Theoretical and Applied Aerodynamics

J.J. Chattot · M.M. Hafez

Theoretical and Applied Aerodynamics

and Related Numerical Methods

Springer

J.J. Chattot
Department of Mechanical
and Aerospace Engineering
University of California
Davis, CA
USA

M.M. Hafez
Department of Mechanical
and Aerospace Engineering
University of California
Davis, CA
USA

ISBN 978-94-017-7793-3 ISBN 978-94-017-9825-9 (eBook)
DOI 10.1007/978-94-017-9825-9

Springer Dordrecht Heidelberg New York London

Softcover reprint of the hardcover 1st edition 2015

Printed on acid-free paper

Springer Science+Business Media B.V. Dordrecht is part of Springer Science+Business Media (www.springer.com)

To our families for their love and support

Preface

The purpose of this book is to expose students to the classical theories of aerodynamics to enable them to apply the results to a wide range of projects, from aircraft to wind turbines and propellers. Most of the tools are analytical, but computer codes are also available and are used by the students to carry out seven to eight projects during the course of a quarter. These computer tools can be found at http://mae.ucdavis.edu/chattot/EAE127/ along with the project statements.

The main focus is on aircraft and the theories and codes that help in estimating the forces and moments acting on profiles, wings, wing-tail and fuselage configurations, appropriate to the flow regime, i.e., subsonic, transonic, supersonic, viscous or inviscid, depending on the Mach number and Reynolds number.

The book culminates with a study of the longitudinal equilibrium of a glider and its static stability, a topic that is not usually found in an aerodynamics but in a stability and controls book. This chapter reflects the expertise of one of the authors (JJC), who has been involved for several years in the SAE Aero Design West competition, as faculty advisor for a student team, (http://students.sae.org/competitions/aerodesign/west/) and has developed the tools and capabilities enabling students to develop their own designs and perform well in the competition. As all airplane modelers know, placing the center of gravity in the correct location is critical to the viability of an aircraft, and a statically stable remote controlled model is a requirement for human piloting.

The material is presented in a progressive way, starting with plane, two-dimensional flow past cylinders of various cross sections and then by mid-quarter, moving to three-dimensional flows past finite wings and slender bodies. In a similar fashion, inviscid incompressible flow is followed by compressible flow and transonic flow, the latter requiring the numerical solution of the nonlinear transonic small disturbance equation (TSD). Viscous effects are discussed and also, due to nonlinear governing equations, numerical simulation is emphasized.

A set of problems with solutions is placed in Part III. It corresponds to final examinations given over the last 10 years or so that the students have 2 hours to complete.

Finally, the reader is assumed to have the basic knowledge in fluid mechanics that can be found in standard textbooks on this topic, in particular as concerns the physical properties of fluids (density, pressure, temperature, equation of state, viscosity, etc.) and the conservation theorems using control volumes. The reader is also assumed to master undergraduate mathematics (calculus of single variables, vector calculus, linear algebra, and differential equations). Three appendices are included in the book, summarizing the material relevant to the subject of interest.

Aerodynamics has a long history and it has reached a mature status during the last century. There are at least 20 books written on aerodynamics in the last 20 years (see references). Some of these are excellent textbooks and some are outdated or out of print. All of the existing texts are based, however, on small disturbance theories. These theories are essential to gain understanding of the physical phenomena involved and the corresponding structure of the flow fields. They also provide good approximations for some simple cases. For practical problems, however, there is a demand for accurate solutions using modern computer simulation. Small disturbance theories can still provide special solutions to test the computer codes. More important perhaps, they can provide a guideline to construct accurate and efficient algorithms for practical flow simulations. They are also used to develop the far field behavior required for the numerical solution of the boundary value problems. In general, the linearized boundary conditions and the restriction to Cartesian grids are no longer sufficient. Grid generation algorithms for complete airplanes, although still a major task in a simulation, are nowadays used routinely in industry. Hence, small disturbance approximations are no longer necessary and indeed full nonlinear potential flow codes, developed over the last two decades, are available everywhere. While it is argued that the corrections to potential flow solutions due to vorticity generated at the shocks can be ignored for cruising speed at design conditions, the viscous effects are definitely important to assess. Again, boundary layer approximations can be useful as a guideline to construct effective viscous/inviscid interaction procedures.

In the book we adopt this view in contrast to a complete CFD approach based on the solution of the Navier-Stokes equations everywhere in the field for more than one reason: it is more attractive, from an educational viewpoint, to use potential flow model and viscous correction. It is also more practical, since Euler and hence Navier-Stokes codes are more expensive and subject to errors due to artificial viscosity as a result of the discrete approximations. A simple example is the accurate capturing of the wake of a wing and the calculation of induced drag, still a challenge today; for the same reasons, the simulations of propellers and helicopter rotor flows are in continuing development, let alone, the problem of turbulence.

In the text, the formulation and the numerics are developed progressively to allow for both small disturbances and full nonlinear potential flows with viscous/inviscid interactions. Only a few existing books (two or three) address these issues and we hope to cover this material in a thorough and simple manner.

The book contains an extensive list of references on aerodynamics including textbooks, advanced and specialized books, classical and old books, flight mechanics books as well as references cited in the text.

Davis, California
November 2013

J.J. Chattot
M.M. Hafez

Contents

Part I
Fundamental Aerodynamics

Chapter 1
Introduction

1.1 Definitions and Notations

Aerodynamics is the study of the forces on a body in a relative motion with air. Aerodynamics is a subset of fluid mechanics. Other related subjects are hydrodynamics and gas dynamics where the medium is water or gas (at high temperature). In this book we are mainly interested in the aerodynamics of airplanes, also rockets, propellers and wind mills, Fig. 1.1.

Other applications include aerodynamics of cars, trains, ships, sails, buildings and bridges. Aerodynamics applications in technology are of course in turbomachinery (compressors and turbines) and in heating, drying and mixing processes (for example aerodynamics of combustion).

Aerodynamics in nature is manifested in flying birds, also in atmospheric boundary layers, storms, tornadoes and hurricanes and their effects on objects like trees as well as surface erosion and land desertification.

There are three ingredients in any aerodynamic analysis: the geometry of the configuration or the model, the medium or the fluid, and the relative motion or the flow.

Consider an airplane flying in the air and in particular the wing. The geometry of a typical wing can be characterized by the maximum camber d and maximum thickness e of a cross section (airfoil) with maximum chord c and span b, see Fig. 1.2.

The medium (air) can be characterized by certain properties: viscosity coefficient (μ), coefficient of heat conductivity (k) and the specific heats c_p and c_v under constant pressure and constant volume respectively.

For steady flight, as in cruising speed conditions, the flow features including the forces on the body are the same as those obtained if the wing is fixed and air is blown on it. What counts is the relative motion between the body and the particles of the air. Figure 1.3 demonstrates the two cases. This observation is the basis of wind tunnel testing. It is easier for analytical and numerical analysis to fix the body and solve the steady state equations in this frame where the variables do not change with time!

The flow can be described in terms of the local velocity vector (**V**) and the thermodynamic variables: pressure (p), friction (τ), density (ρ) temperature (T)

J.J. Chattot and M.M. Hafez, *Theoretical and Applied Aerodynamics*,
DOI 10.1007/978-94-017-9825-9_1

Fig. 1.1 Seaplane, space launcher and wind turbine (from http://en.wikipedia.org/wiki/File:DeHavilland_Single_Otter_Harbour_Air.jpg Author: BW Bell, http://images.ksc.nasa.gov/photos/2000/high/KSC-00PD-5028.jpg Author: NASA, http://images.nrel.gov/viewphoto.php?imageId=6326944 Author: HC Sorensen)

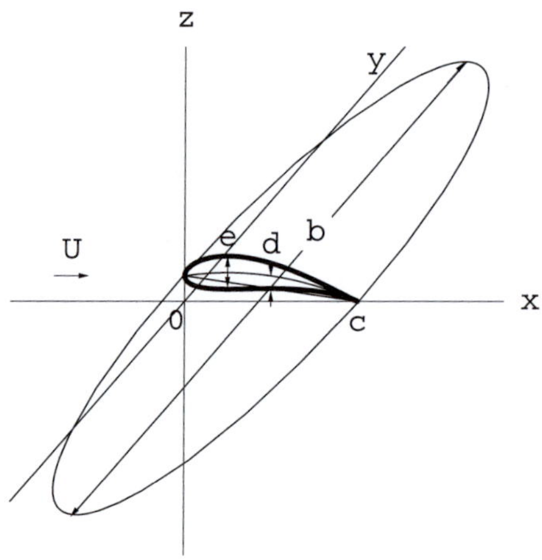

Fig. 1.2 Typical wing and cross section parameters

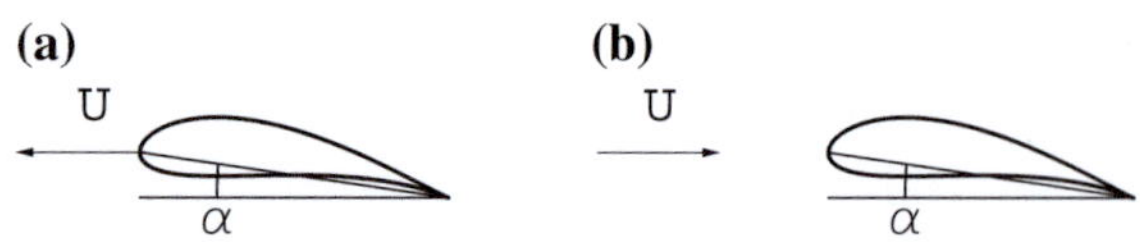

Fig. 1.3 **a** Wing is moving in air at rest, **b** wing is fixed and air is blown on it

and pulsation (ω) when oscillations are present. The velocity has magnitude and direction. Again what counts is the relative angle between the body axis and the wind (i.e. the same results are obtained if the wind is in the horizontal direction and the wing at angle of attack or the wing is horizontal and the wind inclined with the same angle as shown in Fig. 1.4). In the case of a cambered airfoil, the angle of attack and the camber ratio are the parameters indicative of the flow asymmetry.

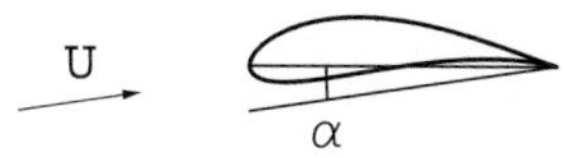

Fig. 1.4 Wing at angle of attack

In Table 1.1, the above 13 (or 14) dimensional quantities are summarized together with the corresponding nondimensional parameters. The nondimensional parameters are identified once reference quantities are chosen. For example, the relative camber $\frac{d}{c}$, the thickness ratio $\frac{e}{c}$ and the aspect ratio $AR = \frac{b^2}{S} \propto \frac{b}{c}$ where $S \propto bc$ is the area of the wing. The incoming flow velocity U is chosen as reference velocity. The ratio of the specific heats $\gamma = \frac{c_p}{c_v}$ and the angle of attack α are obviously nondimensional. On the other hand the viscosity and heat conductivity coefficients have different units. A nondimensional parameter of the ratio of viscosity to heat conductivity is introduced as $P_r = \frac{\mu c_p}{k}$ and is called the Prandtl number.

The ratio of the speed of the body (or the wind) U to the speed of sound a is called the Mach number, $M = \frac{U}{a}$ (where $a = \sqrt{\gamma RT} = \sqrt{\frac{\gamma p}{\rho}}$ for a perfect gas). The Mach number represents the compressibility effects.

The ratio of the normal stress (pressure p) to the tangential stress (friction τ) is related to the Reynolds number $Re \propto \frac{p}{\tau}$. The conventional definition is $Re = \frac{\rho U c}{\mu}$. The Reynolds number represents the ratio of inertia to viscous stress as well. The flow oscillations are characterized by the Strouhal number $St = \frac{\omega c}{U}$. The result for pressure or shear stress depends on the parameters via a dimensional relationship of the form

$$p\,(or\ \tau) = f(b, c, d, e, c_p, c_v, k, \mu, T, U, \rho, \omega, \alpha)$$

The study of aerodynamics is mainly to find the nondimensional stresses:

- the pressure coefficient $C_p = \frac{p - p_\infty}{\frac{1}{2}\rho U^2}$
- the skin friction coefficient $C_f = \frac{\tau}{\frac{1}{2}\rho U^2}$

as functions of the nondimensional parameters of flight representing the geometry, the medium and the motion, i.e.

$$C_p\,(or\ C_f) = F\left(AR, \frac{d}{c}, \frac{e}{c}, \gamma, P_r, M, Re, S_t, \alpha\right)$$

Table 1.1 Dimensional quantities and nondimensional parameters

	Dimensional quantities	Nondimensional parameters
Model	$b,\ c,\ d,\ e$	$AR,\ \frac{d}{c},\ \frac{e}{c}$
Medium	$c_p,\ c_v,\ k,\ \mu$	$\gamma,\ P_r$
Motion	$T,\ V,\ \rho,\ \omega,\ p,\ (or\ \tau)$	$M,\ Re,\ St,\ \alpha,\ C_p,\ (or\ C_f)$

Notice in the definition of C_p, p_∞ indicates the atmospheric pressure and $\frac{1}{2}\rho U^2$ is called the dynamic pressure. (C_p should not be confused with c_p, the latter being the specific heat under constant pressure.)

The relation given above is consistent with the Buckingham PI theorem. Since we have four basic units (length, time, mass and temperature) and 13 (or 14) dimensional quantities, hence there are at least 9 (or 10) nondimensional parameters, in fact 10 (or 11) since α is already a dimensionless parameter.

Integration of surface stresses provides the moment about a point (or an axis) and the forces, the lift, and the drag (the force normal and in the flow direction respectively).

Our study is limited to small angles of attack, relative cambers and thickness ratios, i.e. $|\alpha| \ll 1$, $|\frac{d}{c}| \ll 1$ and $\frac{e}{c} \ll 1$. Moreover, γ and P_r are constants. For air, under normal conditions, $\gamma = 1.4$ and $P_r = 0.72$.

Three types of wings are of interest: high and low aspect ratio wings and their limits, two dimensional and axisymmetric bodies, as well as wings of aspect ratio of order one.

The Mach number range includes constant density, incompressible flow ($M = 0$ not because V is zero, but because a is large) and compressible subsonic $M < 1$, transonic $M \simeq 1$ and supersonic flows $M > 1$. A short introduction of hypersonic flows, $M \gg 1$ is presented but rarefied gas dynamics is not included in this book.

Only high Reynolds number flows are of interest in conventional aerodynamics including laminar, transitional and turbulent flows, depending on critical values of Reynolds number.

1.2 Discussion of Mathematical Models

In general, modeling of fluid motion is a complicated subject particularly with high speed turbulent flows. It is argued that simple models based on small disturbance potential flow equation, with boundary layer corrections are attractive, at least for educational purposes, to identify the difficulties of relevant phenomena at design conditions and to study successfully practical solutions for these problems.

The students should be aware of the limitations of these simple models and the range of their validity. In the following chart, different formulations are listed with their relations, Fig. 1.5.

In the common commercial and military flight regimes, continuum mechanics assumptions are acceptable and the Navier-Stokes equations are considered the high fidelity model. For high Reynolds number flows, the viscous stresses and the heat conduction process are ignored outside a thin layer adjacent to the solid surfaces and in the wake.

Assuming inviscid and adiabatic flows, the Navier-Stokes equations reduce to the Euler equations, representing conservation of mass, momentum and energy for a perfect gas.

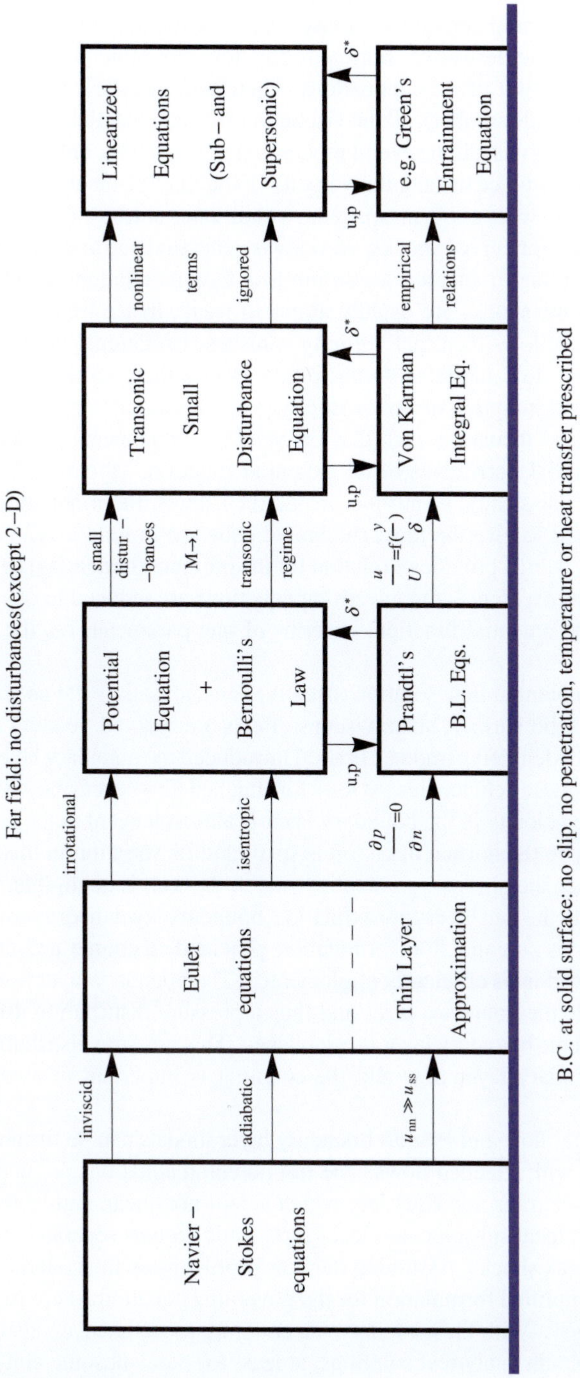

Fig. 1.5 From Navier-Stokes to linearized models

Adjacent to the solid surface, the thin layer approximation is based on ignoring the second order tangential derivatives in viscous and heat conduction terms, compared to the corresponding derivatives normal to the flow direction. Notice that the Euler equations are a subset of these equations. Therefore, outside a thin layer (its size depends on Reynolds number), all the viscous and heat conduction terms are ignored.

The Euler equations admit several modes, namely acoustic, entropy and vorticity modes (vorticity is twice the angular velocity of the fluid element around its center). For steady flows, with uniform upstream conditions, entropy is generated across shock waves, and for curved shocks, vorticity is generated due to variation of entropy from one streamline to another, according to a famous relation due to Crocco. At design conditions, shocks are usually assumed weak, hence isentropic conditions can be used. It follows then that vorticity vanishes. Irrotational flows are easier to handle since the fluid particles do not rotate around their centers as in rotational flows. With the assumption of zero vorticity, a potential function can be introduced (thanks to Stokes' theorem), such that the velocity components are represented in terms of the partial derivatives of the potential function. Moreover, an integral of motion exists. For steady, isentropic, irrotational flows, the momentum equations can be integrated to give the most celebrated equation in fluid mechanics, namely Bernoulli's law, which provides a relation for the pressure (or density) in terms of the velocity magnitude. Hence, the governing equations are reduced to one (nonlinear) equation for the potential function, in terms of one parameter, i.e. the free stream Mach number.

The steady potential flow solution (usually) does not satisfy the no slip boundary condition at a solid surface, since viscous effects are ignored, leading to zero skin friction drag (d'Alembert paradox). Prandtl introduced the boundary layer concept in 1904, to rectify these deficiencies. At least for attached flows over smooth streamlined thin bodies, the velocity in the boundary layer is almost tangent to the body surface, i.e. the velocity in the normal direction is small and its variation in this direction is also small, hence the pressure gradient normal to the wall is negligible.

Together with the thin layer approximation, boundary layer theory provides a powerful tool to fix the potential flow formulation, provided, of course, a viscous/inviscid interaction procedure is efficiently implemented. The boundary layer flow pushes up the flow outside the boundary layer and thus a pressure distribution different from the one without the boundary layer is established. This pressure distribution in return affects the boundary layer flow and the coupling is important to obtain the right interaction.

Both potential flow and Prandtl boundary layer models can be further simplified for thin bodies with attached flows. The full potential equation can be reduced to a small disturbance transonic equation, which is still nonlinear, and is able to model flows with both local subsonic and local supersonic regions separated by sonic line and possibly weak shocks. Assuming the flow profile inside the boundary layer, one can obtain a simplified formulation for the governing equations there following von Karman's analysis. With these models, the coupling procedures are also simplified.

Finally, to obtain analytical solutions, at least for pure subsonic and pure supersonic flows, the nonlinear small disturbance equation is linearized. Ignoring the nonlinear terms can be justified away from the transonic regime. Superposition techniques can be used to establish solutions for realistic shapes. Also, empirical formulae can be used to solve the von Karman equation, in the boundary layer, for both laminar and turbulent flows.

In Fig. 1.5, the boundary conditions at the solid surface are the no slip and no penetration conditions for the velocity together with the specification of wall temperature or heat flux. In the far field, disturbances die out (except in some two dimensional cases).

1.3 Description of the Book Content

In the text, the theory is developed for different regimes and the numerics are explained for some problems. In this respect, some computer code are available for full potential and transonic small disturbance equations. Boundary layer codes for laminar and turbulent flows are also available.

In Fig. 1.6, a road map for the materials of the book is described. The block in the middle covers steady two-dimensional incompressible inviscid flows. The generation of lift around a rotating cylinder (Magnus effect) is mathematically modeled as

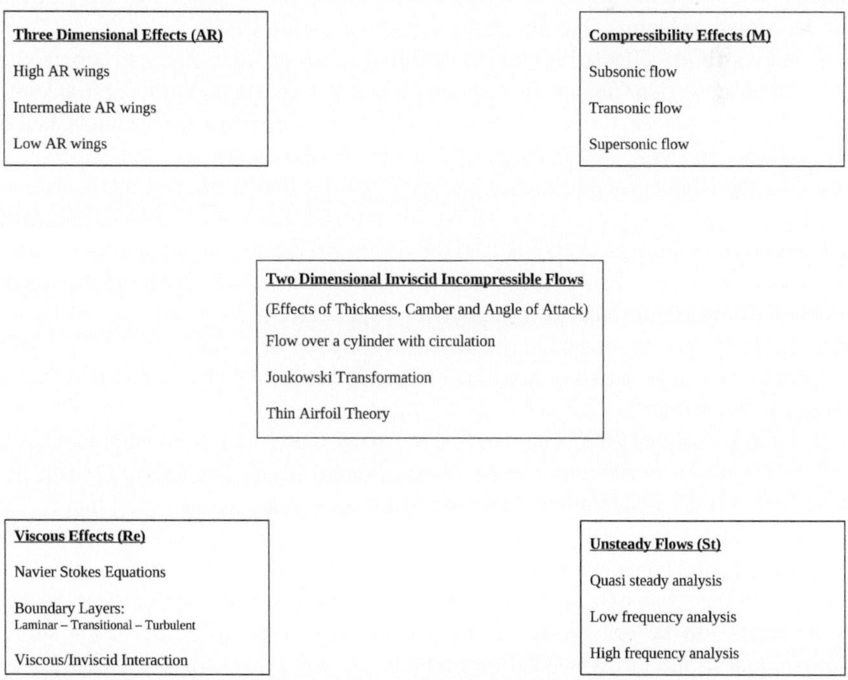

Fig. 1.6 Road map for the book

a potential flow with circulation. In the text, the important notion of circulation is discussed in details from both physical and mathematical views. Without circulation, there is no lift (because of symmetry). The circulation increases the velocity on one side and retards it on the other side. The difference in velocity is associated with a difference in pressure (Bernoulli's law), hence a lifting force is generated. An analytical closed form solution is obtained for this model. The lift is related to the asymmetry generated by the circulation and a relation between lift and the angle of the flow separation is derived. Also, the Kutta-Joukowski lift theorem is established.

Next, the Joukowski transformation is introduced to relate the flow over a cylinder to a flow over Joukowski airfoils (including the flat plate and circular arc as special cases). The celebrated formula for the lift coefficient of flat plates (or for that matter any thin airfoil) is derived. Similarly, a formula for the camber effect is obtained from the circular arc. In all cases, the Kutta-Joukowski condition at the sharp trailing edge is applied, which states that the flow leaves the airfoil smoothly to mimic the real viscous flow. The physical source of circulation (and hence lift) can be explained in terms of the vorticity of the fluid elements in the boundary layer, due to shear stresses between the layers of the fluid. The general Joukowski airfoil results are also obtained (without using complex variables) and the effect of thickness on lift is delineated.

Next, the thin airfoil theory of Glauert and Munk is discussed in details, allowing the calculation of lift (and moment) for general thin airfoils using superposition of fundamental solutions (in terms of sources for the thickness problems and vortices for the lifting problems). The above topics are covered in Chaps. 2 and 3.

Compressibility effects for two-dimensional subsonic, transonic and supersonic flows are discussed in Chap. 4. Both the linearized theory and the nonlinear transonic small disturbance theory are treated in detail. For subsonic flows, the Prandtl-Glauert rule relating subsonic and incompressible flow results is derived. For supersonic flows, Ackeret theory for thin airfoils provides the lift and wave drag estimates, as well as an explanation for why we have this type of drag. For transonic flows, the failure of the linearized theories is explained and the characteristic and shock relations, based on small disturbance theory, are given. Numerical schemes to solve the nonlinear equation using the Murman-Cole method are explained and the numerical results are presented. This chapter is one of the unique features of this book.

Chapter 5 deals with two-dimensional unsteady cases for both incompressible and compressible flows.

Three-dimensional effects are studied in Chaps. 6 and 7 for high and low aspect ratio (AR) wings. For incompressible flows, Prandtl lifting line theory (1918), the Weissinger vortex lattice method and the Munk and Jones slender wing theory are covered. The concept of induced drag (or vortex drag), also called drag due to lift, is explained and it is shown that for given lift, an elliptic planform (without twist) is optimum in the sense that the induced drag is minimum (the induced drag is related to the added kinetic energy stored in the cross flow induced by the vortex sheet, that is equal to the thrust needed to maintain the wing motion). The extension to compressible flows, with low and high Mach numbers, is followed for both bodies of revolution and slender wings. Optimum shapes for supersonic projectiles are discussed, together with the transonic and supersonic area rules, as well as conical flows. Transonic lifting line theory, swept and oblique wings are also covered.

Finally, viscous effects or Reynolds number effects, are dealt with in Chaps. 8 and 9, starting with Navier-Stokes equations and incompressible and compressible boundary layer theory, including viscous/inviscid interaction procedures, and the calculation of skin friction drag and form drag (that is pressure drag due to the boundary layer interaction with the inviscid flow, which is important in the case of separation. In general, there are four types of drag, skin friction drag, induced drag, wave drag and form drag. It is difficult to separate their contributions, particularly if the shock penetrates the boundary layer).

The book contains also four special topics, including wind turbines, airplane design projects, hypersonic flows and flow analogies (electric and hydraulic). Another special feature of the book is three appendices, the first two consist of ten exams and their solutions, and the last deals with mathematical methods in aerodynamics, including the method of complex variables, method of characteristics and the conservation laws with shock waves and contact discontinuities.

There are already many text books available on aerodynamics (see Refs. [1–12] below) and the authors would like to add one more. The authors hope that the reader will enjoy this book as much as the authors did.

References

1. Anderson, J.: Fundamentals of Aerodynamics. McGraw Hill, New York (2007)
2. Bertin, J.: Aerodynamics for Engineers. Prentice Hall, Upper Saddle River (2002)
3. Houghton, E.L., Carpenter, P.W.: Aerodynamics for Engineering Students. Elsevier, Amsterdam (2003)
4. Keuthe, A.M., Chow, C.Y.: Foundations of Aerodynamics. Wiley, New York (1998)
5. McCormick, B.: Aerodynamics, Aeronautics and Flight Mechanics. Wiley, New York (1995)
6. Moran, J.: An Introduction to Theoretical and Computational Aerodynamics. Wiley, New York (1984). Also Dover (2005)
7. Smetana, F.O.: Introductory Aerodynamics and Hydrodynamics of Wings and Bodies. AIAA (1997)
8. Katz, J., Plotkin, A.: Low Speed Aerodynamics. Cambridge University Press, Cambridge (2002)
9. Flandro, G., McMahon, H., Roach, R.: Basic Aerodynamics—Incompressible Flow. Cambridge University Press, New York (2012)
10. Pope, A.: Basic Wing and Airfoil Theory. Dover Publisher, New York (2009)
11. Ashley, H., Landhal, M.: Aerodynamics of Wings and Bodies. Addison Wesley, Boston (1965) Also Dover (1985)
12. Jones, R.T.: Wing Theory. Princeton, New Jersey (1990)

Chapter 2
Inviscid, Incompressible Flow Past Circular Cylinders and Joukowski Airfoils

2.1 Background

2.1.1 Notation

In two-dimensions, the obstacles are cylinders of various cross sections, whose axis is perpendicular to the plane in which the flow is taking place (Fig. 2.1). When a wing will be considered, the wing span will be described by the variable y. Hence, the flow past cylinders will take place in the (x, z) plane, see Fig. 2.2. In this chapter we will consider steady, 2-D, inviscid, incompressible, adiabatic and irrotational flow, also called *potential flow*. The influence of gravity will be neglected.

2.1.2 Governing Equations

Lets (u, w) represent the perturbation velocity or deviation from the uniform flow $\mathbf{V}_\infty = (U, 0)$. At any point in the flow field the velocity is $\mathbf{V} = (U + u, w)$, the density is $\rho = const.$ and the pressure p. The conservation of mass and irrotationality condition form a system of two linear first-order partial differential equations (PDEs) for (u, w) as

$$\begin{cases} \frac{\partial u}{\partial x} + \frac{\partial w}{\partial z} = 0 \\ \frac{\partial u}{\partial z} - \frac{\partial w}{\partial x} = 0 \end{cases} \tag{2.1}$$

In local polar coordinates (r, θ), the system reads in terms of $\mathbf{V} = (V_r, V_\theta)$

$$\begin{cases} \frac{\partial r v_r}{\partial r} + \frac{\partial v_\theta}{\partial \theta} = 0 \\ \frac{\partial r v_\theta}{\partial r} - \frac{\partial v_r}{\partial \theta} = 0 \end{cases} \tag{2.2}$$

J.J. Chattot and M.M. Hafez, *Theoretical and Applied Aerodynamics*,
DOI 10.1007/978-94-017-9825-9_2

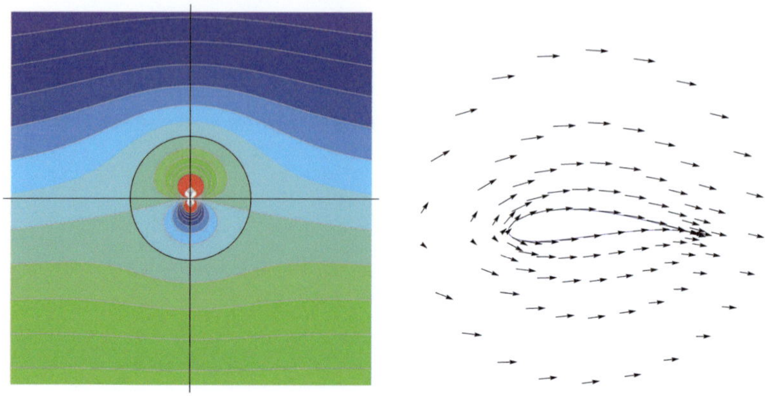

Fig. 2.1 Flow past a circular cylinder with circulation and mapping to a Joukowski airfoil

Fig. 2.2 Coordinate system and notation

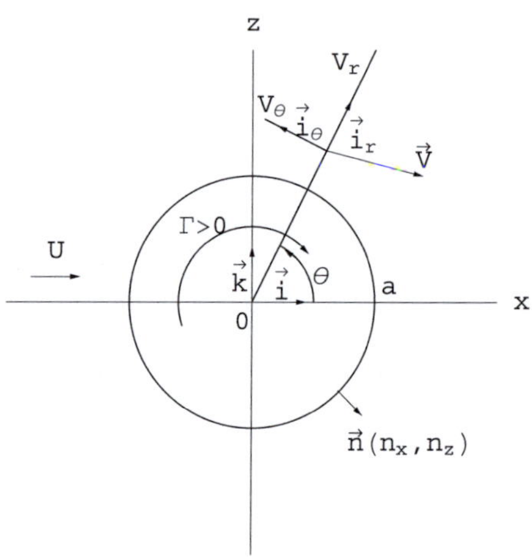

where $V_r = U \cos\theta + v_r$ and $V_\theta = -U \sin\theta + v_\theta$ are the total velocity components, including the contribution of the uniform flow. The irrotationality condition is the consequence of conservation of momentum for inviscid, adiabatic, incompressible flow. In other words, a solution to the above system also satisfies the momentum equations. The energy equation (conservation of mechanical energy) is also satisfied, under the stated assumptions.

The pressure is obtained from Bernoulli's equation, again a consequence of the momentum conservation (it can be interpreted also as a conservation of mechanical energy per unit mass, where the potential energy is constant):

$$\frac{p}{\rho} + \frac{1}{2}V^2 = \frac{p_\infty}{\rho} + \frac{1}{2}\mathbf{V}_\infty^2 = \frac{p_\infty}{\rho} + \frac{1}{2}U^2 = const. \tag{2.3}$$

Note that the Bernoulli's equation is nonlinear. However, as we will see, it can be linearized under the assumption of small perturbation.

2.1.3 Boundary Conditions

The boundary conditions determine the solution of the PDEs. They correspond to the necessary and sufficient conditions for the well-posedness of solutions to the PDEs, i.e. to exist, be unique and depend continuously on the data. There are two types of boundary conditions traditionally associated with the above system, the tangency condition and the asymptotic condition in the far field.

The tangency condition expresses the fact that in an inviscid flow, the fluid is expected to slip along a solid, impermeable surface. This translates to

$$(\mathbf{V}.\mathbf{n})_{obstacle} = V_n = 0 \tag{2.4}$$

where V_n represents the normal relative velocity with respect to the solid surface.

Other conditions can be prescribed, for example in the case of a porous surface, the normal relative velocity can be given as

$$V_n = g \tag{2.5}$$

where g is a given function along the surface.

The asymptotic condition states that, for a finite obstacle, the flow far away from the body, returns to the uniform, undisturbed flow $\mathbf{V}_\infty$. In terms of u and w this reads

$$u, w \rightarrow 0 \; as \; x^2 + z^2 \rightarrow \infty \tag{2.6}$$

2.1.4 Other Formulations

For irrotational flows, a velocity potential can be introduced: let $\phi(x, z)$ be the perturbation potential. The full potential is given by $\Phi = Ux + \phi(x, z)$. The perturbation velocity components are

$$u = \frac{\partial \phi}{\partial x}, \quad w = \frac{\partial \phi}{\partial z} \tag{2.7}$$

Then the irrotationality condition is identically satisfied, but substituting u and w into the conservation of mass results in

$$\Delta\phi = \frac{\partial^2 \phi}{\partial x^2} + \frac{\partial^2 \phi}{\partial z^2} = 0 \tag{2.8}$$

In polar coordinates this is

$$v_\theta = \frac{1}{r}\frac{\partial \phi}{\partial \theta}, \quad v_r = \frac{\partial \phi}{\partial r} \tag{2.9}$$

$$\frac{\partial}{\partial r}\left(r\frac{\partial \phi}{\partial r}\right) + \frac{1}{r}\frac{\partial^2 \phi}{\partial \theta^2} = 0 \tag{2.10}$$

This is the governing equation for potential flows, the Laplace equation. The perturbation velocity field is given by *grad* $\phi = \left(\frac{\partial \phi}{\partial x}, \frac{\partial \phi}{\partial y}, \frac{\partial \phi}{\partial z}\right)$ (or in short notation $\nabla\phi$). The boundary conditions become

$$\begin{cases} \nabla\Phi.\mathbf{n}\,|_{obstacle} = (U\mathbf{i} + \nabla\phi).\mathbf{n}\,|_{obstacle} = 0 \\ \nabla\phi \to 0 \; as \; x^2 + z^2 \to \infty \end{cases} \tag{2.11}$$

A streamfunction can be introduced: let $\psi(x, z)$ be the perturbation streamfunction. The full stream function is $\Psi = Uz + \psi(x, z)$. The perturbation velocity components are obtained from

$$u = \frac{\partial \psi}{\partial z}, \quad w = -\frac{\partial \psi}{\partial x} \tag{2.12}$$

then the equation of conservation of mass is identically satisfied and substituting u and w in the irrotationality condition yields

$$\Delta\psi = \frac{\partial^2 \psi}{\partial x^2} + \frac{\partial^2 \psi}{\partial z^2} = 0 \tag{2.13}$$

In polar coordinates this is also

$$v_r = \frac{1}{r}\frac{\partial \psi}{\partial \theta}, \quad v_\theta = -\frac{\partial \psi}{\partial r} \tag{2.14}$$

$$\frac{\partial}{\partial r}\left(r\frac{\partial \psi}{\partial r}\right) + \frac{1}{r}\frac{\partial^2 \psi}{\partial \theta^2} = 0 \tag{2.15}$$

This is the governing equation for the streamfunction. The streamfunction also satisfies Laplace equation. The boundary conditions, however, read

$$\begin{cases} \Psi\,|_{obstacle} = (Uz + \psi)\,|_{obstacle} = const. \\ \left(\frac{\partial \psi}{\partial z}, -\frac{\partial \psi}{\partial x}\right) \to 0 \; as \; x^2 + z^2 \to \infty \end{cases} \tag{2.16}$$

The first condition results from the identity $\nabla\Psi.\mathbf{t}|_{obstacle} = 0$, where $\mathbf{t}$ represents a unit tangent vector to the solid surface ($\mathbf{n}.\mathbf{t} = 0$), which proves that $\Psi\,|_{obstacle} = const.$, the value of this constant is however not known a priori. The

$\Psi = const.$ lines are the streamlines of the flow. Solid obstacles are streamlines and conversely, streamlines can be materialized to represent a body surface.

When both, Φ and Ψ exist, simultaneously, potential and stream functions are called conjugate harmonic functions. It can be easily shown that the curves $\Phi = const.$ are orthogonal to the curves $\Psi = const.$

2.2 Elementary Solutions

2.2.1 Uniform Flow

The uniform flow $\mathbf{V}_\infty = (U, 0)$ is represented by the leading term of the full potential and stream functions:

$$\Phi = Ux, \ \Psi = Uz \tag{2.17}$$

A sketch of the solution is shown in Fig. 2.3.

The uniform flow $\mathbf{W}_\infty = (0, W)$ is represented by the leading term of the full potential and stream functions:

$$\Phi = Wz, \ \Psi = -Wx \tag{2.18}$$

A sketch of the solution is shown in Fig. 2.4.

A uniform flow in an arbitrary direction can be obtained by the superposition of these two elementary solutions.

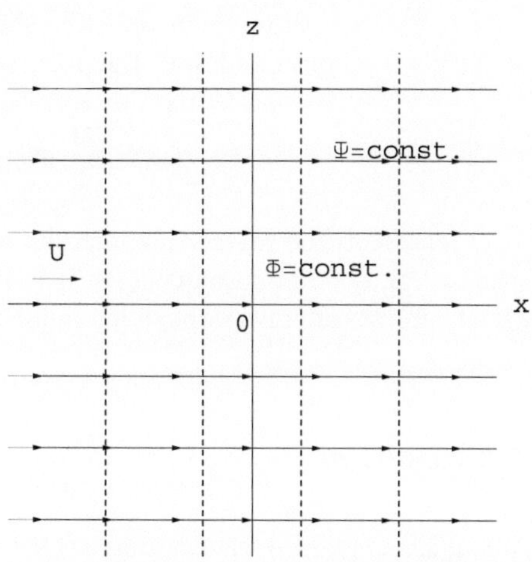

Fig. 2.3 Uniform flow: equipotential lines and streamlines

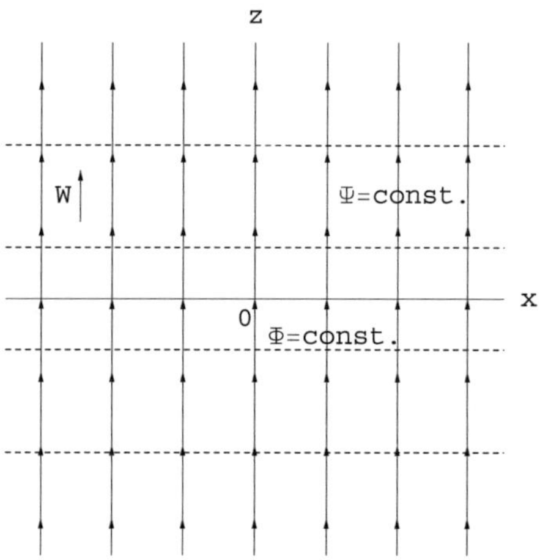

Fig. 2.4 Uniform flow: equipotential lines and streamlines

2.2.2 *Source and Sink*

A line source and sink is the building block to represent profile thickness as will be seen later. If a source or a sink is placed in the plane, it is convenient to define a polar coordinate system with origin at the source or sink location. Let (r, θ) represent the distance and polar angle of a point to the source/sink.

The velocity components are

$$v_r = \frac{Q}{2\pi}\frac{1}{r}, \quad v_\theta = 0 \tag{2.19}$$

Note that the perturbation velocity vanishes in the far field, which is consistent with the asymptotic condition. The potential and stream functions are given by

$$\phi = \frac{Q}{2\pi}\ln r, \quad \psi = \frac{Q}{2\pi}\theta \tag{2.20}$$

Q represents the source/sink intensity and has unit of volume flow rate per unit span (m^2/s). $Q > 0$ is a source, $Q < 0$ a sink. The equipotential lines are circles and the streamlines are rays through the origin. See Fig. 2.5.

2.2.3 *Doublet*

A doublet is obtained when a source $(Q, -a)$ and a sink $(-Q, a)$ located symmetrically along the x-axis are merged at the origin following the limiting process

$$lim_{a\to 0,\ Q\to\infty}\,(Source(Q, -a) + Sink(-Q, a)) \tag{2.21}$$

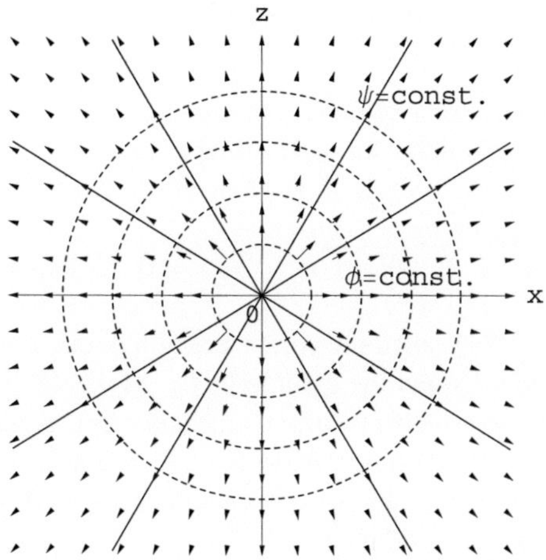

Fig. 2.5 Source: equipotental lines and streamlines

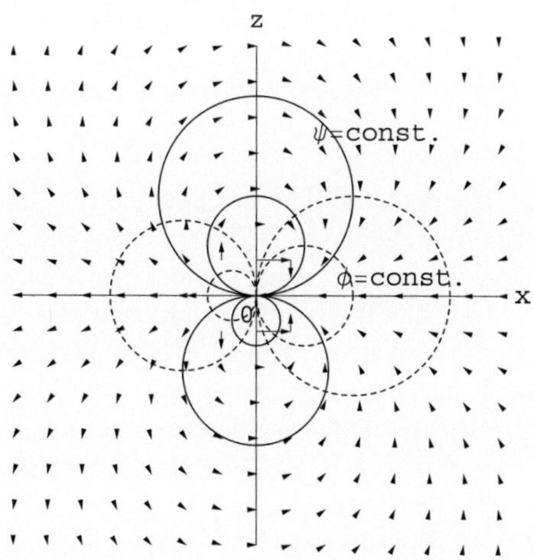

Fig. 2.6 Doublet: equipotential lines and streamlines

The results for the velocity components are

$$v_r = \frac{-D\cos\theta}{r^2}, \ v_\theta = \frac{-D\sin\theta}{r^2} \tag{2.22}$$

where D is the strength of the doublet, $D = lim_{a\to 0,\ Q\to\infty}\ aQ$

The results for the potential and stream functions are

$$\phi = D\frac{\cos\theta}{r}, \ \psi = -D\frac{\sin\theta}{r} \tag{2.23}$$

The equipotential lines and streamlines are circles, tangent at the origin to the z-axis for the former and to the x-axis for the latter. See Fig. 2.6.

2.2.4 Potential Vortex

Consider a cylinder rotating clockwise in a fluid at rest at infinity. The particles will be entrained, through viscous shear forces, to move around the cylinder in concentric circles. It can be shown that, at steady-state, such a flow is irrotational and the velocity is given by

$$v_r = 0, \ 2\pi r v_\theta = const. = -\Gamma \tag{2.24}$$

Γ is called the vortex strength or circulation and has unit (m^2/s). Such a potential flow around the cylinder is called a potential vortex (this terminology may be confusing since the vorticity in such a flow is zero outside the cylinder otherwise it is not a potential flow). Now, the cylinder radius can be vanishingly small. The potential flow solution however is singular at $r = 0$, a point inside the cylinder, no matter how small the radius is.

Notice that a potential vortex corresponds to the flow obtained when the potential and stream lines of a source/sink flow are exchanged. The solution is given by

$$\phi = -\frac{\Gamma}{2\pi}\theta, \ \psi = \frac{\Gamma}{2\pi}\ln r \tag{2.25}$$

The potential lines are now the rays and the streamlines the circles centered at the origin. See Fig. 2.7. The vortex solution is a building block for lifting airfoils to represent incidence and camber as will be seen later.

Note that the velocity components (u, w) (obtained from ϕ and ψ by taking the partial derivatives with respect to x and z when using the Cartesian coordinates, or $\frac{\partial}{\partial r}$ and $\frac{1}{r}\frac{\partial}{\partial \theta}$ when using polar coordinates), decay in the far field as $\frac{1}{r}$, which is consistent with the asymptotic condition.

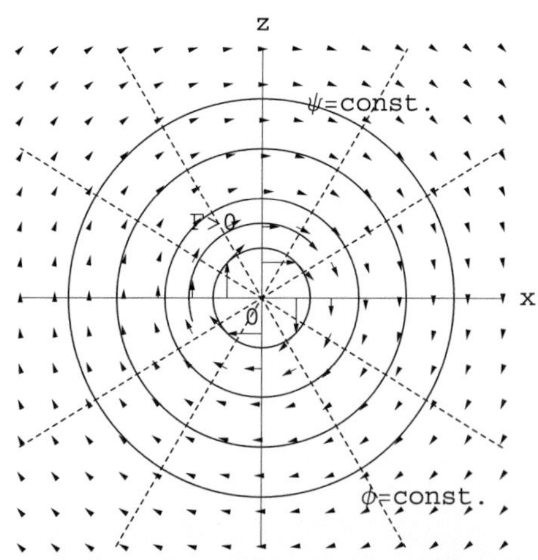

Fig. 2.7 Potential vortex: potential lines and streamlines

2.3 Superposition of Elementary Solutions

From the property of linear governing equations, if ϕ_1 and ϕ_2 are each solution of the Laplace equation, $\phi_1 + \phi_2$ or any linear combination will be a solution of the Laplace equation. The same is true for the stream functions. This is the basis of the principle of superposition. Note that the elementary solutions for ϕ and ψ are singular at their centers (infinite velocity), hence they are also called singular solutions and the superposition of singular solutions is called the method of singularities. These singular solutions are regular everywhere in the flow field when the centers are excluded, for example by enclosing them within the obstacle.

Mathematically one can prove that it is possible to solve for the flow past an arbitrary obstacle by superposing elementary solutions, but, in general, the resulting pressure field is not the sum of the individual pressure fields, due to the nonlinearity of Bernoulli's equation. For small disturbances, the Bernoulli equation can be linearized, allowing for superposition of pressure fields as well.

2.3.1 Global Integrals

Let $Q = \oint_C (\mathbf{V}.\mathbf{n})dl$ and $\Gamma = \oint_C (\mathbf{V}.\mathbf{dl})$, where $\mathbf{dl}$ represents a small oriented element of the contour and $dl = |\mathbf{dl}|$ is the length of the element. The first integral represents the net volume flow rate out of contour C and the second, the circulation along the same contour, taken in the clockwise direction. It is easy to show that if the contour C contains sources and sinks of intensity $Q_1, Q_2, \ldots$ the result will be $Q = Q_1 + Q_2 + \cdots$ the algebraic sum of the sources and sinks inside the contour. If the contour C contains potential vortices of circulation $\Gamma_1, \Gamma_2, \ldots$ the result will be $\Gamma = \Gamma_1 + \Gamma_2 + \cdots$ the algebraic sum of vortices inside the contour.

2.3.2 Example of Superposition: Semi-infinite Obstacle

The superposition of the uniform flow and a source ($Q > 0$) at the origin produces a semi-infinite body, called Rankine body. Stagnation points and stagnation streamlines play an important role in the description of the flow topology. Here, the velocity field is given by

$$V_r = U\cos\theta + \frac{Q}{2\pi}\frac{1}{r}, \quad V_\theta = -U\sin\theta \tag{2.26}$$

At a stagnation point $V_r = V_\theta = 0$. The stagnation point S for this flow is found to be

$$\theta_s = \pi, \ r_s = \frac{Q}{2\pi U} \tag{2.27}$$

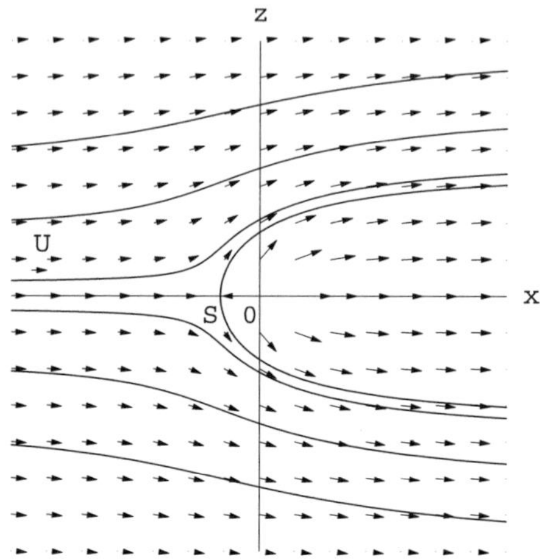

Fig. 2.8 Half-body and flow vector field

The main flow features can be studied with the streamfunction

$$\Psi = Ur\sin\theta + \frac{Q}{2\pi}\theta \tag{2.28}$$

A remarkable streamline is that which corresponds to the half-body. It corresponds to $\Psi = \frac{Q}{2}$. It is also the stagnation streamline. It is made of several pieces, the negative x-axis and the half-body itself. The flow is sketched in Fig. 2.8.

Note that any streamline can be materialized as a solid surface. The flow inside the half body is also represented, however it is singular at the origin and cannot represent a realistic flow near that point. The maximum thickness of the half-body is $\frac{Q}{\pi U}$. One can calculate the force on a solid surface coinciding with the separation streamline using the momentum theorem and Bernoulli's equation applied to a contour composed of the half-body and a large circle of radius R to close the contour. The lift vanishes due to symmetry, but the axial force on the half-body is given by $D = -\rho U Q$, a thrust.

2.4 Flow Past a Circular Cylinder

Finite bodies require that the sum of all sources and sinks be zero. Doublets, as limit of a source and a sink of equal and opposite intensity, have zero net volume flow. The superposition of a uniform flow and a doublet ($D > 0$) produces a flow field given by

$$V_r = U\cos\theta - D\frac{\cos\theta}{r^2},\ \ V_\theta = -U\sin\theta - D\frac{\sin\theta}{r^2} \tag{2.29}$$

The velocity potential and streamfunction are

$$\Phi = Ur\cos\theta + D\frac{\cos\theta}{r} = \left(Ur + \frac{D}{r}\right)\cos\theta \tag{2.30}$$

$$\Psi = Ur\sin\theta - D\frac{\sin\theta}{r} = \left(Ur - \frac{D}{r}\right)\sin\theta \tag{2.31}$$

By inspection of the streamfunction, one can see that $\Psi = 0$ for $r = a = \sqrt{\frac{D}{U}}$. This solution represents the flow past a circular cylinder of radius a. At the surface of the cylinder, the tangency condition, $V_r = 0$, is satisfied. We have also seen earlier that the potential vortex admits streamlines which are circles centered at the origin. In fact, the most general solution of the flow past a circular cylinder of radius a is given by the superposition of the uniform flow, a doublet of intensity $D = Ua^2$ and a potential vortex. The corresponding velocity field is

$$V_r = U\left(1 - \frac{a^2}{r^2}\right)\cos\theta,\ \ V_\theta = -U\left(1 + \frac{a^2}{r^2}\right)\sin\theta - \frac{\Gamma}{2\pi}\frac{1}{r} \tag{2.32}$$

Note that the circulation Γ is arbitrary. The tangency condition is satisfied by $V_r(a, \theta) = 0$. The asymptotic condition is also satisfied as $\mathbf{V} \to \mathbf{V}_\infty$ when $r \to \infty$.

Stagnation points are found on the cylinder by satisfying the condition

$$V_\theta(a, \theta_s) = -2U\sin\theta_s - \frac{\Gamma}{2\pi}\frac{1}{a} = 0 \tag{2.33}$$

This admits two solutions provided $|\Gamma| \le 4\pi Ua$, namely $\theta_s = \arcsin\left(-\frac{\Gamma}{4\pi Ua}\right)$ and $\pi + \arcsin\left(\frac{\Gamma}{4\pi Ua}\right)$, where $-\frac{\pi}{2} \le \theta_s \le \frac{\pi}{2}$. The stagnation points are points A and B in Fig. 2.9. The streamlines are given by

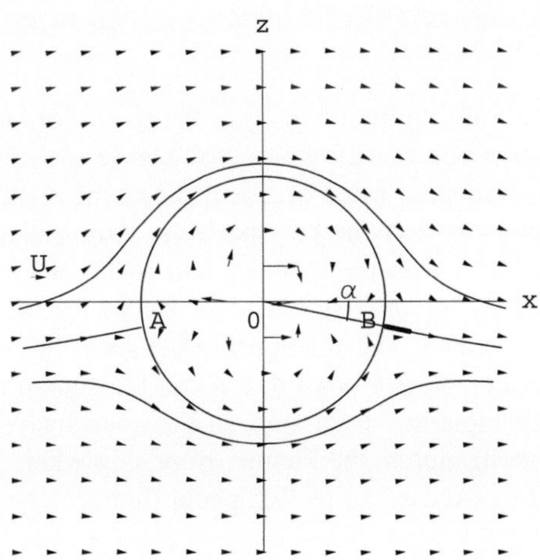

Fig. 2.9 Flow past a circular cylinder with circulation

$$\Psi = U\left(r - \frac{a^2}{r}\right)\sin\theta + \frac{\Gamma}{2\pi}\ln r = const. \tag{2.34}$$

A remarkable streamline is the stagnation streamline, $\Psi = \frac{\Gamma a}{2\pi}$. It is made of the cylinder and the two branches attached to the stagnation points.

The question of how to determine the value for Γ is important. Within the inviscid flow framework, a possible mechanism for controlling the circulation is the use of a small flap, a thin flat plate placed perpendicular to the cylinder at θ_s forcing the rear stagnation point to locate at point B, as shown in Fig. 2.9. Once the rear stagnation point is established, the front stagnation point A will settle at $\pi - \theta_s$. In a sense, θ_s relates to the "incidence" $\alpha = -\theta_s$ of the cylinder. The previous equation for the stagnation points can be written as

$$\Gamma(\alpha) = 4\pi U a \sin\alpha \tag{2.35}$$

which reflects the dependency of the circulation on the flap location.

The pressure field is obtained from the Bernoulli equation

$$p(r,\theta) = p_\infty + \frac{1}{2}\rho(U^2 - V_r^2 - V_\theta^2) \tag{2.36}$$

On the cylinder surface this reduces to

$$p(\theta) = p_\infty + \frac{1}{2}\rho\left[U^2 - \left(\frac{\Gamma}{2\pi a}\right)^2 - \frac{2U\Gamma}{\pi a}\sin\theta - 4U^2\sin^2\theta\right] \tag{2.37}$$

Knowing the surface pressure distribution, one can calculate the lift and drag over the cylinder

$$L' = -\int_0^{2\pi}(p(\theta) - p_\infty)\sin\theta a d\theta,\ D' = -\int_0^{2\pi}(p(\theta) - p_\infty)\cos\theta a d\theta \tag{2.38}$$

Substituting for $p(\theta)$ in the above formulae yields $L' = \rho U\Gamma$ and $D' = 0$, the latter being an obvious consequence of symmetry.

Another, more practical means of controlling the circulation is by giving the cylinder a rotational speed Ω. Any real fluid having viscosity, the particles in contact with the cylinder will rotate at speed Ω, thus entraining the nearby particles through viscous shear forces to rotate also. As a result, a circulatory flow is created. The relationship between the rotation speed Ω and the circulation Γ is not straightforward, but can be studied numerically or experimentally. See for example the discussion, in the book by White [1], of the Magnus effect and the application to the Flettner rotor ship where the sails are replaced by rotating cylinders. According to Wikipedia (http://en.wikipedia.org): "Assisted by Albert Betz,

Fig. 2.10 Buckau Flettner rotor ship (from https://en.wikipedia.org/wiki/File:Buckau_Flettner_Rotor_Ship_LOC_37764u.jpg)

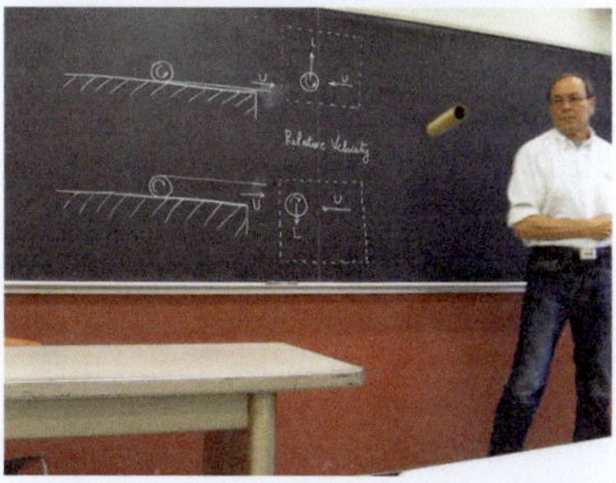

Fig. 2.11 JJC demonstrating the Magnus effect

Jacob Ackeret and Ludwig Prandtl, Flettner constructed an experimental rotor vessel, and in October 1924 the Germaniawerft finished construction of a large two-rotor ship named Buckau. The vessel was a refitted schooner which carried two cylinders (or rotors) about 15 m (50 ft) high, and 3 m (10 ft) in diameter, driven by an electric propulsion system of 50 hp (37 kW) power," Fig. 2.10.

A simple experiment with a small cardboard cylinder and the flat surface of a table demonstrates the Magnus effect. A small string is wound around the cylinder and its extremity is pulled briskly to accelerate the cylinder both in translation and rotation. Depending on the sense of rotation, the cylinder will fall quickly to the ground, or fly for a short time well above the table, see Fig. 2.11.

2.5 Flow Past Arbitrary Airfoils

2.5.1 Kutta-Joukowski Lift Theorem

The theory of conformal mapping tells us that it is possible to map any simply connected profile onto a circular cylinder. In the transformation, the flow past the profile maps into the flow past a circular cylinder derived earlier. The far field flow is preserved. The inverse transformation will therefore describe the exact solution of the flow past the original profile. One such transformation is the Joukowski transformation

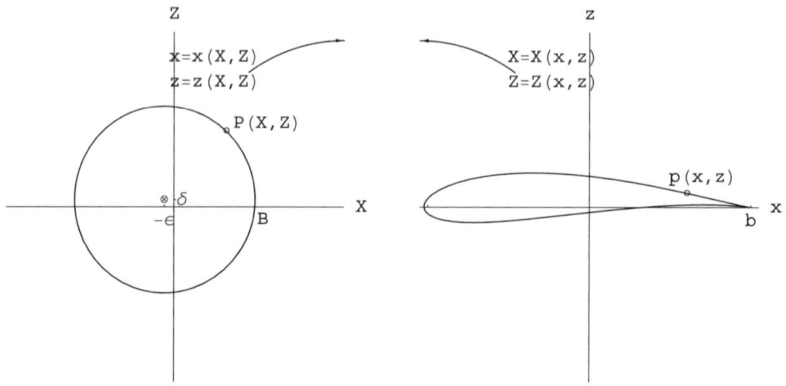

Fig. 2.12 Cylinder plane (X, Z) and physical plane (x, z)

$$x = X\left(1 + \frac{a^2}{X^2 + Z^2}\right), \quad z = Z\left(1 - \frac{a^2}{X^2 + Z^2}\right) \tag{2.39}$$

where (x, z) represent the physical plane of the profile and (X, Z) the plane of the cylinder. Notice that the transformation preserves the far field (i.e. $x \to X$ and $z \to Z$ far from the origin). A cylinder of radius r_0, where $r_0^2 = (a + \epsilon)^2 + \delta^2$ centered at $X = -\epsilon,\ Z = \delta$ and passing through point B located at $X = a,\ Z = 0$ maps onto a family of Joukowski airfoils depending on ϵ and δ. The profile has a cusp at the trailing edge, point b, image of point B. See Fig. 2.12. In the figure, p is the image of P through the transformation.

The polar coordinate representation of the cylinder is

$$X = r(\theta)\cos\theta,\ Z = r(\theta)\sin\theta \tag{2.40}$$

where $r(\theta) = -(\epsilon\cos\theta - \delta\sin\theta) + \sqrt{a^2 + 2\epsilon a + (\epsilon\cos\theta - \delta\sin\theta)^2}$.

Through the mapping, the parametric representation of the Joukowski profile is obtained

$$\begin{cases} x(\theta) = \left(r(\theta) + \frac{a^2}{r(\theta)}\right)\cos\theta \\ z(\theta) = \left(r(\theta) - \frac{a^2}{r(\theta)}\right)\sin\theta \end{cases} \tag{2.41}$$

In the mapping, the doublet and potential vortex contribute to a distribution of singularities (sources, sinks and vorticity distributions) inside the profile. In the far field, however, the leading terms in the perturbation velocity field correspond again to a doublet (closed obstacle) and a potential vortex with the same net circulation Γ. To calculate the force on an arbitrary profile, one can apply the momentum theorem to the domain inside a large circle C of radius R centered at the origin

$$\mathbf{F}' = -\oint_C \left[\rho\mathbf{V}(\mathbf{V}.\mathbf{n}) + p\mathbf{n}\right] R d\theta \tag{2.42}$$

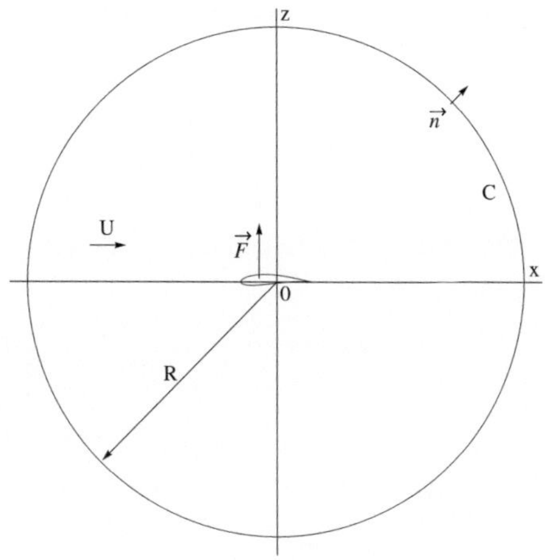

Fig. 2.13 Control surface for application of the momentum theorem

where $\mathbf{F'}$ represents the force per unit span applied by the fluid to the profile. See Fig. 2.13.

We will assume that any solution near C can be expanded asymptotically as

$$v_r = U\cos\theta + O\left(\frac{1}{R^2}\right), \quad v_\theta = -U\sin\theta - \frac{\Gamma}{2\pi R} + O\left(\frac{1}{R^2}\right) \tag{2.43}$$

Let $\mathbf{n} = (\cos\theta, \sin\theta)$ then

$$\mathbf{V}.\mathbf{n} = v_r = U\left(1 - \frac{a^2}{R^2}\right)\cos\theta = U\cos\theta + O\left(\frac{1}{R^2}\right) \tag{2.44}$$

The force will be evaluated in the Cartesian coordinate system. The Cartesian form of $\mathbf{V}$ is

$$\mathbf{V} = \left(U + \frac{\Gamma}{2\pi R}\sin\theta + O\left(\frac{1}{R^2}\right)\right)\mathbf{i} + \left(-\frac{\Gamma}{2\pi R}\cos\theta + O\left(\frac{1}{R^2}\right)\right)\mathbf{k} \tag{2.45}$$

where $\mathbf{i}$ and $\mathbf{k}$ are the unit vectors of the x and z-axis respectively. Hence, the first term in the integrand Eq. (2.42) reads

$$\rho\mathbf{V}(\mathbf{V}.\mathbf{n}) = \rho\left[\left(U + \frac{\Gamma}{2\pi R}\sin\theta + O\left(\frac{1}{R^2}\right)\right)\mathbf{i} + \left(-\frac{\Gamma}{2\pi R}\cos\theta + O\left(\frac{1}{R^2}\right)\right)\mathbf{k}\right] \times \left(U\cos\theta + O\left(\frac{1}{R^2}\right)\right) \tag{2.46}$$

When integrated on the circle, the $\cos\theta$ and $\sin\theta\cos\theta$ terms vanish and there remains only

$$\oint_C \rho \mathbf{V}(\mathbf{V}.\mathbf{n}) R d\theta = \oint_C \rho \left[O\left(\frac{1}{R}\right)\mathbf{i} + \left(-\frac{U\Gamma}{2\pi}\cos^2\theta + O\left(\frac{1}{R}\right)\right)\mathbf{k}\right] d\theta \quad (2.47)$$

The pressure term can also be expanded as

$$p(R,\theta) = p_\infty - \frac{1}{2}\rho\left(\frac{\Gamma}{2\pi R}\right)^2 - \frac{\rho U \Gamma}{2\pi R}\sin\theta + O\left(\frac{1}{R^2}\right) \quad (2.48)$$

The first two terms which are constant on the circle vanish in the integration and there remains for the second term in the momentum integral

$$\oint_C \left[-\frac{\rho U \Gamma}{2\pi R}\sin\theta + O\left(\frac{1}{R^2}\right)\right] (\cos\theta \mathbf{i} + \sin\theta \mathbf{k})\, R d\theta \quad (2.49)$$

Again, the product $\sin\theta\cos\theta$ vanishes in the integration and we are left with

$$\oint_C p\mathbf{n} R d\theta = \oint_C \left[O\left(\frac{1}{R}\right)\mathbf{i} + \left(-\frac{\rho U\Gamma}{2\pi}\sin^2\theta + O\left(\frac{1}{R}\right)\right)\mathbf{k}\right] \quad (2.50)$$

The results is

$$\mathbf{F}' = O\left(\frac{1}{R}\right)\mathbf{i} + \left(\rho U \Gamma + O\left(\frac{1}{R}\right)\right)\mathbf{k} \quad (2.51)$$

Now letting $R \to \infty$ yields the final result

$$\mathbf{F}' = \rho U \Gamma \mathbf{k} \quad (2.52)$$

This result is known as the Kutta-Joukowski lift theorem:

A uniform flow produces on a cylinder of arbitrary cross-section a lift force per unit span $L' = \rho U \Gamma$ that is proportional to the circulation Γ around the profile. The orientation of the force is obtained by rotating the incoming flow velocity vector 90° opposite to the circulation.

Another way to derive this theorem is to consider a control volume as in Fig. 2.14.

If the top and bottom boundaries of the control volume are far from the obstacle (distance H), the pressure will have the atmospheric value and the velocity will be the undisturbed velocity. The circulation, however, will not vanish as it remains constant over any closed curve enclosing the obstacle. The momentum theorem, Eq. (2.42), in projection on the z-axis reduces to

$$L' = -\int_{-\infty}^{\infty} \rho \left([(U+u)w]_{in} - [(U+u)w]_{out}\right) dz \quad (2.53)$$

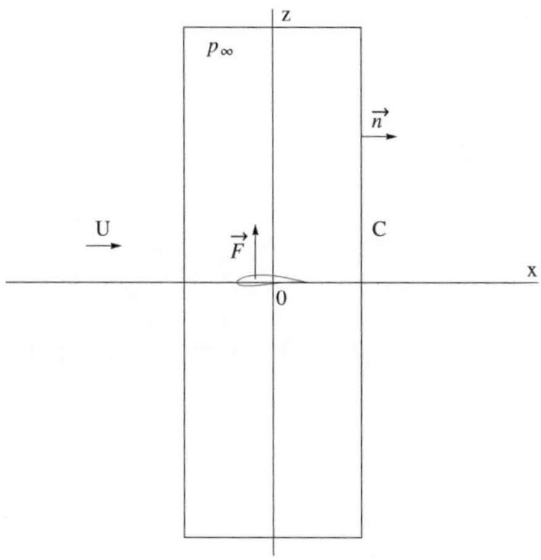

Fig. 2.14 Control volume for the evaluation of lift

where pressure does not contribute because it is atmospheric on the upper and lower surfaces and its action is perpendicular to the z-axis along the side boundaries. Since u and w are of order $O\left(\frac{1}{\sqrt{D^2+H^2}}\right)$ or higher, where D $(H \gg D)$ is the distance of the side boundaries to the z-axis, the result can be expressed as

$$L' = \rho U \Gamma + O\left(\frac{H}{D^2 + H^2}\right) \tag{2.54}$$

The final result is obtained by letting $D \to \infty$. A similar approach, with a control volume where now $D \gg H$ can be used to prove that the drag $D' = 0$. This result is consistent with the calculation of the lift over the cylinder, since with the Joukowski transformation or any transformation of the obstacle to a cylinder that preserves the far field, the lift over the obstacle is the same as the lift over the cylinder at the same angle of incidence. Furthermore, as a consequence of the conservation of momentum, the lift is proportional to the circulation as $L' = \rho U \Gamma$ and the drag $D' = 0$.

2.5.2 The d'Alembert Paradox

The d'Alembert paradox results from the fact that when a profile is moving at uniform velocity in a fluid, the resulting force is perpendicular to the velocity. The drag is zero. This result is consistent with the inviscid, frictionless fluid assumption. But in fact, with all real fluids, there is friction along the surface of the profile and there will be (a small) drag. The analysis of viscous forces in the boundary layer will resolve this paradox.

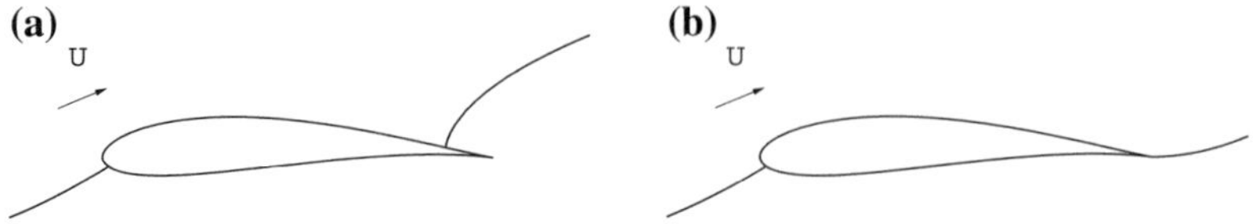

Fig. 2.15 **a** Profile that does not satisfy the K-J condition and **b** Profile that does satisfy the K-J condition

2.6 The Kutta-Joukowski Condition

With the Joukowski transformation, the mapping of the cylinder to the profile is singular at B, $X = a$, $Z = 0$ (the other singular point is at $X = -a$, $Z = 0$). This point corresponds to the cusped trailing edge B of the Joukowski profile, see Fig. 2.12. It has been found experimentally that the flow will leave the profile at a sharp trailing edge (cusp or small angle). This is due to viscosity. As the boundary layer grows from the leading edge to the trailing edge, the fluid particles do not have enough momentum to come around the trailing edge, and hence separate from the profile there. The sharp trailing edge of the profile plays the same role as the flap for the cylinder. This condition makes the inviscid flow solution past a profile unique by fixing the circulation.

The Kutta-Joukowski (K-J) condition states the the flow must leave the profile at the sharp trailing edge "smoothly". See Fig. 2.15.

Revisiting the Joukowski transformation of Sect. 2.5.1, one can see that with an incoming flow making an angle α with the X-axis and with an arbitrary circulation around the cylinder, the rear stagnation point will be located at an arbitrary point such as P in Fig. 2.12, therefore the stagnation streamline will leave the profile at point p and the K-J condition will not be satisfied. The velocity will be infinite at the trailing edge. In order to enforce the K-J condition, the circulation must be adjusted such that the rear stagnation point is located at point B. This uniquely determines the value of the circulation Γ as a function of incidence α.

2.7 Definitions

The chord c of a profile is the radius of the largest circle centered at the trailing edge and touching the leading edge.

The incidence angle or angle of attack is defined as the angle between the profile chord and the incoming flow velocity vector.

It is convenient to use dimensionless coefficients to represent pressure, forces and moments. The pressure coefficient is defined as

$$C_p = \frac{p - p_\infty}{\frac{1}{2}\rho V_\infty^2} = 1 - \left(\frac{V}{V_\infty}\right)^2 \tag{2.55}$$

where the denominator, $\frac{1}{2}\rho V_\infty^2$ is called the dynamic pressure (has dimension of a pressure, unit Pa) and use has been made of the Bernoulli equation.

The lift, drag and moment coefficients per unit span are made dimensionless, in 2-D flow, with the dynamic pressure and a reference length or length squared as

$$C_l = \frac{L'}{\frac{1}{2}\rho V_\infty^2 c}, \quad C_d = \frac{D'}{\frac{1}{2}\rho V_\infty^2 c}, \quad C_{m,o} = \frac{M'_{,o}}{\frac{1}{2}\rho V_\infty^2 c^2} \tag{2.56}$$

where the prime indicates that the force or moment is per unit span and the chord is the reference length in all cases in 2-D. Note that lower case subscripts will be used for two-dimensional coefficients to leave upper-case subscripts notation for wing, tail and complete configurations. Here, the moment coefficient is taken about point O, which in general will represent the leading edge of a profile or the nose of the airplane.

Other important quantities are:

- the lift slope $\frac{dC_l}{d\alpha}$
- the zero incidence lift coefficient $C_{l,0}$
- the moment slope $\frac{dC_{m,o}}{d\alpha}$
- the zero incidence moment coefficient $C_{m,o,0}$.

2.7.1 Center of Pressure—Aerodynamic Center

The center of pressure (c.p.) is the point about which the moment of the aerodynamic forces is zero.

The aerodynamics center (a.c.), also called neutral point, is the point about which the moment of the aerodynamic forces is independent of incidence.

2.7.2 Results for the Circular Cylinder

The reference length for the circular cylinder is the diameter $c = 2a$.

The pressure coefficient on the cylinder can be obtained from the velocity derived earlier

$$C_p(\theta) = 1 - \left(\frac{V(\theta)}{V_\infty}\right)^2 = 1 - \left(\frac{V_\theta}{U}\right)^2 = 1 - \left(2\sin\theta + \frac{\Gamma(\alpha)}{2\pi U a}\right)^2 \tag{2.57}$$

where $\Gamma(\alpha) = 4\pi U a \sin\alpha$. See Fig. 2.16 for the two cases $\alpha = 0$ and $\alpha = \frac{\pi}{2}$.

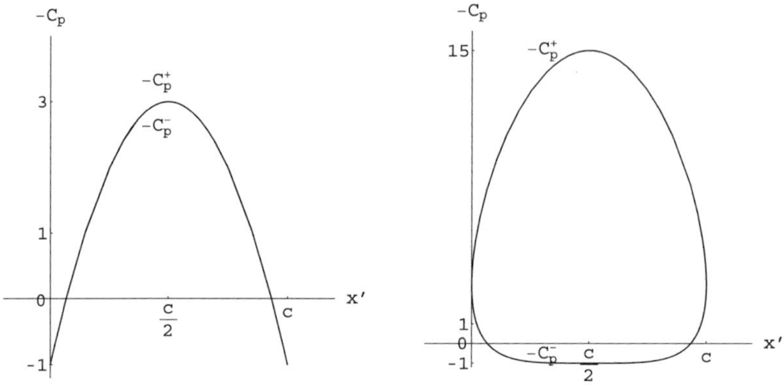

Fig. 2.16 Pressure coefficients for the cylinder, $\Gamma = 0$ (*left*), $\Gamma = 4\pi a$ (*right*)

The lift coefficient is

$$C_l(\alpha) = 4\pi \sin\alpha \tag{2.58}$$

This is a purely theoretical result, as real flow effects will dramatically change the flow field, compared to that shown in Fig. 2.9, even for small incidence angles. From the inviscid flow point of view, however, the cylinder at low incidence (near $\alpha = 0$) has a lift slope

$$\frac{dC_l}{d\alpha} = 4\pi \tag{2.59}$$

which is twice the result we will find in thin airfoil theory. The maximum lift that could be achieved theoretically with the small flap device is $C_{lmax} = 4\pi$, a very large number, when compared with high lift profiles, as we will see. This result corresponds to $\alpha = \frac{\pi}{2}$. See Fig. 2.17.

Fig. 2.17 Ideal cylinder-with-flap lift curve

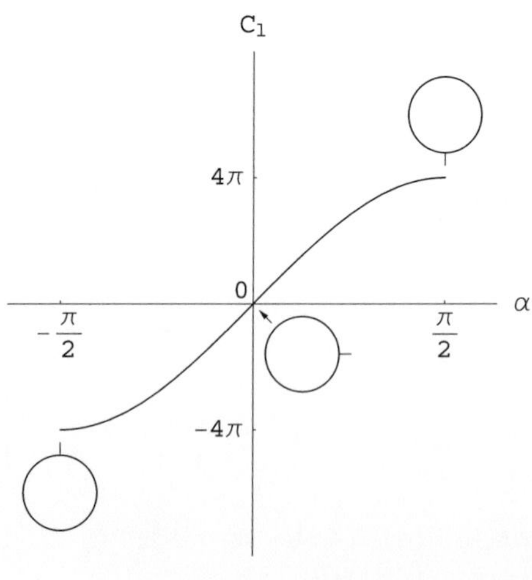

The drag coefficient for the cylinder, as for all airfoils in inviscid, incompressible flow is (d'Alembert paradox)

$$C_d = 0 \tag{2.60}$$

In frictionless flow, the contact force between the fluid and the obstacle is normal to the surface. Hence, the center O of the cylinder is the center of pressure, i.e. $C_{m,o} = 0$ for all α's. This result indicates that O is also the aerodynamic center.

2.8 Special Cases of Joukowski Airfoils

2.8.1 The Ellipse at Zero Incidence

Consider the circle centered at the origin and of radius $b \geq a$ in the cylinder plane. Its image in the physical plane, through the Joukowski transformation, Eq. (2.39), is an ellipse centered at the origin, with main axis along Ox, from $-2a$ to $2a$ and with parametric representation

$$\begin{cases} x = \frac{b^2+a^2}{b} \cos\theta = \frac{c}{2} \cos\theta \\ z = \frac{b^2-a^2}{b} \sin\theta = \frac{e}{2} \sin\theta \end{cases} \tag{2.61}$$

where θ represents the polar angle in the cylinder plane. The ellipse has thickness e (small axis) and chord c (large axis). The potential function in the cylinder plane is

$$\Phi = U\left(r + \frac{b^2}{r}\right)\cos\theta \tag{2.62}$$

and the velocity components are

$$V_r = \frac{\partial \Phi}{\partial r} = U\left(1 - \frac{b^2}{r^2}\right)\cos\theta,\ V_\theta = \frac{1}{r}\frac{\partial \Phi}{\partial \theta} = -U\left(1 + \frac{b^2}{r^2}\right)\sin\theta \tag{2.63}$$

On the surface of the cylinder, this reduces to

$$V_r = 0,\ V_\theta = -2U\sin\theta \tag{2.64}$$

Using the chain rule

$$\begin{cases} \frac{\partial \Phi}{\partial r} = \frac{\partial \Phi}{\partial x}\frac{\partial x}{\partial r} + \frac{\partial \Phi}{\partial z}\frac{\partial z}{\partial r} \\ \frac{\partial \Phi}{\partial \theta} = \frac{\partial \Phi}{\partial x}\frac{\partial x}{\partial \theta} + \frac{\partial \Phi}{\partial z}\frac{\partial z}{\partial \theta} \end{cases} \tag{2.65}$$

and the Joukowski transformation

$$\begin{cases} x = \frac{r^2+a^2}{r}\cos\theta \\ z = \frac{r^2-a^2}{r}\sin\theta \end{cases} \tag{2.66}$$

the partial derivatives $\frac{\partial\Phi}{\partial x}$ and $\frac{\partial\Phi}{\partial z}$ can be solved for, on the ellipse as

$$\begin{cases} \frac{\partial\Phi}{\partial x} = 2U\frac{\left(1+\frac{a^2}{b^2}\right)\sin^2\theta}{\left(1+\frac{a^2}{b^2}\right)^2\sin^2\theta+\left(1-\frac{a^2}{b^2}\right)^2\cos^2\theta} \\ \frac{\partial\Phi}{\partial z} = -2U\frac{\left(1-\frac{a^2}{b^2}\right)\sin\theta\cos\theta}{\left(1+\frac{a^2}{b^2}\right)^2\sin^2\theta+\left(1-\frac{a^2}{b^2}\right)^2\cos^2\theta} \end{cases} \tag{2.67}$$

Upon eliminating θ in terms of x along the ellipse and a and b in terms of e and c, one obtains

$$\begin{cases} \frac{\partial\Phi}{\partial x} = U + u = U\frac{(1+\frac{e}{c})\left(1-(\frac{2x}{c})^2\right)}{1-\left(\frac{2x}{c}\right)^2+(\frac{e}{c})^2\left(\frac{2x}{c}\right)^2} \\ \frac{\partial\Phi}{\partial z} = w = \pm U\frac{(1+\frac{e}{c})\frac{e}{c}\frac{2x}{c}\sqrt{1-\left(\frac{2x}{c}\right)^2}}{1-\left(\frac{2x}{c}\right)^2+(\frac{e}{c})^2\left(\frac{2x}{c}\right)^2} \end{cases} \tag{2.68}$$

where the minus (plus) sign corresponds to the upper (lower) surface.

The above results reduce to the solutions for a circle ($\frac{e}{c} = 1$) and for the flat plate ($\frac{e}{c} = 0$). The surface speed on the ellipse is given by

$$\frac{V}{U} = \frac{\left(1+\frac{e}{c}\right)}{\sqrt{1+\left(\frac{e}{c}\right)^2\frac{\left(\frac{2x}{c}\right)^2}{1-\left(\frac{2x}{c}\right)^2}}} = \left(1+\frac{e}{c}-\frac{1}{2}\left(\frac{e}{c}\right)^2\frac{\left(\frac{2x}{c}\right)^2}{1-\left(\frac{2x}{c}\right)^2}+\cdots\right) \tag{2.69}$$

where the result has been expanded in terms of the small parameter $\frac{e}{c}$. The first order term in $\frac{e}{c}$ is constant (independent of x) and the same as the result of thin airfoil theory, as will be discussed in the next chapter. It is clear that any higher order term is singular at the leading edge (and trailing edge) for this blunt obstacle. The expansion is valid in the region $1-\left(\frac{2x}{c}\right)^2 > \frac{\left(\frac{e}{c}\right)^2}{1+\left(\frac{e}{c}\right)^2}$.

2.8.2 The Ellipse at Incidence

We proceed in the same way as before. The potential, with the flow coming at an angle α from the X-axis, reads

$$\Phi = U\left(r + \frac{b^2}{r}\right)\cos(\theta - \alpha) - \frac{\Gamma}{2\pi}(\theta - \alpha) \tag{2.70}$$

In this case, on the cylinder, the velocity components are

$$V_r = 0,\ V_\theta = -2U\sin(\theta - \alpha) - \frac{\Gamma}{2\pi b} \tag{2.71}$$

The circulation corresponding to a stagnation point at $\theta = 0$ is given by $\Gamma = 4Ub\sin\alpha$. On the surface of the ellipse, after some algebra, one finds

$$\left\{\begin{array}{l} \frac{\partial\Phi}{\partial x} = 2U\frac{\left(1+\frac{a^2}{b^2}\right)\left(\sin(\theta-\alpha)+\frac{\Gamma}{4Ub}\right)\sin\theta}{\left(1+\frac{a^2}{b^2}\right)^2\sin^2\theta+\left(1-\frac{a^2}{b^2}\right)^2\cos^2\theta} \\ \frac{\partial\Phi}{\partial z} = -2U\frac{\left(1-\frac{a^2}{b^2}\right)\left(\sin(\theta-\alpha)+\frac{\Gamma}{4Ub}\right)\cos\theta}{\left(1+\frac{a^2}{b^2}\right)^2\sin^2\theta+\left(1-\frac{a^2}{b^2}\right)^2\cos^2\theta} \end{array}\right. \tag{2.72}$$

Since the ellipse has a rounded trailing edge, there is no mechanism to control the circulation in inviscid flow and an ellipse at incidence will in general produce a flow with zero circulation. For $\Gamma = 0$ the two stagnation points are located at $\theta = \alpha$ and $\theta = \pi + \alpha$, see Fig. 2.18.

The pressure coefficient is given by

$$C_p(\theta) = 1 - \frac{4\sin^2(\theta - \alpha)}{\left(1 + \frac{a^2}{b^2}\right)^2\sin^2\theta + \left(1 - \frac{a^2}{b^2}\right)^2\cos^2\theta} \tag{2.73}$$

This is represented in Fig. 2.19 as $-C_p^+$ and $-C_p^-$ for the upper and lower surfaces, respectively.

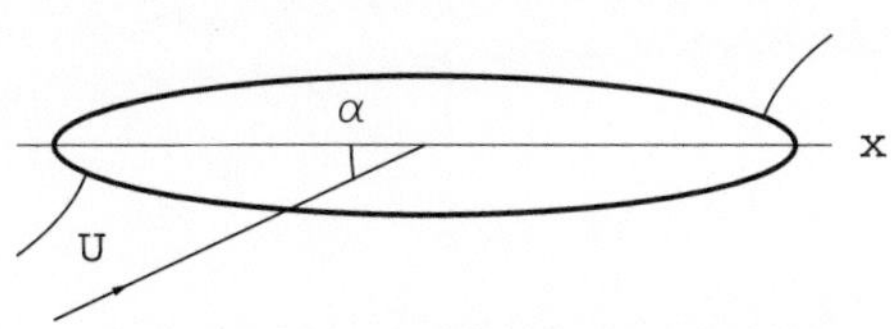

Fig. 2.18 Ellipse at 25° incidence ($\Gamma = 0$)

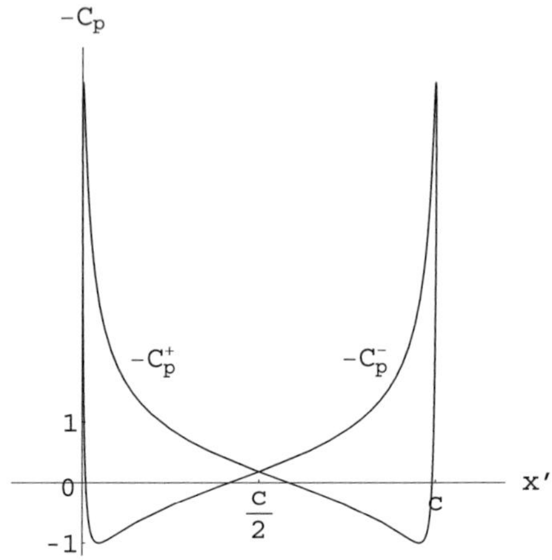

Fig. 2.19 Pressure coefficients for the ellipse at incidence ($\Gamma = 0$)

From anti-symmetry, the lift is zero. The moment however is not. It is obtained from

$$C_{m,o} = -\left(1 - \left(\frac{e}{c}\right)^2\right) \int_0^{2\pi} \frac{\sin^2(\theta - \alpha)\sin\theta\cos\theta}{\left(1 + \frac{a^2}{b^2}\right)^2 \sin^2\theta + \left(1 - \frac{a^2}{b^2}\right)^2 \cos^2\theta} d\theta \qquad (2.74)$$

where O stands for the center of the ellipse, where the moment is calculated. After lengthy algebra, the result reads

$$C_{m,o} = \frac{\pi}{2}\left(1 - \left(\frac{e}{c}\right)^2\right)\sin\alpha\cos\alpha \qquad (2.75)$$

Again, the results for the circle and the flat plate are obtained for $\frac{e}{c} = 1$ and $\frac{e}{c} = 0$, respectively.

In order to force the flow to stagnate at $\theta = 0$, a small flap can be added at the trailing edge of the ellipse. In this case, the circulation corresponding to a stagnation point at $\theta = 0$ is given by $\Gamma = 4\pi U b \sin\alpha$ and the velocity components can be written as

$$\begin{cases} \frac{\partial \Phi}{\partial x} = 4U \frac{\left(1+\frac{a^2}{b^2}\right)\sin\frac{\theta}{2}\cos(\frac{\theta}{2}-\alpha)\sin\theta}{\left(1+\frac{a^2}{b^2}\right)^2\sin^2\theta+\left(1-\frac{a^2}{b^2}\right)^2\cos^2\theta} \\ \frac{\partial \Phi}{\partial z} = -4U \frac{\left(1-\frac{a^2}{b^2}\right)\sin\frac{\theta}{2}\cos(\frac{\theta}{2}-\alpha)\cos\theta}{\left(1+\frac{a^2}{b^2}\right)^2\sin^2\theta+\left(1-\frac{a^2}{b^2}\right)^2\cos^2\theta} \end{cases} \qquad (2.76)$$

The two stagnation points are now located at $\theta = 0$ and $\theta = \pi + 2\alpha$. The corresponding flow is sketched in Fig. 2.20.

The surface pressure distribution for the ellipse with a small flap is shown in Fig. 2.21.

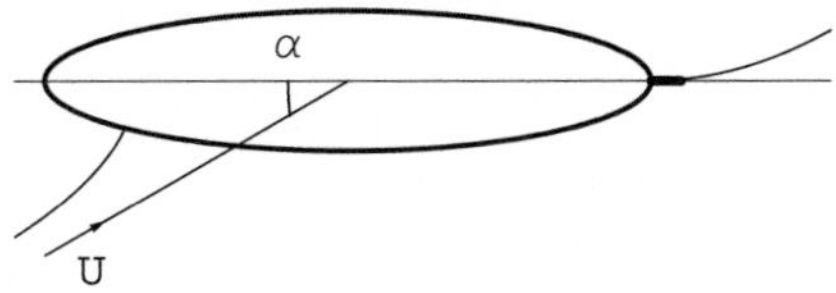

Fig. 2.20 Flow past an ellipse with a small flap at the trailing edge

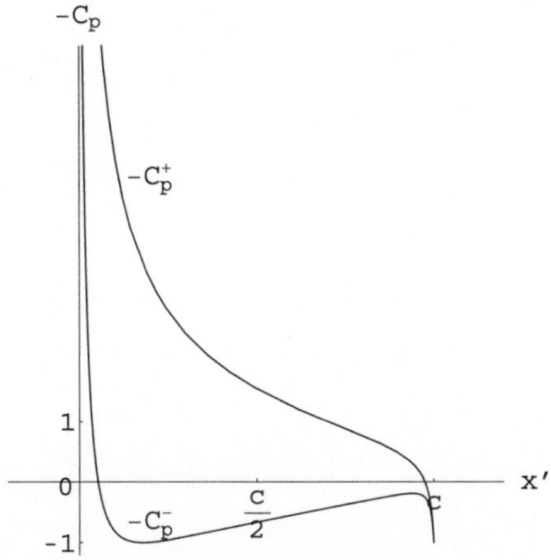

Fig. 2.21 Pressure distribution on the ellipse with small flap at the trailing edge

The lift coefficient is given by the Kutta-Joukowski lift theorem

$$C_l = \frac{2\Gamma}{Uc} \tag{2.77}$$

where $c = 2(b^2 + a^2)/b$. Substituting the value of the circulation gives

$$C_l = 4\pi \frac{b^2}{b^2 + a^2} \sin\alpha = 2\pi \left(1 + \frac{e}{c}\right) \sin\alpha \tag{2.78}$$

The results for the cylinder and the flat plate are obtained for $e/c = 1$ and $e/c = 0$, respectively. In small perturbation theory, the term e/c can be neglected as a second-order term when multiplied by $sin\alpha \simeq \alpha$.

2.8.3 The Flat Plate at Incidence

For this case, Joukowski transformation has $\epsilon = \delta = 0$. The flat plate is mapped into a circle with its center at the origin in the (X, Z)-plane, Fig. 2.22. The mapping can be interpreted as a conformal transformation from the cylinder plane to the physical plane such that $x = \Phi/U$, $z = \Psi/U$, where Φ and Ψ are the velocity potential and stream function of the flow past a circular cylinder without circulation. Indeed, Eqs. (2.30) and (2.31) written in terms of X and Z read

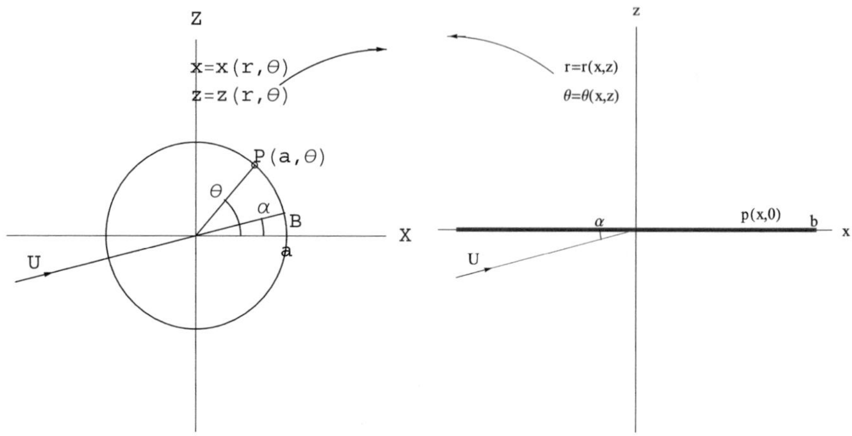

Fig. 2.22 Flat plate transformation

$$x = \frac{\Phi}{U} = \left(r + \frac{a^2}{r}\right)\cos\theta = X\left(1 + \frac{a^2}{X^2 + Z^2}\right) \tag{2.79}$$

$$z = \frac{\Psi}{U} = \left(r - \frac{a^2}{r}\right)\sin\theta = Z\left(1 - \frac{a^2}{X^2 + Z^2}\right) \tag{2.80}$$

In this transformation, the cylinder of radius a centered at the origin, which corresponds to the stagnation streamline $\Psi = 0$, is mapped onto the x-axis ($z = 0$), in the interval $[-2a, 2a]$.

This case corresponds to $b = a = \frac{c}{4}$ in Sects. 2.8.1 and 2.8.2, and the results derived previously can be used.

In the mapping, the flat plate parametric representation is given by

$$x = 2acos\theta = \frac{c}{2}cos\theta,\ z = 0 \tag{2.81}$$

and the velocity components in the physical plane are

$$\begin{cases} \frac{\partial\Phi}{\partial x} = 2U\frac{\left(\sin(\theta-\alpha)+\frac{\Gamma}{4\pi a}\right)}{\sin\theta} \\ \frac{\partial\Phi}{\partial z} = 0 \end{cases} \tag{2.82}$$

Forcing the circulation to be zero yields an anti-symmetric flow that does not satisfy the Kutta-Joukowski condition, Fig. 2.23. The lift is zero, but the moment coefficient about the mid-plate reduces to

$$C_{m,o} = \frac{\pi}{2}\sin\alpha\cos\alpha \tag{2.83}$$

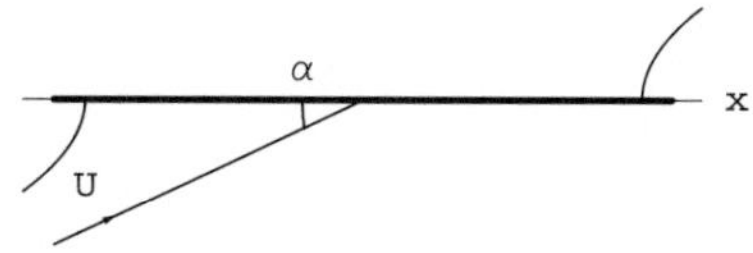

Fig. 2.23 Flat plate at 25° incidence ($\Gamma = 0$)

Making use of the mapping, the velocity on the flat plate now reads

$$\begin{cases} \frac{\partial \Phi}{\partial x} = 2U\left(\cos\alpha \mp \sin\alpha \frac{\frac{2x}{c}}{\sqrt{1-(\frac{2x}{c})^2}}\right) \\ \frac{\partial \Phi}{\partial z} = 0 \end{cases} \quad (2.84)$$

where the minus (plus) sign corresponds to the upper (lower) surface. The stagnation points are at $x = \pm\frac{c}{2}\cos\alpha$, $z = 0$.

When the Kutta-Joukowski condition is applied, the velocity on the flat plate becomes

$$\frac{\partial \Phi}{\partial x} = U\frac{\cos\left(\frac{\theta}{2} - \alpha\right)}{\cos\frac{\theta}{2}}, \quad \frac{\partial \Phi}{\partial z} = 0 \quad (2.85)$$

The streamline in this case leaves the trailing edge "smoothly", Fig. 2.24.

Note that the velocity is infinite at $\theta = \pi$ (leading edge) in all cases except for $\alpha = 0$. The stagnation point corresponds to $\theta = \pi + 2\alpha$, i.e. $x = -\frac{c}{2}cos2\alpha$. It is also worthy to note that the exact solution is consistent with the thin airfoil theory result that will be derived in the next chapter, since the flat plate is the ultimate thin airfoil, having zero thickness and zero camber. Another assumption that will be made is that of small α. If we map the flat plate to the segment $[0, c]$ with $x' = \frac{c}{2}\left(1 + \frac{x}{2a}\right)$ the horizontal velocity component becomes

$$\frac{\partial \Phi}{\partial x} = U\left(cos\alpha \pm sin\alpha\sqrt{\frac{c-x'}{x'}}\right) \simeq U\left(1 \pm \alpha\sqrt{\frac{c-x'}{x'}}\right) \quad (2.86)$$

The plus (minus) sign corresponds to the upper (lower) surface. This last result is the thin airfoil result with a perturbation u above and below the plate that is symmetric about the x'-axis.

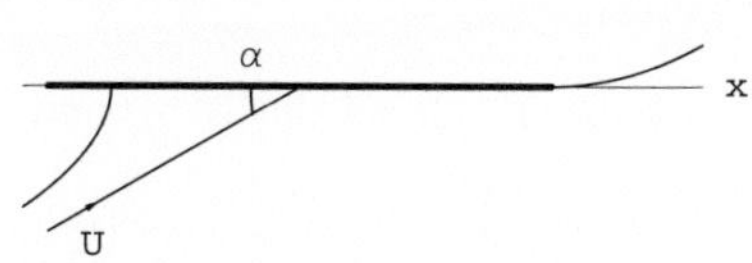

Fig. 2.24 Flat plate at 25° incidence ($\Gamma = 4\pi a sin\alpha$)

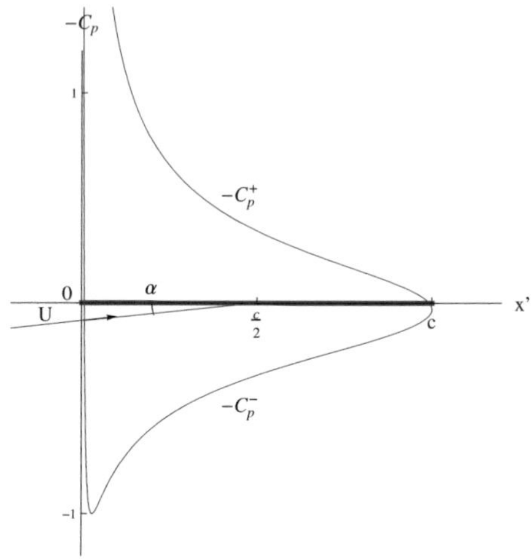

Fig. 2.25 Pressure coefficient along the plate at $\alpha = 10°$

The pressure coefficient is given by

$$C_p(\theta) = 1 - \left(\frac{cos(\frac{\theta}{2} - \alpha)}{cos\frac{\theta}{2}} \right)^2 \tag{2.87}$$

The result for $-C_p$ is displayed in Fig. 2.25. It is not symmetric w.r.t. the Ox'-axis. If we anticipate again on thin airfoil theory, the linearized C_p will be symmetric, i.e. $C_p^- = -C_p^+$.

The lift coefficient is obtained from the Kutta-Joukowski lift theorem

$$C_l = \frac{\rho U \Gamma}{\frac{1}{2}\rho U^2 4a} = 2\pi sin\alpha \tag{2.88}$$

It is interesting, however, to consider the integration of the pressure along the flat plate directly. The result is expected to be normal to the plate, i.e. in the z-direction. Indeed, integrating in θ the elementary forces due to pressure around the plate yields the pressure contribution $\mathbf{F'_p} = (0, F'_p)$, where

$$F'_p = 4\pi\rho U^2 a sin\alpha\, cos\alpha \tag{2.89}$$

On the other hand, the lift per unit span, as given by the Kutta-Joukowski theorem is

$$L' = 4\pi\rho U^2 a sin\alpha \tag{2.90}$$

acting in a direction perpendicular to the incoming velocity vector as $\mathbf{L'} = (-L' sin\alpha, L' cos\alpha)$. The difference between these results is due to a horizontal suction force $\mathbf{F'_s} = (F'_s, 0)$ acting at the leading edge, in the direction of the negative x-axis, that is due to the leading edge singularity. This missing component, parallel to the camber line and of magnitude

$$F'_s = -4\pi\rho U^2 a sin^2\alpha \tag{2.91}$$

points to the fact that integration of pressure does not capture the localized force that exists at a sharp leading edge.

The moment w.r.t. the leading edge is the sum of elementary moments

$$dM_{,o} = -\frac{1}{2}\rho U^2 C_p(\theta)(x(\theta) + 2a)dx(\theta) \tag{2.92}$$

Here, O stands for the leading edge. The moment is given by

$$M_{,o} = 2\rho U^2 a^2 \int_0^{2\pi} \left(1 - \left(\frac{cos(\frac{\theta}{2} - \alpha)}{cos\frac{\theta}{2}}\right)^2\right)(1 + cos\theta)sin\theta\, d\theta \tag{2.93}$$

Some simplifications occur when the half-angle $\frac{\theta}{2}$ is used. The result is

$$M_{,o} = -2\pi\rho U^2 a^2 sin2\alpha \tag{2.94}$$

and the moment coefficient reads

$$C_{m,o} = -\frac{\pi}{4} sin2\alpha \tag{2.95}$$

The moment of the aerodynamic forces can be calculated at an arbitrary point D along the x'-axis as $C_{m,D} = C_{m,o} + \frac{x'_D}{c} C_l$. The center of pressure satisfies $C_{m,c.p.} = 0$, i.e.

$$\frac{x'_{c.p.}}{c} = -\frac{C_{m,o}}{C_l} = \frac{1}{4} cos\alpha \simeq \frac{1}{4} \tag{2.96}$$

For small values of α, the center of pressure for the flat plate is located at the quarter-chord.

Taking the derivative w.r.t. α of the same general formula for the moment at point D and now setting the result to zero, gives the aerodynamics center

$$\frac{\partial C_{m,D}}{\partial \alpha} = -\frac{\pi}{2} cos2\alpha + \frac{x'_D}{c} 2\pi cos\alpha = 0 \tag{2.97}$$

For small values of α one obtains the small perturbation result derived in the next chapter

$$\frac{x'_{a.c.}}{c} = \frac{1}{4} \tag{2.98}$$

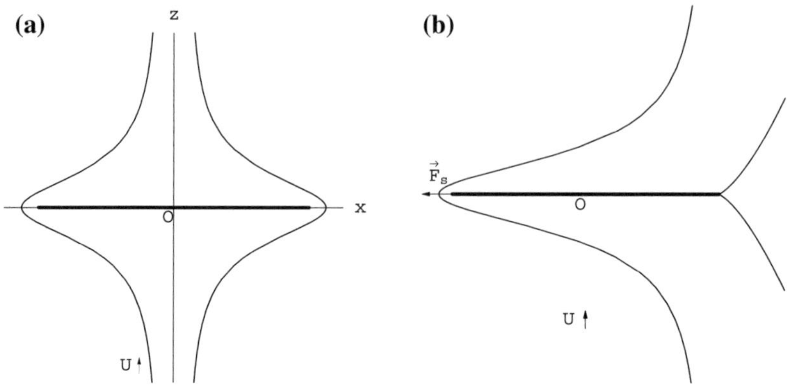

Fig. 2.26 Flat plate perpendicular to incoming flow **a** with zero circulation and **b** with trailing edge that satisfies the K-J condition, and suction force

To close this section, the limiting case of the flow coming perpendicular to the plate ($\alpha = 90°$) is considered, with the plate having zero circulation or satisfying the Kutta-Joukowski condition. The two results are shown in Fig. 2.26.

In Fig. 2.26a, the flow is perfectly symmetric about the mid-plate and the net resulting force and moments are zero. Again, this is an idealized flow representation and viscosity will radically change the picture by having separation at the sharp edges of the plate. The same can be said of Fig. 2.26b in terms of realism of the picture, but it is intriguing to observe that there is circulation, hence lift, and the force is not perpendicular to the flat plate, but aligned with the plate and to the left. The occurrence of such a force, again a suction force, will be discussed in more details in the next chapter.

2.8.4 Circular Arc

For this case, $\epsilon = 0$ and $\delta > 0$. The polar coordinates are centered at $(0, \delta)$. The potential function reads

$$\Phi = 2Ua' \cos(\theta - \alpha) - \frac{\Gamma}{2\pi}(\theta - \alpha) \tag{2.99}$$

where $a'^2 = a^2 + \delta^2$. The azimuthal component of velocity on the cylinder is

$$V_\theta = -2U \sin(\theta - \alpha) - \frac{\Gamma}{2\pi a'} \tag{2.100}$$

The circulation, that places the stagnation point at $(0, a)$ is

$$\Gamma = 4\pi U\,(a \sin\alpha + \delta \cos\alpha) \tag{2.101}$$

and the lift coefficient as given by the Kutta-Joukowski lift theorem

$$C_l = 2\pi \left(\sin\alpha + \frac{\delta}{a} \cos\alpha \right) \tag{2.102}$$

The most interesting situation for this airfoil corresponds to $\alpha = 0$ where a symmetry w.r.t. the z-axis (and Z-axis) is obtained. In this case, the surface velocity reduces to a simple expression

$$\frac{V}{U} = \frac{a^2 + 2\delta \sin\theta \sqrt{a^2 + (\delta \sin\theta)^2}}{a^2 + \delta^2} \tag{2.103}$$

The pressure coefficients are shown in Fig. 2.27 for $\alpha = 0$, $\frac{d}{c} = 0.1$ and $\frac{d}{c} = 0.2$.

A consequence of symmetry is that the leading edge, as the trailing edge, satisfies a Kutta-Joukowski condition. This is only possible for $\alpha = 0$. The leading edge is "adapted". $\alpha = 0$ is the ideal angle of attack for the circular arc. The lift coefficient in this case is $C_l = 2\pi \frac{\delta}{a}$. The lift per unit span

$$L' = 4\pi\rho U^2 \delta = 2\pi\rho U^2 d \tag{2.104}$$

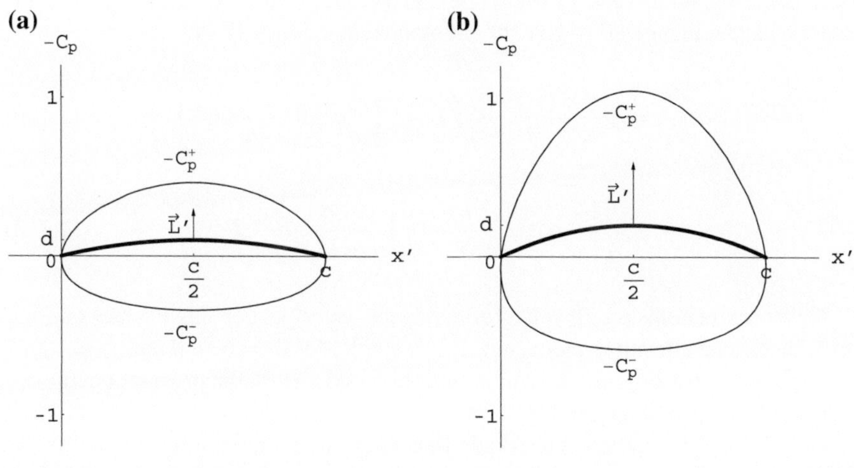

Fig. 2.27 C_p distributions for the circular arc at $\alpha = 0$: **a** $\frac{d}{c} = 0.1$, **b** $\frac{d}{c} = 0.2$

only depends on d, where $d = 2\delta$ represents the height of the circular arc in the physical plane (the chord is $c = 4a$). This result was first obtained by Kutta in 1902. The center of pressure is at the center of the arc, i.e.

$$\frac{x'_{c.p.}}{c} = \frac{1}{2} \tag{2.105}$$

2.8.5 Joukowski Airfoil at Incidence

The same result for the potential function is obtained in this case with $\epsilon \geq 0$ and $\delta \geq 0$

$$\Phi = 2Ua' \cos(\theta - \alpha) - \frac{\Gamma}{2\pi}(\theta - \alpha) \tag{2.106}$$

where now $a'^2 = (a + \epsilon)^2 + \delta^2$, with the following formula for the circulation that enforces the Kutta-Joukowski condition

$$\Gamma = 4\pi U\,((a + \epsilon)\sin\alpha + \delta\cos\alpha) \tag{2.107}$$

and lift coefficient

$$C_l = 2\pi\left(\left(1 + \frac{\epsilon}{a}\right)\sin\alpha + \frac{\delta}{a}\cos\alpha\right) \tag{2.108}$$

The exact solution is quite complicated and is best calculated numerically. However, assuming small values for thickness and camber, i.e. $\frac{\epsilon}{a} \ll 1$ and $\frac{\delta}{a} \ll 1$, the formula can be linearized to give the x-component of velocity as

$$\frac{\partial\Phi}{\partial x} = U\frac{\cos\left(\frac{\theta}{2} - \alpha\right)}{\cos\left(\frac{\theta}{2}\right)}\left(1 - 2\left\{\frac{\epsilon}{a}\cos\theta - \frac{\delta}{a}\sin\theta\right\}\right. \tag{2.109}$$

$$\left. + \left\{\frac{\epsilon}{a}\cos\alpha - \frac{\delta}{a}\sin\alpha\right\}\frac{\cos\left(\frac{\theta}{2}\right)}{\cos\left(\frac{\theta}{2} - \alpha\right)}\right) \tag{2.110}$$

This formula holds for all values of incidence, $|\alpha| \leq \frac{\pi}{2}$. For small values of α the small disturbance result is recovered. For example, for $\alpha = \delta = 0$ and $\frac{\epsilon}{a} = 0.077$ corresponding to a 10 % thick symmetric Joukowski airfoil, the pressure coefficient at zero incidence is shown in Fig. 2.28.

The thin airfoil solution is a straight line, since the perturbation velocity in x' is linear

$$U + u(x') = U\left(1 + \frac{4}{3\sqrt{3}}\frac{e}{c}\left(3 - 4\frac{x'}{c}\right)\right) \tag{2.111}$$

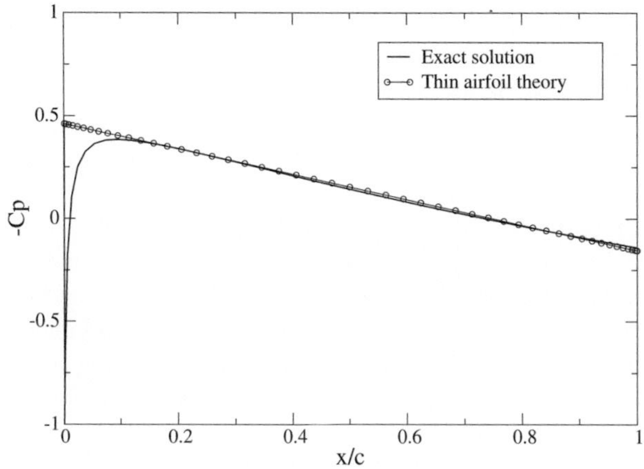

Fig. 2.28 Comparison of exact and numerical (small disturbance) C_p distributions

To find the solution in the physical plane and in particular the surface velocity for arbitrary Joukowski airfoils, the values of the potential function at certain points on the circle, image of the profile, are first calculated. The same values are assigned to the corresponding points on the airfoil which can be found from the transformation formulae. The surface velocity can be evaluated numerically as

$$V = \frac{\Delta \Phi}{\Delta s} \tag{2.112}$$

where $\Delta \Phi$ is the difference between the values of Φ at the two close points on the surface and Δs is the length of the airfoil segment between these two points. The surface pressure distribution is readily obtained from Bernoulli's law, i.e.

$$C_p = 1 - \left(\frac{V}{U}\right)^2 \tag{2.113}$$

2.9 Summary of Chapter 2

In this Chapter, mathematical solutions are found analytically for a theoretical model of steady, two-dimensional, inviscid, incompressible, adiabatic flow with uniform upstream conditions. This flow has no vorticity. The governing first order partial differential equations (Cauchy/Riemann system) for the velocity components are linear, representing conservation of mass and the irrotationality condition. The pressure is decoupled and is obtained from Bernoulli's law.

The Cauchy/Riemann system can be replaced by a single potential or stream-function equation which satisfies Laplace equation. The uniform flow is a trivial solution of these equations. Other fundamental solutions are found (i.e. source/sink and potential vortex) and more general solutions are constructed via superposition, for example doublet and half-body flows.

The solution for the flow over a circular cylinder is of particular interest. The boundary conditions are the tangency condition at the solid surface and uniform flow in the far field. From symmetry, the cylinder has zero lift and zero drag. The latter result is referred to d'Alembert Paradox. In the real flow, there is drag due to friction and due to the difference in pressure on the front and the back of the cylinder because of separation. However, this mathematical solution is useful since it provides the pressure distributions over thin bodies at small angle of attack via mapping techniques. For example, the Joukowski transformation provides analytical solutions for a family of symmetric shapes. In such cases, the flow is attached and the viscous effects are confined to the thin boundary layers, hence the mathematical model of inviscid flow yields a reasonable approximation.

A more interesting case is the flow with lift. Flow asymmetry is necessary to generate lift. One can think of a rotating cylinder. In the real flow, the effect of the rotation of the cylinder is transmitted through a viscous layer, augmenting the velocity on one side and retarding it on the other, thus creating a circulatory flow and lift. The inviscid flow model however, admits a slip velocity at the surface and has no mechanism to control the asymmetry other than adding a small flap to force flow stagnation at a specified point on the surface of the cylinder. The mathematical solution is obtained as uniform flow plus a doublet and a potential vortex. The doublet strength is determined to satisfy the tangency condition at the cylinder surface. In general the mathematical solution is not realistic since there is flow separation (unless the cylinder rotates so fast that the boundary layer is confined to a thin annulus attached to the surface). Through the Joukowski transformation however, the solution can be mapped to a flow over a lifting airfoil. To obtain a solution consistent with the experimental results, at least for small angles of attack when the flow is attached and the viscous effects are confined to thin boundary layers, the circulation in the mathematical model is chosen such that the flow leaves the trailing edge smoothly, in a direction bisecting the trailing edge angle. This celebrated Kutta-Joukowski condition renders the mathematical solution unique by choosing the most physically relevant solution which has no flow around the trailing edge, since in a real flow, the viscous effects will not allow such a behavior.

Once the vortex strength is determined, the lift is calculated via the Kutta-Joukowski lift theorem which is based on momentum balance. The same result can be obtained by integrating the surface pressure.

The Kutta-Joukowski transformation has allowed us to obtain exact solutions for the flow past ellipses, flat plates and Joukowski airfoils. The circles in the cylinder plane and their images in the physical plane are shown for example in Fig. 2.29.

Finally, the main formulae in this chapter are the Kutta-Joukowski lift theorem result $L' = \rho U \Gamma$ and the relation between the strength of the potential vortex and the position of the stagnation points on the cylinder. Through the Joukowski transformation, these formulae reveal the dependence of the lift on the angle of attack for a family of airfoils.

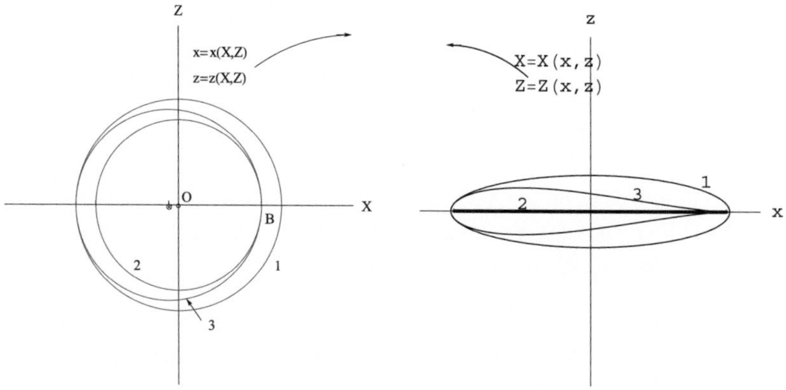

Fig. 2.29 Circles *1*, *2* and *3* in the cylinder plane and their images: *1* ellipse, *2* flat plate, *3* Joukowski profile

In the next chapter, thin airfoil theory is introduced. The same results can be obtained numerically, using for example the method of singularities.

More accurate analyses, where the viscous effects in the boundary layer are fully accounted for, are presented later. Also, compressibility and three-dimensional effects are dealt with in separate chapters.

2.10 Problems

2.10.1

A 2-D model for tornadoes can be described as a core cylinder of fluid with solid body rotation and a potential vortex flow outside the core. Plot the velocity distribution and streamlines. Plot the lines of constant potentials where they exist. What is Bernoulli's law for such a flow? Calculate the pressure distribution and in particular the pressure at the center (Hint: use normal momentum equation to calculate the pressure in the core).

2.10.2

A 2-D model for hurricanes can be described by a potential vortex and a source for the flow outside the core. Show that the streamlines are spirals. Find the velocity and pressure distributions.

2.10.3

Derive the formula for velocity distributions due to a doublet using summation of velocity vectors of a sink and a source of strength Q and a distance $2a$ apart, in the limit of $a \to 0$ and $Q \to \infty$, such that $aQ = D$.

2.10.4

Show that the velocity due to a doublet with its axis in the x-direction is the derivative with respect to x of the velocity field due to a source.

2.10.5

Show that a doublet can be produced by a pair of counter rotating potential vortices of strength Γ and $-\Gamma$ at a distance $2a$, along the z-axis, in the limit $a \to 0$ and $\Gamma \to \infty$, such that $a\Gamma$ is finite (Fig. 2.30).

2.10.6

Show that the velocity due to a doublet with its axis in the x-direction is the derivative w.r.t. z of the velocity field due to a potential vortex.

2.10.7

Obtain the velocity and pressure distributions around a Rankine body described by the closed streamline due to a source and a sink of equal strength and at a distance $2a$ apart in a uniform stream.

2.10.8

Obtain the velocity and pressure distributions around a Kelvin body described by the closed streamline due to a pair of potential vortices of equal strength and opposite direction in a uniform stream, perpendicular to their axis.

2.10.9

Show that the Joukowski transformation with $\epsilon = \delta = 0$, maps a circle to a flat plate with $c = 4a$, hence for a flat plate at incidence produces a lift $C_l = 2\pi \sin\alpha$. What is the drag coefficient C_d? Explain d'Alembert paradox, considering that the pressure acting normal to the flat plate will produce a force normal to the plate and not normal to the flow direction!

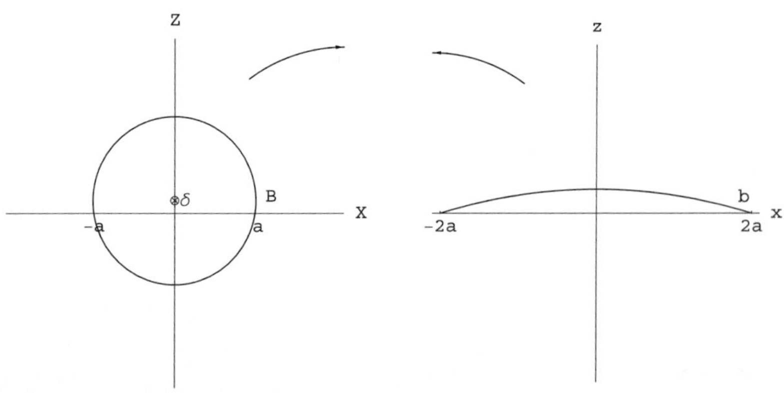

Fig. 2.30 Problem 2.10.10

2.10.10

Show that the Joukowski transformation with $\epsilon = 0$ and $0 < \delta \ll 1$ maps a circle to a shallow circular arc as in Fig. 2.27. Find the pressure distribution over the circular arc and calculate C_l and C_d. Where is the center of pressure? Where is the aerodynamic center?

2.10.11

Consider an ellipse at incidence. Show that the lift is zero. Consider next an ellipse with a small flap at the trailing edge. Find the lift (Hint: use Joukowski transformation).

Construct a symmetric airfoil with a blunt leading edge and a sharp trailing edge (Hint: use a parabola or a circle for the nose and a cubic or a quartic for the rest of the airfoil). Plot the airfoil shape.

Reference

1. White, F.M.: Fluid Mechanics, 7th edn. McGraw Hill, New York (2009)

Chapter 3
Inviscid, Incompressible Flow Past Thin Airfoils

3.1 Introduction

The solution of the inviscid, incompressible and irrotational (potential) two-dimensional flow past an arbitrary profile with a sharp trailing edge exists and can be obtained using complex variables and mapping techniques. However, this approach is very difficult in general, except for particular classes of profiles, such as the Joukowski airfoils.

A more practical approach consists in simplifying the model one step further by assuming what is called a "thin airfoil", an airfoil that creates only a small disturbance to the uniform incoming flow. It will then be possible to simplify the tangency condition, linearize the expression of the pressure coefficient and construct the solution of the flow past an arbitrary airfoil, using superposition of elementary solutions.

3.1.1 Definition of a Thin Airfoil

The thin airfoil geometry is defined with the profile chord aligned with the x-axis, from $x = 0$ to $x = c$, which corresponds to zero incidence. The (smooth) leading

J.J. Chattot and M.M. Hafez, *Theoretical and Applied Aerodynamics*,
DOI 10.1007/978-94-017-9825-9_3

edge is tangent to the z-axis at the origin, as in Fig. 3.2. The sharp trailing edge is located at $(c, 0)$.

The equations of the upper and lower surfaces of the airfoil are given by $f^+(x)$ and $f^-(x)$, respectively. The camber distribution $d(x)$ and thickness distribution $e(x) \geq 0$ are defined as follows

$$d(x) = \frac{1}{2}\left(f^+(x) + f^-(x)\right), \; e(x) = f^+(x) - f^-(x) \tag{3.1}$$

Note that, with the above definition of the airfoil geometry, $f^{\pm}(0) = f^{\pm}(c) = 0$. Hence, $d(0) = e(0) = d(c) = e(c) = 0$. With these distributions, the profile upper and lower surfaces are obtained from the superposition of camber and thickness as

$$f^+(x) = d(x) + \frac{1}{2}e(x), \; f^-(x) = d(x) - \frac{1}{2}e(x) \tag{3.2}$$

The curve $z = d(x)$ is called the mean camber line. Let d be its maximum absolute value. The maximum value of $e(x)$, e is called maximum thickness.

The relative camber of the airfoil is $\frac{d}{c}$ and $\frac{e}{c}$ is the relative thickness.

A thin airfoil is defined as an airfoil with small relative camber and relative thickness

$$\frac{d}{c} \ll 1, \; \frac{e}{c} \ll 1 \tag{3.3}$$

The Wright Brothers glider wings can be considered as equipped with thin airfoils as can be seen in Fig. 3.1.

3.1.2 *Profile at Incidence*

The thin profile is set at incidence α by performing a rotation of angle α in the clockwise direction, about the trailing edge. The exact transformation of a point $M(x, z)$ of the profile into $M'(x', z')$ reads

Fig. 3.1 View of Wright Brothers glider (from NASA, http://wright.nasa.gov/airplane/kite00.html)

$$\begin{cases} x' = x + (x - c)(\cos\alpha - 1) + z\sin\alpha \\ z' = z - (x - c)\sin\alpha + z(\cos\alpha - 1) \end{cases} \tag{3.4}$$

In thin airfoil theory, we will further assume that the incidence angle (in radians) is small

$$\alpha \ll 1 \tag{3.5}$$

With this assumption, the above transformation can be simplified to first order in $\frac{d}{c}$, $\frac{e}{c}$ and α as

$$\begin{cases} x' = x \\ z' = z - (x - c)\alpha \end{cases} \tag{3.6}$$

which corresponds to a shearing transformation. This is a great simplification, because the equation of the profile is not modified by the transformation other than by the addition of a term proportional to α, since the ordinate of the profile at incidence now reads $z^{\pm}(x) = f^{\pm}(x) + \alpha(c - x)$. See Fig. 3.2.

The question arises as to how small these coefficients have to be for the solution to be of practical use? In our experience, good results are obtained with relative cambers of the order of 6 %, relative thicknesses of up to 12 % and angles of incidence between -10 and $+10°$. As will be seen, when a profile is used at its "design condition", the viscous effects are minimized and the inviscid result will be representative of the real flow as far as the lift and moment coefficients are concerned. In off-design conditions, one should be more cautious with the bounds given above.

3.1.3 *Examples of Camber and Thickness Distributions*

A symmetric profile has zero camber, i.e. $d(x) = 0$. The simplest camber distribution is the parabolic camber line:

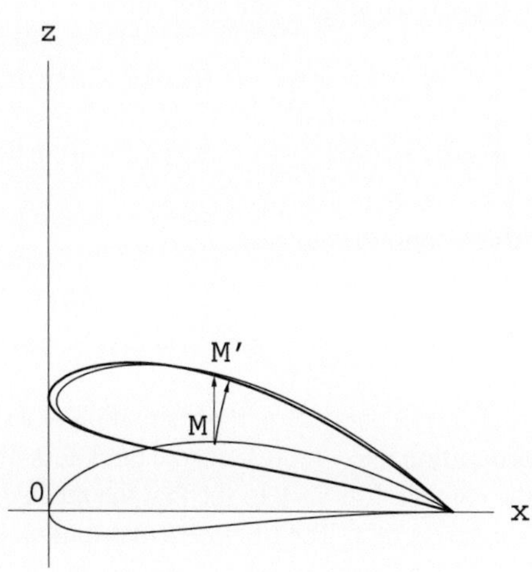

Fig. 3.2 Rotation and shearing transformation

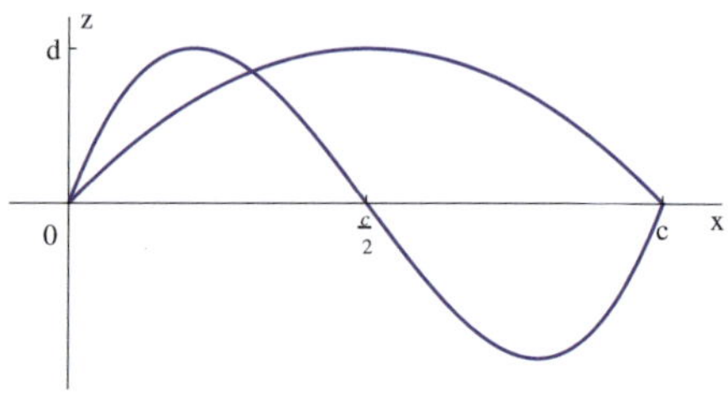

Fig. 3.3 Parabolic and cubic camber lines

$$d(x) = 4d\frac{x}{c}\left(1 - \frac{x}{c}\right) \tag{3.7}$$

A cubic distribution can represent a small windmill if an axis is placed at the mid-chord

$$d(x) = 6\sqrt{3}d\frac{x}{c}\left(1 - 2\frac{x}{c}\right)\left(1 - \frac{x}{c}\right) \tag{3.8}$$

see the camber distributions in Fig. 3.3, where the z-coordinate has been stretched.

A simple thickness distribution is represented by a semi-cubic

$$e(x) = \frac{3\sqrt{3}}{2}e\sqrt{\frac{x}{c}}\left(1 - \frac{x}{c}\right) \tag{3.9}$$

which gives a finite trailing edge angle. The maximum thickness is at $x = \frac{c}{3}$.

The Joukowski profile described in Sect. 2.5.1 can be linearized by expanding its analytic expression to first order in ϵ and δ, where (ϵ,δ) is related to the center of the cylinder image of the profile and also to thickness and camber. See Fig. 2.11. For $\epsilon \ll 1$ and $\delta \ll 1$, to first order the radius reads

$$r(\theta) = -(\epsilon\cos\theta - \delta\sin\theta) + a\left(1 + \frac{\epsilon}{a}\right) \tag{3.10}$$

Substituting this into the Cartesian coordinates of the profile and expanding in terms of the small coefficients gives

$$\begin{cases} x(\theta) = \left(a(1 + \frac{\epsilon}{a}) - (\epsilon\cos\theta - \delta\sin\theta) + a(1 - \frac{\epsilon}{a} + \frac{\epsilon}{a}\cos\theta - \frac{\delta}{a}\sin\theta\right)\cos\theta \\ z(\theta) = \left(a(1 + \frac{\epsilon}{a}) - (\epsilon\cos\theta - \delta\sin\theta) - a(1 - \frac{\epsilon}{a} + \frac{\epsilon}{a}\cos\theta - \frac{\delta}{a}\sin\theta\right)\sin\theta \end{cases} \tag{3.11}$$

which simplifies to give

$$\begin{cases} x(\theta) = 2a\cos\theta \\ z(\theta) = 2\epsilon(1 - \cos\theta)\sin\theta + 2\delta\sin^2\theta \end{cases} \tag{3.12}$$

This is a parametric representation of a Quasi-Joukowski airfoil. In $z(\theta)$, the term proportional to ϵ represents the thickness distribution and the term proportional to δ, the camber. It is easy to see that the mean camber line is parabolic by elimination of θ in terms of x. The thickness distribution can also be expressed in terms of x as

Fig. 3.4 Semi-cubic, Quasi-Joukowski and other thickness distributions

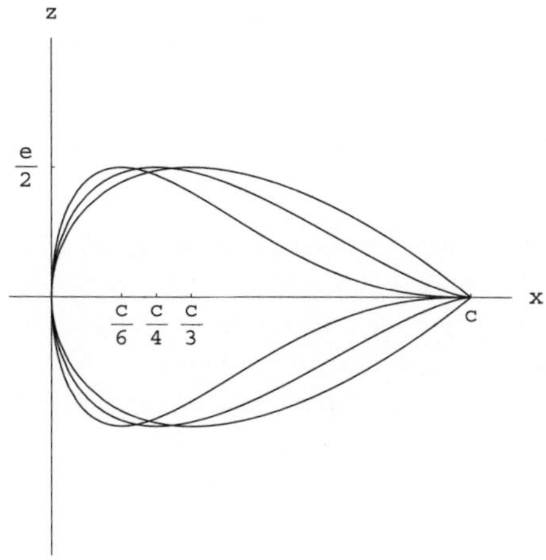

Fig. 3.5 Quasi-Joukowski airfoil at incidence

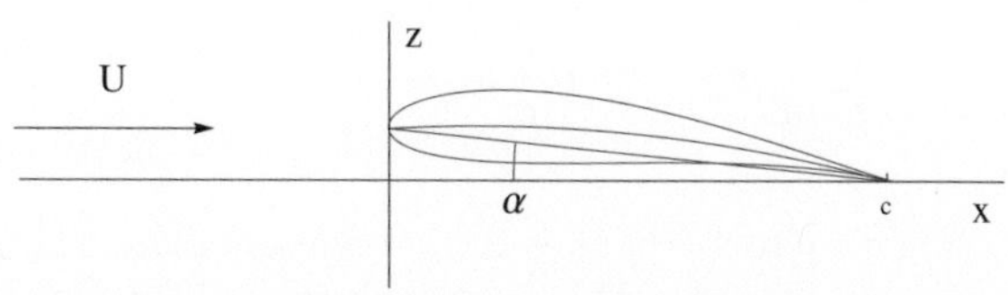

$$e(x) = \frac{16}{3\sqrt{3}} e \sqrt{\frac{x}{c}} \left(1 - \frac{x}{c}\right)^{\frac{3}{2}} \tag{3.13}$$

The Quasi-Joukowski airfoil has a cusped trailing edge. The maximum thickness is at $x = \frac{c}{4}$.

The following thickness distribution has a maximum at $x = \frac{c}{6}$. As will be discussed with viscous effects, having the maximum thickness far forward helps the boundary layer recovery as the flow slows down towards the trailing edge. This thickness distribution may be interesting for that reason:

$$e(x) = \frac{216}{25\sqrt{5}} e \sqrt{\frac{x}{c}} \left(1 - \frac{x}{c}\right)^{\frac{5}{2}} \tag{3.14}$$

See Fig. 3.4 for the different thickness distributions, where the scale has been stretched in the z-direction.

A Quasi-Joukowski airfoil at incidence, with the chord and camber lines is shown in Fig. 3.5.

3.2 Small Disturbance Linearization Method

As mentioned earlier, a unique solution exists for the inviscid, incompressible flow past an airfoil with sharp trailing edge. The solution satisfies the PDEs (2.1), the tangency (2.4) and the asymptotic boundary conditions (2.6). It is of the general form

$$\mathbf{V} = \left(U + u\left(x, z; \frac{d}{c}, \frac{e}{c}, \alpha\right), w\left(x, z; \frac{d}{c}, \frac{e}{c}, \alpha\right)\right) \tag{3.15}$$

where the dependency on the geometrical parameters, coming from the tangency condition, is indicated. u and w are the exact perturbation velocity components and, in general, their analytical form cannot be obtained in closed form. However, since the solution depends continuously on the data, it is possible to expand u and w in a Taylor series with respect to the small parameters $\frac{d}{c}$, $\frac{e}{c}$ and α. If we limit ourselves to the first order terms, we find

$$\mathbf{V} = \left(U + \frac{d}{c}u_d(x, z) + \frac{e}{c}u_e(x, z) + \alpha u_\alpha(x, z) + O\left(\frac{d}{c} + \frac{e}{c} + \alpha\right)^2, \right.$$
$$\left. \frac{d}{c}w_d(x, z) + \frac{e}{c}w_e(x, z) + \alpha w_\alpha(x, z) + O\left(\frac{d}{c} + \frac{e}{c} + \alpha\right)^2\right) \tag{3.16}$$

where the terms $O(\frac{d}{c} + \frac{e}{c} + \alpha)^2$ represent the remainders in the Taylor expansions and are of second and higher order in the small parameters or products of them. They can be neglected if the parameters are small enough. Note that in the limit

$$\frac{d}{c}, \ \frac{e}{c}, \ \alpha \to 0 \tag{3.17}$$

the profile becomes a flat plate aligned with the incoming flow, and the uniform flow is recovered. This is the process we have used to derive the parametric representation of the Quasi-Joukowski profile geometry.

The governing equations are already linear. The next step is the linearization of the tangency condition.

3.2.1 Linearization of the Tangency Condition

The tangency condition reads $\mathbf{V}.\mathbf{n}|_{obstacle} = 0$. Consider a small element of the profile surface $(dx, dz) = \mathbf{t}dl$ where $\mathbf{t}$ is the tangent unit vector and dl the length of the element. A normal vector to the profile is $\mathbf{n}dl = (dz, -dx)$. The above condition can be written

$$(U + u, w).(dz, -dx)|_{obstacle} = ((U + u)dz - wdx)\,|_{obstacle} = 0 \tag{3.18}$$

Dividing by dx ($dx \neq 0$) and applying the condition on the obstacle

$$(U + u(x, z(x)))\,\frac{dz}{dx} - w(x, z(x)) = 0 \tag{3.19}$$

Using the fact that u, w, z and $\frac{dz}{dx}$ are all of order $O(\frac{d}{c}+\frac{e}{c}+\alpha)$, the above expression can be simplified and expanded to first order as

$$U\frac{dz}{dx} - w(x,0) + u(x,0)\frac{dz}{dx} - z(x)\frac{\partial w}{\partial z}(x,0) + O\left(\frac{d}{c}+\frac{e}{c}+\alpha\right)^2 = 0 \quad (3.20)$$

Note that the third and fourth terms are already second order terms and can be included in the remainder. Finally, keeping only the leading terms results in the new form of the tangency condition

$$w(x,0) = U\frac{dz}{dx}(x) \quad (3.21)$$

To be more specific, one can formulate the tangency condition in terms of the profile shape and incidence as

$$w(x,0^{\pm}) = U\left(f'^{\pm}(x) - \alpha\right) \quad (3.22)$$

This operation is called the transfer of the tangency condition from the profile surface to the x-axis. Note that, as regard the mathematical treatment of the tangency condition, the profile is replaced by a slit along the x-axis with an upper (0^+) and a lower (0^-) edge.

Note also that the tangency condition is expressed only in terms of w.

3.2.2 Linearization of the Pressure Coefficient

This is the last element of the theory that needs to be linearized. Using Cartesian coordinates, the exact C_p can be expanded as

$$C_p = 1 - \left(\frac{V}{U}\right)^2 = 1 - \frac{U^2 + 2Uu + u^2 + w^2}{U^2} = -2\frac{u}{U} + O\left(u^2 + w^2\right) \quad (3.23)$$

Again, neglecting second and higher order terms, the pressure coefficient becomes

$$C_p = -2\frac{u}{U} \quad (3.24)$$

This expression allows for the superposition of solutions as will be needed later.

Note that the pressure coefficient only dependent on u.

Figure 3.6 depicts the situation for the small perturbation analysis: a thin airfoil will disturb the uniform flow only slightly, so that, the velocity vector $\mathbf{V}$ in the domain surrounding the profile (with the exception of singular points) will point inside a small ball centered at the end of the undisturbed velocity vector $\mathbf{U}$. The

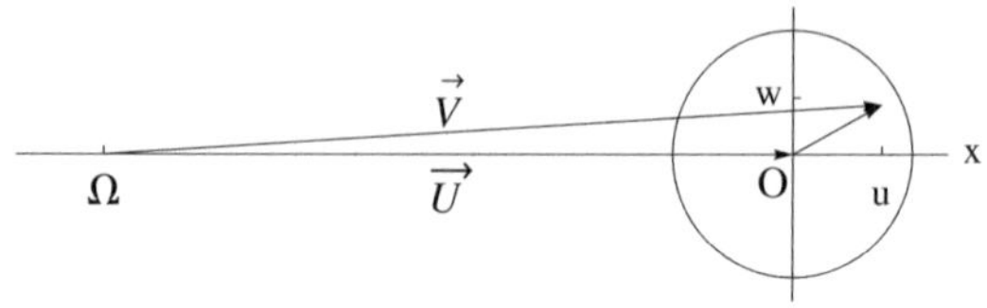

Fig. 3.6 Effect of a small perturbation (u, w) on the orientation and modulus of **V**

direction of the velocity vector is controlled primarily by w (tangency condition), whereas the magnitude of the vector is controlled primarily by u (pressure).

Note also, that stagnation will correspond to a point located near point Ω, which implies large negative values of u. In fact, the theory will allow u to go to $-\infty$ at a stagnation point.

3.3 Decomposition into Symmetric and Lifting Problems

The small disturbance assumption has allowed for the complete linearization of the problem of the flow past a thin airfoil. It is therefore possible to use linear superposition of solutions. The general problem of the flow past a thin cambered airfoil at incidence can be viewed as the sum of a symmetric problem corresponding to the flow past a symmetric profile having the same thickness distribution at zero incidence, and a lifting problem corresponding to the flow past a thin cambered plate having the same camberline and incidence. This is depicted in Fig. 3.7.

The symmetric problem will be solved by distributing sources and sinks on a cut along the x-axis, from zero to c. Due to the symmetry of the problem, there will be no net force in the z-direction and no net moment. The drag will also be zero (d'Alembert paradox).

The lifting problem will be handled by a distribution of vortices on a cut along the x-axis, from zero to c. In general, the flow will produce lift and moment, but again zero drag.

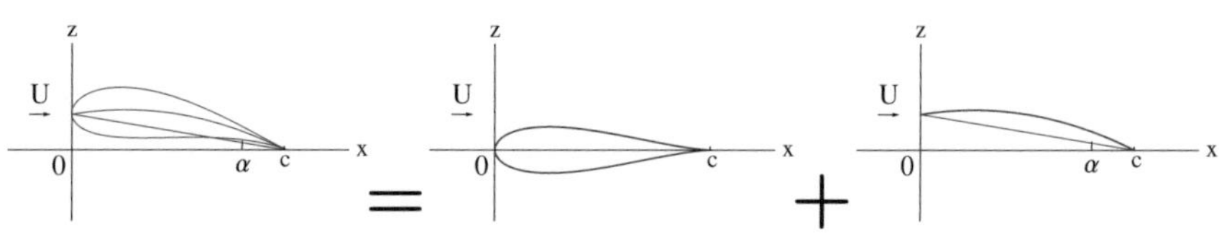

Fig. 3.7 Decomposition into symmetric and lifting problems

3.4 The Symmetric Problem

In the following, the streamfunction will be chosen as an intermediate function to solve for the flow velocity field. Similar results would be obtained with the velocity potential.

We consider a small element along the x-axis of length $d\xi$ with a distribution of source/sink $dQ = Q'(\xi)d\xi$. The contribution of this element to the perturbation streamfunction at a point $M(x, z)$ is given in Cartesian coordinates by

$$d\psi = \frac{Q'(\xi)d\xi}{2\pi} \arctan\left(\frac{z}{x-\xi}\right) \tag{3.25}$$

Upon integration, this becomes

$$\psi(x, z) = \frac{1}{2\pi}\int_0^c Q'(\xi) \arctan\left(\frac{z}{x-\xi}\right) d\xi \tag{3.26}$$

The perturbation velocity components are obtained by taking derivatives with respect to x and z. Note that the derivatives will be carried out under the integral sign, which requires some regularity of the integrand and convergence of the integral. We shall assume that these conditions are fulfilled.

$$u(x, z) = \frac{\partial\psi}{\partial z} = \frac{1}{2\pi}\int_0^c \frac{Q'(\xi)(x-\xi)}{(x-\xi)^2+z^2} d\xi \tag{3.27}$$

$$w(x, z) = -\frac{\partial\psi}{\partial x} = \frac{1}{2\pi}\int_0^c \frac{Q'(\xi)z}{(x-\xi)^2+z^2} d\xi \tag{3.28}$$

Note that u is even and w is odd in z. This is consistent with the symmetry of the flow about the x-axis. For that reason, u will be continuous across the cut $[0, c]$, whereas w will be allowed to jump. Let $< a >= a^+ - a^-$ represent the jump of a across the cut, then $< u >= 0$ and $< w >\neq 0$. Since the pressure is related to u, the pressure will be continuous across the cut, and w will have a jump corresponding to the change of sign of the profile slope ($\pm\frac{e'(x)}{2}$) between the upper and lower surfaces. See Fig. 3.8.

The tangency condition on the upper surface is applied by letting $z \to 0^+$ while $0 \leq x \leq c$ and

$$w(x, 0^+) = Uf'^+(x) = \frac{1}{2}Ue'(x) \tag{3.29}$$

which will also, by symmetry, enforce the tangency condition on the lower surface. In the process, the integral goes to zero, almost everywhere, except at $\xi = x$, where the integrand is of the form $\frac{0}{0}$. One can show that the limit is finite and equal to

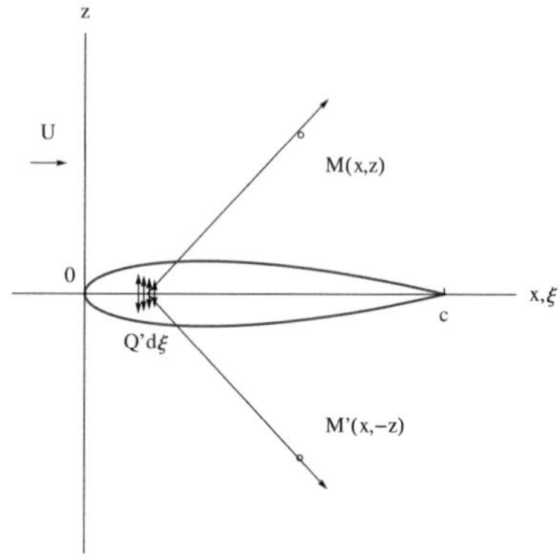

Fig. 3.8 Symmetric flow induced by a source element

$$\lim_{z \to 0^+} w(x,z) = \frac{1}{2} Q'(x) \tag{3.30}$$

The proof for this result is obtained as follows. The integral is split into three pieces

$$w(x,z) = \frac{1}{2\pi} \left\{ \int_0^{x-\epsilon} + \int_{x-\epsilon}^{x+\epsilon} + \int_{x+\epsilon}^{c} \frac{Q'(\xi) z}{(x-\xi)^2 + z^2} d\xi \right\} \tag{3.31}$$

where $\epsilon > 0$ is a small number. The first and last integrands are regular and $\frac{z}{(x-\xi)^2+z^2} \to 0$ as $z \to 0$, in other words

$$\lim_{z \to 0} \frac{1}{2\pi} \left\{ \int_0^{x-\epsilon} + \int_{x+\epsilon}^{c} \frac{Q'(\xi) z}{(x-\xi)^2 + z^2} d\xi \right\} = 0, \ \forall \epsilon > 0 \tag{3.32}$$

In the middle integral, the dummy variable ξ is replaced by t s.t. $\xi = x + zt$. The integral becomes

$$w_\epsilon(x,z) = \frac{1}{2\pi} \int_{-\frac{\epsilon}{z}}^{\frac{\epsilon}{z}} \frac{Q'(x+zt)}{(1+t^2)} dt \tag{3.33}$$

Now, if Q' is a smooth function, it can be expanded in Taylor series as $Q'(x+zt) = Q'(x) + ztQ''(x) + O(\epsilon^2)$. Note that $-\epsilon \le zt \le \epsilon$. Substitution of this last result into the integral yields

$$w_\epsilon(x,z) = \frac{Q'(x)}{2\pi} \int_{-\frac{\epsilon}{z}}^{\frac{\epsilon}{z}} \frac{dt}{1+t^2} + O(\epsilon) = \frac{Q'(x)}{2\pi} [\arctan t]_{-\frac{\epsilon}{z}}^{\frac{\epsilon}{z}} + O(\epsilon) \tag{3.34}$$

Letting $z \to 0^+$, followed by $\epsilon \to 0$, gives the anticipated result.

By identification with the previous equation, the source/sink strength is found to be $Q'(x) = Ue'(x)$. When the thickness increases, sources are used and when it decreases, sinks are used. For a closed profile the net sum of sources and sinks is zero, a consequence of $e(0) = e(c) = 0$.

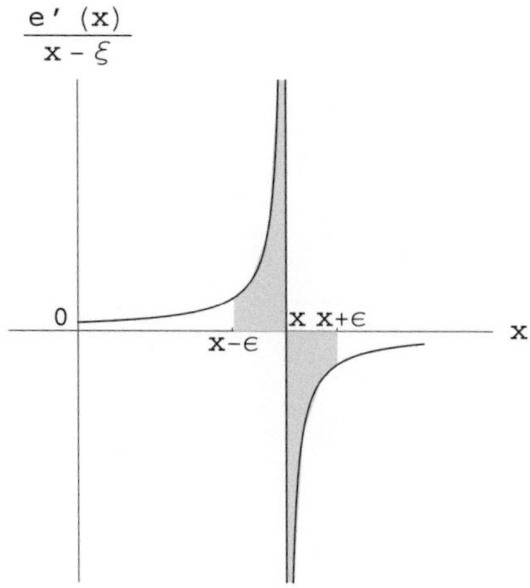

Fig. 3.9 Limiting process for principal value integral

Now that the singularity distribution is known, the perturbation velocity field can be calculated using the above equations for $u(x, z)$ and $w(x, z)$. We are especially interested by the velocity on the obstacle. We know w already and u is given by

$$u(x, 0) = \frac{U}{2\pi} \int_0^c \frac{e'(\xi)}{x - \xi} d\xi \tag{3.35}$$

Note that the integrand is singular at $\xi = x$. The singularity is a "simple pole". A finite value can be obtained according to the limiting process

$$u(x, 0) = \lim_{\epsilon \to 0} \frac{U}{2\pi} \left\{ \int_0^{x-\epsilon} \frac{e'(\xi)}{x - \xi} d\xi + \int_{x+\epsilon}^c \frac{e'(\xi)}{x - \xi} d\xi \right\} \tag{3.36}$$

In this process, the two diverging contributions cancel out because the areas, as depicted in Fig. 3.9, are equal and opposite. Such integrals are called "principal value" integrals.

The proof is easily obtained by studying the following limit for the remainder

$$R = \lim_{\epsilon \to 0} \int_{x-\epsilon}^{x-\epsilon^2} \frac{e'(\xi)}{x - \xi} d\xi + \int_{x+\epsilon^2}^{x+\epsilon} \frac{e'(\xi)}{x - \xi} d\xi \tag{3.37}$$

and making use of $\epsilon^2 \ll \epsilon$ and of the Taylor expansion

$$e'(\xi) = e'(x) + (\xi - x)e''(x) + \frac{(\xi - x)^2}{2!} e'''(x) + O(\epsilon^3) \tag{3.38}$$

where ξ has been written as $\xi = x + (\xi - x)$.

The first term in R integrates as

$$\int_{x-\epsilon}^{x-\epsilon^2} \frac{e'(\xi)}{x-\xi} d\xi = e'(x)\left(\ln\epsilon - \ln\epsilon^2\right) - e''(x)\left(\epsilon - \epsilon^2\right) + O(\epsilon^3) \tag{3.39}$$

The second term gives

$$\int_{x+\epsilon^2}^{x+\epsilon} \frac{e'(\xi)}{x-\xi} d\xi = e'(x)\left(\ln\epsilon^2 - \ln\epsilon\right) - e''(x)\left(\epsilon - \epsilon^2\right) + O(\epsilon^3) \tag{3.40}$$

Adding the two results, gives

$$R = -2e''(x)\left(\epsilon - \epsilon^2\right) + O(\epsilon^3) \tag{3.41}$$

which vanishes in the limit $\epsilon \to 0$.

Consider as example, the flow past the symmetric wedge of semi-angle θ at zero incidence. The thickness distribution is $e(x) = 2\theta x$ for small values of θ. This results in $Q'(x) = 2U\theta = const$. The velocity along the wedge is

$$u(x,0) = \frac{U\theta}{\pi}\int_0^c \frac{d\xi}{x-\xi} \tag{3.42}$$

Going through the limiting process in detail, let

$$\begin{aligned} u_\epsilon(x,0) &= \frac{U\theta}{\pi}\left(\int_0^{x-\epsilon}\frac{d\xi}{x-\xi} + \int_{x+\epsilon}^{c}\frac{d\xi}{x-\xi}\right) \\ &= \frac{U\theta}{\pi}\left(\ln x - \ln\epsilon + \ln\epsilon - \ln(c-x)\right) \end{aligned} \tag{3.43}$$

Note that the terms in $\ln\epsilon$ cancel out and the limit for $\epsilon \to 0$ is simply

$$u(x,0) = \frac{U\theta}{\pi}\ln\frac{x}{c-x} \tag{3.44}$$

The pressure coefficient, $C_p = -2u(x,0)/U$ is shown in Fig. 3.10. Note that the areas under the curve are equal and of opposite signs. For this reason, the resulting pressure force in the x-direction (drag) is zero.

To close this section, it is worth mentioning that, although thickness does not contribute to forces and moments, it will be beneficial in mitigating adverse viscous effects due to sharp leading edges.

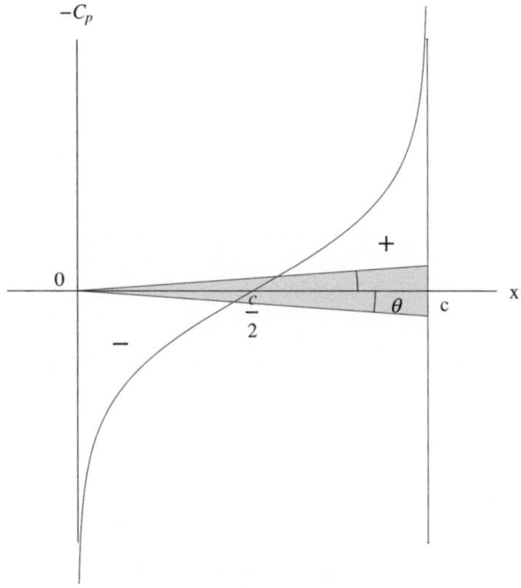

Fig. 3.10 Pressure coefficient along the wedge

3.5 Lifting Problem

We consider a small element along the x-axis of length $d\xi$ with a distribution of vorticity $d\Gamma = \Gamma'(\xi)d\xi$. The contribution of this element to the perturbation stream-function at a point $M(x, z)$ is given in Cartesian coordinates by

$$d\psi = \frac{\Gamma'(\xi)d\xi}{2\pi}\frac{1}{2}\ln\left[(x-\xi)^2 + z^2\right] \tag{3.45}$$

Upon integration, this becomes

$$\psi(x, z) = \frac{1}{4\pi}\int_0^c \Gamma'(\xi)\ln\left[(x-\xi)^2 + z^2\right]d\xi \tag{3.46}$$

From this we obtain the perturbation velocity field as

$$u(x, z) = \frac{\partial\psi}{\partial z} = \frac{1}{2\pi}\int_0^c \frac{\Gamma'(\xi)z}{(x-\xi)^2 + z^2}d\xi \tag{3.47}$$

$$w(x, z) = -\frac{\partial\psi}{\partial x} = -\frac{1}{2\pi}\int_0^c \frac{\Gamma'(\xi)(x-\xi)}{(x-\xi)^2 + z^2}d\xi \tag{3.48}$$

Note that u is odd and w is even in z, hence $< u > \neq 0$ and $< w >= 0$ across the cut $[0, c]$. This is consistent with the vortex singularity, see Fig. 3.11.

For the application of the tangency condition on the thin cambered plate at incidence, we let $z \to 0$, while $0 \le x \le c$. The condition becomes

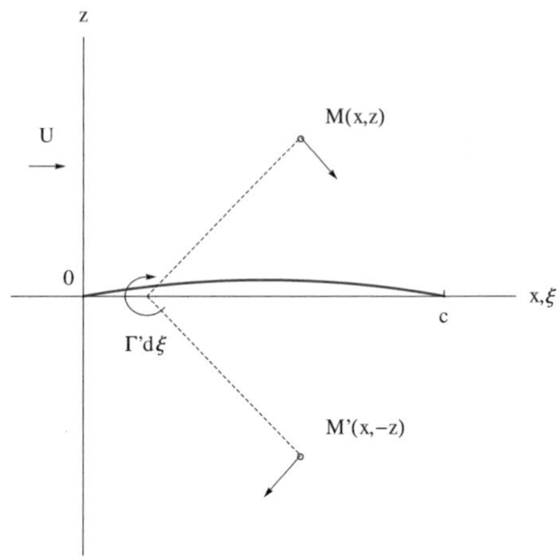

Fig. 3.11 Lifting flow induced by a vortex element

$$w(x,0) = -\frac{1}{2\pi}\int_0^c \frac{\Gamma'(\xi)}{x-\xi}d\xi = U\left(d'(x)-\alpha\right) \tag{3.49}$$

The equation above is the fundamental integral equation of thin airfoil theory. In general the airfoil shape $d(x)$ is given as well as the flow conditions U, α. The unknown vorticity $\Gamma'(x)$ is the solution of an integral equation. Once Γ' is found, which satisfies the Kutta-Joukowski condition, $\Gamma'(c) = 0$, one can obtain the velocity field everywhere and in particular $u(x, 0^{\pm})$ and hence the pressure coefficients on the thin cambered plate, $C_p^{\pm}(x)$. Using the result derived earlier with the same singular integrand, the limiting process yields

$$\lim_{z\to 0^+} u(x,z) = \frac{1}{2\pi}\int_0^c \frac{\Gamma'(\xi)z}{(x-\xi)^2+z^2}d\xi = \frac{\Gamma'(x)}{2} = u(x,0^+) \tag{3.50}$$

It is easy to show by symmetry that $u(x,0^-) = -\frac{\Gamma'(x)}{2}$. Across the thin cambered plate (or cut along the x-axis) the u-component has a jump, $< u(x) >= u(x,0^+) - u(x,0^-) = \Gamma'(x)$. There is also a pressure (or C_p) jump, except at the trailing edge, where the flow leaves the profile "smoothly". In contrast, the leading edge sustains in general an infinite velocity ($\Gamma'(0) = \pm\infty$), the only exception being when the leading edge is "adapted", in which case the flow attaches "smoothly" and satisfies a $K-J$ condition there ($\Gamma'(0) = 0$), see Fig. 3.12 for typical flow fields at the leading and trailing edges.

The mathematical form of the Kutta-Joukowski condition can be explained by the fact that, if the flow leaves the trailing edge smoothly, the pressures above and below the trailing edge must be equal, hence $u^+(c) = u^-(c)$ which implies that $\Gamma'(c) = 0$.

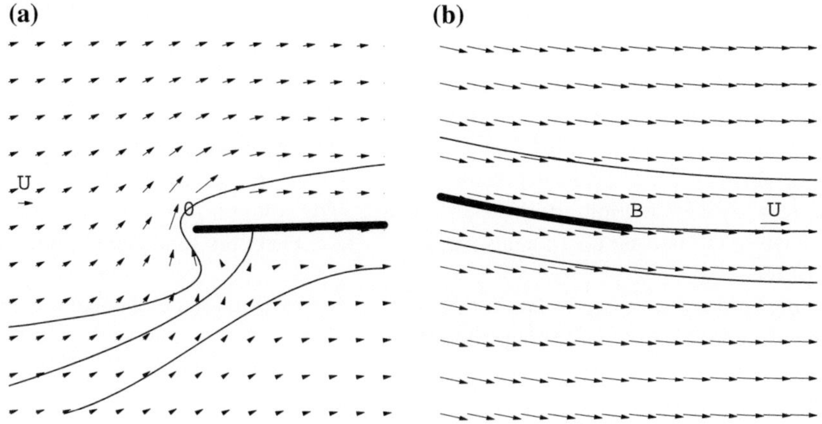

Fig. 3.12 Local flow conditions: **a** leading edge flow, **b** trailing edge flow

3.5.1 Solution of the Fundamental Integral Equation

The solution to the fundamental integral equation of thin airfoil theory has been obtained by using a change of variable that plays an important role in this theory, as well as in the Prandtl lifting line theory, as will be seen later. It is due to Glauert [1] and reads

$$\begin{cases} x = \frac{c}{2}(1 - \cos t), \ \ 0 \leq t \leq \pi \\ \xi = \frac{c}{2}(1 - \cos \theta), \ \ 0 \leq \theta \leq \pi \end{cases} \tag{3.51}$$

This transformation maps a uniform distribution of points on a half circle (with steps Δt or $\Delta\theta$) by projection onto its diameter (x or ξ). The new distribution clusters the points near the leading and trailing edges. See Fig. 3.13.

The vorticity distribution is assumed to be of the form

$$\Gamma'[\xi(\theta)] = 2U\left\{A_0 \frac{1+\cos\theta}{\sin\theta} + \sum_{n=1}^{\infty} A_n \sin n\theta\right\} \tag{3.52}$$

Fig. 3.13 Geometric interpretation of the change of variable

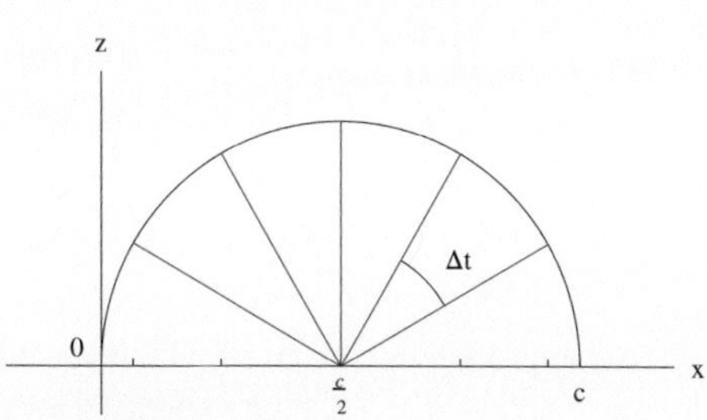

The first term in the bracket is singular at $\theta = 0$ (leading edge) when $A_0 \neq 0$. It represents the leading edge singularity. The first term by itself represents the flow past a flat plate, as will be seen shortly. Notice that this term is regular at the trailing edge as $\Gamma'(\theta) \to 0$ as $\theta \to \pi$. The infinite summation, the Fourier series, is the superposition of regular solutions that satisfy two $K - J$ conditions, at the leading and trailing edges, since all the terms vanish there. In other words, the solution proposed for Γ' enforces, by construction, the $K - J$ condition. The solution procedure consists in calculating the Fourier coefficients A_0, A_1,...,A_n,...using the fundamental integral equation.

Substitution of the expression for Γ' into the fundamental integral equation, and using the change of variable results in $w(x, 0)$

$$-\frac{1}{2\pi}\int_0^{\pi} \frac{2U\left\{A_0\frac{1+\cos\theta}{\sin\theta} + \sum_{n=1}^{\infty} A_n \sin n\theta\right\}\frac{c}{2}\sin\theta}{\frac{c}{2}\left(\cos\theta - \cos t\right)} d\theta = U\left(d'[x(t)] - \alpha\right) \tag{3.53}$$

or, after some rearranging

$$-\frac{U}{\pi}A_0\int_0^{\pi}\frac{1+\cos\theta}{\cos\theta - \cos t}d\theta - \frac{U}{\pi}\sum_{n=1}^{\infty}A_n\int_0^{\pi}\frac{\sin\theta\sin n\theta}{\cos\theta - \cos t}d\theta = U\left(d'[x(t)] - \alpha\right) \tag{3.54}$$

The two principal value integrals that appear in the above expression can be found in books on integrals. The first one is equal to π and the second one to $-\pi\cos nt$ so that the fundamental integral equation, after some simplification, reads

$$d'[x(t)] - \alpha = -A_0 + \sum_{n=1}^{\infty} A_n \cos nt, \quad 0 \le t \le \pi \tag{3.55}$$

The left-hand-side is well defined for $0 \le t \le \pi$. The right-hand-side is a Fourier series of period 2π. It is a series of cosines, hence even in t. Knowing its value in $[0, \pi]$ is sufficient to define it for all values of its argument $-\infty < t < \infty$. Consider, for example, a parabolic plate at incidence α. The left-hand-side is

$$d'[x(t)] - \alpha = 4\frac{d}{c}\left(1 - 2\frac{x}{c}\right) - \alpha = 4\frac{d}{c}\cos t - \alpha \tag{3.56}$$

This is shown in Fig. 3.14 for $\frac{d}{c} = 0.03$ and $\alpha = 5° = 0.087$ rd.
In this particular case, the Fourier coefficients are easily found by inspection:

$$A_0 = \alpha, \; A_1 = 4\frac{d}{c}, \; A_n = 0, \; n \ge 2 \tag{3.57}$$

This approach can be used whenever the profile camber is a polynomial of low degree, because $d'(x)$ can be expressed as a sum of cosines by elimination of x with t through $x = \frac{c}{2}(1 - \cos t)$ and use of some identities to replace $\cos^2 t$ by $\cos 2t$,

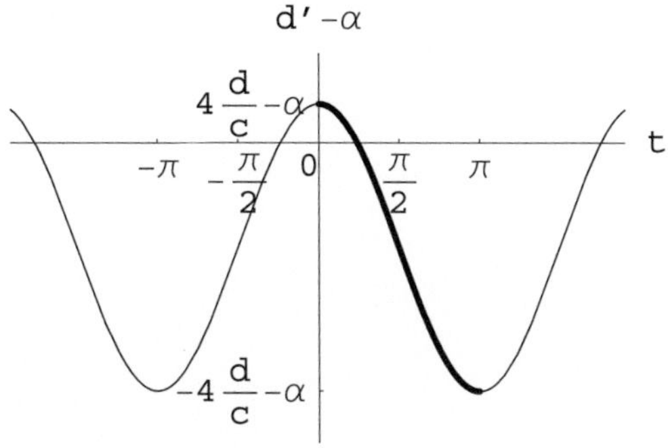

Fig. 3.14 Fourier series representation for a thin parabolic cambered plate at incidence

etc…If the profile camber is not a polynomial, or is defined by piecewise continuous functions, the Fourier coefficients have to be calculated using the general method of integration, using orthogonality of the modes as

$$A_0 = \alpha - \frac{1}{\pi}\int_0^c d'[x(t)]dt \tag{3.58}$$

$$A_n = \frac{2}{\pi}\int_0^c d'[x(t)]\cos nt dt \tag{3.59}$$

Note that A_0 is the only coefficient which depends on α, and it is of the form

$$A_0 = \alpha - \alpha_{adapt} \tag{3.60}$$

where $\alpha_{adapt} = \frac{1}{\pi}\int_0^c d'[x(t)]dt$ is a geometry related constant. When $\alpha = \alpha_{adapt}$, the singular term vanishes ($A_0 = 0$) and the leading edge is "adapted", in other words satisfies the $K - J$ condition. α_{adapt} is called the *angle of adaptation* or *ideal angle of attack.*

3.5.2 Example: Flat Plate

For the flat plate, $d'(x) = 0$. The above equations for the Fourier coefficients reduce to $A_0 = \alpha$, $A_1 = \ldots = A_n = 0, n \geq 1$. The vorticity distribution reduces to the singular term

$$\Gamma'[x(t)] = 2U A_0 \frac{1 + \cos t}{\sin t} = 2U\alpha\sqrt{\frac{c - x}{x}} \tag{3.61}$$

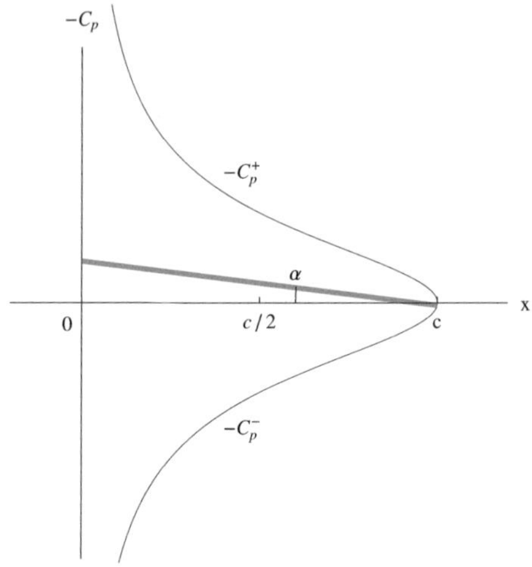

Fig. 3.15 Cp distribution for a flat plate at incidence

The leading edge of the flat plate is singular for $\alpha \neq 0$. The angle of adaptation is $\alpha_{adapt} = 0$, which corresponds to the uniform flow (zero perturbation). The result for $C_p^{\pm}(x)$ at $\alpha = 5°$ is shown in Fig. 3.15.

3.5.3 Example: Parabolic Plate

With $d(x) = 4d\frac{x}{c}(1 - \frac{x}{c})$ and $d'(x) = 4\frac{d}{c}(1 - 2\frac{x}{c}) = 4\frac{d}{c}\cos t$, as seen earlier, the Fourier coefficients are easily identified to be

$$A_0 = \alpha,\ A_1 = 4\frac{d}{c},\ A_n = 0,\ n \geq 2 \tag{3.62}$$

hence, the vorticity distribution reads

$$\Gamma'[x(t)] = 2U\left\{\alpha\frac{1+\cos t}{\sin t} + 4\frac{d}{c}sint\right\} \tag{3.63}$$

Upon elimination of t in terms of x, the result becomes

$$\Gamma'[x(t)] = 2U\left\{\alpha\sqrt{\frac{c-x}{x}} + 8\frac{d}{c}\frac{\sqrt{x(c-x)}}{c}\right\} \tag{3.64}$$

Note that this result can be interpreted as the superposition of the flow past a flat plate at incidence (first term proportional to α) plus a parabolic cambered plate at zero incidence (second term proportional to $\frac{d}{c}$). It is interesting also to realize that the

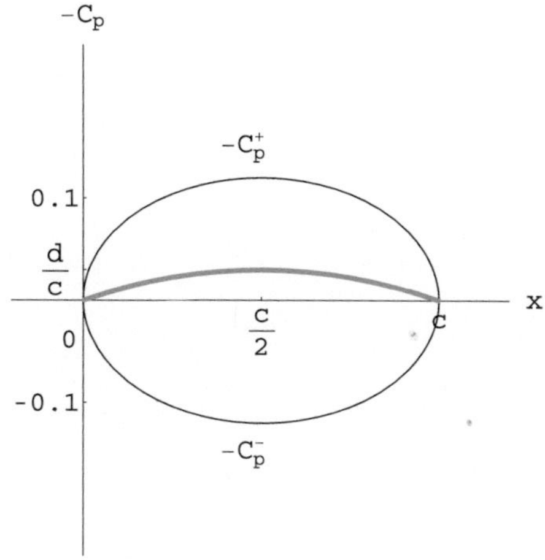

Fig. 3.16 Cp distribution for the parabolic plate at zero incidence

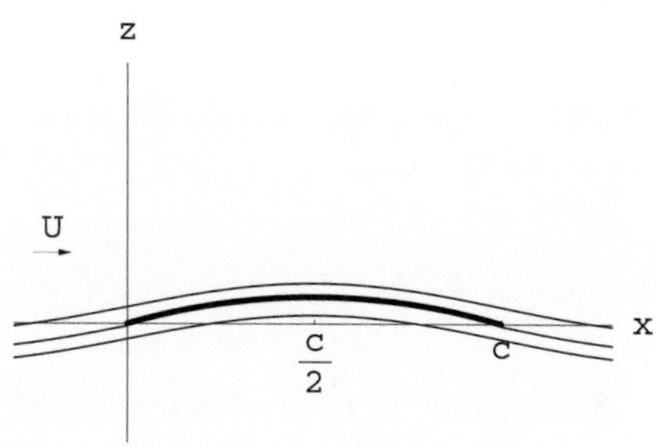

Fig. 3.17 Streamlines of the flow past a parabolic plate at zero incidence

parabolic plate at zero incidence satisfies two $K - J$ conditions, one at the trailing edge but also one at the leading edge. Indeed, the angle of adaptation is $\alpha_{adapt} = 0$. At zero incidence, the parabolic plate has lift and the C_p distributions are half ellipses, as shown in Fig. 3.16.

The flow in this case can be depicted as in Fig. 3.17.

Due to the flow symmetry w.r.t the axes $x = \frac{c}{2}$, the lift force will be located at $x = x_{c.p.} = \frac{c}{2}$, the center of pressure. When the incidence is positive, the flow will go around the leading edge clockwise and the stagnation point will be located on the lower surface. The opposite is true if the incidence is negative.

The angle of adaptation has a practical interest for thin cambered plates: at the ideal angle of attack, the viscous effects are minimized because the flow does not have to go around the leading edge with very high velocities and then slow down abruptly, which often provokes separation. However, since it is not possible to control perfectly the incidence of the profile of a wing, thickness is added in the leading edge region that allows the stagnation point to move about the ideal position, without major penalties in viscous effects. See the sketch in Fig. 3.18.

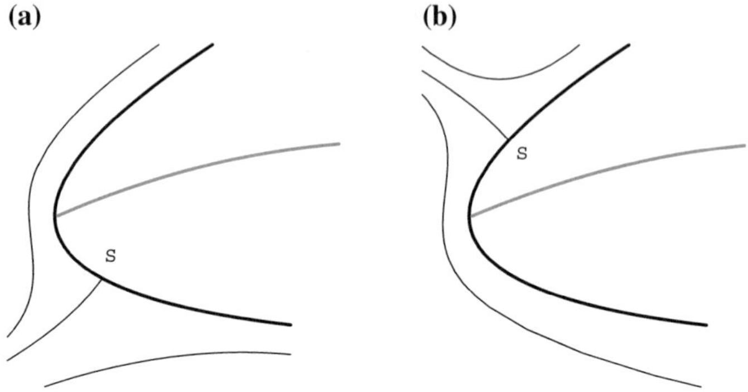

Fig. 3.18 Movement of stagnation point with incidence: **a** $\alpha > \alpha_{adapt}$, **b** $\alpha < \alpha_{adapt}$

3.5.4 Forces

The lift force is obtained by application of the Kutta-Joukowski lift theorem, where one accounts for the net resulting circulation

$$L' = \rho U \Gamma = \rho U \int_0^c \Gamma'(x) dx \tag{3.65}$$

Substituting the expansion for Γ' and integrating with the variable t yields

$$L' = \rho U \int_0^\pi 2U \left\{ A_0 \frac{1 + \cos t}{\sin t} + \sum_{n=1}^{\infty} A_n \sin nt \right\} \frac{c}{2} \sin t dt \tag{3.66}$$

With the elimination of $\sin t$ in the first term and using orthogonality of the sines in the second, there remains only two terms

$$L' = \pi \rho U^2 c \left(A_0 + \frac{A_1}{2} \right) \tag{3.67}$$

The lift coefficient reads

$$C_l = \frac{L'}{\frac{1}{2}\rho U^2 c} = \frac{2\Gamma}{Uc} = 2\pi \left(A_0 + \frac{A_1}{2} \right) \tag{3.68}$$

The important result is that the lift coefficient only depends on the first two modes, A_0 which depends on α and α_{adapt}, and A_1 which reflects the contribution to the camber of a parabolic plate, mode 1.

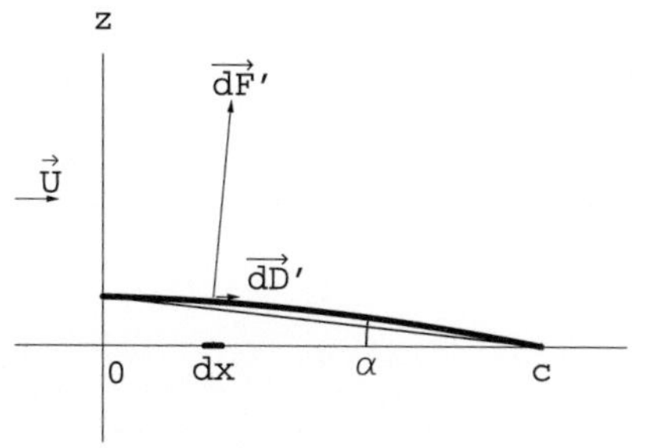

Fig. 3.19 Drag calculation

Application:

- the lift coefficient of a flat plate is $C_l = 2\pi\alpha$
- the lift coefficient of a parabolic plate is $C_l = 2\pi(\alpha + 2\frac{d}{c})$

Note, that at incidence of adaptation, the lift of the flat plate is zero, but it is $4\pi\frac{d}{c}$ for the parabolic plate.

The drag is zero (d'Alembert paradox). This seems quite surprising when considering the case of a flat plate at incidence α, since the integration of pressure will produce a force F' normal to the plate, hence a drag $D' = F' \sin\alpha \simeq F'\alpha$.

The proof of zero drag can be carried out in the general case as follows. A small element of a cambered plate is subject to a pressure difference, hence a small force dF' perpendicular to the surface element

$$\mathbf{dF}' = \frac{1}{2}\rho U^2 \left(C_p^-(x) - C_p^+(x)\right) \mathbf{n}dl \tag{3.69}$$

where $\mathbf{n}dl = (-dz, dx)$. See Fig. 3.19.

The projection of the force onto the x-axis gives the drag contribution

$$\begin{aligned} dD' = dF'_x &= -\frac{1}{2}\rho U^2 \left(C_p^-(x) - C_p^+(x)\right) dz \\ &= -\frac{1}{2}\rho U^2 \left(C_p^-(x) - C_p^+(x)\right) \frac{dz}{dx} dx \end{aligned} \tag{3.70}$$

Now using $C_p = -2\frac{u}{U}$, $< u >= \Gamma'$ and $\frac{dz}{dx} = \frac{w(x,0)}{U}$, one can write

$$dD' = -\rho\Gamma'(x)w(x,0)dx \tag{3.71}$$

Upon integration and replacing $w(x,0)$ by its expression in terms of Γ' one gets

$$\begin{aligned} D' &= -\rho \int_0^c \Gamma'(x) \left\{ -\frac{1}{2\pi} \int_0^c \frac{\Gamma'(\xi)}{x - \xi} d\xi \right\} dx \\ &= -\frac{\rho}{2\pi} \int_0^c \int_0^c \frac{\Gamma'(x)\Gamma'(\xi)}{x - \xi} d\xi dx \end{aligned} \tag{3.72}$$

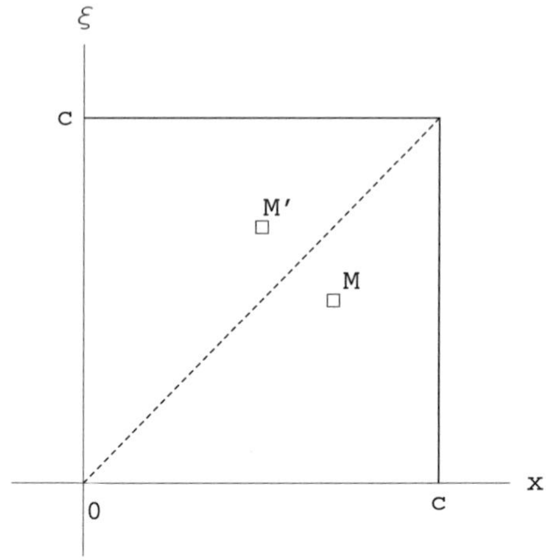

Fig. 3.20 Antisymmetry property of drag integrand (*...singular line*)

The last expression corresponds to the integration on a square domain $[0, c] \times [0, c]$ of an antisymmetric function, i.e. replacing x by ξ and vice-versa changes the integrand in its opposite, as shown in Fig. 3.20 for M and M'. The result is $D' = 0$ or $C_d = 0$ for all thin cambered plates.

Now the question remains concerning the result of the pressure integration along the flat plate and the apparent drag calculated to be

$$D' = -\rho U \int_0^c \Gamma'(x) \{-\alpha\} \, dx = \rho U \Gamma \alpha = L'\alpha \tag{3.73}$$

where $w(x, 0)$ has been replaced by $-\alpha U$. This apparent mystery is explained by the presence of a *suction force* acting at the leading edge of thin cambered plates in the direction of the tangent to the camberline, whenever the solution is singular at the leading edge. This suction force is not accounted for in the integration of pressure along the plate, but is found in a momentum balance on any contour that excludes the leading edge singularity (for example by using a small circle around it). The suction force $\mathbf{F}'_s$ balances exactly the "drag" from the pressure integration, i.e. $F'_s = -L'\alpha$, see Fig. 3.21.

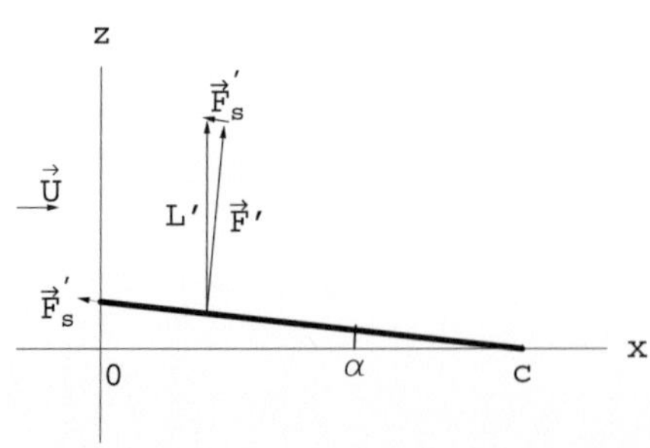

Fig. 3.21 Suction force and resulting lift force

3.5.5 Moment

The elementary moment of a small aerodynamic force located at x about the leading edge reads (Fig. 3.19)

$$dM'_{,o} = -dF'x = -\rho U \Gamma'(x) x dx \tag{3.74}$$

the minus sign is needed for the "nose up" moment to be positive. Replacing Γ' by its expansion and integrating in t from zero to π gives

$$M'_{,o} = -\rho U \int_0^{\pi} 2U \left\{ A_0 \frac{1+\cos t}{\sin t} + \sum_{n=1}^{\infty} A_n \sin nt \right\} \frac{c}{2}(1-\cos t)\frac{c}{2} \sin t dt \tag{3.75}$$

Rearranging and using orthogonality of sine modes one gets

$$M'_{,o} = -\frac{1}{2}\rho U^2 c^2 \left\{ A_0 \int_0^{\pi} \sin^2 t dt + A_1 \int_0^{\pi} \sin^2 t dt - \frac{1}{2} A_2 \int_0^{\pi} \sin^2 2t dt \right\} \tag{3.76}$$

The result is

$$M'_{,o} = -\frac{\pi}{4}\rho U^2 c^2 \left(A_0 + A_1 - \frac{A_2}{2} \right) \tag{3.77}$$

The moment coefficient is

$$C_{m,o} = \frac{M',o}{\frac{1}{2}\rho U^2 c^2} = -\frac{\pi}{2} \left(A_0 + A_1 - \frac{A_2}{2} \right) \tag{3.78}$$

Note that the moment depends on one more mode, mode 2, than lift. Application:

- moment coefficient for the flat plate: $C_{m,o} = -\frac{\pi}{2}\alpha$
- moment coefficient for the parabolic plate: $C_{m,o} = -\frac{\pi}{2}\left(\alpha + 4\frac{d}{c}\right)$

3.5.6 Center of Pressure

Let A and B be two points located at x_A and x_B along the x-axis. $M'_{,A}$ and $M'_{,B}$ are the moments and L' the resulting force. The following formula will be used to transfer the moment from one point to the other

$$M'_{,B} = M'_{,A} + (x_B - x_A)L' \tag{3.79}$$

or in terms of the dimensionless coefficients

$$C_{m,B} = C_{m,A} + \left(\frac{x_B}{c} - \frac{x_A}{c}\right) C_l \tag{3.80}$$

Now consider an arbitrary point B, the moment coefficient at this point can be calculated from the moment at the nose (A)

$$C_{m,B} = C_{m,o} + \left(\frac{x_B}{c} - \frac{x_o}{c}\right) C_l = C_{m,o} + \frac{x_B}{c} C_l \tag{3.81}$$

By definition, B will be the center of pressure if $C_{m,B} = 0$, i.e.

$$\frac{x_B}{c} = \frac{x_{c.p.}}{c} = -\frac{C_{m,o}}{C_l} \tag{3.82}$$

In terms of the Fourier coefficients, this becomes

$$\frac{x_{c.p.}}{c} = \frac{A_0 + A_1 - \frac{A_2}{2}}{4\left(A_0 + \frac{A_1}{2}\right)} \tag{3.83}$$

Since A_0 depends on α, in general the center of pressure location will vary with incidence.

Application:

- location of center of pressure for flat plate: $\frac{x_{c.p.}}{c} = \frac{A_0}{4A_0} = \frac{1}{4}$. The center of pressure is at the quarter-chord for all symmetric airfoils.
- location of center of pressure for parabolic plate: $\frac{x_{c.p.}}{c} = \frac{\alpha + 4\frac{d}{c}}{4\left(\alpha + 2\frac{d}{c}\right)}$

This last result is sketched in Fig. 3.22. The wide range of location for the center of pressure puzzled the early aerodynamicists who studied cambered profiles for their favorable lifting capabilities. It is worth noting that for large incidences, the center of pressure will be located near the quarter chord.

3.5.7 Aerodynamic Center

Starting from the expression of the aerodynamic moment coefficient at an arbitrary point B, we take the derivative with respect to α as

$$\frac{\partial C_{m,B}}{\partial \alpha} = \frac{\partial C_{m,o}}{\partial \alpha} + \frac{x_B}{c} \frac{\partial C_l}{\partial \alpha} = -\frac{\pi}{2} + 2\pi \frac{x_B}{c} \tag{3.84}$$

where we have made use of the fact that none of the Fourier coefficients depend on α except A_0 and $\frac{\partial A_0}{\partial \alpha} = 1$. By definition, since $\frac{\partial C_{m,a.c.}}{\partial \alpha} = 0$. The aerodynamic center will be located at

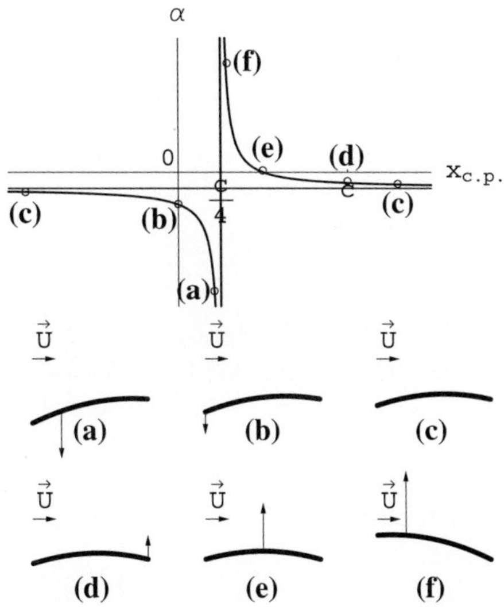

Fig. 3.22 Location of center of pressure versus α for parabolic cambered plate

$$\frac{x_{a.c.}}{c} = \frac{1}{4} \tag{3.85}$$

The aerodynamic center is located at quarter-chord for all thin airfoils.

The moment at the aerodynamic center is

$$C_{m,a.c.} = C_{m,o} + \frac{1}{4}C_l = -\frac{\pi}{4}(A_1 - A_2) \tag{3.86}$$

independent of α.

Application:

- flat plate $C_{m,a.c.} = 0$
- parabolic plate $C_{m,a.c.} = -\pi\frac{d}{c}$.

3.5.8 Example of Design Problem

Design the simplest airfoil (thin cambered plate) such that the center of pressure stays at the quarter-chord at all incidences (small α's) and the design lift $(C_l)_{des} > 0$ is obtained at the angle of adaptation α_{adapt}. By simplest airfoil, we mean one that has the smallest number of non zero Fourier coefficients. This has a practical interest for a "flying wing" configuration, where the long fuselage providing a moment from the tail is not available to balance the pitching moment at different incidences.

The condition reads

$$\frac{x_{c.p.}}{c} = \frac{A_0 + A_1 - \frac{A_2}{2}}{4\left(A_0 + \frac{A_1}{2}\right)} = \frac{1}{4} \tag{3.87}$$

Upon multiplying by the denominator (not zero, since $C_l \neq 0$), one finds

$$A_2 = A_1 \tag{3.88}$$

which is nothing else but $C_{m,a.c.} = 0$.

One such airfoil is the flat plate with $A_0 = \alpha,\ A_1, \ldots, A_n = 0,\ n \geq 1$. This is not a very interesting airfoil, because at adaptation the lift is zero.

The next candidate is such that $A_1 = A_2 = A \neq 0,\ A_3, \ldots, A_n = 0,\ n \geq 3$. It is a combination of a flat plate, a parabolic and a cubic plate. The slope is obtained from the relation

$$d'[x(t)] = \alpha - A_0 + A\,(\cos t + \cos 2t) = \alpha - A_0 + A\left(\cos t + 2\cos^2 t - 1\right) \tag{3.89}$$

Upon elimination of t one finds

$$d'(x) = \alpha - A_0 + A\left(2 - 10\frac{x}{c} + 8\frac{x^2}{c^2}\right) \tag{3.90}$$

Integrating from 0 to x gives

$$d(x) = \int_0^x d'(\xi)d\xi = (\alpha - A_0)\,x + A\left(2x - 5\frac{x^2}{c} + \frac{8}{3}\frac{x^3}{c^2}\right) \tag{3.91}$$

The condition that $d(c) = 0$ determines $A_0 = \alpha - \frac{A}{3}$. Substitution of this result in the expression of the lift coefficient results in $C_l = 2\pi\left(\alpha + \frac{A}{6}\right)$. Finally, the equation of the camberline can be factorized as

$$d(x) = \frac{A}{3}x\left(7 - 8\frac{x}{c}\right)\left(1 - \frac{x}{c}\right) \tag{3.92}$$

From the previous result, the ideal angle of attack is $\alpha_{adapt} = \frac{A}{3}$ and the associated lift $(C_l)_{adapt} = \pi A = (C_l)_{des}$. This determines the unknown coefficient A:

$$A = \frac{(C_l)_{des}}{\pi} \tag{3.93}$$

A sketch of the thin cambered plate at incidence of adaptation is shown in Fig. 3.23.

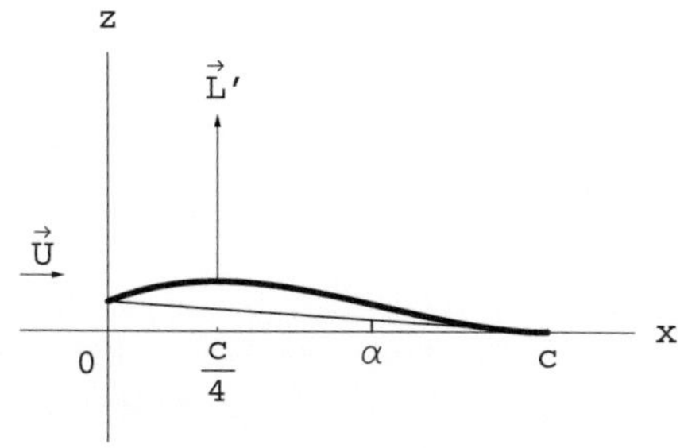

Fig. 3.23 Thin cambered plate at design conditions

3.6 A Family of Profiles with Minimum Pressure Gradient

The Society of Automotive Engineering (SAE) organizes every year a collegiate design competition which encourages the students to design, build and fly a remote controlled model airplane, capable of carrying a maximum payload, and that can take-off within 200 ft from a start line, make four-90° turns, touch down and roll within 400 ft from a prescribed line. Besides these take-off and landing constraints, other requirements are imposed such as not losing parts at any time during the run, using an unmodified specified gas engine with provided fuel, as well as some geometric constraints that vary from year to year (imposed maximum span or maximum surface area of the lifting elements, etc.)

From these rules, it is clear that one would like to use a wing profile having high lift capabilities C_{lmax}. Indeed, assuming that the best propeller has been chosen for the engine during the acceleration phase on the ground, the velocity is primarily governed by the total airplane weight. In order for the airplane to rotate and take-off with a maximum payload, a high lift coefficient is needed.

From the thin airfoil theory and inviscid flow view point, all airfoils are equivalent in the sense of being capable of giving a C_l value

$$C_l = 2\pi \left(\alpha - \alpha_{adapt} + 2\frac{d}{c} \right) \tag{3.94}$$

by a combination of interchangeable incidence and camber. Of course, in reality, viscous effects at a given Reynolds number, will limit the maximum lift coefficient to a value that depends on the airfoil camber and thickness distributions. The problem of optimization seems therefore irrelevant, unless one accounts for viscous effects. One idea is to try to minimize the negative impact of viscosity by searching for the profile that will create the "least disturbance" to an inviscid flow.

Potential flows have been shown to minimize the integral of pressure over the domain surrounding the profile (Bateman principle), i.e. to minimize

$$B = \int_{\Omega} p(x, z) dx dz \tag{3.95}$$

where Ω represents the domain of the flow.

Here we propose to minimize the integral of the pressure gradient

$$I = \frac{1}{(\rho U^2)^2} \int_{\Omega} \left\{ \left(\frac{\partial p}{\partial x} \right)^2 + \left(\frac{\partial p}{\partial z} \right)^2 \right\} dxdz \tag{3.96}$$

with the understanding, as will be seen in a later chapter, that very large negative pressure gradients increase shear stress and friction drag, whereas very large positive pressure gradients trigger boundary-layer separation and pressure drag. Using the small perturbation assumption $p(x, z) = p_\infty - \rho U u$, the integral can be written as

$$I = \frac{1}{U^2} \int_{\Omega} \left\{ \left(\frac{\partial u}{\partial x} \right)^2 + \left(\frac{\partial u}{\partial z} \right)^2 \right\} dxdz \tag{3.97}$$

The square of the first derivatives can be replaced by the equivalent expression, say for the x-derivative

$$\left(\frac{\partial u}{\partial x} \right)^2 = \frac{\partial}{\partial x} \left(u \frac{\partial u}{\partial x} \right) - u \frac{\partial^2 u}{\partial x^2} \tag{3.98}$$

and similarly for the z-derivative, so that the integral becomes

$$\begin{aligned} I = &\frac{1}{U^2} \int_{\Omega} \left\{ \frac{\partial}{\partial x} \left(u \frac{\partial u}{\partial x} \right) + \frac{\partial}{\partial z} \left(u \frac{\partial u}{\partial z} \right) \right\} dxdz \\ &- \frac{1}{U^2} \int_{\Omega} u \left(\frac{\partial^2 u}{\partial x^2} + \frac{\partial^2 u}{\partial z^2} \right) dxdz \end{aligned} \tag{3.99}$$

The last term vanishes because u, like w, ϕ and ψ are solution of Laplace equation. The divergence theorem is applied to the first term and results in a contour integral that vanishes at infinity, but contributes on the profile contour to the term

$$I = \frac{1}{U^2} \int_{\partial\Omega} u \left(\frac{\partial u}{\partial x}, \frac{\partial u}{\partial z} \right) .\mathbf{n}dl \tag{3.100}$$

$\partial\Omega$ is the profile boundary. $\mathbf{dl} = (dx, dz)$ represents an element tangent to the profile surface in the clockwise contour integral and the normal element $\mathbf{n}dl = (dz, -dx)$ points into the profile. Making use of the two governing PDEs, the integral can be written as

$$I = \frac{1}{U^2} \int_{\partial\Omega} u \left(-\frac{\partial w}{\partial z}, \frac{\partial w}{\partial x} \right) .\mathbf{n}dl = -\frac{1}{U^2} \int_{\partial\Omega} u \frac{dw}{dx} dx \tag{3.101}$$

Now, we can use the following expressions for u and w on the upper and lower surfaces

$$u^+(x) = u_e(x) + \frac{\Gamma'(x)}{2}, \; u^-(x) = u_e(x) - \frac{\Gamma'(x)}{2} \tag{3.102}$$

$$w^+(x) = w_d(x) + U\frac{e'(x)}{2}, \; w^-(x) = w_d(x) - U\frac{e'(x)}{2} \tag{3.103}$$

where $u_e(x)$ represents the horizontal component of velocity induced by the distribution of sources and sinks (thickness problem) and $w_d(x)$ is the vertical component induced by the vorticity distribution (lifting problem). Replacing the contour integral by a simple integral from zero to c, substituting the above expressions and after some simplification, one gets

$$\begin{aligned} I &= -\frac{1}{U^2}\int_0^c \left(u^+\frac{dw^+}{dx} - u^-\frac{dw^-}{dx}\right)dx \\ &= -\frac{1}{U^2}\int_0^c \left(Uu_e e'' + \Gamma'\frac{dw_d}{dx}\right)dx = I_e + I_d \end{aligned} \tag{3.104}$$

This result indicates that the minimization problem can also be decomposed into symmetric (I_e) and lifting (I_d) problems.

The symmetric problem reads

$$I_e = -\frac{1}{U^2}\int_0^c Uu_e e'' dx, \; u_e(x) = \frac{U}{2\pi}\int_0^c \frac{e'(\xi)}{x-\xi}d\xi \tag{3.105}$$

We will proceed as was done for the lifting problem. Let

$$\begin{cases} e'[x(t)] = 2\left\{E_0\frac{1+\cos t}{\sin t} + \sum_{n=1}^{\infty} E_n \sin nt\right\} \\ \qquad x(t) = \frac{c}{2}(1-\cos t) \end{cases}, \; 0 \le t \le \pi \tag{3.106}$$

Note that we have kept a singular term which tends to infinity as $t \to 0$ (i.e. $x \to 0$). The behavior near the leading edge is

$$e'[x(t)] \sim \frac{4E_0}{\sin t} \sim \frac{2E_0}{\sqrt{\frac{x}{c}}} \tag{3.107}$$

which corresponds to a parabolic blunt nose. The trailing edge is cusped. The horizontal velocity component is

$$u_e(x) = U\left\{E_0 - \sum_{n=1}^{\infty} E_n \cos t\right\} \tag{3.108}$$

and e'' is given by

$$e''[x(t)] = \frac{4}{c}\left\{-E_0\frac{1+\cos t}{\sin^3 t} + \sum_{n=1}^{\infty} E_n \frac{n\cos nt}{\sin t}\right\} \tag{3.109}$$

After some algebra, the final result is

$$I_e = \pi\left\{2E_0\sum_{n=1}^{\infty} E_n + \sum_{n=1}^{\infty} nE_n^2\right\} \tag{3.110}$$

In order for the profile to close at the trailing edge, the condition $\int_0^c e'(x)dx = 0$ must be satisfied. This is found to be the case when $E_1 = -2E_0$. Therefore the simplest profile corresponds to modes zero and mode 1, i.e.

$$e'[x(t)] = 2E_0\left\{\frac{1+\cos t}{\sin t} - 2\sin t\right\} \tag{3.111}$$

Upon integration, the thickness distribution is obtained

$$e[x(t)] = E_0 c \sin t\,(1+\cos t) = 4E_0 c\sqrt{\frac{x}{c}}\left(1-\frac{x}{c}\right)^{\frac{3}{2}} \tag{3.112}$$

which is the Quasi-Joukowski thickness distribution, with $E_0 = \frac{4}{3\sqrt{3}}\frac{e}{c}$.

Adding an extra term, mode 2, with $E_2 = -\frac{3}{4}E_0$ produces a profile with a far-forward maximum thickness

$$e[x(t)] = E_0 c \sin t\,(1+\cos t) - \frac{3}{2}E_0 c \sin^3 t = 4E_0 c\sqrt{\frac{x}{c}}\left(1-\frac{x}{c}\right)^{\frac{5}{2}} \tag{3.113}$$

which is the third thickness distribution introduced earlier, with $E_0 = \frac{54}{25\sqrt{5}}\frac{e}{c}$.

With the lifting problem, we will assume that the leading edge is adapted, since we remarked that a singular leading edge will not behave well in presence of viscous effects and will probably trigger separation, Chattot [2]. Proceeding as was done earlier in the chapter, we let

$$\Gamma'[x(t)] = 2U\sum_{n=1}^{\infty} A_n \sin nt \tag{3.114}$$

using only regular terms in the series, which means that the flat plate is not an element of this family. Hence we have

$$w_d(x) = U \sum_{n=1}^{\infty} A_n \cos nt \tag{3.115}$$

After some algebra one gets in this case

$$I_d = \pi \sum_{n=1}^{\infty} n A_n^2 \tag{3.116}$$

Keeping, for example, only the first mode results in the parabolic cambered plate at zero incidence, Fig. 3.17.

More interesting results have been obtained with this approach in the design of multiple element airfoil camberlines. For instance, a double element airfoil can be designed for minimum pressure gradient and a given high lift coefficient, say $C_l = 2.3$. A number of design parameters have to be chosen: the relative length of each element, the gap width between the two elements. Here we used a 50–50 chord percentage for each element as overlap was not advantageous with the SAE rules. The inviscid result indicates that the first element has $C_{l1} = 1.145$ and the second element $C_{l2} = 1.155$. See Fig. 3.24 for the profile geometry where 12 % thickness distributions have been added to the camberline design. The inviscid C_p distributions are also shown. The gap width was set by analyzing several designs with a viscous code, MSES for multiple element airfoils by Drela [3] and selecting the best result. The final design geometry and C_p distributions are shown in Fig. 3.25. Note that the second element thickness was reduced to 7 % as it was found that the stagnation point was not moving with incidence, whereas the thickness of the first element was increased to 17 % to allow for large displacements of the leading element stagnation point. The maximum lift coefficient was found to be $C_l = 3.1$, (see Fig. 11.6).

A wind tunnel model of a low aspect ratio half-wing has been built with foam core and carbon fiber skin which validated the high lift capabilities of the design by comparison with the numerical model, Fig. 3.26.

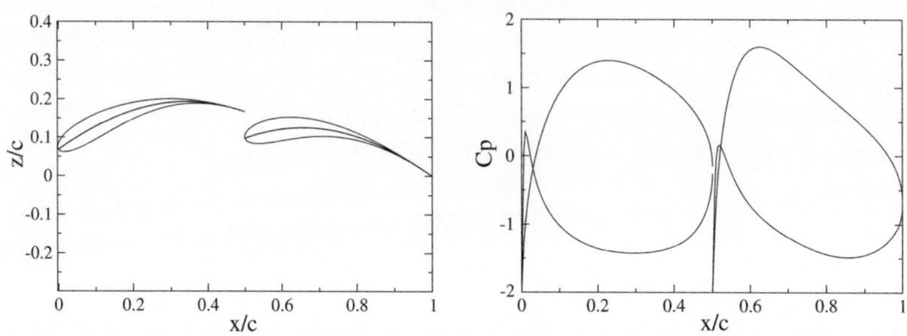

Fig. 3.24 Double element geometry and inviscid C_p distributions at design conditions

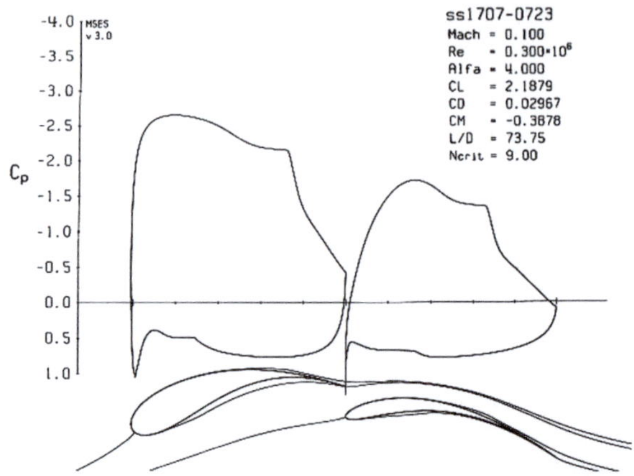

Fig. 3.25 Double element geometry and viscous C_p distribution at design conditions

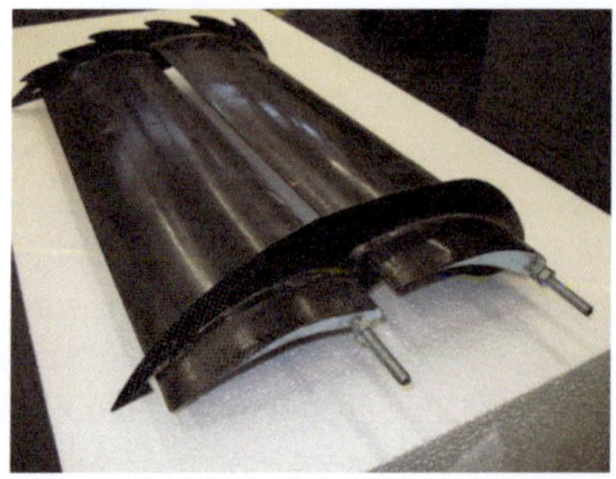

Fig. 3.26 Double element wind tunnel test model

3.7 Numerical Solution of the Fundamental Integral Equation

Numerical solution of the fundamental integral equation can be obtained cheaply and accurately for arbitrary geometries. Such simulation tools replace advantageously the analytical method and allow to spend more time analyzing the results and understanding the physics, than computing integrals. The benefit of the theory is undeniable in demonstrating the dependency of lift and moment on solely a few modes and in helping understand the suction force and the notion of ideal angle of attack. The fundamental integral equation of thin airfoil theory

$$w(x,0) = -\frac{1}{2\pi}\int_0^c \frac{\Gamma'(\xi)}{x-\xi}d\xi = U\left(d'(x) - \alpha\right) \tag{3.117}$$

is solved using finite differences. One important feature in the discretization of the problem is the use of the Glauert transformation, which allows for rapid convergence and high accuracy. Let the distribution of control points be given by

$$x_i = \frac{c}{2}(1 - \cos t_i), \; i = 1, 2, \ldots, ix \tag{3.118}$$

where $t_i = (i-1)\Delta t = \pi \frac{i-1}{ix-1}$, and ix is the maximum number of control points. Dummy points are introduced between the control points as

$$\xi_j = \frac{c}{2}(1 - \cos\theta_j), \; j = 1, 2, \ldots, ix - 1 \tag{3.119}$$

where $\theta_j = (j-1)\Delta t + \frac{1}{2}\Delta t = \pi \frac{2j-1}{2ix-2}$, $j = 1, 2, \ldots, ix - 1$.

Note that there is one less dummy points than control points. The integral is replaced by the summation

$$w_i = -\frac{1}{2\pi}\sum_{j=1}^{ix-1} \frac{\Gamma_{j+1} - \Gamma_j}{x_i - \xi_j} = U\left(f_i' - \alpha\right), \; i = 2, 3, \ldots, ix - 1 \tag{3.120}$$

The principal value integral limiting process is avoided by having the x_i's and ξ_j's at different locations, such that $x_i - \xi_j \neq 0$. Let's denote by $a_{i,j} = -\frac{1}{2\pi}\left(\frac{1}{x_i - \xi_{j-1}} - \frac{1}{x_i - \xi_j}\right)$, the coefficient of Γ_j in the ith equation. The terms involving ξ_0 and ξ_{ix} are not included.

Two boundary conditions are added. At the leading edge, $i = 1$, the circulation is zero, i.e. $\Gamma_1 = 0$. At the trailing edge, $\Gamma_{ix}' = 0$. This is implemented with the formula

$$b_{ix-2}\Gamma_{ix-2} + b_{ix-1}\Gamma_{ix-1} + \Gamma_{ix} = 0 \tag{3.121}$$

where

$$\begin{aligned} b_{ix-2} &= \frac{(x_{ix} - x_{ix-1})^2}{(x_{ix} - x_{ix-2})^2 - (x_{ix} - x_{ix-1})^2} \\ b_{ix-1} &= -\frac{(x_{ix} - x_{ix-2})^2}{(x_{ix} - x_{ix-2})^2 - (x_{ix} - x_{ix-1})^2} \end{aligned} \tag{3.122}$$

which are obtained by fitting a parabola through the last three mesh points, with horizontal tangent at $x_{ix} = c$.

The fundamental integral equation in discrete form, including boundary conditions, can be written as a linear system of equations for the Γ_i's

$$\begin{pmatrix} 1 & 0 & 0 & \ldots & \ldots & 0 \\ a_{2,1} & a_{2,2} & \ldots & \ldots & \ldots & a_{2,ix} \\ a_{3,1} & a_{3,2} & a_{3,3} & \ldots & \ldots & a_{3,ix} \\ . & & & & & \\ a_{ix-1,1} & \ldots & \ldots & \ldots & a_{ix-1,ix-1} & a_{ix-1,ix} \\ 0 & \ldots & \ldots & b_{ix-2} & b_{ix-1} & 1 \end{pmatrix} \begin{pmatrix} \Gamma_1 \\ \Gamma_2 \\ . \\ \Gamma_i \\ . \\ \Gamma_{ix} \end{pmatrix} = \begin{pmatrix} 0 \\ U(f_2' - \alpha) \\ . \\ U(f_i' - \alpha) \\ . \\ 0 \end{pmatrix} \tag{3.123}$$

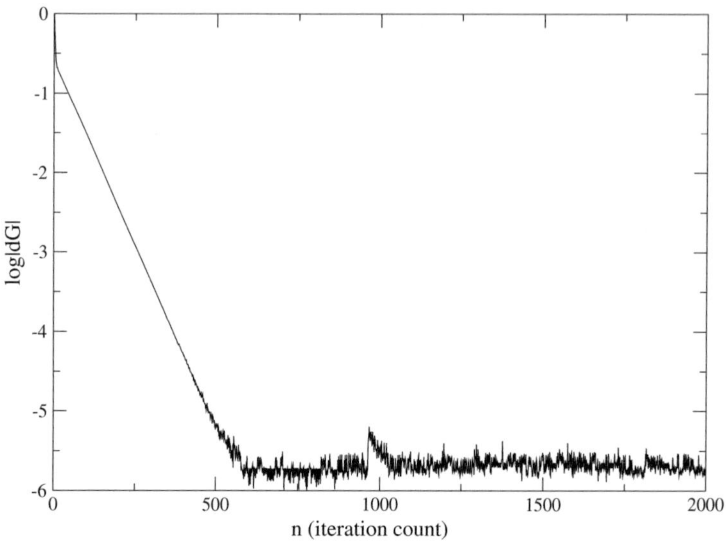

Fig. 3.27 Convergence history

This very large, practically full matrix, can be solved efficiently by relaxation as

$$a_{i,i}\Delta\Gamma_i = U(f_i' - \alpha) - \sum_{j=2}^{ix} a_{i,j}\Gamma_j^{n,n+1}, \ i = 2, 3, \dots, ix - 1 \tag{3.124}$$

$$\Delta\Gamma_{ix} = -b_{ix-2}\Gamma_{ix-2}^{n+1} - b_{ix-1}\Gamma_{ix-1}^{n+1} - \Gamma_{ix}^{n} \tag{3.125}$$

and $\Gamma_i^{n+1} = \Gamma_i^n + \omega\Delta\Gamma_i$. n represent the iteration index and the upper index $(n, n+1)$ indicates that the value may be "old" or "new". The relaxation factor [4], $0 < \omega < 2$, can be chosen as high as $\omega = 1.8$.

The convergence history with a fine mesh of $ix = 101$ points is shown in Fig. 3.27. It takes about 500 iterations to converge to machine accuracy.

Results for the parabolic plate at zero incidence are presented in Fig. 3.28 and compared with the analytic solution for mesh systems of $ix = 6, 11, 21$ and 51 points. As can be seen, results are sufficiently accurate with $ix \geq 21$.

The convergence with mesh refinement is shown in Fig. 3.29, where the following error norms have been used

$$\epsilon_1 = \frac{1}{ix}\sum_{i=1}^{ix} |\Gamma_i - \Gamma_{exact}(x_i)| \tag{3.126}$$

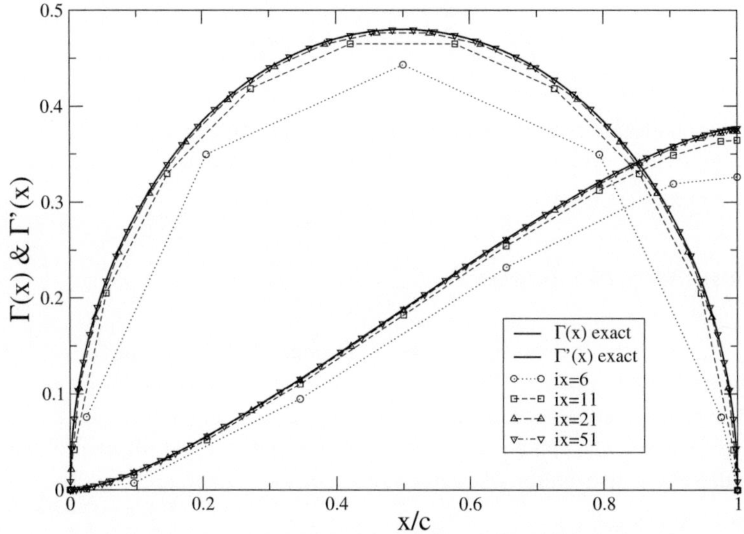

Fig. 3.28 Comparison with analytic solution of Γ and Γ'

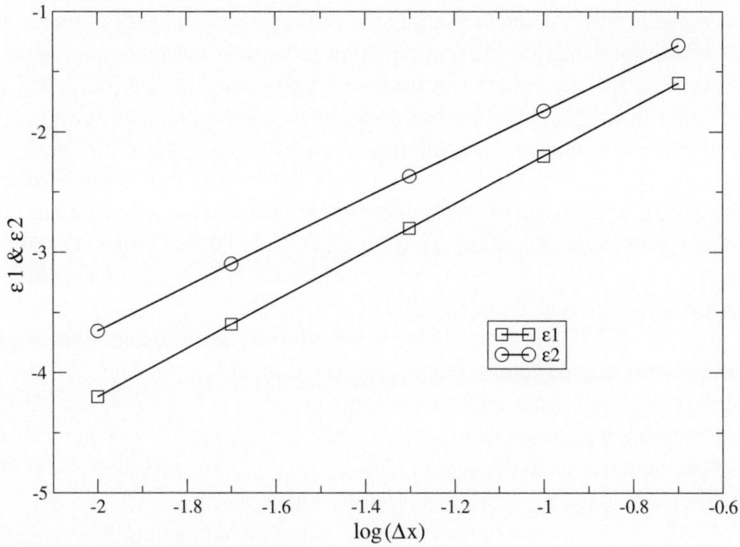

Fig. 3.29 Rate of convergence of Γ and Γ'

$$\epsilon_2 = \frac{1}{ix-1} \sum_{i=1}^{ix-1} \left| \frac{\Gamma_{i+1} - \Gamma_i}{x_{i+1} - x_i} - \Gamma'_{exact}(\xi_i) \right| \tag{3.127}$$

The circulation Γ converges with a slope of 2. The rate of convergence of 1.8 for Γ' is slightly less, probably due to the singularity in slope at $x = 0$ and $x = c$.

3.8 Summary of Chapter 3

In this chapter, small disturbance theory associated with thin airfoils is used to simplify and obtain linearized formulations of the tangency condition and pressure field. The general problem of inviscid, incompressible potential flow past thin airfoils is now fully linear and can be decomposed into symmetric and lifting problems that are handled by a distribution of sources/sinks for the former, and of vortices for the latter.

The symmetric problem does not contribute to lift or moment, however, thickness affects the pressure distribution. The lifting problem leads to the fundamental integral equation of thin airfoil theory, which relates the vorticity distribution Γ' to the induced velocity component w along the profile, via the tangency condition. The solution of the fundamental integral equation is worked out using a singular term plus a Fourier series of regular terms for the vorticity along the cut $[0, c]$. It is shown how the singular term and the Fourier coefficients can be obtained for a given thin cambered plate at incidence. The interpretation of the singular term coefficient as $A_0 = \alpha - \alpha_{adapt}$ points to the existence, for each thin airfoil of an angle of adaptation, that depends purely on the airfoil geometry, such that the leading edge satisfies also a Kutta-Joukowski condition. This has a practical interest since separation at a sharp leading edge is likely to occur in real fluid when $\alpha \neq \alpha_{adapt}$. The positive role of thick leading edge is also emphasized in this regard.

The forces and moments are obtained for arbitrary airfoils and from these, the center of pressure can be found. In all cases, the drag is zero (d'Alembert paradox). In general, as the incidence varies from large negative to large positive values, the center of pressure will travel from $-\infty$ to $+\infty$. At very high or low incidences, the center of pressure moves to the quarter chord, $\frac{x_{c.p.}}{c} = \frac{1}{4}$. The aerodynamic center for all thin airfoils is found to be at the quarter chord, $\frac{x_{a.c.}}{c} = \frac{1}{4}$.

Application of the theory to the design of a thin airfoil exemplifies the practical aspects and usefulness of the results derived in this chapter. It is also shown that some particular thickness and camber distributions can be related to the concept of minimum pressure gradient field surrounding the airfoil. This idea has been applied to the design of double element airfoils, which have proved to perform well in practice.

Finally, numerical aspects of the solution of the fundamental integral equation of thin airfoil theory are given, which, properly implemented, contribute to fast and accurate analysis tools for the designer.

3.9 Problems

3.9.1

Calculate the horizontal perturbation velocity $u(x, 0)$ due to an ellipse at zero incidence. The thickness distribution of the ellipse is given by $e(x) = 2e\sqrt{\frac{x}{c}\left(1-\frac{x}{c}\right)}$. Hint: use Glauert change of variable. Why was this thickness distribution not included in the family of profiles with minimum pressure gradient?

3.9.2

Calculate the horizontal perturbation velocity $u(x, 0)$ due to a biconvex airfoil at zero incidence. The thickness distribution of the biconvex profile is given by $e(x) = 4e\frac{x}{c}\left(1-\frac{x}{c}\right)$. Hint: use Glauert change of variable. Sketch the pressure coefficient $-C_p$.

3.9.3

Calculate the horizontal perturbation velocity $u(x, 0)$ due to a double wedge airfoil at zero incidence. The thickness distribution of the double wedge profile is given by $e(x) = 2\theta x$, $0 \le x \le \frac{c}{2}$ and $e(x) = 2\theta(c - x)$, $\frac{c}{2} \le x \le c$. Hint: no change of variable is necessary. Sketch the pressure coefficient $-C_p$.

3.9.4

Consider a thin flat plate with a break in slope at $x = \frac{c}{2}$. The equation of each piece is given by $d(x) = \theta x$, $0 \le x \le \frac{c}{2}$ and $d(x) = \theta(c - x)$, $\frac{c}{2} \le x \le c$. Calculate the horizontal perturbation velocity $u(x, 0)$ and $C_p^{\pm}$. Hint: Use the integral formulae given for the Fourier coefficients. Find the angle of adaptation α_{adapt}. Give the values for $C_l(\alpha)$ and $C_{m,o}(\alpha)$. Sketch $-C_p^{\pm}$ at α_{adapt}.

3.9.5

A double wedge airfoil at zero incidence ($\alpha = 0$) is moving with velocity U in a uniform atmosphere. The geometry of the airfoil is sketched in Fig. 3.30. The half-angle is $\theta << 1rd$. The equation of the upper surface is $z^+(x) = \theta x$, $0 < x < c/2$, $z^+(x) = \theta(c - x)$, $c/2 < x < c$. Find the perturbation velocity (u, w) on the upper surface, given that $u(x, 0) = \frac{U}{\pi}\int_0^c \frac{z'^+(\xi)d\xi}{x-\xi}$. Give the C_p distribution at $\alpha = 0$ and make a sketch of it. Will there be a lift force? A drag force? The incidence is changed to $\alpha > 0$. Give the new C_p distribution and make a sketch of it. What will the lift coefficient be? The drag coefficient?

3.9.6

A thin airfoil problem is characterized by small, dimensionless parameters of order $O \ll 1$. Give the definition of these small parameters. Give the definition of the Center of Pressure. In inviscid, incompressible flow, give the formulae for the lift coefficient C_l, drag coefficient C_d and moment coefficient $C_{m,o}$ of a thin parabolic

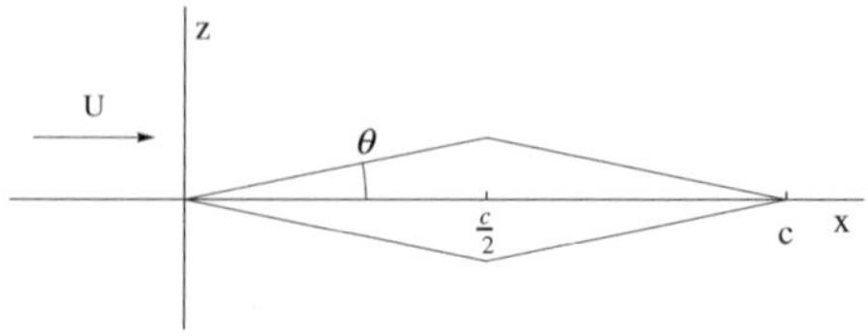

Fig. 3.30 Problem 3.9.5: Double wedge airfoil

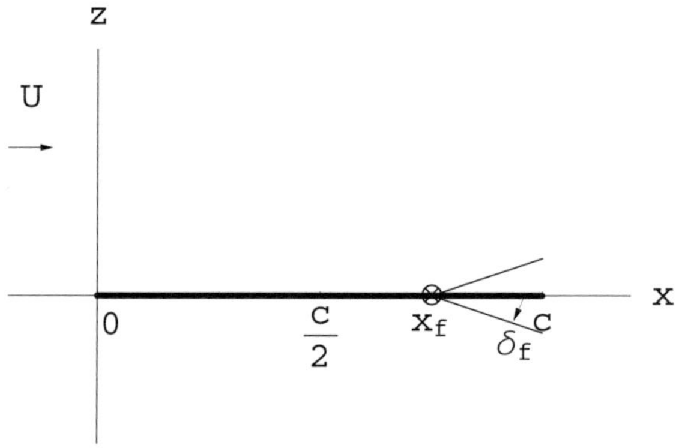

Fig. 3.31 Problem 3.9.7: Flat plate with flap

plate in terms of the small, dimensionless parameters. If thickness is added to the plate, how do these coefficients vary?

If this thin profile is allowed to rotate freely about an axis placed at the leading edge, find the equilibrium incidence. Hint: the moment of the aerodynamic forces is zero (neglect weight). Is the equilibrium statically stable?

3.9.7

A thin flat plate has a hinge located at $(x_f, 0)$, $0 \leq x_f \leq c$, about which the end of the plate can rotate by an angle δ_f (positive down) and that plays the role of a flap. The equation of the thin plate is given by:

$$f^{\pm}(x) = d(x) = \begin{cases} 0, \ 0 \leq x \leq x_f \\ \delta_f(x_f - x), \ x_f \leq x \leq c \end{cases}$$

Calculate the slope $d'(x)$ of the plate (check your algebra as other results depend on this). See Fig. 3.31. Let t_f be the value of the parameter t corresponding to the flap hinge in the classical parametric representation of the thin plate as $x_f = x(t_f) = \frac{c}{2}(1 - \cos t_f)$, $0 \leq t_f \leq \pi$. Using the formula derived in class $A_0 = \alpha - \frac{1}{\pi}\int_0^{\pi} d'[x(t)]dt$, calculate A_0. Check your result for $t_f = \pi$ $(x_f = c)$. Conclude. Check your result for $t_f = 0$ $(x_f = 0)$. Conclude.

What is the angle of adaptation $\alpha_{adapt}(t_f, \delta_f)$? Check your result for $t_f = \pi$ $(x_f = c)$. Conclude. Check your result for $t_f = 0$ $(x_f = 0)$. Conclude. Calculate the incidence of adaptation for $x_f = 3c/4$ and $\delta_f = 10°$.

Sketch the plate and, qualitatively, the flow at the calculated incidence of adaptation, in particular the streamlines near the leading edge and the trailing edge.

Calculate A_1 in terms of the average camber $\frac{d}{c}$, t_f and δ_f. Check your result for $t_f = \pi$ ($x_f = c$). Conclude. Give the value of the aerodynamic coefficients C_l and C_d in terms of α, t_f and δ_f. If thickness is added to the thin plate, how will these coefficients be affected?

References

1. Glauert, H.: The Elements of Aerofoil and Airscrew Theory, 2nd edn. Cambridge University Press, Cambridge (1948)
2. Chattot, J.-J.: Optimization in applied aerodynamics. Comput. Fluid Dyn. J. **9**(3), 306–311 (2000)
3. Drela, M.: MSES—Multi-element airfoil design/analysis software. http://raphael.mit.edu/research.html (1994)
4. Chattot, J.-J.: Computational Aerodynamics and Fluid Dynamics: An Introduction. Scientific Computation. Springer, Berlin (2004)

Chapter 4
Inviscid, Compressible Flow Past Thin Airfoils

4.1 Introduction

Compressibility effects are significant when the incoming Mach numbers or the local Mach numbers take values that are no longer small compared to unity. Indeed, one often considers that if the Mach number is larger than 0.3 the flow can no longer be treated as incompressible. Such a situation can occur at very low incoming Mach number, if the geometry is such that the incompressible solution reaches infinite velocities at some point. This is the case, as we have seen in Chap. 3, at a sharp leading edge, Fig. 3.12a. This would also be the case at the shoulder of a double wedge profile, Fig. 3.30, even at zero incidence. In both cases, the incompressible flow solution admits an infinite velocity. If an inviscid, compressible flow model were used, the velocity would not be infinite, but the flow would accelerate beyond sonic speeds and a supersonic bubble terminated by a shock would appear in the midst of the overall subsonic flow field. Although these are interesting theoretical problems, we will not focus our attention on this type of compressibility effects induced by discontinuities in the geometry, that can be eliminated by introducing a blunt nose or a rounded shoulder. In the framework of thin airfoil theory, compressibility effects will be the result of a combination of thickness, camber, angle of incidence and Mach number such that, as the Mach number is progressively increased from low subsonic to supersonic values, all the other parameters being held fixed, the solution deviates progressively more and more from the incompressible flow solution, exhibiting in general shock waves, such as the detached shock in Fig. 4.1.

4.2 Governing Equations

Assuming 2-D, steady, inviscid and irrotational compressible flow with uniform conditions at infinity, the governing equations of conservation of mass, the irrotationality condition and energy equation (generalized Bernoulli) read in terms of the perturbation velocity components (u, w)

J.J. Chattot and M.M. Hafez, *Theoretical and Applied Aerodynamics*,
DOI 10.1007/978-94-017-9825-9_4

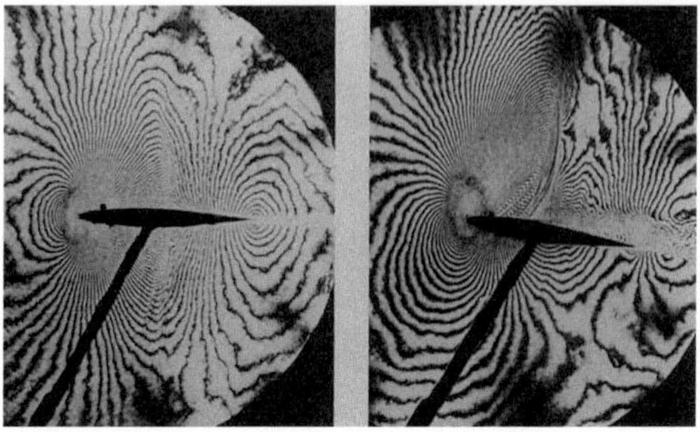

Fig. 4.1 Holographic interferometry of transonic flow fields, (from history.nasa.gov)

$$\frac{\partial \rho(U+u)}{\partial x} + \frac{\partial \rho w}{\partial z} = 0 \tag{4.1}$$

$$\frac{\partial w}{\partial x} - \frac{\partial u}{\partial z} = 0 \tag{4.2}$$

$$\frac{\gamma p}{(\gamma-1)\rho} + \frac{V^2}{2} = H_0 = const. \tag{4.3}$$

where $\gamma = \frac{c_p}{c_v}$ is the ratio of specific heats, $\overrightarrow{V} = (U+u, w)$ the total velocity and H_0 is the uniform stagnation enthalpy for isoenergetic flow. The entropy for such flows is uniform and can be written

$$\frac{p}{\rho^\gamma} = \frac{p_0}{\rho_0^\gamma} = const. \tag{4.4}$$

For air, the gas index is $\gamma = 1.4$. The subscript "0" stands for undisturbed, incoming flow quantities.

Elimination of pressure between the energy equation and the entropy condition yields a density-velocity relation

$$\frac{\rho}{\rho_0} = \left[1 + \frac{\gamma-1}{2}\left(M_0^2 - \left(\frac{V}{a_0}\right)^2\right)\right]^{\frac{1}{\gamma-1}} \tag{4.5}$$

where the incoming Mach number and speed of sound are $M_0 = \frac{U}{a_0}$ and $a_0 = \sqrt{\frac{\gamma p_0}{\rho_0}}$, respectively.

4.3 Linearized Compressible Flow Potential Equation

As before, we will assume that the uniform flow is only slightly disturbed by a thin airfoil at incidence, where the incidence is also a small parameter, $\alpha \ll 1$. The above density-velocity relation can be expanded for small values of u and w to read

$$\frac{\rho}{\rho_0} = \left[1 + \frac{\gamma - 1}{2}\left(M_0^2 - \frac{U^2 + 2Uu + u^2 + w^2}{a_0^2}\right)\right]^{\frac{1}{\gamma-1}}$$

$$= 1 + \frac{1}{2}\left(-\frac{2Uu}{a_0^2}\right) + O\left(\frac{u^2 + w^2}{a_0^2}\right) \simeq 1 - M_0^2 \frac{u}{U} \tag{4.6}$$

The relative change in density is a more meaningful number and takes the form

$$\frac{\rho - \rho_0}{\rho_0} = -M_0^2 \frac{u}{U} \tag{4.7}$$

which states that the Mach number is an amplification factor of the compressibility effects: the larger the Mach number, the larger the relative changes in density for a given longitudinal perturbation in velocity. Note that the incompressible limit is recovered for $M_0 = 0$. The changes in density are considered negligible when $M_0 \leq 0.3$. On the other hand, when $M_0 > 0.3$, compressibility effects are expected.

The linearized C_p can also be derived from the density-velocity relation. Fist notice that the pressure ratio of local pressure to the undisturbed pressure is to first order

$$\frac{p}{p_0} = \left(\frac{\rho}{\rho_0}\right)^{\gamma} = 1 - \gamma M_0^2 \frac{u}{U} \tag{4.8}$$

and using the identity $\gamma p_0 M_0^2 = \rho_0 U^2$, simplifies the pressure coefficient to read, as before

$$C_p = -2\frac{u}{U} \tag{4.9}$$

in compressible, 2-D flow.

The tangency condition is also unaffected by the changes in density and remains

$$w(x, 0^{\pm}) = U\left(f'^{\pm}(x) - \alpha\right) \tag{4.10}$$

The equation of conservation of mass is nonlinear. The last step is to linearize it using the result just obtained for the density. The mass flux components are expanded as

$$\begin{cases} \rho(U+u) = \rho_0\left(1 - M_0^2 \frac{u}{U}\right) U\left(1+\frac{u}{U}\right) = \rho_0 U\left(1+(1-M_0^2)\frac{u}{U}+\cdots\right) \\ \rho w = \rho_0 U\left(\frac{w}{U}+\cdots\right) \end{cases} \tag{4.11}$$

where the second and higher order terms have been omitted. Substitution in the mass conservation yields a linear system of two first order PDEs

$$(1-M_0^2)\frac{\partial u}{\partial x} + \frac{\partial w}{\partial z} = 0 \tag{4.12}$$

$$\frac{\partial w}{\partial x} - \frac{\partial u}{\partial z} = 0 \tag{4.13}$$

The irrotationality condition allows to introduce a velocity potential for the perturbation velocity such that $u = \frac{\partial \phi}{\partial x}$ and $w = \frac{\partial \phi}{\partial z}$. The conservation of mass is transformed into a second order PDE

$$(1-M_0^2)\frac{\partial^2 \phi}{\partial x^2} + \frac{\partial^2 \phi}{\partial z^2} = 0 \tag{4.14}$$

called the subsonic linearized compressible flow potential equation when $M_0 < 1$ and the potential equation for linearized supersonic flow when $M_0 > 1$. The change of sign of the coefficient $1 - M_0^2$ is significant in associating different physical phenomena with different equation types, elliptic for subsonic flow with perturbation decay in all directions, and hyperbolic for supersonic flow with subdomains separated by wave fronts. This can be illustrated by a moving source emitting sound waves at equal intervals and examining the development of the wave fronts for Mach number $\frac{U}{a_0} < 1$ and $\frac{U}{a_0} > 1$, Fig. 4.2. In subsonic flow, the waves move away from the source and the source influence is felt everywhere. The perturbations of the supersonic source are confined within a wedge (a cone in 3-D).

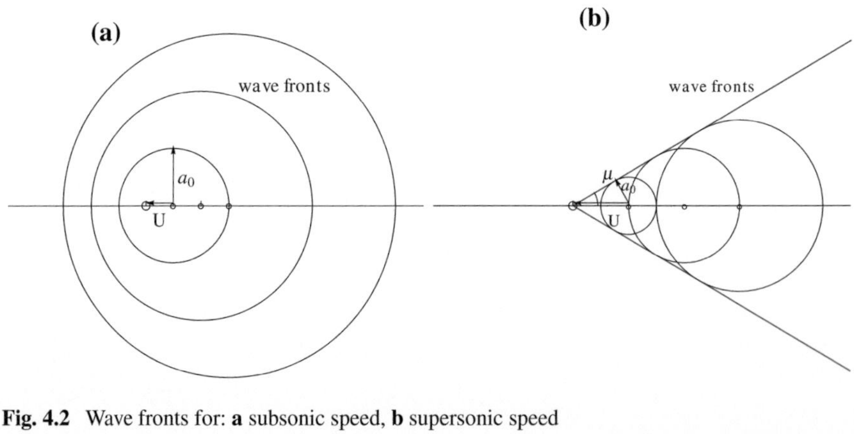

Fig. 4.2 Wave fronts for: **a** subsonic speed, **b** supersonic speed

In supersonic flow, the wave fronts admit envelopes which make an angle μ, called the Mach angle, with the flow direction. It is easy to see from Fig. 4.2 that the Mach angle is given by

$$\sin\mu = \frac{a_0}{U} = \frac{1}{M_0}, \quad \tan\mu = \frac{1}{\sqrt{M_0^2 - 1}} \tag{4.15}$$

4.4 Prandtl-Glauert Transformation

In subsonic flow, $M_0 < 1$, the linearized potential flow equation is elliptic and the Prandtl-Glauert transformation reduces it to Laplace equation. Let $\beta = \sqrt{1 - M_0^2}$ and define new dependent and independent variables as

$$\begin{cases} \widetilde{\phi}(\widetilde{x}, \widetilde{z}) = \beta\phi(x, z) \\ \widetilde{x} = x \\ \widetilde{z} = \beta z \end{cases} \tag{4.16}$$

Then, the governing equation becomes

$$\beta^2 \frac{1}{\beta}\frac{\partial^2\widetilde{\phi}}{\partial\widetilde{x}^2} + \frac{1}{\beta}\frac{\partial^2\widetilde{\phi}}{\partial\widetilde{z}^2}\beta^2 = \beta\left(\frac{\partial^2\widetilde{\phi}}{\partial\widetilde{x}^2} + \frac{\partial^2\widetilde{\phi}}{\partial\widetilde{z}^2}\right) = 0 \tag{4.17}$$

The transformation of the boundary conditions reads

$$\nabla\widetilde{\phi} \to 0, \ as \ \widetilde{x}^2 + \widetilde{z}^2 \to \infty \tag{4.18}$$

$$\frac{1}{\beta}\frac{\partial\widetilde{\phi}(\widetilde{x}, 0^{\pm})}{\partial\widetilde{z}}\beta = \frac{\partial\widetilde{\phi}(\widetilde{x}, 0^{\pm})}{\partial\widetilde{z}} = U\left(f'^{\pm}(\widetilde{x}) - \alpha\right) \tag{4.19}$$

The airfoil slope is unaffected by the transformation and the solution to this problem is equivalent to solving for the incompressible flow past the same airfoil. For this reason, except for a finite number of singular points such as leading edge and other sharp edges, the subsonic compressible flow solution past a profile is smooth and does not admit lines of discontinuity.

Let the subscript 0 represent the incompressible solution, i.e. at Mach $M_0 = 0$. The pressure coefficient is modified as follows

$$C_p(x) = -2\frac{u(x, 0)}{U} = -\frac{2}{U}\frac{1}{\beta}\frac{\partial\widetilde{\phi}(\widetilde{x}, 0)}{\partial\widetilde{x}} = \frac{1}{\beta}\left(C_p\right)_0 = \frac{\left(C_p\right)_0}{\sqrt{1 - M_0^2}} \tag{4.20}$$

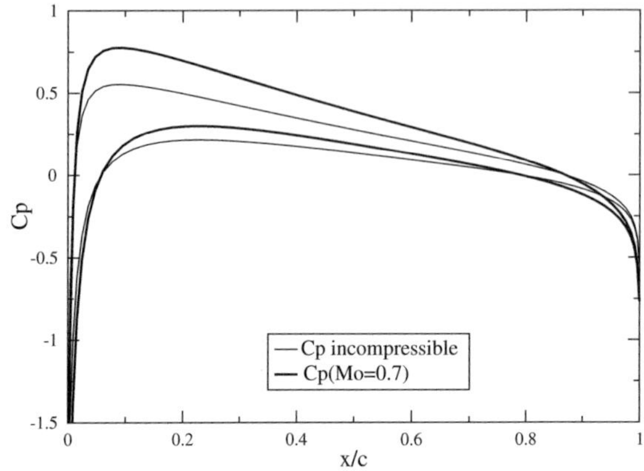

Fig. 4.3 Prandtl-Glauert correction applied to C_p

where $\left(C_p\right)_0 = -2\frac{\tilde{u}}{U}$, see Fig. 4.3. The forces and moments at Mach number M_0 will also be divided by $\sqrt{1 - M_0^2}$, which is the Prandtl-Glauert rule

$$C_l = \frac{(C_l)_0}{\sqrt{1 - M_0^2}}, \quad C_{m,o} = \frac{\left(C_{m,o}\right)_0}{\sqrt{1 - M_0^2}}, \quad C_d = 0 \tag{4.21}$$

Note that the drag is still zero. To be complete, one can add that the center of pressure and the aerodynamic center coincide with their incompressible counterparts.

4.5 Linearized Supersonic Flow

Let $\beta = \sqrt{M_0^2 - 1}$, when $M_0 > 1$. The potential equation for linearized supersonic flow is of hyperbolic type and models wave propagation. It is analogous to the wave equation that models the vibrations of a string. The PDE and boundary conditions read:

$$\begin{cases} -\beta^2 \frac{\partial^2 \phi}{\partial x^2} + \frac{\partial^2 \phi}{\partial z^2} = 0 \\ \frac{\partial \phi(x, 0^\pm)}{\partial z} = U(f'^{\pm}(x) - \alpha) \\ \phi(x, z) \to 0, \ x^2 + z^2 \to \infty, \ \textit{almost everywhere} \end{cases} \tag{4.22}$$

The general solution of the PDE, the d'Alembert solution, is given by

$$\phi(x, z) = F(x - \beta z) + G(x + \beta z) \tag{4.23}$$

where $F(\xi)$ and $G(\zeta)$ are arbitrary functions of a single argument. These unknown functions are uniquely determined by the boundary conditions. It is easy to verify

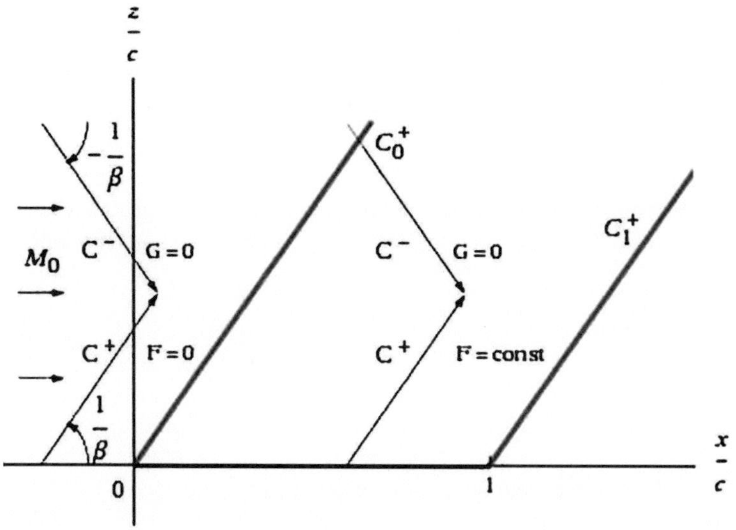

Fig. 4.4 D'Alembert solution for the *upper* surface

that such a solution satisfies the PDE with any functions F and G. The families of straight lines, $\xi = x - \beta z = const.$ and $\zeta = x + \beta z = const.$ are the characteristic lines of the PDE. They represent, in this approximation, wave fronts of infinitesimal perturbations as well as shocks and expansion waves of finite strength. They have slopes $dz/dx = \pm 1/\beta$.

Consider a profile placed along the x-axis, $0 \le x \le c$. The supersonic flow is from left to right. Upstream of the profile leading edge the flow is undisturbed. Perturbations will occur downstream of the two characteristics passing through the origin. Consider first the upper surface. The C_0^+ characteristic through O and C_1^+ through the trailing edge delimit a semi-infinite strip where the perturbation due to the profile affects the flow solution, see Fig. 4.4.

The solution will be obtained once the expressions for F and G are found. In the undisturbed flow region, upstream of the C_0^+ characteristic, the solution for the perturbation potential is $\phi = const. = 0$ (the potential is defined up to an arbitrary constant). In other words:

$$F(x - \beta z) + G(x + \beta z) = 0, \quad \forall x, z, \quad z \ge 0, \ \xi = x - \beta z \le 0 \qquad (4.24)$$

which implies that $F = G = 0$. Furthermore, since G remains constant along the $\zeta = x + \beta z = const.$, $G = 0$ in the semi-infinite strip $0 \le \xi = x - \beta z \le c$. The solution there reduces to $\phi(x, z) = F(x - \beta z)$. In order to apply the tangency condition, one needs to calculate the w-component along the x-axis. It is

$$\frac{\partial \phi(x, z)}{\partial z} = F'(\xi)(-\beta) \quad \Rightarrow \quad \frac{\partial \phi(x, 0^+)}{\partial z} = -\beta F'(x) = U(f'^{+}(x) - \alpha) \quad (4.25)$$

The last equation can be readily integrated as an ODE to give

$$F(x) = -\frac{U}{\beta}(f^+(x) - \alpha x) + const. \tag{4.26}$$

Finally, the constant is found by enforcing continuity of the potential at $x = 0$, hence the constant is zero. The above equation defines the functional form of F. Replacing x with ξ extends the definition of ϕ from the axis to the semi-infinite strip as

$$\phi(x, z) = -\frac{U}{\beta}\left(f^+(x - \beta z) - \alpha(x - \beta z)\right), \quad z \geq 0,\ 0 \leq \xi = x - \beta z \leq c \tag{4.27}$$

Note that, above the airfoil, the solution is only a function of $\xi = x - \beta z$. It is also easy to see that the tangency condition for $w(x, 0^+)$ is satisfied. There remains to calculate the u-component on the profile and the pressure coefficient. The former is now obtained:

$$\frac{\partial \phi(x, z)}{\partial x} = -\frac{U}{\beta}\left(f'^+(x - \beta z) - \alpha\right), \quad \Rightarrow \quad \frac{\partial \phi(x, 0^+)}{\partial x} = -\frac{U}{\beta}\left(f'^+(x) - \alpha\right) \tag{4.28}$$

The pressure coefficient on the profile is, as usual

$$C_p^+(x) = -2\frac{u(x, 0^+)}{U} = \frac{2}{\beta}\left(f'^+(x) - \alpha\right) \tag{4.29}$$

Proceeding in a similar way, one would find that, below the profile, the potential is $\phi(x, z) = G(x + \beta z)$ (i.e. $F = 0$ for $z \leq 0,\ 0 \leq \zeta = x + \beta z \leq c$) so that, upon satisfying the tangency condition $w(x, 0^-) = U(f'^-(x) - \alpha)$, the perturbation potential is found to be

$$\phi(x, z) = \frac{U}{\beta}\left(f^-(x + \beta z) - \alpha(x + \beta z)\right), \quad z \leq 0,\ 0 \leq \zeta = x + \beta z \leq c \tag{4.30}$$

The u-component of velocity on the profile is now

$$\frac{\partial \phi(x, z)}{\partial x} = \frac{U}{\beta}\left(f'^-(x + \beta z) - \alpha\right), \quad \Rightarrow \quad \frac{\partial \phi(x, 0^-)}{\partial x} = \frac{U}{\beta}\left(f'^-(x) - \alpha\right) \tag{4.31}$$

and the C_p

$$C_p^-(x) = -2\frac{u(x, 0^-)}{U} = -\frac{2}{\beta}\left(f'^-(x) - \alpha\right) \tag{4.32}$$

The expressions for $C_p^{\pm}(x)$ are known as Ackeret formulae.

Note that there was no reference to a Kutta-Joukowski condition, even when there is circulation and lift. Indeed, due to the wave character of supersonic flow, the solution, hence the forces and moment on the profile, do not depend on the solution downstream of the trailing edge. It is interesting, however, to discuss the solution downstream of the profile. At the trailing edge the potential is discontinuous. Indeed, the solution in the two semi-infinite strips above and below the profile reads:

$$\begin{cases} \phi(x,z) = -\frac{U}{\beta}\left(f^+(x-\beta z) - \alpha(x-\beta z)\right), & z \geq 0,\ 0 \leq \xi = x - \beta z \leq c \\ \phi(x,z) = \frac{U}{\beta}\left(f^-(x+\beta z) - \alpha(x+\beta z)\right), & z \leq 0,\ 0 \leq \zeta = x + \beta z \leq c \end{cases} \tag{4.33}$$

At the trailing edge the jump of potential is $< \phi(c,0) >= \phi(c,0^+) - \phi(c,0^-) = \Gamma(c) = 2\frac{U}{\beta}\alpha c$. The potential function for a lifting problem is multi-valued and a cut is introduced to regularize it. Although it could be any line $\xi \geq c, \varsigma \geq c$, originating at the trailing edge, it is convenient to align the cut with the x-axis. The cut does not represent a physical surface, streamline or wall, therefore pressure and velocity are continuous across it. The continuity of pressure or $u(x,z)$ is simply $\Gamma'(x) =< u(x,0) >= u(x,0^+) - u(x,0^-) = 0,\ x \geq c$. This condition can be integrated along the x-axis from $x = c$ to give $< \phi(x,0) >= \phi(x,0^+) - \phi(x,0^-) = \Gamma(c)$. The same result can be obtained when the cut is an arbitrary line. Above the cut the solution is still of the form $\phi(x,z) = F(x - \beta z)$, whereas below the cut it is of the form $\phi(x,z) = G(x + \beta z)$. The condition along the cut is therefore

$$F(x) - G(x) = \Gamma(c) \tag{4.34}$$

The continuity of the w-component leads to

$$\frac{\partial\phi(x,0^+)}{\partial z} = -\beta F'(x) = \frac{\partial\phi(x,0^-)}{\partial z} = \beta G'(x), \quad \Rightarrow F'(x) + G'(x) = 0 \tag{4.35}$$

This last ODE can be integrated to give

$$F(x) + G(x) = const. = 0 \tag{4.36}$$

where the constant is obtained for $x = c$. From these conditions the solution for F and G is obtained:

$$\begin{cases} F = const. = \frac{\Gamma(c)}{2} \\ G = const. = -\frac{\Gamma(c)}{2} \end{cases} \tag{4.37}$$

The solution reads

$$\begin{cases} \phi(x,z) = \frac{U}{\beta}\alpha c, & z \geq 0,\ c \leq \xi = x - \beta z \\ \phi(x,z) = -\frac{U}{\beta}\alpha c, & z \leq 0,\ c \leq \zeta = x + \beta z \end{cases} \tag{4.38}$$

The flow field, downstream of the profile is the uniform, undisturbed flow $\mathbf{V} = (U, 0)$.

4.5.1 Jump Conditions: Shock and Expansion Waves

The jump conditions associated with the system of first-order PDEs

$$\begin{cases} (1 - M_0^2)\frac{\partial u}{\partial x} + \frac{\partial w}{\partial z} = 0 \\ \frac{\partial w}{\partial x} - \frac{\partial u}{\partial z} = 0 \end{cases} \tag{4.39}$$

are found to be

$$\begin{cases} (1 - M_0^2) < u > n_x + < w > n_z = 0 \\ < w > n_x - < u > n_z = 0 \end{cases} \tag{4.40}$$

where $\mathbf{n} = (n_x, n_z)$ is a unit vector normal to the jump line. Upon elimination of the normal vector, the *shock polar* is obtained:

$$(M_0^2 - 1) < u >^2 - < w >^2 =$$

$$\left(\sqrt{M_0^2 - 1} < u > + < w >\right)\left(\sqrt{M_0^2 - 1} < u > - < w >\right) = 0 \tag{4.41}$$

The shock polar is made of two straight branches in the hodograph plane $(< u >, < w >)$, $J^+ : \beta < u > + < w > = 0$ and $J^- : \beta < u > - < w > = 0$, see Fig. 4.5. The origin of the hodograph plane corresponds to infinitely weak waves, i.e. characteristic lines, the half-plane $< u > < 0$ represents shock waves and the half-plane $< u \gg 0$ represents expansion waves.

The slope of the jump lines J^+ and J^- in the physical plane are obtained by selecting a root and using one of the jump relations

$$\frac{dz}{dx} = -\frac{n_x}{n_z} = -\frac{< u >}{< w >} \quad \Rightarrow$$

$$\left(\frac{dz}{dx}\right)_{J^+} = \frac{1}{\beta} = \left(\frac{dz}{dx}\right)_{C^+}, \quad \left(\frac{dz}{dx}\right)_{J^-} = -\frac{1}{\beta} = \left(\frac{dz}{dx}\right)_{C^-} \tag{4.42}$$

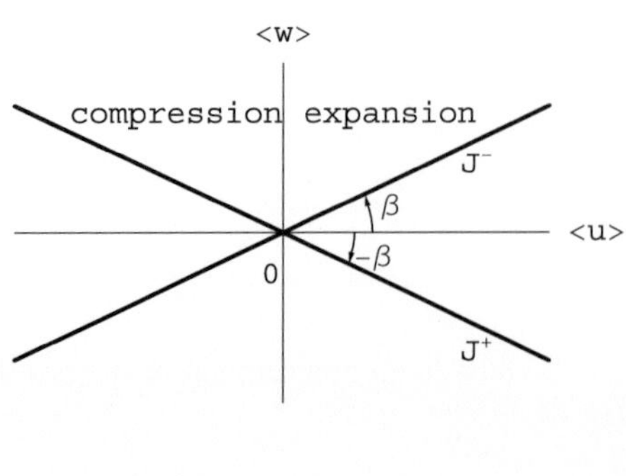

Fig. 4.5 Shock polar for linearized supersonic flow

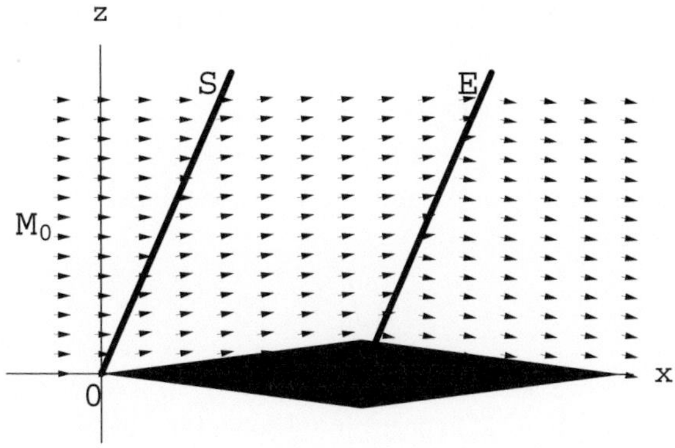

Fig. 4.6 Supersonic flow past a double wedge. S is a shock, E is an expansion "fan"

Note that in linear theory the jump lines coincide with the C^+ and C^- characteristics, Fig. 4.4, and are perpendicular to their images J^+ and J^- in the hodograph plane, respectively. Shock waves as well as expansion fans are represented by characteristics lines across which the velocity vector is discontinuous. Shocks can be differentiated from expansion waves by the fact that the u-component decreases (pressure increases) across them when following a particle, whereas the u-component increases (pressure decreases) as a particle crosses an expansion line. Figure 4.6 depicts the flow past a double wedge profile at zero incidence. Note also, that in linear theory, an expansion fan collapses on a single characteristic line which acts as an "expansion shock."

4.5.2 Forces

The force contribution per unit span is given by

$$d\vec{F}' = -\frac{1}{2}\rho U^2 \left(C_p^-(x)(\vec{n}\,dl)^- + C_p^+(x)(\vec{n}\,dl)^+\right) \tag{4.43}$$

where $(\vec{n}\,dl)^- = (dz^-, -dx)$ and $(\vec{n}\,dl)^+ = (-dz^+, dx)$, see Fig. 4.7.

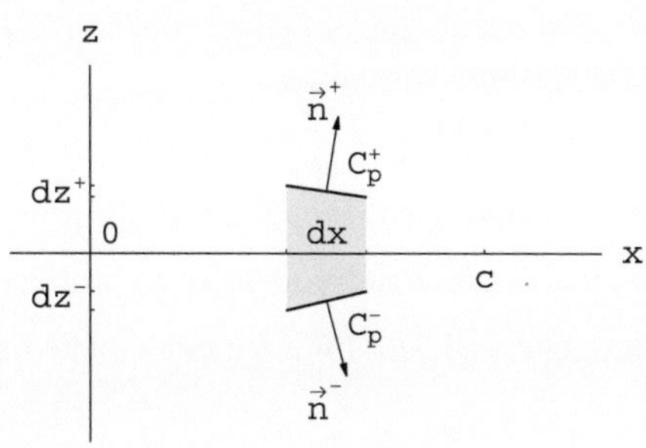

Fig. 4.7 Schematic for calculation of forces

The lift corresponds to the z-component of the force, i.e.

$$dL' = \frac{1}{2}\rho U^2 \left(C_p^-(x) - C_p^+(x)\right) dx \quad \Rightarrow \quad dC_l = \left(C_p^-(x) - C_p^+(x)\right) \frac{dx}{c} \tag{4.44}$$

With the help of Ackeret formulae, one finds

$$C_l = \frac{2}{\beta}\int_0^c (\alpha - f^{'-}(x) + f^{'+}(x) - \alpha)\frac{dx}{c} = 4\frac{\alpha}{\beta} \tag{4.45}$$

where we accounted for $f^{\pm}(0) = f^{\pm}(c) = 0$. The result is remarkable in that, contrary to what happens in subsonic flow, camber does not contribute to lift.

The drag is the x-component with

$$dD' = -\frac{1}{2}\rho U^2 \left(C_p^-(x)dz^- - C_p^+(x)dz^+\right) \quad \Rightarrow$$

$$dC_d = -\left(C_p^-(x)\frac{dz^-}{dx} - C_p^+(x)\frac{dz^+}{dx}\right)\frac{dx}{c} \tag{4.46}$$

Again, using Ackeret formulae, one can evaluate an inviscid drag which is called the "wave drag" with coefficient given by

$$\begin{aligned} C_d &= \frac{2}{\beta}\int_0^c \left((f^{'-}(x) - \alpha)^2 + (f^{'+}(x) - \alpha)^2\right)\frac{dx}{c} \\ &= 4\frac{\alpha^2}{\beta} + \frac{2}{\beta}\int_0^c \left((f^{'-}(x))^2 + (f^{'+}(x))^2\right)\frac{dx}{c} \end{aligned} \tag{4.47}$$

Let $(C_d)_{\alpha=0}$ be the integral term in the above equation. Introducing camber and thickness, it can be rewritten as

$$(C_d)_{\alpha=0} = \frac{4}{\beta}\int_0^c \left((d^{'}(x))^2 + \left(\frac{e^{'}(x)}{2}\right)^2\right)\frac{dx}{c} \tag{4.48}$$

The result shows that in supersonic flow, camber, thickness and incidence, all contribute to wave drag.

Application:

- drag coefficient for a flat plate: $C_d = 4\frac{\alpha^2}{\beta}$
- drag coefficient for a biconvex profile of thickness $\frac{e}{c}$: $C_d = 4\frac{\alpha^2}{\beta} + \frac{16}{3\beta}(\frac{e}{c})^2$
- drag coefficient for a parabolic plate of camber $\frac{d}{c}$: $C_d = 4\frac{\alpha^2}{\beta} + \frac{4}{3\beta}(\frac{4d}{c})^2$

Note, that in linearized supersonic flow theory, the flow at the leading edge of the profile is still governed by a wave-type hyperbolic equation and the upper and lower surfaces do not communicate. Hence, the flow does not go around the leading edge as it does in subsonic flow, and there does not exist a suction force.

4.5.3 Moment

The moment about the leading edge is given by

$$dM_{,o} = -xdF_z = -\frac{1}{2}\rho U^2 \left(C_p^-(x) - C_p^+(x)\right) xdx \quad \Rightarrow$$

$$dC_{m,o} = -\left(C_p^-(x) - C_p^+(x)\right) \frac{x}{c}\frac{dx}{c} \tag{4.49}$$

Using again Ackeret results, the moment coefficient can be written

$$C_{m,o} = \frac{2}{\beta}\int_0^c (f^{'-}(x) - \alpha + f^{'+}(x) - \alpha)\frac{x}{c}\frac{dx}{c} =$$

$$-2\frac{\alpha}{\beta} + \frac{2}{\beta}\int_0^c (f^{'-}(x) + f^{'+}(x))\frac{x}{c}\frac{dx}{c} \tag{4.50}$$

Let $\left(C_{m,o}\right)_{\alpha=0}$ be the integral term. Introducing camber and thickness reduces the equation to

$$\left(C_{m,o}\right)_{\alpha=0} = \frac{4}{\beta}\int_0^c d^{'}(x)\frac{x}{c}\frac{dx}{c} = -\frac{4}{\beta}\int_0^c \frac{d(x)}{c}\frac{dx}{c} \tag{4.51}$$

where we have performed an integration by parts and used $d(0) = d(c) = 0$. The integral represents the dimensionless area under the camberline. It is worth noting that, although lift does not depend on camber, the aerodynamic moment in general does.

Application:

- moment coefficient of a flat plate or symmetric profile: $C_{m,o} = -2\frac{\alpha}{\beta}$
- moment coefficient of a parabolic plate of camber $\frac{d}{c}$: $C_{m,o} = -2\frac{\alpha}{\beta}\text{-}\frac{8}{3\beta}\frac{d}{c}$.

4.5.4 Center of Pressure

Let D be an arbitrary point along the chord, the moment at D reads

$$C_{m,D} = C_{m,o} + \frac{x_D}{c}C_l = \left(C_{m,o}\right)_{\alpha=0} - 2\frac{\alpha}{\beta} + 4\frac{x_D}{c}\frac{\alpha}{\beta} \tag{4.52}$$

Setting the result to zero gives the location of the center of pressure, i.e.

$$\frac{x_{c.p.}}{c} = \frac{1}{2} - \frac{\beta \left(C_{m,o}\right)_{\alpha=0}}{4\alpha} = \frac{1}{2} + \frac{1}{\alpha} \int_0^c \frac{d(x)}{c} \frac{dx}{c} \tag{4.53}$$

In general, the center of pressure will go to $\pm\infty$ when α goes to zero, and will be close to the mid-chord for large incidences.

Application:

- location of the center of pressure for a flat plate and symmetric profiles: $\frac{x_{c.p.}}{c} = \frac{1}{2}$
- location of the center of pressure for a parabolic plate of camber $\frac{d}{c}$: $\frac{x_{c.p.}}{c} = \frac{1}{2} - \frac{2}{3}\frac{d}{c}\frac{1}{\alpha}$.

4.5.5 Aerodynamic Center

Taking the partial derivative $\frac{\partial}{\partial \alpha}$ of $C_{m,D}$ and setting the result to zero, gives the location of the aerodynamic center, that is

$$\frac{x_{a.c.}}{c} = \frac{1}{2} \tag{4.54}$$

this is in sharp contrast to subsonic flow, where the aerodynamic center is located at the quarter-chord. This phenomena has been responsible for many accidents in the early days of supersonic flights. The Franco-British Concorde transport aircraft was handling the change of location of the aerodynamic center, during the transonic acceleration, by moving fuel from a tank located in the front fuselage area to a tank located in the tail area of the fuselage and doing the reverse for return to subsonic and landing, Fig. 4.8.

4.6 Limit of Validity of Linearized Theories

The above linear theories all fail when the incoming Mach number approaches unity. Indeed, the Prandtl-Glauert correction shows a dependency of the local and global coefficients with incoming Mach number through the multiplication factor $1/\sqrt{1-M_0^2}$, whereas, the linearized supersonic flow theory shows the corresponding dependency through the factor $1/\sqrt{M_0^2-1}$. In both cases, the solution becomes infinite at $M_0 = 1$. This is in contradiction with experiments that indicate, for example, that the lift and drag coefficients are finite and reach a maximum near $M_0 = 1$. In graphic form, this is shown in Fig. 4.9.

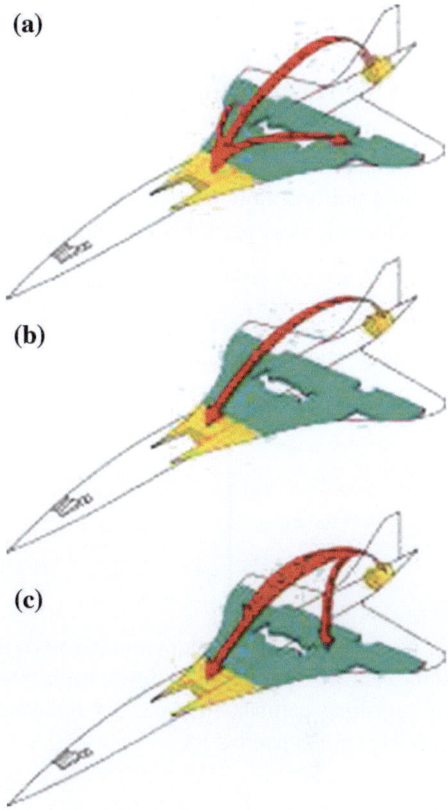

Fig. 4.8 Fuel pitch trim. **a** trimming fuel during transonic acceleration **b** trimming fuel during supersonic cruise **c** trimming fuel during transonic deceleration (from https://en.wikipedia.org/wiki/File:Concorde_fuel_trim.svg Author: steal88)

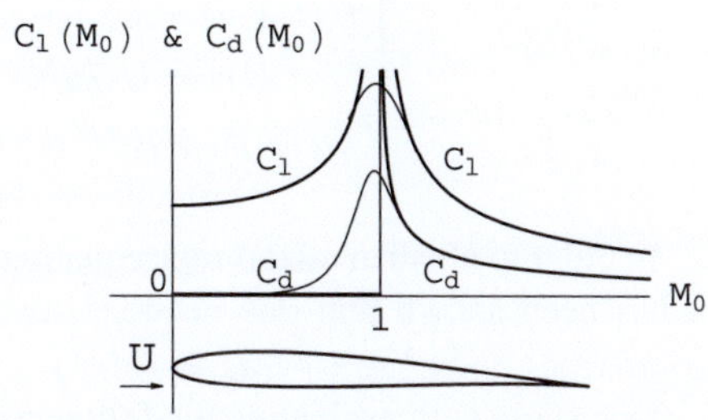

Fig. 4.9 Linear theories and nonlinear transonic results near $M_0 = 1$

The linearized potential equation does not allow for continuous description of flow fields from subsonic to supersonic flow regimes. In particular, it cannot represent flow fields in which regions of supersonic flow coexist with regions of subsonic flow, called mixed-type flows. Such complex flow fields, which may include sonic line, expansion wave and shock wave, require a mixed-type equation, one that is elliptic in regions of subsonic flow and hyperbolic where the flow is supersonic. In order to change type, such an equation has to be nonlinear. It will be presented in the next section. Note that near Mach one, the aerodynamic coefficients, as given by the nonlinear theory, reach an extremum. This is called "sonic freezing."

4.7 Transonic Small Disturbance Theory

4.7.1 Governing Equation

Assuming inviscid, steady, 2-D flow of perfect gas, the Euler equations are perturbed about a uniform flow near Mach one. In this section for practical reasons, we will use dimensionless perturbation velocity components (u, w) defined as

$$V_x = U(1+u), \quad V_z = Uw \tag{4.55}$$

where $\overrightarrow{V}$ is the total velocity. With this notation, the Euler equations read

$$\begin{cases} \frac{\partial \rho U(1+u)}{\partial x} + \frac{\partial \rho U w}{\partial z} = 0 \\ \frac{\partial \rho U^2(1+u)^2+p}{\partial x} + \frac{\partial \rho U^2(1+u)w}{\partial z} = 0 \\ \frac{\partial \rho U^2(1+u)w}{\partial x} + \frac{\partial \rho U^2 w^2+p}{\partial z} = 0 \\ H_0 = \frac{\gamma p}{(\gamma-1)\rho} + \frac{U^2((1+u)^2+w^2)}{2} = const. \end{cases} \tag{4.56}$$

Note that the energy equation has been replaced by the constant enthalpy condition, valid for isoenergetic flows, even when shocks are present. For a thin airfoil of vanishing camber, thickness or incidence of order $\epsilon \ll 1$, the flow dependent variables are expanded as

$$\begin{cases} \rho = \rho_0(1 + \epsilon^{\frac{2}{3}}\rho^{(1)} + \epsilon^{\frac{4}{3}}\rho^{(2)} + \cdots) \\ U(1+u) = U(1 + \epsilon^{\frac{2}{3}}u^{(1)} + \epsilon^{\frac{4}{3}}u^{(2)} + \cdots) \\ Uw = U(\epsilon w^{(1)} + \epsilon^{\frac{5}{3}}w^{(2)} + \cdots) \\ p = p_0 + \rho_0 U^2(\epsilon^{\frac{2}{3}}p^{(1)} + \epsilon^{\frac{4}{3}}p^{(2)} + \cdots) \end{cases} \tag{4.57}$$

In order to obtain a mixed-type equation in a transformed plane $(\xi = x, \zeta = \epsilon^{\frac{1}{3}}z)$, it has been shown that one needs to assume that, as $\epsilon \to 0$, the Mach number approaches unity, i.e. $1 - M_0^2 = O(\epsilon^{\frac{2}{3}})$.

One starts with evaluating the following groupings:

$$\begin{cases} \rho U(1+u) = \rho_0 U \left(1 + \epsilon^{\frac{2}{3}}(\rho^{(1)} + u^{(1)}) + \epsilon^{\frac{4}{3}}(\rho^{(1)}u^{(1)} + \rho^{(2)} + u^{(2)}) + \cdots\right) \\ \rho U w = \rho_0 U \left(\epsilon w^{(1)} + \epsilon^{\frac{5}{3}}(\rho^{(1)}w^{(1)} + w^{(2)}) + \cdots\right) \\ \rho U^2(1+u)^2 = \rho_0 U^2 \\ \times \left(1 + \epsilon^{\frac{2}{3}}(\rho^{(1)} + 2u^{(1)}) + \epsilon^{\frac{4}{3}}(2\rho^{(1)}u^{(1)} + u^{(1)2} + \rho^{(2)} + 2u^{(2)}) + \cdots\right) \\ \rho U^2(1+u)w = \rho_0 U^2 \left(\epsilon w^{(1)} + \epsilon^{\frac{5}{3}}(\rho^{(1)}w^{(1)} + u^{(1)}w^{(1)} + w^{(2)}) + \cdots\right) \\ \rho U^2 w^2 = \rho_0 U^2 \left(\epsilon^2 w^{(1)2} + \cdots\right) \end{cases} \tag{4.58}$$

When these expressions are substituted in the Euler equations, at the lowest order one finds the following results.

For conservation of mass:

$$\epsilon^{\frac{2}{3}}\left\{\frac{\partial}{\partial\xi}(\rho^{(1)}+u^{(1)})\right\}=0 \tag{4.59}$$

For the x-momentum:

$$\epsilon^{\frac{2}{3}}\left\{\frac{\partial}{\partial\xi}(\rho^{(1)}+2u^{(1)}+p^{(1)})\right\}=0 \tag{4.60}$$

For the z-momentum:

$$\epsilon\left\{\frac{\partial w^{(1)}}{\partial\xi}+\frac{\partial p^{(1)}}{\partial\zeta}\right\}=0 \tag{4.61}$$

The first two PDEs can be integrated as

$$\rho^{(1)}+u^{(1)}=F(\zeta),\quad \rho^{(1)}+2u^{(1)}+p^{(1)}=G(\zeta) \tag{4.62}$$

where F and G are arbitrary functions of ζ. However, since the perturbations vanish at infinity for all values of ζ, $F=G=0$. Hence $\rho^{(1)}=-u^{(1)}$ and $p^{(1)}=-u^{(1)}$. The z-momentum equation yields a first PDE for $(u^{(1)}, w^{(1)})$, so that we now have

$$\begin{cases}\frac{\partial w^{(1)}}{\partial\xi}-\frac{\partial u^{(1)}}{\partial\zeta}=0\\ \rho^{(1)}=-u^{(1)}\\ p^{(1)}=-u^{(1)}\end{cases} \tag{4.63}$$

The PDE states that the flow is irrotational at order one, i.e. there exists $\phi^{(1)}(\xi,\zeta)$ such that $\nabla\phi^{(1)}=(u^{(1)}, w^{(1)})$. In order to find a second PDE for $(u^{(1)}, w^{(1)})$, we need to carry out the expansions to the next order for the PDEs and for the equation of constant enthalpy. Consider first the enthalpy equation. To lowest orders one finds

$$\frac{\gamma}{\gamma-1}\frac{p_0}{\rho_0}+\frac{U^2}{2}=H_0 \tag{4.64}$$

$$\epsilon^{\frac{2}{3}}\frac{a_0^2}{(\gamma-1)}\left\{\gamma M_0^2p^{(1)}-\rho^{(1)}+(\gamma-1)M_0^2u^{(1)}\right\} \tag{4.65}$$

where we have made use of $a_0^2=\gamma p_0/\rho_0$ and $M_0=U/a_0$. When $p^{(1)}$ and $\rho^{(1)}$ are eliminated in terms of $u^{(1)}$, the terms do not balance to zero, but there remains

$$\epsilon^{\frac{2}{3}}\frac{a_0^2}{(\gamma-1)}\left\{(1-M_0^2)u^{(1)}\right\}=O(\epsilon^{\frac{4}{3}}) \tag{4.66}$$

since, by hypothesis, $(1 - M_0^2) = O(\epsilon^{\frac{2}{3}})$. This remainder is added to the next term in the expansion. Let

$$\frac{(1 - M_0^2)}{\epsilon^{\frac{2}{3}}} = K \tag{4.67}$$

K is of order unity and is called the *transonic similarity parameter*. It is defined up to a multiplicative factor M_0^κ, $\kappa > 0$. Two definitions are commonly used:

$$\begin{cases} K = \frac{1-M_0^2}{\epsilon^{\frac{2}{3}}}, & (Cole) \\ K = \frac{1-M_0^2}{M_0^2 \epsilon^{\frac{2}{3}}}, & (Spreiter) \end{cases} \tag{4.68}$$

At the next order we now obtain

$$\epsilon^{\frac{4}{3}} \frac{a_0^2}{(\gamma - 1)} \left\{ Ku^{(1)} + \gamma M_0^2 p^{(2)} - \gamma M_0^2 \rho^{(1)} p^{(1)} - \rho^{(2)} + \rho^{(1)2} \right.$$
$$\left. + \frac{\gamma - 1}{2} M_0^2 (u^{(1)2} + 2u^{(2)}) \right\} = O(\epsilon^2) \tag{4.69}$$

The right-hand-side must be $O(\epsilon^2)$ or higher. Again, eliminating $p^{(1)}$ and $\rho^{(1)}$ in terms of $u^{(1)}$ yields for the unknowns of the level 2 approximation:

$$\gamma M_0^2 p^{(2)} - \rho^{(2)} + (\gamma - 1) M_0^2 u^{(2)} + Ku^{(1)} + (1 - \gamma M_0^2) u^{(1)2} + \frac{\gamma - 1}{2} M_0^2 u^{(1)2} = O(\epsilon^{\frac{2}{3}}) \tag{4.70}$$

Returning now to the conservation of mass and x-momentum

$$\begin{cases} \epsilon^{\frac{4}{3}} \left\{ \frac{\partial}{\partial \xi} (\rho^{(1)} u^{(1)} + \rho^{(2)} + u^{(2)}) + \frac{\partial w^{(1)}}{\partial \zeta} \right\} = 0 \\ \epsilon^{\frac{4}{3}} \left\{ \frac{\partial}{\partial \xi} (2\rho^{(1)} u^{(1)} + u^{(1)2} + \rho^{(2)} + 2u^{(2)} + p^{(2)}) + \frac{\partial w^{(1)}}{\partial \zeta} \right\} = 0 \end{cases} \tag{4.71}$$

Subtracting one from the other and substituting $\rho^{(1)} = -u^{(1)}$ yields

$$\epsilon^{\frac{4}{3}} \left\{ \frac{\partial}{\partial \xi} (u^{(2)} + p^{(2)}) \right\} = 0 \tag{4.72}$$

which can be integrated to give $p^{(2)} = -u^{(2)}$. Now, solving the energy equation for $\rho^{(2)}$ in terms of $u^{(2)}$ and first-order terms provides the following result:

$$\rho^{(2)} = -M_0^2 u^{(2)} + Ku^{(1)} + \left(1 - \frac{\gamma + 1}{2} M_0^2\right) u^{(1)2} + O(\epsilon^{\frac{2}{3}}) \tag{4.73}$$

Final substitution of these expressions in the first PDE above, reduces it to

$$\epsilon^{\frac{4}{3}}\left\{\frac{\partial}{\partial\xi}\left[(1-M_0^2)u^{(2)}+Ku^{(1)}-\frac{\gamma+1}{2}M_0^2u^{(1)\,2}+O(\epsilon^{\frac{2}{3}})\right]+\frac{\partial w^{(1)}}{\partial\zeta}\right\}=0 \tag{4.74}$$

Note that the first and last terms in the inner bracket are of higher order and can be dropped to give at $O(\epsilon^{\frac{4}{3}})$:

$$\epsilon^{\frac{4}{3}}\left\{\frac{\partial}{\partial\xi}\left[Ku^{(1)}-\frac{\gamma+1}{2}M_0^2u^{(1)\,2}\right]+\frac{\partial w^{(1)}}{\partial\zeta}\right\}=0 \tag{4.75}$$

In the second-order approximation, the level-2 unknowns $\rho^{(2)}$, $u^{(2)}$, $w^{(2)}$ and $p^{(2)}$ do not appear, but the nonlinear contribution involving level-1 unknowns remains.

Reverting to the physical variables (x, z), with Cole's definition of the transonic similarity parameter, the results are the following for the velocity, density and pressure:

$$\begin{cases} U(1+\epsilon^{\frac{2}{3}}u^{(1)}) = U(1+u) \\ U\epsilon w^{(1)} = Uw \\ \rho = \rho_0(1+\epsilon^{\frac{2}{3}}\rho^{(1)}) = \rho_0(1-u) \\ p = p_0+\rho_0U^2\epsilon^{\frac{2}{3}}p^{(1)} = p_0-\rho_0U^2u \end{cases} \tag{4.76}$$

Note that now $C_p = -2u$. The governing equations for (u, w) read

$$\begin{cases} \frac{\partial}{\partial x}\left[(1-M_0^2)u-\frac{\gamma+1}{2}M_0^2u^2\right]+\frac{\partial w}{\partial z}=0 \\ \frac{\partial w}{\partial x}-\frac{\partial u}{\partial z}=0 \end{cases} \tag{4.77}$$

a first-order system of PDEs in conservation form. If we introduce the perturbation potential $\phi(x, z)$ such that $\nabla\phi = (u, w)$, the system is transformed into a second-order PDE

$$\frac{\partial}{\partial x}\left[(1-M_0^2)\frac{\partial\phi}{\partial x}-\frac{\gamma+1}{2}M_0^2\left(\frac{\partial\phi}{\partial x}\right)^2\right]+\frac{\partial^2\phi}{\partial z^2}=0 \tag{4.78}$$

called the Transonic Small Disturbance equation (TSD).

In non-conservative form, it reads

$$\left[(1-M_0^2)-(\gamma+1)M_0^2\frac{\partial\phi}{\partial x}\right]\frac{\partial^2\phi}{\partial x^2}+\frac{\partial^2\phi}{\partial z^2}=0 \tag{4.79}$$

Let $u^* = (\partial\phi/\partial x)^* = (1-M_0^2)/\left((\gamma+1)M_0^2\right)$ represent the critical perturbation velocity. The TSD equation is a nonlinear equation of mixed-type. When $u = \partial\phi/\partial x < u^*$, the equation is of elliptic type, which corresponds to subsonic flow. When $u = \partial\phi/\partial x > u^*$, the equation is of hyperbolic type, which corresponds to supersonic flow. The sonic line is given by $u = \partial\phi/\partial x = u^*$.

The linearized Mach number-u relation is given by

$$M(u) = M_0 + (1 - M_0)\frac{u}{u^*} = M_0 + \frac{\gamma + 1}{1 + M_0} M_0^2 u \tag{4.80}$$

In the supersonic subdomain, characteristic lines exist. The rigorous full theory of characteristics is beyond the scope of this book (see Appendix C2 for a brief introduction). For further information the reader is referred to Courant and Hilbert, Ref. [1], or also to JJC, [2].

The slopes of the C^+ and C^- characteristics can be found to be:

$$\left(\frac{dz}{dx}\right)_{C+} = \frac{1}{\sqrt{(\gamma + 1)M_0^2\sqrt{u - u^*}}}, \quad \left(\frac{dz}{dx}\right)_{C-} = -\frac{1}{\sqrt{(\gamma + 1)M_0^2\sqrt{u - u^*}}} \tag{4.81}$$

The slopes of the characteristic lines vary with the local value of u and their shapes depend on the solution. Note that their slopes are infinite at the sonic line, Fig. 4.10. In the lower part of the figure, a visualization of the characteristic lines was obtained fortuitously when pieces of tape were placed on the profile to cover the pressure taps.

4.7.2 Boundary Conditions

The boundary conditions associated with the TSD equation are the same as those of the linear models:

$$\begin{cases} \frac{\partial\phi(x,0^\pm)}{\partial z} = U\left(f'^{\pm}(x) - \alpha\right) \\ \nabla\phi(x,z) \to 0, \; x^2 + z^2 \to \infty \end{cases} \tag{4.82}$$

Note that the tangency condition is applied on the x-axis, which is consistent with the small disturbance assumption. For a lifting airfoil, the Kutta-Joukowski condition reads

$$\Gamma'(c) = U < u(c, 0) >= U(u(c, 0^+) - u(c, 0^-)) = 0 \tag{4.83}$$

where $\Gamma(x) = U < \phi(x, 0) >= U(\phi(x, 0^+) - \phi(x, 0^-))$ represents the circulation distribution inside the profile.

4.7.3 Jump Conditions

The TSD is written as a system of two first-order PDEs in conservation form.

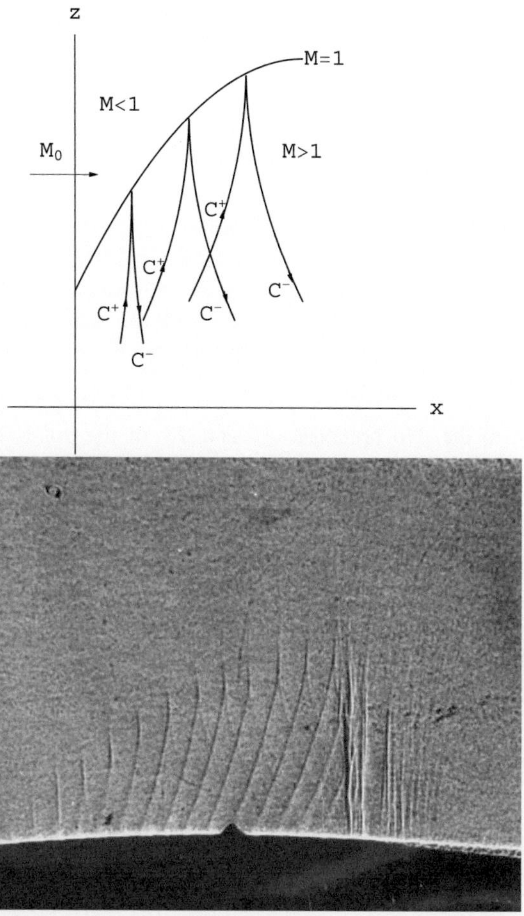

Fig. 4.10 Characteristic lines near the sonic line. *Top*, sketch. *Bottom*, shadowgraph, courtesy W.J. Bannink, TU-Delft

$$\left\{ \begin{array}{c} \frac{\partial}{\partial x}\left[(1-M_0^2)u - \frac{\gamma+1}{2}M_0^2 u^2\right] + \frac{\partial w}{\partial z} = 0 \\ \frac{\partial w}{\partial x} - \frac{\partial u}{\partial z} = 0 \end{array} \right. \tag{4.84}$$

The jump conditions are given by

$$\left\{ \begin{array}{c} < (1-M_0^2)u - \frac{\gamma+1}{2}M_0^2 u^2 > n_x + < w > n_z = 0 \\ < w > n_x - < u > n_z = 0 \end{array} \right. \tag{4.85}$$

Upon elimination of $\mathbf{n} = (n_x, n_z)$, the unit vector normal to the jump line, the shock polar is obtained

$$< (1-M_0^2)u - \frac{\gamma+1}{2}M_0^2 u^2 >< u > + < w >^2 = 0 \tag{4.86}$$

The term in the first bracket can be written differently by using the identity $< u^2 >= 2\bar{u} < u >$ where $\bar{u} = (u_1 + u_2)/2$ represents the average value of u across the shock. This results in

$$\left(1 - M_0^2 - (\gamma + 1)M_0^2\bar{u}\right) < u >^2 + < w >^2 = 0 \tag{4.87}$$

Given the upstream state (u_1, w_1), if $\bar{u} < u^*$, the term in the bracket is positive and the only solution is the trivial solution $< u >=< w >= 0$. No jump is allowed and $(u_2, w_2) = (u_1, w_1)$. The flow is locally subsonic. If $\bar{u} = u^*$, $< w >= 0$. This corresponds to a smooth transition through the sonic line if $< u >= 0$, or to the normal shock which does not deviate the flow since $w_2 = w_1$, if $< u >\neq 0$. If $\bar{u} > u^*$, two solutions exist $< w >= \pm\sqrt{(\gamma + 1)M_0^2}\sqrt{u_1 - u^* + \frac{<u>}{2}} < u >$ corresponding to the two branches J^+ and J^- of the polar. The maximum flow deviation through a shock occurs when $< u >= 4(u^* - u_1)/3$ and is given by

$$|< w >|_{max} = \frac{4}{3\sqrt{3}}\sqrt{(\gamma + 1)M_0^2}(u_1 - u^*)^{\frac{3}{2}} \tag{4.88}$$

The shock polar is depicted in Fig. 4.11.

The slopes of the shock waves in the physical plane are given by

$$\frac{dz}{dx} = -\frac{n_x}{n_z} = -\frac{< u >}{< w >} \Rightarrow \left(\frac{dz}{dx}\right)_{J^+} = \frac{1}{\sqrt{(\gamma + 1)M_0^2}\sqrt{u_1 - u^* + \frac{<u>}{2}}},$$

$$\left(\frac{dz}{dx}\right)_{J^-} = -\frac{1}{\sqrt{(\gamma + 1)M_0^2}\sqrt{u_1 - u^* + \frac{<u>}{2}}} \tag{4.89}$$

Note that the slopes of the lines that connect the downstream states to the double point O on the shock polar are perpendicular to the images of the shocks in the physical plane. In the limit of infinitely weak shocks, the tangents to the double point O are perpendicular to the characteristic lines representing them, Fig. 4.12.

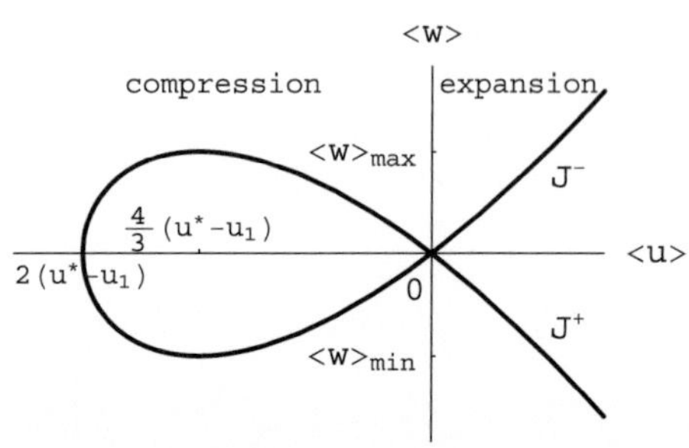

Fig. 4.11 TSD polar

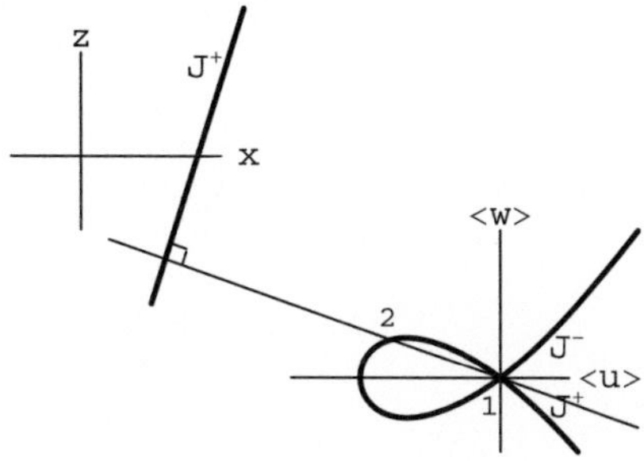

Fig. 4.12 Polar in hodograph plane and shock in physical plane

4.7.4 Riemann Invariants

The Riemann invariants associated with the TSD are found to be

$$\begin{cases} \lambda = -\frac{2}{3}\sqrt{(\gamma+1)M_0^2}(u-u^*)^{\frac{3}{2}} + w \\ \mu = -\frac{2}{3}\sqrt{(\gamma+1)M_0^2}(u-u^*)^{\frac{3}{2}} - w \end{cases} \tag{4.90}$$

The Riemann invariants verify the following first-order PDEs, called *compatibility relations*:

$$\begin{cases} CR^+ : \frac{\partial\lambda}{\partial x} + \frac{1}{\sqrt{(\gamma+1)M_0^2}\sqrt{u-u^*}}\frac{\partial\lambda}{\partial z} = 0 \\ CR^- : \frac{\partial\mu}{\partial x} - \frac{1}{\sqrt{(\gamma+1)M_0^2}\sqrt{u-u^*}}\frac{\partial\mu}{\partial z} = 0 \end{cases} \tag{4.91}$$

The compatibility relations state that the Riemann invariants λ and μ are constant along the C^+ and C^- characteristics, respectively. This property is exploited in the numerical technique called the *method of characteristics*, that is used to solve supersonic flows (see Appendix C2).

4.7.5 Forces and Moment

The nonlinearity of the governing equation does not allow for superposition of elementary solutions nor the decomposition of the problem into a symmetric and lifting problems. However, the symmetric problem still has zero lift and moment, but wave drag may occur when supersonic regions are present in the flow.

The lift is given by the integral seen earlier

$$C_l = \int_0^c \left(C_p^-(x) - C_p^+(x)\right)\frac{dx}{c} = \int_0^c \left(2u^+(x) - 2u^-(x)\right)\frac{dx}{c} =$$

$$2\int_0^c < u(x) > \frac{dx}{c} = \frac{2}{U}\int_0^c \frac{d\Gamma(x)}{dx}\frac{dx}{c} = 2\frac{\Gamma(c)}{Uc} \tag{4.92}$$

This result is consistent with the Kutta-Joukowski lift theorem.
The drag is also calculated as before

$$C_d = -\int_0^c \left(C_p^-(x)\frac{dz^-}{dx} - C_p^+(x)\frac{dz^+}{dx} \right)\frac{dx}{c} =$$

$$\int_0^c \left(2u^-(x)\frac{dz^-}{dx} - 2u^+(x)\frac{dz^+}{dx} \right)\frac{dx}{c} \tag{4.93}$$

One can substitute the following expressions from the tangency conditions

$$\frac{dz^+}{dx} = w^+(x), \quad \frac{dz^-}{dx} = w^-(x) \tag{4.94}$$

to obtain

$$C_d = -2\int_0^c < uw > \frac{dx}{c} \tag{4.95}$$

Another approach to evaluating the drag is based on the x-momentum theorem applied to a control volume with exterior unit normal, whose boundary $\delta\Omega$ is made of a large circle of radius R, connected by a cut C to the profile P and shock(s) S_k that are excluded, see Fig. 4.13. Inside the domain bounded by $\delta\Omega$, the flow is smooth. Application of the divergence theorem to the x-momentum Euler equation leads to the contour integral

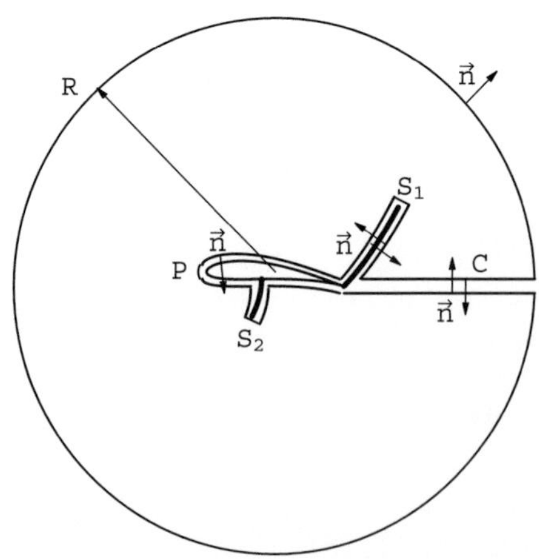

Fig. 4.13 Contour for evaluation of the drag integral

$$\int_{\Omega}\left\{\frac{\partial}{\partial x}\left[\rho U^2(1+u)^2+p\right]+\frac{\partial}{\partial z}\left[\rho U^2(1+u)w\right]\right\}dxdz=$$

$$\int_{\partial\Omega}\left[\rho U(1+u)(\vec{V}.\vec{n})+pn_x\right]dl=0 \tag{4.96}$$

On the circle R, the perturbation velocity (u, w) decays as $R \to \infty$ resulting in no contribution. Along the profile P the contribution is precisely the drag force D'. Along the cut C the flow is continuous and again no contribution will occur, but along the shock(s) S_k the flow is discontinuous. For the full Euler equations, though, there would be no net result from the shocks, because the jump conditions would be satisfied there and more precisely

$$<\rho(\vec{V}.\vec{n})>=0, \quad <\rho U(1+u)(\vec{V}.\vec{n})+pn_x>=0 \tag{4.97}$$

from conservation of mass and x-momentum. This is no longer true when the solution is approximated by its leading terms $(u^{(1)}, w^{(1)})$. The only two non zero contributions are therefore along the profile and along the shock(s).

Mass and momentum are conserved to second order ($O(\epsilon^{\frac{4}{3}})$), but to third order this is no longer the case. Indeed, expanding the conservation of mass and x-momentum to third order gives

$$\begin{cases} \epsilon^2\left\{\frac{\partial}{\partial\xi}(Ku^{(2)}+\rho^{(3)}+u^{(1)}\rho^{(2)}+\rho^{(1)}u^{(2)})+\frac{\partial}{\partial\zeta}(w^{(2)}+\rho^{(1)}w^{(1)})\right\}=0 \\ \epsilon^2\left\{\frac{\partial}{\partial\xi}\left(Ku^{(2)}+\rho^{(3)}+2u^{(3)}+p^{(3)}+2u^{(1)}\rho^{(2)}+\right.\right. \\ \quad 2\rho^{(1)}u^{(2)}+2u^{(1)}u^{(2)}+\rho^{(1)}u^{(1)2}) \\ \quad \left.+\frac{\partial}{\partial\zeta}(w^{(2)}+\rho^{(1)}w^{(1)}+u^{(1)}w^{(1)})\right\}=0 \end{cases} \tag{4.98}$$

Subtracting the first equation from the second yields a combined mass-momentum equation

$$\epsilon^2\left\{\frac{\partial}{\partial\xi}(u^{(3)}+p^{(3)}+u^{(1)}\rho^{(2)}+\rho^{(1)}u^{(2)}+2u^{(1)}u^{(2)}+\rho^{(1)}u^{(1)2})+\right.$$

$$\left.\frac{\partial}{\partial\zeta}(u^{(1)}w^{(1)})\right\}=0 \tag{4.99}$$

Upon replacing $\rho^{(1)}$ and $\rho^{(2)}$ from previous results, this reduces to

$$\epsilon^2\left\{\frac{\partial}{\partial\xi}(u^{(3)}+p^{(3)}+Ku^{(1)2}-\frac{\gamma+1}{2}M_0^2u^{(1)3}+(1-M_0^2)u^{(1)}u^{(2)})+\right.$$

$$\left.\frac{\partial}{\partial\zeta}(u^{(1)}w^{(1)})\right\}=0 \tag{4.100}$$

The last term in the ξ-derivative is of higher order and can be removed, while the ζ-derivative term can be expanded as

$$w^{(1)}\frac{\partial u^{(1)}}{\partial \zeta}+u^{(1)}\frac{\partial w^{(1)}}{\partial \zeta}=w^{(1)}\frac{\partial w^{(1)}}{\partial \xi}-u^{(1)}\frac{\partial}{\partial \xi}\left(Ku^{(1)}-\frac{\gamma+1}{2}u^{(1)2}\right) \tag{4.101}$$

using irrotationality and the governing equation at the lower order. Finally, this reduces to a single derivative term

$$\epsilon^2\left\{\frac{\partial}{\partial \xi}\left(u^{(3)}+p^{(3)}+\frac{Ku^{(1)2}}{2}-\frac{\gamma+1}{6}M_0^2u^{(1)3}+\frac{w^{(1)2}}{2}\right)\right\}=0 \tag{4.102}$$

This is an equation for $p^{(3)}$ in the whole domain

$$p^{(3)}=-u^{(3)}-\frac{Ku^{(1)2}}{2}+\frac{\gamma+1}{6}M_0^2u^{(1)3}-\frac{w^{(1)2}}{2} \tag{4.103}$$

Substituting this result into the combined mass-momentum equation yields, after some simplification and elimination of higher order terms

$$\epsilon^2\left\{\frac{\partial}{\partial \xi}\left(K\frac{u^{(1)2}}{2}-\frac{\gamma+1}{3}M_0^2u^{(1)3}-\frac{w^{(1)2}}{2}\right)+\frac{\partial}{\partial \zeta}(u^{(1)}w^{(1)})\right\}=0 \tag{4.104}$$

This equation is equivalent to the previously derived governing equation for smooth flow, since it can be written as

$$\epsilon^2u^{(1)}\left\{\frac{\partial}{\partial \xi}\left(Ku^{(1)}-\frac{\gamma+1}{2}M_0^2u^{(1)2}\right)+\frac{\partial w^{(1)}}{\partial \zeta}\right\}=0 \tag{4.105}$$

However, when the flow is discontinuous, such as at a shock, the solution $(u^{(1)}, w^{(1)})$ to the low order governing equations does not satisfy the jump conditions associated with this new conservation equation. In fact, the mass and x-momentum are not conserved at third order. This is consistent with the fact that the flow is irrotational to the first two orders but not to third order. Hence drag appears at third order $(O(\epsilon^2))$. If we revert to the physical variables, the drag reads

$$D'=\rho U^2\int_{S_k} < \left((1-M_0^2)\frac{u^2}{2}-\frac{\gamma+1}{3}M_0^2u^3-\frac{w^2}{2}\right)n_x+uw\,n_z > dl \tag{4.106}$$

where S_k represents the shock line(s) with arc length l and $\vec{n}$ is the normal unit vector to S_k oriented in the flow direction. This determines how the jump is calculated, i.e. downstream values minus upstream ones. Here we have used the combined mass-momentum equation. The reason for using this equation is that mass is not conserved to third order, therefore, this imbalance will add momentum in the flow direction in the amount corresponding to $< \rho U(\overrightarrow{V}.\overrightarrow{n}) >$ which must be subtracted from the x-momentum given by $< \rho U(1+u)(\overrightarrow{V}.\overrightarrow{n}) + pn_x >$.

The above expression for D' can be simplified further by using the shock conditions derived earlier. As a first step, one eliminates the n_z component from the irrotationality condition

$$n_z = \frac{< w >}{< u >} n_x \tag{4.107}$$

we also make use of the identities $< a^2 >= 2\overline{a} < a >$ where $\overline{a} = \frac{a_1+a_2}{2}$ is the average value across the discontinuity, $< ab >= \overline{a} < b > + \overline{b} < a >$ and $< a^3 >= (\overline{a^2} + 2\overline{a}^2) < a >$. The jump term in D' now reads

$$\Big[(1-M_0^2)\overline{u} < u > -\frac{\gamma+1}{3}M_0^2(\overline{u^2}+2\overline{u}^2) < u > -\overline{w} < w > +$$

$$(\overline{u} < w > + \overline{w} < u >)\frac{< w >}{< u >}\Big] n_x \tag{4.108}$$

The final step consists in using the shock polar relation multiplied by $\frac{\overline{u}}{<u>}n_x$ and subtract it from the above equation

$$\left[\left((1-M_0^2)\overline{u} - (\gamma+1)M_0^2\overline{u}^2\right) < u > + \overline{u}\frac{< w >^2}{< u >}\right] n_x = 0 \tag{4.109}$$

One finds the following result

$$\left[-\frac{\gamma+1}{3}M_0^2(\overline{u^2}+2\overline{u}^2) < u > + (\gamma+1)M_0^2\overline{u}^2 < u >\right] n_x \tag{4.110}$$

This can be rearranged to read

$$-\frac{\gamma+1}{12}M_0^2 < u >^3 n_x \tag{4.111}$$

The drag is thus

$$D' = -\frac{\gamma+1}{12}M_0^2\rho U^2 \int_{S_k} < u >^3 n_x dl$$

or the drag coefficient

$$C_d = -\frac{\gamma+1}{6} M_0^2 \int_{S_k} < u >^3 n_x \frac{dl}{c} \tag{4.112}$$

This makes it clear that if there are no shock waves in the flow, the wave drag is zero.

The moment is obtained from

$$M,o = \frac{1}{2}\rho U^2 \int_0^c (C_p^+(x) - C_p^-(x)) x dx = -\rho U^2 \int_0^c (u^+(x) - u^-(x)) x dx, \quad \Rightarrow$$

$$M,o = -\rho U^2 \int_0^c \left(\frac{\Gamma'(x)}{U}\right) x dx \tag{4.113}$$

and the moment coefficient

$$C_{m,o} = -\frac{2}{U}\int_0^c \Gamma'(x) \frac{x}{c}\frac{dx}{c} \tag{4.114}$$

4.7.6 Murman-Cole Scheme

Murman and Cole introduced in 1970 a "type-dependent" scheme and a line relaxation algorithm to solve the TSD equation. It was the first time that such a scheme was employed, in which different discretizations are used at subsonic points (centered) and supersonic points (upwind). Although they started from the conservative form of the PDE (Eq. 4.78), the first scheme did not conserve mass across a shock wave, resulting in jump conditions not being satisfied. The reason was that in the switching back of the scheme from an upwind- to a centered-scheme, a loss of conservation occurred. This was later corrected by Murman with a fully conservative four-operator scheme [3]. The sonic-point operator of Murman was revisited by Chattot [4] along with extension of the scheme to the solution of the equations of gas dynamics, a system of three first order PDEs. The four-operator scheme with the modified sonic point treatment is presented here. Lets assume for simplicity that the solution is sought on a Cartesian mesh system with uniform mesh steps Δx and Δz. In this type-dependent scheme, one defines the local flow regime by comparing the x-component of the perturbation $u_{i,j}$ with the sonic perturbation u^*, where the "switches" are defined as

$$u_{i,j} = \frac{\phi_{i+1,j} - \phi_{i-1,j}}{2\Delta x} \tag{4.115}$$

and the most recent values of the potential ϕ is used in the evaluation.

(i) $u_{i-1,j} > u^*$ and $u_{i,j} > u^*$, (supersonic point)

$$\left((1-M_0^2)-(\gamma+1)M_0^2 u_{i-1,j}\right)\frac{\phi_{i,j}^{n+1}-2\phi_{i-1,j}^{n+1}+\phi_{i-2,j}^{n+1}}{\Delta x^2}+$$

$$\frac{\phi_{i,j+1}^{n+1}-2\phi_{i,j}^{n+1}+\phi_{i,j-1}^{n+1}}{\Delta z^2}=0 \tag{4.116}$$

(ii) $u_{i-1,j} > u^*$ and $u_{i,j} < u^*$, (shock point)

$$\left((1-M_0^2)-(\gamma+1)M_0^2 u_{i-1,j}\right)\frac{\phi_{i,j}^{n+1}-2\phi_{i-1,j}^{n+1}+\phi_{i-2,j}^{n+1}}{\Delta x^2}+$$

$$\left((1-M_0^2)-(\gamma+1)M_0^2 u_{i,j}\right)\frac{\phi_{i+1,j}^{n}-2\phi_{i,j}^{n+1}+\phi_{i-1,j}^{n+1}}{\Delta x^2}+$$

$$\frac{\phi_{i,j+1}^{n+1}-2\phi_{i,j}^{n+1}+\phi_{i,j-1}^{n+1}}{\Delta z^2}=0 \tag{4.117}$$

(iii) $u_{i-1,j} < u^*$ and $u_{i,j} > u^*$, (sonic point)

$$\left((1-M_0^2)-(\gamma+1)M_0^2\frac{\phi_{i,j}^{n+1}-\phi_{i-1,j}^{n+1}}{\Delta x}\right)\frac{\phi_{i+1,j}^{n}-\phi_{i,j}^{n}-\phi_{i-1,j}^{n+1}+\phi_{i-2,j}^{n+1}}{2\Delta x^2}+$$

$$\frac{\phi_{i,j+1}^{n+1}-2\phi_{i,j}^{n+1}+\phi_{i,j-1}^{n+1}}{\Delta z^2}=0 \tag{4.118}$$

(iv) $u_{i-1,j} < u^*$ and $u_{i,j} < u^*$, (subsonic point)

$$\left((1-M_0^2)-(\gamma+1)M_0^2 u_{i,j}\right)\frac{\phi_{i+1,j}^{n}-2\phi_{i,j}^{n+1}+\phi_{i-1,j}^{n+1}}{\Delta x^2}+$$

$$\frac{\phi_{i,j+1}^{n+1}-2\phi_{i,j}^{n+1}+\phi_{i,j-1}^{n+1}}{\Delta z^2}=0 \tag{4.119}$$

The upper index n indicates an "old" (n) or a "new" ($n+1$) value in the iteration process.

A few comments about the scheme: the shock-point operator is not consistent with the PDE but is conservative, which is more important (consistency is not defined when the solution is discontinuous). In the fully conservative scheme of Murman, the sonic-point scheme sets the x-derivative term to zero. Here the sonic point operator is consistent but not conservative, although the conservation error can be shown to be of order $O(\Delta x)$. Note that the sonic-point operator is implicit via the coefficient of the second derivative in x, not the second derivative in x itself ($\phi_{i,j}^{n+1}$ is in the bracket not in the second partial derivative). The latter can be written

$$\frac{\phi_{i+1,j}^{n} - \phi_{i,j}^{n} - \phi_{i-1,j}^{n+1} + \phi_{i-2,j}^{n+1}}{2\Delta x^2} =$$

$$\frac{1}{\Delta x}\left(\frac{\phi_{i+1,j}^{n} - \phi_{i-1,j}^{n+1}}{2\Delta x} - \frac{\phi_{i,j}^{n} - \phi_{i-2,j}^{n+1}}{2\Delta x}\right) = \frac{u_{i,j} - u_{i-1,j}}{\Delta x} > 0$$

The flow accelerates, hence, the coefficient of $\phi_{i,j}^{n+1}$ from the x-derivative term has the right sign and contributes to reinforcing the diagonal of the tridiagonal matrix.

Results are shown in Fig. 4.14 of the flow past a 10 % thick Quasi-Joukowski airfoil of unit chord with leading edge at the origin of the coordinate system, at incidence $\alpha = 10°$, near sonic conditions. The mesh system uses stretching and extends from $x = -2$ to $x = 5$ and from $z = -10$ to $z = 50$. Red lines represent

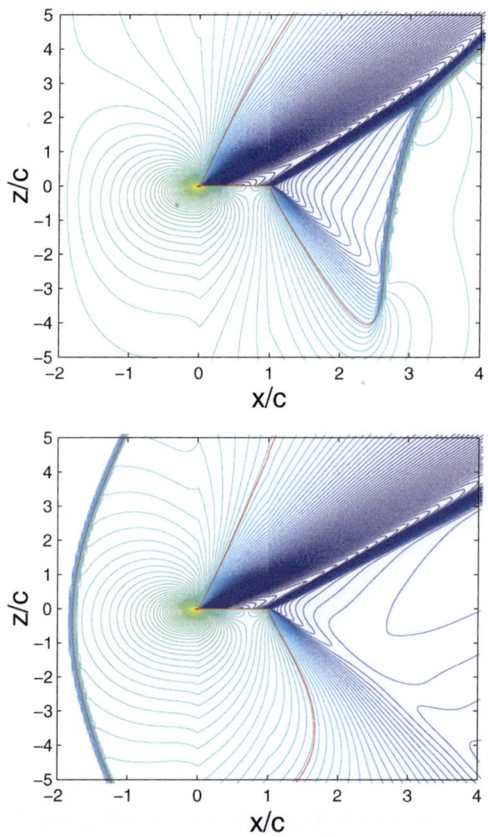

Fig. 4.14 Iso-*Mach* lines of the near sonic flows: *Top*, $M_0 = 0.9$; *Bottom*, $M_0 = 1.1$

sonic lines. The pressure distributions on the profile for the two cases are very similar, as can be seen from the Mach lines near the airfoil. This is associated with sonic freezing. Indeed, behind the bow shock in the $M_0 = 1.1$ flow case, the Mach number will be close to $M_1 = 0.9$, which explains why the flow features near the profile are very similar.

Commercial jet aircrafts cruise at Mach number in the vicinity of $M = 0.85$. In such condition, the profile will see a fairly large supersonic bubble on the upper surface terminated, in general, by a shock wave. In order to mitigate the wave drag, supercritical airfoil sections have been developed by Whitcomb and Clark [5]. Supercritical profiles, also called "peaky" profiles, tend to have a blunt nose followed by a rapid change of slope and a fairly flat upper surface to achieve a moderate shock strength by limiting the maximum value of the Mach number in the supersonic region. Shock-free airfoils have also been developed with inviscid codes for which the supersonic region is terminated by a smooth recompression. In practice, however, these shock-free airfoils may see a weak shock occur in place of the smooth transition from supersonic to subsonic, due to viscous/inviscid interaction as well as off-design flight conditions. For more details, see Boerstoel [6].

4.7.7 A Useful Nozzle Flow Solution

The design of transonic nozzles can benefit from an exact solution of the TSD equation that represents near-sonic flow conditions with a linear acceleration along the plane or axis of symmetry. Consider the two disturbance potential functions:

$$\phi(x, z) = \frac{x^2}{2} + \frac{\gamma+1}{2} x z^2 + \frac{(\gamma+1)^2}{24} z^4 \tag{4.120}$$

$$\phi(x, z) = \frac{x^2}{2} + \frac{\gamma+1}{4} x z^2 + \frac{(\gamma+1)^2}{64} z^4 \tag{4.121}$$

The reference Mach number has been set to $M_0 = 1$. The first function satisfies the above TSD 4.78 (see also Cole and Cook [7]) , whereas the second is a solution of the TSD equation for axisymmetric flow

$$-(\gamma+1)\frac{\partial \phi}{\partial x}\frac{\partial^2 \phi}{\partial x^2} + \frac{\partial^2 \phi}{\partial z^2} + \frac{1}{z}\frac{\partial \phi}{\partial z} = 0 \tag{4.122}$$

where z represents the distance from the axis of revolution. These solutions are similar and it will suffice to describe the flow features of one of them. The axisymmetric solution is more relevant to the design of a transonic nozzle for aerospace applications. The perturbation velocity field is given by

$$\begin{cases} u = x + \frac{\gamma+1}{4} z^2 \\ w = \frac{\gamma+1}{2} xz + \frac{(\gamma+1)^2}{16} z^3 \end{cases} \tag{4.123}$$

Some remarkable lines are the streamlines and the iso-C_p (or iso-u) lines. The first are obtained by numerical integration of the ODE

$$\frac{dz}{dx} = w = \frac{\gamma+1}{2} xz + \frac{(\gamma+1)^2}{16} z^3 \tag{4.124}$$

using a fourth-order Runge-Kutta algorithm. This is an approximation, consistent with the small disturbance tangency condition. It is also possible to introduce a perturbation stream-function that satisfies

$$\frac{\partial \psi}{\partial x} = -\rho w z = -z \frac{\partial \phi}{\partial z}, \quad \frac{\partial \psi}{\partial z} = \rho u z = -\frac{\gamma+1}{2} z \left(\frac{\partial \phi}{\partial x} \right)^2 \tag{4.125}$$

Upon integration, one finds

$$\psi(x, z) = -\frac{\gamma+1}{4} x^2 z^2 - \frac{(\gamma+1)^2}{16} x z^4 - \frac{(\gamma+1)^3}{192} z^6 \tag{4.126}$$

For the 2-D, planar solution, the corresponding stream-function is given by

$$\psi(x, z) = -\frac{\gamma+1}{2} x^2 z - \frac{(\gamma+1)^2}{6} x z^3 - \frac{(\gamma+1)^3}{40} z^5 \tag{4.127}$$

Then, the streamlines of the solution can be more accurately obtained by solving $\Psi = \rho^* a^* \left(\frac{z^2}{2} + \psi(x, z) \right) = const.$ or for planar, 2-D flow $\Psi = \rho^* a^* \left(z + \psi(x, z) \right) = const.$ The stream-function can be integrated simultaneously with the potential

The Mach number is given by the linear relation

$$M(u) = 1 + \frac{\gamma+1}{2} u \tag{4.128}$$

The iso-C_p (or iso-Mach) lines are simply $u = const.$ lines of equation

$$x = u - \frac{\gamma+1}{4} z^2 \tag{4.129}$$

which are parabolae with axis along the x-axis where the tangent is vertical. The sonic line corresponds to $u = 0$ and passes through the origin. Note that the sonic line does not correspond to the nozzle throat. One streamline has been selected to represent the nozzle wall (thicker line), Fig. 4.15.

In the supersonic subdomain one can also calculate the characteristic lines. As mentioned, they admit a cusp on the sonic line. Note also that the "line of throats" corresponds to the minimum of each streamline and is given by $w = 0$, i.e.

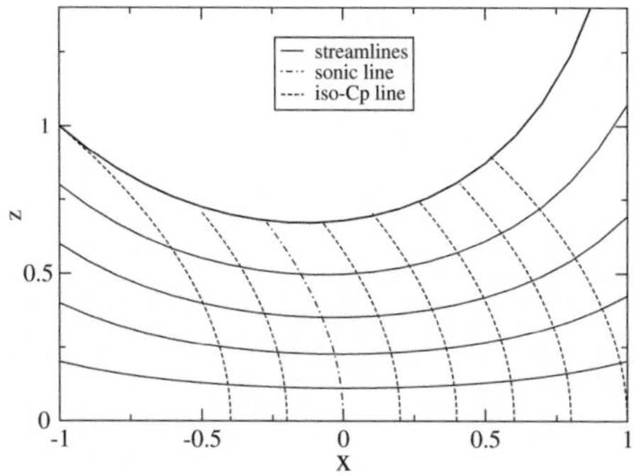

Fig. 4.15 *Streamlines* and *iso-C_p lines* near a nozzle throat

$$x = -\frac{\gamma+1}{8}z^2 \tag{4.130}$$

The radius of curvature R at the throat is an important design parameter. It varies with the streamline that is used as a nozzle wall and is found to be

$$\left(\frac{1}{R}\right)_{throat} = \left(\frac{d\theta}{ds}\right)_{throat} = \frac{\gamma+1}{2}z_{throat} \tag{4.131}$$

Also, the C^- characteristic that touches the sonic line on the x-axis is sometimes called the "frontier of transonic flow" because any small perturbation made downstream of it does not affect the incoming flow. This is depicted in Fig. 4.16.

This solution is very accurate near the sonic line since the perturbation is vanishingly small there. It can be used to design a supersonic nozzle producing a uniform flow at Mach number M_1. If the Mach number is too large to use the TSD equation, the solution can be matched and continued downstream of the frontier of transonic flow with the full potential equation to achieve an accurate design. For a low supersonic Mach number, say $M_1 = 1.48$, the small disturbance model is acceptable and the calculation can proceed readily as described below.

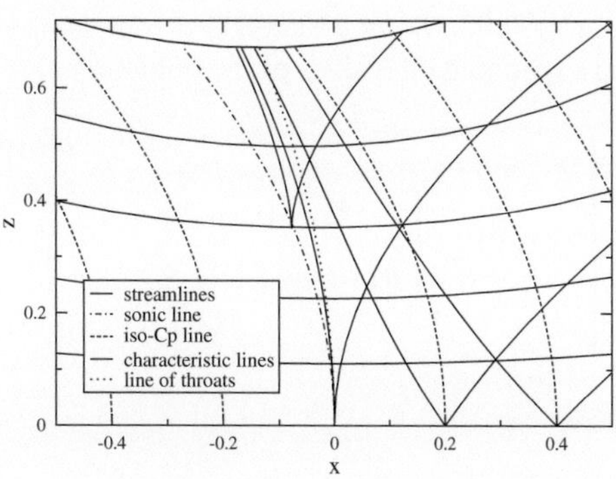

Fig. 4.16 *Characteristic lines* and *line of throats* near a nozzle throat

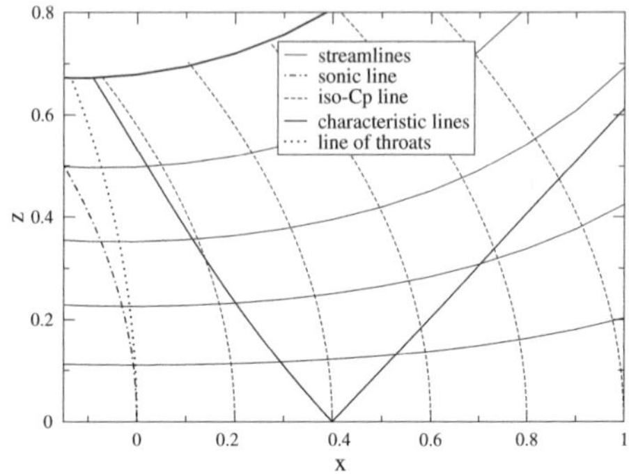

Fig. 4.17 Characteristic initial value problem for a nozzle design providing a uniform flow $u = u_1$, $w = 0$

The following initial value problem will provide the design of a uniform supersonic flow. Given M_1 the corresponding value of u_1 is determined. With the choice $M_1 = 1.48$ one finds $u_1 = 0.4$. From the point $(x_1 = u_1, 0)$ one draws the left-running C_1^- characteristic along which the potential $\phi(x, z)$ as well as (u, v) are known from the exact solution, and the desired right-running straight C_1^+ characteristic line of slope $dz/dx = 1/\sqrt{(\gamma+1)u_1}$ that is bounding the uniform flow region of pertubation velocity $(u_1, 0)$ on its right side. On Fig. 4.17, these characteristics have been superposed to the analytic solution developed earlier. The potential is also known along the C_1^+ to be, by continuity, $\phi(x, z) = u_1 x - u_1^2/2$. In the triangular region between these two characteristics, the characteristic initial value problem is well posed and a solution exists that will replace the analytical solution there. Note that the solution is integrated by marching perpendicular to the flow direction, for example using the method of characteristics.

The design is finalized by integrating the equation for the nozzle wall, starting from the point on the line of throats that provides the desired throat radius of curvature and marching in the flow direction until the C_1^+ characteristic is reached.

Another approach consists in integrating the TSD directly, marching it away from the axis. This will be possible only in the supersonic subdomain. The first step will consist in setting up a Cartesian mesh system (for simplicity) with mesh points $x_i = (i-1)\Delta x,\ i = 1, ix,\ z_j = (j-1)\Delta z,\ j = 1, jx$ with for example $\Delta x = 0.02,\ ix = 51$ and $\Delta z = \Delta x/\sqrt{(\gamma+1)u_1}$. Initial and boundary conditions are specified on 3 sides of the domain, i.e.

$$\begin{cases} \phi(0, z) = \frac{(\gamma+1)^2}{64} z^4, \quad z \geq 0 \\ \phi(x, 0) = \begin{cases} \frac{x^2}{2}, & 0 \leq x \leq x_1 \\ u_1 x - \frac{u_1^2}{2}, & x_1 \leq x \leq 1 \end{cases}, \quad w(x, 0) = \frac{\partial \phi}{\partial z}(x, 0) = 0, \quad 0 \leq x \leq 1 \\ \phi(1, z) = u_1 - \frac{u_1^2}{2}, \quad z \geq 0 \end{cases} \tag{4.132}$$

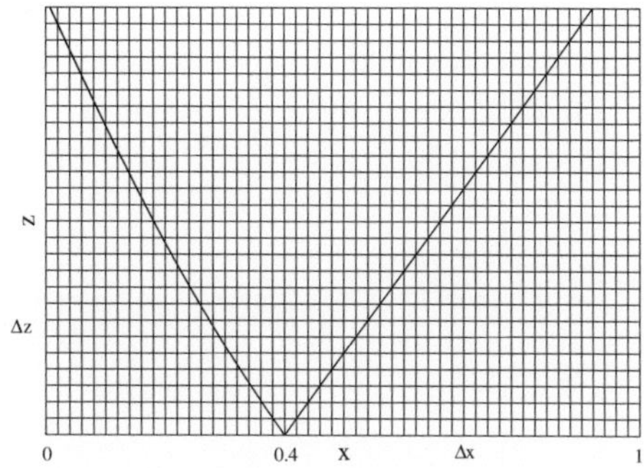

Fig. 4.18 Initial-boundary value problem for a nozzle design providing a uniform flow $u = u_1$, $w = 0$ and superimposed C_1^+ and C_1^- characteristics

See Fig. 4.18.

The numerical scheme is an explicit, second-order accurate, centered scheme for $\phi_{i,j+1}$ that reads

$$-(\gamma+1)\frac{\phi_{i+1,j}-\phi_{i-1,j}}{2\Delta x}\frac{\phi_{i+1,j}-2\phi_{i,j}+\phi_{i-1,j}}{\Delta x^2}+$$

$$\frac{\phi_{i,j+1}-2\phi_{i,j}+\phi_{i,j-1}}{\Delta z^2}+\frac{1}{z_j}\frac{\phi_{i,j+1}-\phi_{i,j-1}}{2\Delta z}=0, \quad j=2, jx-1 \tag{4.133}$$

In order to initiate the marching procedure, two lines of data are needed. For $j = 1$ the values of the potential are given by $\phi_{i,1} = x_i^2/2,\ 0 \le x_i \le x_1$ or $\phi_{i,1} = u_1 x_i - \frac{u_1^2}{2},\ x_1 \le x_i \le 1$. The symmetry condition can be implemented simply by accounting for $\phi_{i,2} = \phi_{i,0}$, where the index 0 corresponds to a fictitious line at $z = -\Delta z$, and for the limiting value of $\frac{1}{z}\frac{\partial \phi}{\partial z}(x, 0)$, which leads to the particular form of the scheme that can be solved for $\phi_{i,2}$

$$-(\gamma+1)\frac{\phi_{i+1,1}-\phi_{i-1,1}}{2\Delta x}\frac{\phi_{i+1,1}-2\phi_{i,1}+\phi_{i-1,1}}{\Delta x^2}+$$

$$2\frac{\phi_{i,2}-\phi_{i,1}}{\Delta z^2}+\left\{\begin{array}{l}\frac{\gamma+1}{2}x_i, \quad 0 < x_i < x_1\\ \quad 0, \quad x_1 < x_i < 1\end{array}\right\}=0 \tag{4.134}$$

The scheme is explicit and as such must satisfy a stability condition (CFL). However, in contrast to the analysis of transonic flow in a given nozzle, the inverse problem of designing a transonic nozzle is much easier. This is due to the fact that the solution is marched in the direction normal to the flow. Stability is ensured because the most restrictive marching step Δz corresponds to the highest value of the axial component of perturbation, here $u = u_1$, and the choice $\Delta z = \Delta x/\sqrt{(\gamma+1)u_1}$ is the maximum allowable value for the whole domain, according to the CFL criterion.

Before solving the design problem, the scheme was used to integrate that particular exact solution of the TSD equation to validate the scheme's accuracy. The boundary conditions are in this case

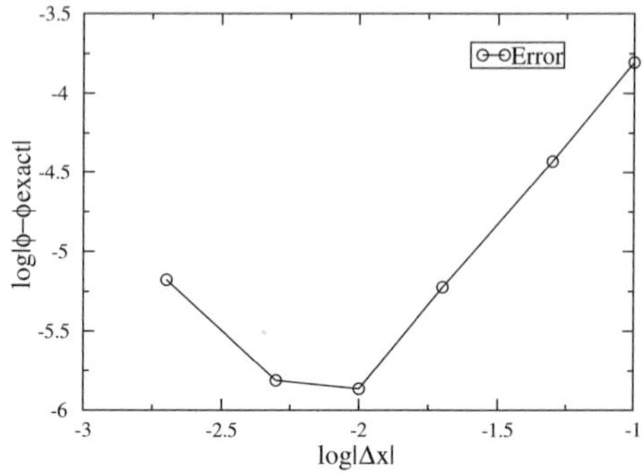

Fig. 4.19 Convergence of numerical scheme using exact solution

$$\begin{cases} \phi(0,z) = \frac{(\gamma+1)^2}{64}z^4, \quad z \geq 0 \\ \phi(x,0) = \frac{x^2}{2}, \ w(x,0) = \frac{\partial\phi}{\partial z}(x,0) = 0, \quad 0 \leq x \leq 1 \\ \phi(1,z) = \frac{1}{2} + \frac{\gamma+1}{4}z^2 + \frac{(\gamma+1)^2}{64}z^4, \quad z \geq 0 \end{cases} \tag{4.135}$$

corresponding to $x_1 = 1$. The solution was calculated on several mesh systems with $ix = 11, 21, 51$ and 101 and compared with the exact solution, using as error norm $\epsilon = max_i |\phi(i, jx) - \phi_{exact}(i, jx)|$ where $z_{jx} = 1$. On finer meshes such as $ix = 201$ or higher, round-off errors were dominating and the convergence test became meaningless. A convergence rate of approximately $R = 2.0$ was found. This is expected for a second-order accurate centered scheme. A convergence rate of 2.0 was also found for the planar solution. See Fig. 4.19.

The design of a nozzle producing a uniform supersonic flow at Mach number $M_1 = 1.48$, i.e. corresponding to $u_1 = 0.4$ is shown in Fig. 4.20. The initial-boundary value problem is discretized with a mesh of $ix = 51$ points. The slope of the C^+ characteristic that bounds the uniform flow region is $\left(\frac{dz}{dx}\right)_{C^+} = \frac{1}{\sqrt{(\gamma+1)u_1}} = 1.02$. Hence $\Delta z = 1.02\Delta x$. The solution is marched 97 steps in z to approximately $z = 1$. The wall streamline passes through the throat at $z_{throat} = 0.553$, which gives the throat a radius of curvature $R_{throat} = 1.5$ and an exit radius $z_1 = 0.608$. The area

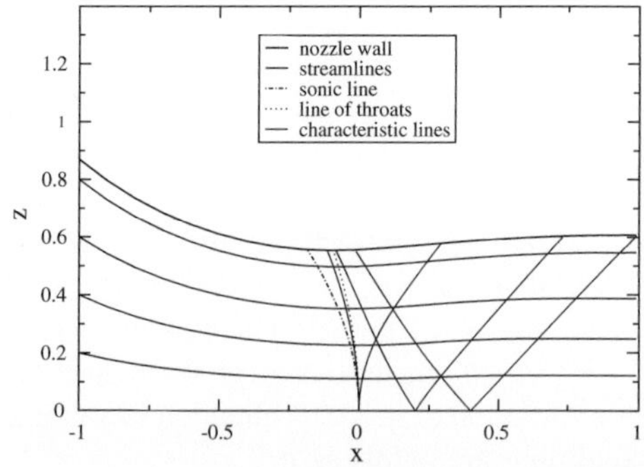

Fig. 4.20 Design of a nozzle providing a uniform flow $u_1 = 0.4$, $w = 0$

ratio $\frac{A_1}{A_{throat}} = \left(\frac{z_1}{z_{throat}}\right)^2 = 1.21$, which is 4 % larger than the value $\frac{A}{A^*} = 1.16$ found in isentropic tables. This discrepancy is probably due to the small perturbation assumption, not the two-dimensional effect that places the sonic line in a larger cross section, upstream of the throat.

4.7.8 Supersonic Flow Adjacent to Uniform Flow Region

We consider the subdomain SW delimited by the C_1^+ which bounds the uniform flow region where the perturbation velocity is $(u_1, 0)$ and the C_1^- characteristic through the point of intersection of the former with the Ox-axis, Fig. 4.21.

Plane Flow

In the case of plane, near sonic flow ($M_0 = 1$), the compatibility relations given earlier are

$$\begin{cases} CR^+ : \frac{d}{dx}\left[-\frac{2}{3}\sqrt{\gamma+1}u^{\frac{3}{2}} + w\right]_{C^+} = 0 \quad along\ C^+ : \left(\frac{dz}{dx}\right)_{C^+} = \frac{1}{\sqrt{(\gamma+1)u}} \\ CR^- : \frac{d}{dx}\left[-\frac{2}{3}\sqrt{\gamma+1}u^{\frac{3}{2}} - w\right]_{C^-} = 0 \quad along\ C^- : \left(\frac{dz}{dx}\right)_{C^-} = -\frac{1}{\sqrt{(\gamma+1)u}} \end{cases} \tag{4.136}$$

We consider the change of independent variables from (x, z) to (ξ, ζ) where $\xi = const$ along C^+ characteristics and $\zeta = const$ along C^- characteristics. One such transformation is given by

$$\begin{cases} d\xi = dx - \sqrt{(\gamma+1)u}dz = \frac{\partial\xi}{\partial x}dx + \frac{\partial\xi}{\partial z}dz \\ d\zeta = dx + \sqrt{(\gamma+1)u}dz = \frac{\partial\zeta}{\partial x}dx + \frac{\partial\zeta}{\partial z}dz \end{cases} \tag{4.137}$$

Let $u(x, z) = U(\xi, \zeta)$ and $w(x, z) = W(\xi, \zeta)$. Using the chain rule, the total derivative along the C^+ characteristic can be written

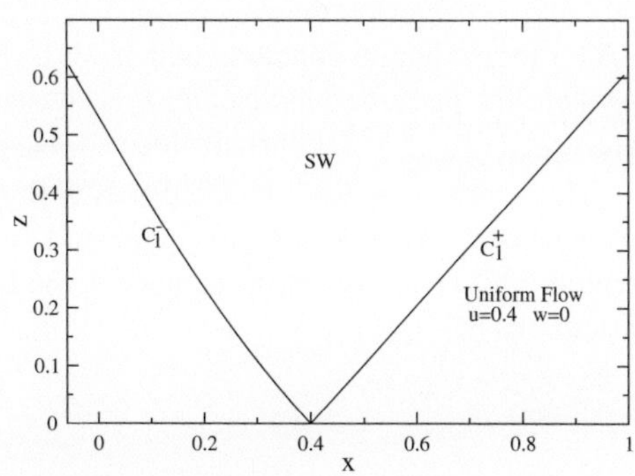

Fig. 4.21 Subdomain SW adjacent to uniform flow region

$$\frac{d}{dx}[]_{C^+} = \frac{\partial}{\partial x} + \frac{1}{\sqrt{(\gamma+1)u}}\frac{\partial}{\partial z} = \left(\frac{d\xi}{dx}\right)_{C^+}\frac{\partial}{\partial \xi} + \left(\frac{d\zeta}{dx}\right)_{C^+}\frac{\partial}{\partial \zeta} = 2\frac{\partial}{\partial \zeta} \quad (4.138)$$

The CR^+ now reads

$$\frac{\partial}{\partial \zeta}\left[-\frac{2}{3}\sqrt{\gamma+1}U^{\frac{3}{2}} + W\right] = 0 \quad (4.139)$$

This PDE can be integrated to give

$$-\frac{2}{3}\sqrt{\gamma+1}U^{\frac{3}{2}} + W = f(\xi),\ \forall\, \xi, \zeta\ in\ SW \quad (4.140)$$

where f is an arbitrary function of a single argument. Similarly, with CR^- one finds

$$\frac{2}{3}\sqrt{\gamma+1}U^{\frac{3}{2}} + W = h(\zeta),\ \forall\, \xi, \zeta\ in\ SW \quad (4.141)$$

This last relation can be fully resolved by taking into account that all C^- characteristics in SW emanate from the uniform flow region, therefore

$$h(\zeta) = \frac{2}{3}\sqrt{\gamma+1}u_1^{\frac{3}{2}} = const,\ \forall\, \xi, \zeta\ in\ SW \quad (4.142)$$

One can now solve for $W(\xi, \zeta)$ as

$$W(\xi, \zeta) = \frac{2}{3}\sqrt{\gamma+1}\left(u_1^{\frac{3}{2}} - U^{\frac{3}{2}}\right) \quad (4.143)$$

This result also indicates that the data (Cauchy data) on the C_1^- characteristic (ξ, ζ_1) cannot be specified arbitrarily. Substitution in CR^+ yields

$$\frac{2}{3}\sqrt{\gamma+1}\left(u_1^{\frac{3}{2}} - 2U^{\frac{3}{2}}\right) = f(\xi),\ \forall\, \xi, \zeta\ in\ SW \quad (4.144)$$

which forces one to conclude that $U = U(\xi)$ only, hence $W = W(\xi)$ as well. As a result, the perturbation velocity u is constant along the C^+ characteristics, which are straight lines with constant slope $\frac{1}{\sqrt{(\gamma+1)u}}$.

The subdomain SW is called a simple wave region and we can state that in plane supersonic flow, a region adjacent to a uniform flow region is a simple wave region. This result has been extended to the Euler equations in 2-D, see Courant and Friedrichs [8].

Axisymmetric Flow

The compatibility relations now read

$$\begin{cases} CR^+ : \frac{d}{dx}\left[-\frac{2}{3}\sqrt{\gamma+1}u^{\frac{3}{2}} + w\right]_{C^+} = -\frac{w}{z} \quad along\ C^+ : \left(\frac{dz}{dx}\right)_{C^+} = \frac{1}{\sqrt{(\gamma+1)u}} \\ CR^- : \frac{d}{dx}\left[-\frac{2}{3}\sqrt{\gamma+1}u^{\frac{3}{2}} - w\right]_{C^-} = -\frac{w}{z} \quad along\ C^- : \left(\frac{dz}{dx}\right)_{C^-} = -\frac{1}{\sqrt{(\gamma+1)u}} \end{cases} \tag{4.145}$$

We will use the same change of independent variables and verify that a simple wave solution exists in SW by assuming that $u(x,z) = U(\xi),\ \forall\ \xi, \zeta\ in\ SW$. The CR^+ now reads

$$2\frac{\partial}{\partial\zeta}\left[-\frac{2}{3}\sqrt{\gamma+1}U^{\frac{3}{2}} + W\right] = 2\frac{\partial W}{\partial\zeta} = \frac{d}{dx}[w]_{C^+} = -\frac{w}{z} \tag{4.146}$$

This relation holds on a C^+ characteristic of equation $\xi = x - z\sqrt{(\gamma+1)U(\xi)} = const$, hence one can write

$$\frac{dw}{w} = -\frac{dx}{x-\xi}\sqrt{(\gamma+1)U(\xi)} \tag{4.147}$$

Upon integration one obtains $\ln|w| = -\sqrt{(\gamma+1)U(\xi)}\ln|x-\xi| + f(\xi)$, where f is an arbitrary function of a single argument. Lets define the C_1^- characteristic with parametric representation $(x_1(\xi), z_1(\xi))$ and characteristic Cauchy data $(U(\xi), W_1(\xi))$ on C_1^-. One finds

$$W(\xi,z) = W_1(\xi)\left(\frac{z_1(\xi)}{z}\right)^{\sqrt{(\gamma+1)U(\xi)}}, \quad along\ C^+ \tag{4.148}$$

Substitution in CR^- yields

$$\frac{d}{dx}\left[-\frac{2}{3}\sqrt{\gamma+1}U(\xi)^{\frac{3}{2}} - W_1(\xi)\left(\frac{z_1(\xi)}{z}\right)^{\sqrt{(\gamma+1)U(\xi)}}\right]_{C^-} = -\frac{W}{z} \tag{4.149}$$

The left-hand-side term can be expanded as

$$2\frac{\partial}{\partial\xi}\left[-\frac{2}{3}\sqrt{\gamma+1}U(\xi)^{\frac{3}{2}} - W(\xi,z)\right] - \frac{1}{\sqrt{(\gamma+1)U(\xi)}}\frac{\partial}{\partial z}[W(\xi,z)] = -\frac{W}{z} \tag{4.150}$$

where the second term is exactly $-\frac{W}{z}$. The RC^- reduces to

$$\frac{\partial}{\partial\xi}\left[-\frac{2}{3}\sqrt{\gamma+1}U(\xi)^{\frac{3}{2}} - W_1(\xi)\left(\frac{z_1(\xi)}{z}\right)^{\sqrt{(\gamma+1)U(\xi)}}\right] = 0 \tag{4.151}$$

which, upon integration, yields

$$-\frac{2}{3}\sqrt{\gamma+1}U(\xi)^{\frac{3}{2}} - W_1(\xi)\left(\frac{z_1(\xi)}{z}\right)^{\sqrt{(\gamma+1)U(\xi)}} = h(\zeta) \tag{4.152}$$

along the C^- characteristic. Note that along a C^-, z is only a function of ξ, not of ζ. Making use of the fact that the C^- originate in the uniform flow region, one concludes that $h(\zeta) = const.$, and the relation becomes

$$\frac{2}{3}\sqrt{\gamma+1}\left(u_1^{\frac{3}{2}} - U(\xi)^{\frac{3}{2}}\right) - W_1(\xi)\left(\frac{z_1(\xi)}{z}\right)^{\sqrt{(\gamma+1)U(\xi)}} = 0 \tag{4.153}$$

Along the C_1^-, the characteristic initial data cannot be chosen arbitrarily since $\frac{2}{3}\sqrt{\gamma+1}\left(u_1^{\frac{3}{2}} - U(\xi)^{\frac{3}{2}}\right) - W_1(\xi) = 0$ there. This completes the verification that a simple wave solution exists in the SW subdomain. Although $w(x, z)$ varies along C^+ characteristics, $u(x, z) = U(\xi)$ is constant on a C^+, implying that they are straight lines of slope $\frac{1}{\sqrt{(\gamma+1)u}}$. In this axisymmetric transonic small disturbance approximation, a region adjacent to a uniform flow region is also a simple wave region.

4.8 Summary of Chapter 4

In this chapter, the effects of compressibility, associated with the changing Mach number in inviscid flow, are reviewed. Thin airfoil and small disturbance assumptions are retained.

At low Mach numbers, $0 \leq M_0 \leq 0.3$, the flow is treated as incompressible flow.

Compressibility effects in the range $0.3 \leq M_0 \leq 0.7$ correspond to subsonic compressible flow and the Prandtl-Glauert rule can be applied, in the absence of shock wave, to evaluate forces and moments by solving the incompressible flow past the thin profile, then the resulting aerodynamic coefficients, C_p, C_l and $C_{m,o}$, are magnified by the multiplying factor $1/\sqrt{1-M_0^2}$. The drag coefficient remains $C_d = 0$.

Above $M_0 = 1.3$, the flow is expected to be essentially supersonic and the linear supersonic theory and Ackeret results, are applicable and yield closed form expressions for the coefficients C_p, C_l and $C_{m,o}$. An inviscid wave drag is generally present with $C_d > 0$. The aerodynamic coefficients depend on Mach number via the multiplication factor $1/\sqrt{M_0^2 - 1}$.

Near Mach one, in the range $0.7 \leq M_0 \leq 1.3$, mixed-type flow is experienced with regions of subsonic and supersonic flows coexisting within the flow field. This is the transonic flow regime and the subsonic compressible and linear supersonic theories are not applicable and yield unphysical results at $M_0 = 1$. The nonlinear transonic small disturbance (TSD) equation bridges the gap between the two linear theories. The derivation of the TSD equation is carried out and characteristic lines and jump conditions associated with a hyperbolic equation are studied. Numerical solution of the TSD equation was needed and the breakthrough came with the Murman-Cole four-operator mixed-type scheme, which is discussed in some detail. The pressure coefficient is obtained from the numerical solution and provide the lift and pitching moment coefficients. The wave drag is shown to reduce to an integral over the shock lines impinging on the profile and hence vanishes in their absence. A rare exact solution of the TSD equation, which represents the near-sonic throat flow in a nozzle, is used to construct numerically subsonic-to-supersonic nozzles producing a uniform supersonic exit flow.

The final part of the chapter is devoted to the discussion of simple wave regions of supersonic flow adjacent to a uniform flow region.

4.9 Problems

4.9.1

A thin profile with parabolic camber $\frac{d}{c} = 0.02$ is set at an angle of incidence $\alpha = 5°$. Find the lift coefficient C_l in compressible flow at Mach $M_0 = 0.7$. What is the corresponding drag coefficient C_d. If the profile is attached to a vertical axis located at $x = 0$ (nose of profile), write the equation of equilibrium that will give the equilibrium angle of incidence α_{eq} at $M_0 = 0.7$. Solve for α_{eq}, and sketch the profile at equilibrium. Is the equilibrium stable? At which axis position along the chord would the equilibrium be neutral?

4.9.2

Consider a thin parabolic plate $z = 4d\frac{x}{c}(1 - \frac{x}{c})$ (zero thickness) in supersonic flow $M_0 > 1$. Give the expressions of $C_p^+(x)$ and $C_p^-(x)$ for this profile at $\alpha = 0$. Make a plot of the pressure distributions and deduct from the graph the value of the lift coefficient. Give the formula for $C_l(\alpha)$. Compute the drag coefficient of the thin parabolic plate for arbitrary incidence α. Make a graph of the profile polar $C_l(\alpha)$ vs. $C_d(\alpha)$. Find on the polar the value $\alpha = \alpha_f$ of maximum "finesse", i.e. corresponding to maximum $f = C_l(\alpha)/C_d(\alpha)$. Calculate f_{max}. Does the value depend on M_0? If it is considered desirable to fly the airfoil at maximum finesse at $M_0 = 2$, give the optimum value of camber $\frac{d}{c}$ of the parabolic profile which should be used for a given design lift coefficient $C_{l,design} = 0.1$.

4.9.3

Consider the thin cambered plate $z = \delta c\left(\frac{7}{6}\frac{x}{c} - \frac{5}{2}(\frac{x}{c})^2 + \frac{4}{3}(\frac{x}{c})^3\right)$ (zero thickness), where δ is a small positive number, in a supersonic flow $M_o > 1$. Give the expressions of $C_p^+(x)$ and $C_p^-(x)$ for this profile at $\alpha = 0$. Make a plot of the pressure distributions and deduct from the graph the value of the lift coefficient. Calculate $\left(C_{m,o}\right)_{\alpha=0}$. If an axis is located at $\frac{x}{c} = \frac{1}{2}$ and the thin cambered plate can rotate about it and is released at zero incidence with zero initial velocity, will it rotate with nose up, nose down or stay indifferent (neglect weight)?

4.9.4

The double wedge of Fig. 4.6 of half-angle θ is set at incidence α in an incoming uniform supersonic flow at Mach $M_o > 1$. Give the expressions of $C_p^+(x)$ and $C_p^-(x)$ for this profile at incidence α. Make a plot of the pressure distributions. Give the expression of the lift coefficient $C_l(\alpha)$. Compute the drag and moment coefficients, $C_d(\alpha)$ and $C_{m,o}(\alpha)$. Make a graph of the profile polar $C_l(\alpha)$ vs. $C_d(\alpha)$. Find on the polar the value $\alpha = \alpha_f$ of maximum "finesse", i.e. corresponding to maximum $f = C_l(\alpha)/C_d(\alpha)$. Sketch the waves as is done in Fig. 4.6, indicating shock waves and expansion fans, when the double wedge is at the incidence of maximum finesse α_f.

4.9.5

Consider a flat plate at incidence α in a uniform supersonic flow at Mach $M_o > 1$. Sketch the shock waves and expansion "fans", the latter being represented by expansion shocks (zero thickness). Using the jump conditions derived in this chapter, find the two components of the perturbation velocity on the upper and lower surface of the plate that satisfy the tangency condition. From these results, find the pressure coefficients and the lift coefficient for the flat plate at incidence.

4.9.6

Describe and interpret what you see in Fig. 4.22 of a bullet cruising at Mach one. In particular discuss briefly: -the absence (out-of-frame?) of the bow shock, - the origin

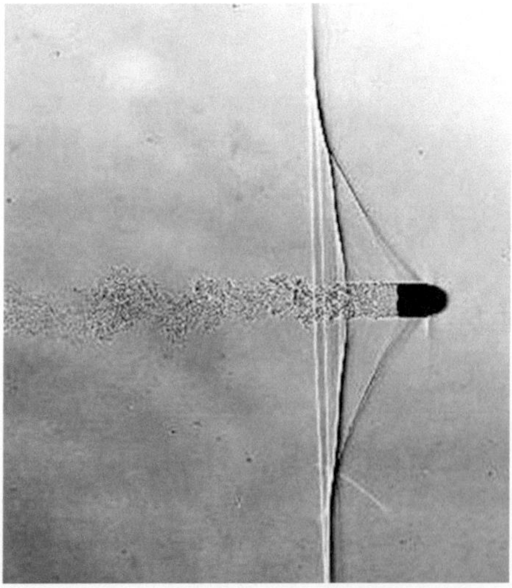

Fig. 4.22 Problem 4.9.6: Shadowgraph of bullet at $Mach = 1$ (from https://en.wikipedia.org/wiki/File:Shockwave.jpg Author: Dpbsmith (Daniel P.B. Smith))

of the wave at the 50 % location from the nose of the bullet, - the lambda shock, - the base flow and how you would handle it in an approximate manner, within the small disturbance theory.

References

1. Courant, R., Hilbert, D.: Methods of Mathematical Physics. Partial Differential Equations, vol. II. Wiley, New York (1989)
2. Chattot, J.-J.: Computational Aerodynamics and Fluid Dynamics: An Introduction. Scientific Computation. Springer, Berlin (2004)
3. Murman, E.M.: Analysis of embedded shock waves calculated by relaxation methods. AIAA J. **12**(5), 626 (1974)
4. Chattot, J.-J.: A conservative box-scheme for the Euler equations. Int. J. Numer. Methods Fluids **31**, 149–158 (1999)
5. Whitcomb, R.T., Clark, L.R.: An Airfoil Shape for Efficient Flight at Supercritical Mach Numbers. NASA Technical Memorandum X-1109 (1965)
6. Boerstoel, J.W.: A transonic hodograph theory for aerofoil design. In: IMA Conference on Computational Methods and Problems in Aeronautical Fluid Dynamics, Manchester, NLR Rep. MP 74024 U (1974)
7. Cole, J.D., Cook, L.P.: Transonic Aerodynamics. North Holland Series in Applied Mathematics and Mechanics, vol. 30 (1986)
8. Courant, R., Friedrichs, K.O.: Supersonic Flow and Shock Waves. Wiley, Pure and Applied Mathematics, vol. I (1967)

Chapter 5
Inviscid, Unsteady Flows Past Airfoils

In this chapter we will study 2-D unsteady incompressible flows such as flow past oscillating profiles or impulsively accelerated airfoils. The flow is assumed to be uniform and potential in the far field. In such cases, the profile contains time dependent bound vorticity and the wake contains shed vorticity that is convected downstream and influences the profile flow conditions. An example of wake shedding vorticity is shown in Fig. 5.1. Unsteady compressible flows are also covered.

5.1 Unsteady Incompressible Flows

5.1.1 Unsteady Flow Past Thin Cambered Plates: Governing Equations

This section is the extension of thin airfoil theory to unsteady flow. Thickness is not considered here as it does not contribute to forces and moment. Consider a uniform incoming flow of velocity U parallel to the x-axis. The total velocity is $\vec{V} = (U + u, w)$, where (u, w) represent the perturbation velocity components. The governing equations are still conservation of mass and irrotationality for the perturbation velocity

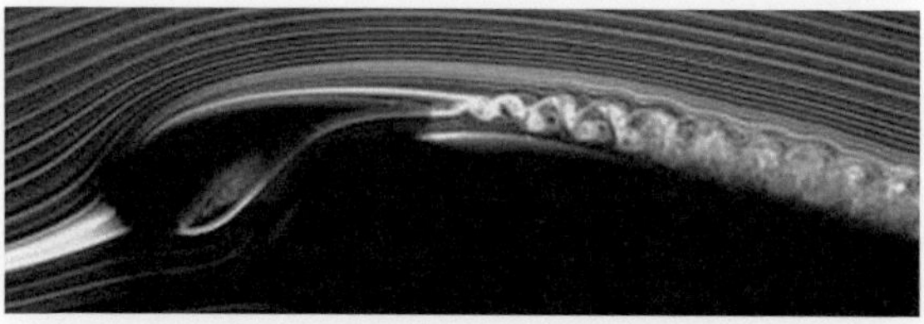

Fig. 5.1 Vortex shedding from a slat trailing edge (courtesy M. Asai, Fluid Dynamics Laboratory, Tokyo Metropolitan University)

J.J. Chattot and M.M. Hafez, *Theoretical and Applied Aerodynamics*,
DOI 10.1007/978-94-017-9825-9_5

$$\begin{cases} \frac{\partial u}{\partial x} + \frac{\partial w}{\partial z} = 0 \\ \frac{\partial w}{\partial x} - \frac{\partial u}{\partial z} = 0 \end{cases} \tag{5.1}$$

Note that this system is identical to that at steady-state. This is because we assume incompressible potential flow $\left(\frac{\partial \rho}{\partial t} = 0\right)$. The unsteady effects will appear through the boundary conditions. Lets assume that the cambered plate can rotate by a small angle $\alpha(t)$ about an axis located at x_Ω along the x-axis and can translate parallel to the z-axis with a small amplitude $h(t)$. The equation of the thin cambered plate at time t is given by

$$z^{\pm}(x,t) = d(x) + (x_\Omega - x)\alpha(t) + h(t) \tag{5.2}$$

The tangency condition reads

$$\left(\vec{V}_r.\vec{n}\right)_{profile} = 0 \tag{5.3}$$

where $\vec{V}_r = \vec{V} - \vec{V}_e$ is the velocity of the fluid in a frame attached to the profile (relative velocity) and $\vec{V}_e$ is the entrainment velocity due to the motion of the profile. The entrainment velocity for a point on the profile is given to first order by

$$\vec{V}_e = \left(0, (x_\Omega - x)\frac{d\alpha}{dt} + \frac{dh}{dt}\right) \tag{5.4}$$

Hence the relative velocity is

$$\vec{V}_r = \left(U + u, w - (x_\Omega - x)\frac{d\alpha}{dt} - \frac{dh}{dt}\right) \tag{5.5}$$

A unit normal vector to the profile is to first order $\vec{n} = (d'(x) - \alpha(t), -1)$.

The tangency condition can be simplified by keeping the first order contributions only and, as was done in steady flow, by transferring the condition to the x-axis between 0 and c as

$$w(x,0,t) = U\left(d'(x) - \alpha(t)\right) + (x_\Omega - x)\frac{d\alpha}{dt} + \frac{dh}{dt} \tag{5.6}$$

Such a flow can be modeled by a distribution of bound vorticity $\Gamma'(x,t)$ along the chord $[0, c]$ representing the vorticity inside the profile (bound vorticity) as well as shed vorticity along the x-axis, $x \geq c$. The equation governing the shed vorticity is the linear convection equation for the circulation

$$\frac{\partial \Gamma}{\partial t} + U\frac{\partial \Gamma}{\partial x} = 0 \tag{5.7}$$

The solution is made unique by requiring the Kutta-Joukowski condition to hold at the trailing edge, i.e.

$$\frac{\partial \Gamma}{\partial x}(c,t) = 0 \tag{5.8}$$

In the far field the perturbation velocity decays, except near the x-axis, when shed vorticity is present

$$u, w \to 0, \quad x^2 + z^2 \to \infty, \quad \textit{almost everywhere} \tag{5.9}$$

The linearization of the mathematical model will be complete with that of the Bernoulli equation. For unsteady, potential flow, Bernoulli equation reads

$$\rho\frac{\partial \phi}{\partial t} + p + \rho\frac{V^2}{2} = p_\infty + \rho\frac{U^2}{2} \tag{5.10}$$

Expanding the velocity term with the small disturbance assumption yields

$$\rho\frac{\partial \phi}{\partial t} + p + \rho U u = p_\infty \tag{5.11}$$

To first order, pressure is given by

$$p(x,z,t) = p_\infty - \rho U u - \rho\frac{\partial \phi}{\partial t} \tag{5.12}$$

5.1.2 *Unsteady Flow Past Thin Cambered Plates: Forces and Moment*

The lift and pitching moment are easily obtained from the pressure contributions in the z-direction, see Fig. 5.2.

From the figure, the elementary lift and moment contributions per unit span are

$$dL' = (p^- - p^+)dx, \quad dM',o = -(p^- - p^+)xdx \tag{5.13}$$

Hence, the global results

$$L' = \rho\int_0^c \left\{U(u^+ - u^-) + \frac{\partial \phi^+}{\partial t} - \frac{\partial \phi^-}{\partial t}\right\} dx \tag{5.14}$$

$$M',o = -\rho\int_0^c \left\{U(u^+ - u^-) + \frac{\partial \phi^+}{\partial t} - \frac{\partial \phi^-}{\partial t}\right\} xdx \tag{5.15}$$

Fig. 5.2 Elementary contribution to forces and pitching moment

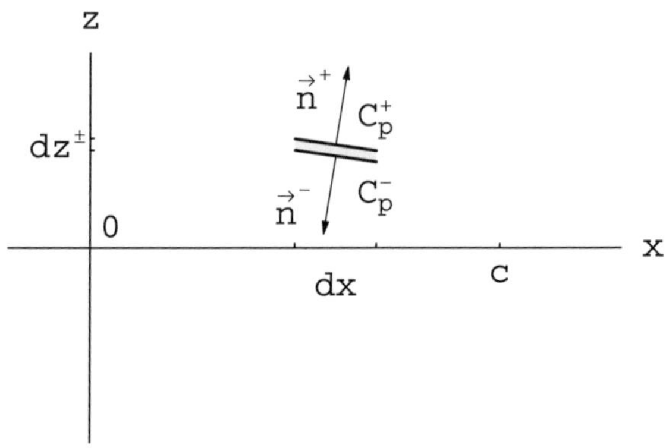

or

$$L' = \rho \int_0^c \left\{ \frac{\partial \Gamma}{\partial t} + U \frac{\partial \Gamma}{\partial x} \right\} dx, \quad M',o = -\rho \int_0^c \left\{ \frac{\partial \Gamma}{\partial t} + U \frac{\partial \Gamma}{\partial x} \right\} x dx \quad (5.16)$$

Here we have made use of the linearized Bernoulli equation and of the relationship between Γ and the jump in ϕ as

$$\Gamma = \phi(x, 0^+, t) - \phi(x, 0^-, t) = < \phi(x, t) > \quad \Rightarrow$$

$$\frac{\partial \Gamma}{\partial x} = u(x, 0^+, t) - u(x, 0^-, t) = < u(x, t) > \quad (5.17)$$

The elementary drag contribution is seen to be

$$dD' = (p^+ - p^-) \frac{dz}{dx} dx \quad (5.18)$$

Upon integrating, one obtains

$$D' = -\rho \int_0^c \left\{ \frac{\partial \Gamma}{\partial t} + U \frac{\partial \Gamma}{\partial x} \right\} \frac{dz}{dx} dx = -\rho \int_0^c \left\{ \frac{\partial \Gamma}{\partial t} + U \frac{\partial \Gamma}{\partial x} \right\} \left\{ d'(x) - \alpha(t) \right\} dx \quad (5.19)$$

It was indicated in Sect. 3.5.4 that the result of pressure integration does not include the suction force that exists at the singular leading edge of a thin plate in steady flow. The same situation exists here, in unsteady flow, and can be handled in the same manner by replacing the slope of the profile, $d'(x) - \alpha(t)$, from the tangency condition as

$$D' = -\frac{\rho}{U} \int_0^c \left\{ \frac{\partial \Gamma}{\partial t} + U \frac{\partial \Gamma}{\partial x} \right\} \left\{ w(x, 0, t) - (x_\Omega - x) \frac{d\alpha}{dt} - \frac{dh}{dt} \right\} dx \quad (5.20)$$

where $w(x, 0, t)$ is made of two parts:

$$w_b(x, 0, t) = -\frac{1}{2\pi}\int_0^c \frac{\partial\Gamma(\xi, t)}{\partial x}\frac{d\xi}{x-\xi}, \quad w_w(x, 0, t) = -\frac{1}{2\pi}\int_c^\infty \frac{\partial\Gamma(\xi, t)}{\partial x}\frac{d\xi}{x-\xi} \tag{5.21}$$

the first part, w_b, which is to be taken in the "principal value" sense, is due to the bound vorticity and the second part, w_w, is the contribution of the "wake" or shed vorticity. Substitution of these contributions in the drag evaluation yields

$$D' = -\frac{\rho}{U}\int_0^c \left\{\frac{\partial\Gamma(x, t)}{\partial t} + U\frac{\partial\Gamma(x, t)}{\partial x}\right\} \times$$

$$\left\{w_b(x, 0, t) + w_w(x, 0, t) - (x_\Omega - x)\frac{d\alpha}{dt} - \frac{dh}{dt}\right\} dx \tag{5.22}$$

Now this can be broken into three pieces, D'_1, D'_2 and D'_3 where

$$D'_1 = -\frac{\rho}{U}\int_0^c \frac{\partial\Gamma(x, t)}{\partial t} w_b(x, 0, t)dx \tag{5.23}$$

$$D'_2 = -\rho\int_0^c \frac{\partial\Gamma(x, t)}{\partial x} w_b(x, 0, t)dx \tag{5.24}$$

$$D'_3 = -\frac{\rho}{U}\int_0^c \left\{\frac{\partial\Gamma(x, t)}{\partial t} + U\frac{\partial\Gamma(x, t)}{\partial x}\right\}\left\{w_w(x, 0, t) - (x_\Omega - x)\frac{d\alpha}{dt} - \frac{dh}{dt}\right\} dx \tag{5.25}$$

D'_1 can be transformed, once we remark that

$$w_b(x, 0, t) = -\frac{1}{2\pi}\frac{\partial}{\partial x}\int_0^c \frac{\partial\Gamma(\xi, t)}{\partial t} \ln|x - \xi|\, d\xi \tag{5.26}$$

and upon integration by parts

$$D'_1 = \frac{\rho}{2\pi}\frac{\partial\Gamma(c, t)}{\partial t}\int_0^c \frac{\partial\Gamma(\xi, t)}{\partial x} \ln|c - \xi|\, d\xi$$

$$-\frac{\rho}{4\pi}\frac{d}{dt}\int_0^c\int_0^c \frac{\partial\Gamma(\xi, t)}{\partial x}\frac{\partial\Gamma(x, t)}{\partial x} \ln|x - \xi|\, d\xi dx \tag{5.27}$$

D'_2 is exactly zero, as in the steady case since it can be written

$$D'_2 = \frac{\rho}{2\pi}\int_0^c\int_0^c \frac{\partial\Gamma(\xi, t)}{\partial x}\frac{\partial\Gamma(x, t)}{\partial x}\frac{d\xi dx}{x-\xi} = 0 \tag{5.28}$$

showing the antisymmetry of the kernel.

D'_3 cannot be simplified much further. In summary, the drag reduces to $D' = D'_1 + D'_3$, that is

$$D' = -\frac{\rho}{U}\int_0^c \frac{\partial \Gamma(x,t)}{\partial t} w_b(x,0,t)dx$$

$$-\frac{\rho}{U}\int_0^c \left\{\frac{\partial \Gamma(x,t)}{\partial t} + U\frac{\partial \Gamma(x,t)}{\partial x}\right\}\left\{w_w(x,0,t) - (x_\Omega - x)\frac{d\alpha}{dt} - \frac{dh}{dt}\right\}dx \tag{5.29}$$

We note that, in the limit of steady flow, $D'_1 = D'_3 = 0$ and $D' = 0$.

5.1.3 Unsteady Flow Past Thin Airfoils: Far Field Condition for Potential

The far field value of the potential is important as boundary condition when solving the unsteady flow potential equation, using finite differences, finite volumes or finite element methods. The leading term of the perturbation potential in the far field is due to the distribution of circulation Γ along the x-axis. It is the only term for thin cambered plates (zero thickness). It reads

$$\phi(x,z,t) = -\frac{1}{2\pi}\int_0^\infty \frac{\partial \Gamma(\xi,t)}{\partial x} \tan^{-1}\left(\frac{z}{x-\xi}\right)d\xi \tag{5.30}$$

where (x, z) are the coordinates of a point where the potential is evaluated. See Fig. 5.3.

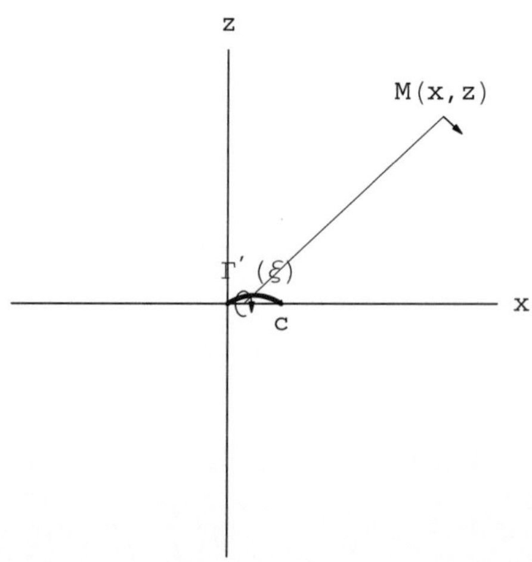

Fig. 5.3 Far field evaluation of potential

The vorticity, $\frac{\partial \Gamma}{\partial x}$, is in general singular at the leading edge $x = 0$ of the airfoil. Hence, in the numerical evaluation of the integral, it may be better to transform the expression by integrating by parts, to obtain

$$\phi(x, z, t) = -\frac{1}{2\pi} \left\{ \left[\Gamma(\xi, t) \tan^{-1} \left(\frac{z}{x - \xi} \right) \right]_0^\infty - \int_0^\infty \Gamma(\xi, t) \frac{z}{(x - \xi)^2 + z^2} d\xi \right\} \tag{5.31}$$

The first term vanishes at both limits, hence there remains

$$\phi(x, z, t) = \frac{1}{2\pi} \int_0^\infty \Gamma(\xi, t) \frac{z}{(x - \xi)^2 + z^2} d\xi \tag{5.32}$$

In steady flow, on a large circle, far from the profile, this becomes

$$\phi(x, z) = -\frac{\Gamma(c)}{2\pi} \tan^{-1} \left(\frac{z}{x} \right) \tag{5.33}$$

5.1.4 Example: Plunging Plate

This classical test case represents a flat plate aligned with the incoming flow that is accelerated impulsively to a uniform velocity $W < 0$ parallel to the z-axis, $\left|\frac{W}{U}\right| \ll 1$. This is equivalent to setting the plate abruptly at incidence $\alpha = -\frac{W}{U}$, which is the option implemented here. Although the incidence α is expected to be small in absolute value, it is possible to set it arbitrarily to the value $\alpha = 1$ rad since the problem is fully linear. For any small value of incidence, the solution obtained needs to be multiplied by α. This problem has been solved numerically and the solution $\Gamma(x, t)$ has been compared with the numerical evaluation of the integral equation that corresponds to the solution past a thin airfoil. The integral equation reads, see JJC [1]:

$$\pi\alpha = \Gamma(t) + \frac{d}{dt} \int_1^{1+t} \frac{\Gamma(1 + t - \xi)(\xi + \sqrt{\xi(\xi - 1)}) d\xi}{\sqrt{\xi(\xi - 1)}(2\xi - 1 + 2\sqrt{\xi(\xi - 1)})} \tag{5.34}$$

Here $\Gamma(t)$ represents the total bound circulation $\Gamma(c, t)$ formulated with dimensionless variables or equivalently setting $U = c = 1$. The tangency condition is treated as indicated in Chap. 3, in the case of steady flow, but accounting now for the induced velocity contribution of the shed vortices. The results for the circulation are shown in Fig. 5.4 for $\Delta t = 0.01$.

The vorticity distribution on the flat plate is displayed in Fig. 5.5. The initial low value of circulation is obtained with an almost perfectly anti symmetrical distribution of vorticity, except for the Kutta-Joukowsky condition which breaks the symmetry. The initial pressure distribution is very different from that at steady-state and the use of a steady profile polar to study unsteady flow (quasi-steady approach) would not give a realistic representation of the pressure nor of the pitching moment at the same lift in this rapidly changing flow.

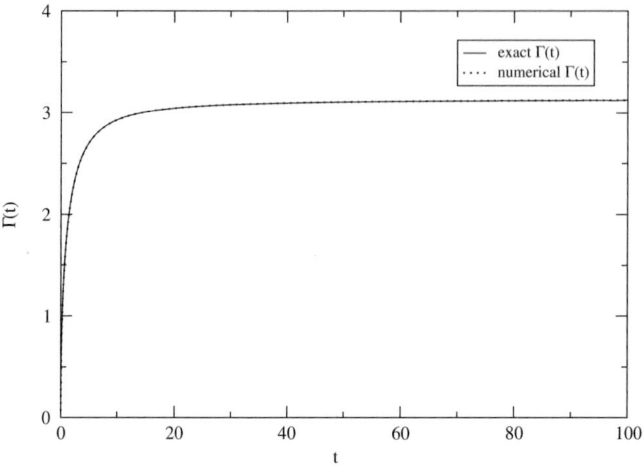

Fig. 5.4 Comparison of circulation $\Gamma(t)$ obtained with thin airfoil theory and integral equation, $\alpha = 1$ rad

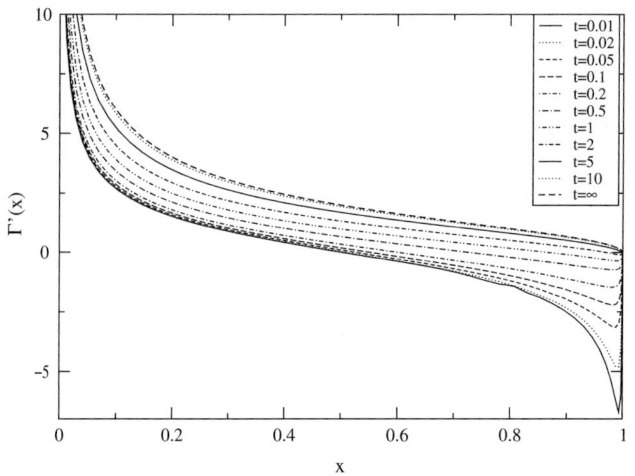

Fig. 5.5 Time evolution of the vorticity distribution on the plate, $\alpha = 1$ rad

One notable feature of the method is the scheme and the mesh for solving the convection equation along the vortex sheet. The exact solution to the convection of vorticity is

$$\Gamma(x,t) = F(x - Ut), \quad x \geq c, \ x - Ut \leq c \tag{5.35}$$

where F is a function of a single argument $\xi = x - Ut$, that satisfies $F(c - Ut) = \Gamma(c,t)$

hence the exact solution is

$$\Gamma(x,t) = \Gamma\left(c, t - \frac{x-c}{U}\right), \quad x \geq c, \ x - Ut \leq c \tag{5.36}$$

This represents a time shift without distortion of the trailing edge circulation along the x-axis. It is possible numerically to achieve the same result with a scheme that has the "perfect shift" property. This is the case with a two-point scheme, at Courant-Friedrich-Lewy (CFL) number one

$$CFL = \frac{U\Delta t}{\Delta x} = 1 \tag{5.37}$$

and with parameter θ set to $\theta = \frac{1}{2}$. A uniform mesh is constructed with $\Delta x = U\Delta t$ to the far field point, say $x = 100c$. The scheme reads

$$\theta \frac{\Gamma_i^{n+\frac{\nu}{\nu+1}} - \Gamma_i^n}{\Delta t} + (1-\theta)\frac{\Gamma_{i-1}^{n+\frac{\nu}{\nu+1}} - \Gamma_{i-1}^n}{\Delta t}$$
$$+U\left\{\theta \frac{\Gamma_i^{n+\frac{\nu}{\nu+1}} - \Gamma_{i-1}^{n+\frac{\nu}{\nu+1}}}{\Delta x} + (1-\theta)\frac{\Gamma_i^n - \Gamma_{i-1}^n}{\Delta x}\right\} = 0 \tag{5.38}$$

This is a combined implicit/explicit scheme, which reduces to Crank-Nicolson scheme for $\theta = \frac{1}{2}$ and is unconditionally stable for $\theta \geq \frac{1}{2}$. The upper index n represents the time step and ν is the inner iteration counter used to converge the circulation so that the tangency condition on the plate as well as the convection equation in the wake are satisfied to a prescribed accuracy at time t^{n+1}.

Results for the lift and drag coefficients are shown in Fig. 5.6.

The lift and drag coefficients tend to infinity near $t = 0$. Numerically, the common value at the first time step $t = \Delta t$ is of order $\frac{1}{\Delta t}$, specifically, with $\Delta t = 0.01$ one

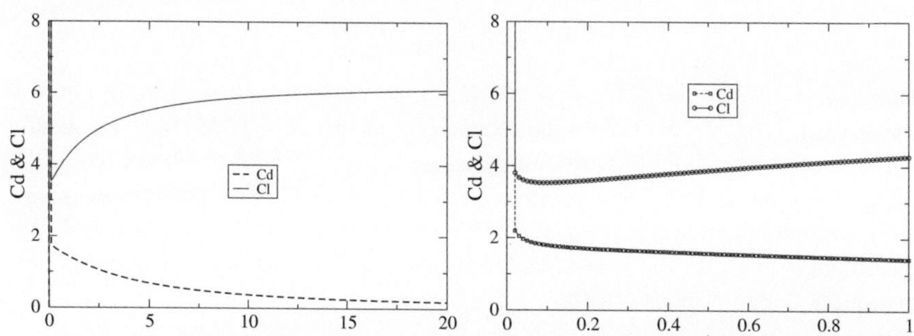

Fig. 5.6 Drag and lift coefficients versus time, $\alpha = 1$ rad (enlarged view right)

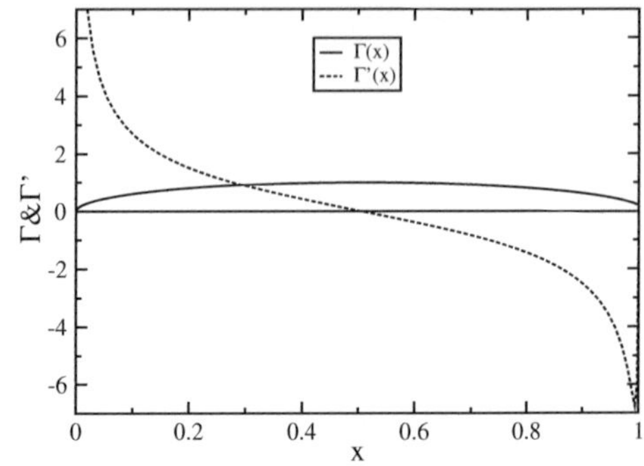

Fig. 5.7 Circulation and vorticity at time $t = \Delta t$

finds $C_l = 160.5$ and $C_d = 158.9$. Then drag decreases monotonically to zero as steady-state is approached, whereas lift decreases then increases to its final value of π. If the term D_2' had been included, the result for drag would have been $C_d(t) = C_l(t)$ since $w(x, 0, t) = -U\alpha = -U$ and $(x_\Omega - x)\frac{d\alpha}{dt} + \frac{dh}{dt} = 0$. Enlarging the scale near $t = 0$ provides a better look at the early evolution of those coefficients, Fig. 5.6. One can see from the numerical results that, at the first time step, the suction force is not present, but with the second time step the suction force already cancels half of the drag.

It is also possible to look at the circulation and vorticity distributions at $t = \Delta t$, Fig. 5.7. It is clear that at a very early time $t = \epsilon > 0$, the circulation must be elliptic since this gives a constant $w(x, \epsilon) = \alpha$.

5.1.5 Example: Pitching NACA0012

We consider now a pitching motion. Thanks to the linearity of the problem, the combined plunging and pitching motion can be obtained by superposition of the solutions of the two motions. The following pitching motion is considered:

$$\alpha(t) = \alpha_0 + \delta\alpha \sin(\omega t) \tag{5.39}$$

where $\alpha_0 = 4°$, $\delta\alpha = 8.3°$, $f = 30\,\text{Hz}$, $U = 98.8\,\text{m/s}$ and the profile chord is $c = 0.1°\,\text{m}$. This corresponds to the reduced frequency $k = fc/U = 0.03$. Results of the numerical simulation are compared with experimental results of Wood [2] in Fig. 5.8. The normal force coefficient is defined as $C_n = C_l \cos\alpha + C_d \sin\alpha$ and represents the component of the force perpendicular to the profile chord. The agreement is good for the lower incidences, $\alpha < 2.5°$ but differences occur at high incidences due to viscous effects.

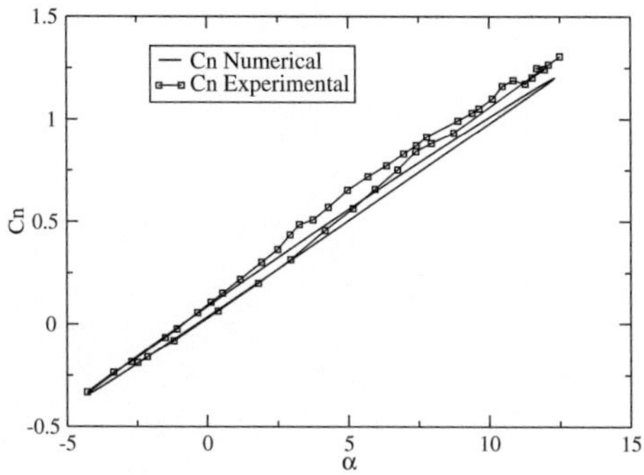

Fig. 5.8 Normal force coefficient comparison

5.2 Unsteady Compressible Flows

5.2.1 The Full Potential Equation

The Euler equations represent conservation of mass, momentum and energy, assuming inviscid and adiabatic flow. Furthermore, for isentropic and irrotational unsteady flows, the governing equations become

$$\frac{\partial \rho}{\partial t} + \frac{\partial}{\partial x}\left(\rho \frac{\partial \Phi}{\partial x}\right) + \frac{\partial}{\partial y}\left(\rho \frac{\partial \Phi}{\partial y}\right) + \frac{\partial}{\partial z}\left(\rho \frac{\partial \Phi}{\partial z}\right) = \frac{\partial \rho}{\partial t} + \nabla . (\rho \nabla \Phi) = 0 \quad (5.40)$$

where

$$\frac{\rho}{\rho_0} = \left[1 - \frac{\gamma - 1}{2} M_0^2 \left(|\nabla \Phi|^2 + 2\frac{\partial \Phi}{\partial t} - 1\right)\right]^{\frac{1}{\gamma-1}} \quad (5.41)$$

The boundary condition on a solid surface defined by $F(x, y, z, t) = 0$, can be written as

$$\frac{dF}{dt} = \frac{\partial F}{\partial t} + \nabla \Phi . \nabla F = 0, \; on \; F = 0 \quad (5.42)$$

where the velocity field of the surface points displacements has been replaced by the fluid velocity field, as both fields should be tangential to the surface.

To guarantee the continuity of the density and pressure across the vortex sheet, which is a slip line allowing for discontinuous tangential velocity only, the following condition must be imposed

$$\frac{d\Gamma}{dt} = \frac{\partial \Gamma}{\partial t} + \bar{\mathbf{V}} . \nabla \Gamma = 0 \quad (5.43)$$

where Γ is the jump in the potential, $\Gamma = \Phi^+ - \Phi^-$ and $\bar{\mathbf{V}}$ is the mean tangential velocity. Note that the normal velocity component on the vortex sheet is continuous.

The Kutta-Joukowski condition at the trailing edge is given by

$$\left(2\frac{\partial \Phi}{\partial t} + |\nabla \Phi|^2\right)^+ = \left(2\frac{\partial \Phi}{\partial t} + |\nabla \Phi|^2\right)^- \tag{5.44}$$

In the far field, a non reflecting, one-dimensional, boundary condition is used [3]

$$\frac{\partial \Phi}{\partial n} = V_n \pm \frac{\frac{\partial \Phi}{\partial t}}{a_0 \pm V_n} \tag{5.45}$$

where V_n is the normal component to the boundary and a_0 is the speed fo sound of the undisturbed flow.

Next, the governing equations are rewritten in non-conservative form and the discussion is limited to 2-D flows

$$\begin{aligned} 2\frac{\partial \Phi}{\partial x}\frac{\partial^2 \Phi}{\partial t \partial x} + 2\frac{\partial \Phi}{\partial y}\frac{\partial^2 \Phi}{\partial t \partial y} + \frac{\partial^2 \Phi}{\partial t^2} &= \left[a^2 - \left(\frac{\partial \Phi}{\partial x}\right)^2\right]\frac{\partial^2 \Phi}{\partial x^2} \\ &- 2\frac{\partial \Phi}{\partial x}\frac{\partial \Phi}{\partial y}\frac{\partial^2 \Phi}{\partial x \partial y} + \left[a^2 - \left(\frac{\partial \Phi}{\partial y}\right)^2\right]\frac{\partial^2 \Phi}{\partial y^2} \end{aligned} \tag{5.46}$$

where a is the speed of sound given by

$$\frac{a^2}{\gamma - 1} + \frac{1}{2}\left[\left(\frac{\partial \Phi}{\partial x}\right)^2 + \left(\frac{\partial \Phi}{\partial y}\right)^2 + 2\frac{\partial \Phi}{\partial t}\right] = \frac{a^2}{\gamma - 1} + \frac{1}{2}U^2 \tag{5.47}$$

The shock jump conditions of the potential formulation is given by

$$< \rho > n_t + < \rho\frac{\partial \Phi}{\partial x} > n_x + < \rho\frac{\partial \Phi}{\partial y} > n_y = 0 \tag{5.48}$$

On the shock, a normal vector is ∇S, where $S(x, y, t)$ is the surface of discontinuity. One finds

$$< \rho > \frac{\partial S}{\partial t} + < \rho\frac{\partial \Phi}{\partial x} > \frac{\partial S}{\partial x} + < \rho\frac{\partial \Phi}{\partial y} > \frac{\partial S}{\partial y} = 0 \tag{5.49}$$

together with the condition of continuity of the potential

$$< \Phi >= 0 \tag{5.50}$$

The characteristics are obtained from the non-conservative form of the full potential equation as follows. Let, $u = \partial \Phi/\partial x$, $v = \partial \Phi/\partial y$ and $w = \partial \Phi/\partial t$, hence

$$2u\frac{\partial u}{\partial t}+2v\frac{\partial v}{\partial t}+\frac{\partial w}{\partial t}=(a^2-u^2)\frac{\partial u}{\partial x}-uv\left(\frac{\partial u}{\partial y}+\frac{\partial v}{\partial x}\right)+(a^2-v^2)\frac{\partial v}{\partial y} \quad (5.51)$$

The system can be written in quasi-linear form as

$$\frac{\partial}{\partial t}\begin{bmatrix} u \\ v \\ w \end{bmatrix}=\begin{bmatrix} 0 & 0 & 1 \\ 0 & 0 & 0 \\ a^2-u^2 & -uv & -2u \end{bmatrix}\frac{\partial}{\partial x}\begin{bmatrix} u \\ v \\ w \end{bmatrix}+\begin{bmatrix} 0 & 0 & 0 \\ 0 & 0 & 1 \\ -uv & a^2-v^2 & -2v \end{bmatrix}\frac{\partial}{\partial y}\begin{bmatrix} u \\ v \\ w \end{bmatrix} \quad (5.52)$$

where we have made use of $\partial u/\partial t = \partial w/\partial x$ and $\partial v/\partial t = \partial w/\partial y$.

The eigenvalues of the two associated matrices are $\lambda = 0$ and $\lambda = u \pm a$, and $\lambda = 0$ and $\lambda = v \pm a$, respectively. The geometric interpretation of this result is that there exist two families of characteristic surfaces. Consider a point in the flow field and let it be the origin of the coordinate system. With respect to that point, one family, corresponding to $\lambda = 0$, are planes that contain the t-axis. The other family are planes through the origin, that make an angle with the t-axis and envelop a cone of equation (see Fig. 5.9),

$$(x-ut)^2+(y-vt)^2=(at)^2 \quad (5.53)$$

The cross section of the cone by a plane $t = const$ is a circle, centered at (ut, vt) and of radius $r = at$.

The origin is inside the circle when $M_0 < 1$, and outside when $M_0 > 1$. In the latter case, the tangents to the circle from the origin are the characteristic lines of the steady flow.

The zero eigenvalues can cause problems in the numerical solutions of the system of first order equations.

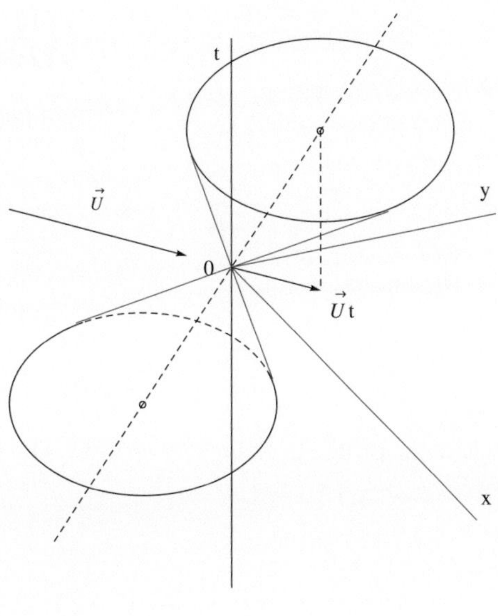

Fig. 5.9 Characteristic cone for the full potential equation

5.2.2 The Transonic Small Disturbance Equation

The unsteady, transonic small disturbance equation (TSD) in Cartesian coordinates is given by

$$A\frac{\partial^2\phi}{\partial t^2}+2B\frac{\partial^2\phi}{\partial t\partial x}=C\frac{\partial^2\phi}{\partial x^2}+\frac{\partial^2\phi}{\partial y^2} \tag{5.54}$$

where

$$A=\frac{k^2M_0^2}{\epsilon^{\frac{2}{3}}},\quad B=\frac{kM_0^2}{\epsilon^{\frac{2}{3}}},\quad C=\frac{1-M_0^2}{\epsilon^{\frac{2}{3}}}-(\gamma+1)M_0^2\frac{\partial\phi}{\partial x}=K-(\gamma+1)M_0^2\frac{\partial\phi}{\partial x} \tag{5.55}$$

K is the transonic similarity parameter. For an airfoil of chord length c, traveling with speed U with an oscillating frequency ω, the reduced frequency is defined as

$$k=\frac{\omega c}{U} \tag{5.56}$$

The perturbation potential in the above equation is scaled by $\epsilon^{\frac{2}{3}}$, where ϵ is the small parameter representing the profile thickness, camber or incidence, while x, y and t are scaled with c, $c/\epsilon^{\frac{1}{3}}$ and $1/\omega$, respectively.

For low reduced frequencies, when $k=O(\epsilon^{\frac{2}{3}})$, the equation becomes

$$2B\frac{\partial^2\phi}{\partial t\partial x}=C\frac{\partial^2\phi}{\partial x^2}+\frac{\partial^2\phi}{\partial y^2} \tag{5.57}$$

The pressure coefficient is

$$C_p\simeq-2\left(\frac{\partial\phi}{\partial x}+k\frac{\partial\phi}{\partial t}\right) \tag{5.58}$$

The cut is placed at $y=0$, $x\geq 1$, and the continuity of pressure requires

$$<\frac{\partial\phi}{\partial x}>+k<\frac{\partial\phi}{\partial t}>=\frac{\partial\Gamma}{\partial x}+k\frac{\partial\Gamma}{\partial t}=0 \tag{5.59}$$

The linearized boundary condition becomes

$$\frac{\partial\phi}{\partial y}(x,0,t)=\frac{\partial f}{\partial x}+k\frac{\partial f}{\partial t},\quad 0\leq x\leq 1 \tag{5.60}$$

where the profile is given by $y=\epsilon f(x,t)$.

The far field non-reflecting downstream boundary condition becomes

$$\frac{\partial \phi}{\partial x} + k\frac{\partial \phi}{\partial t} = 0 \tag{5.61}$$

Notice, this condition is consistent with the treatment of the cut.

The shock jump condition for the unsteady TSD, low frequency approximation, is obtained from the conservation form of the equation [4]

$$2B < \frac{\partial \phi}{\partial x} > n_t =< K\frac{\partial \phi}{\partial x} - \frac{\gamma+1}{2}M_0^2\left(\frac{\partial \phi}{\partial x}\right)^2 > n_x + < \frac{\partial \phi}{\partial y} > n_y \tag{5.62}$$

and from the irrotationality condition $\partial^2\phi/\partial x\partial y = \partial^2\phi/\partial y\partial x$

$$< \frac{\partial \phi}{\partial x} > n_y =< \frac{\partial \phi}{\partial y} > n_x \tag{5.63}$$

upon elimination of n_y and multiplication by $< \partial\phi/\partial x > /n_x$ one finds

$$2B < \frac{\partial \phi}{\partial x} >^2 \frac{n_t}{n_x} = \left(K - (\gamma+1)M_0^2\frac{\bar{\partial \phi}}{\partial x}\right) < \frac{\partial \phi}{\partial x} >^2 + < \frac{\partial \phi}{\partial y} >^2 \tag{5.64}$$

Here, we made use of $< u^2 >= 2\bar{u} < u >$, where $\bar{u} = (u_1 + u_2)/2$, is the average of u across the shock. Also, $n_t/n_x = -dx_S/dt$ is the speed of the shock in the x-direction. With this, the shock equation becomes

$$2B\left(\frac{dx}{dt}\right)_S < \frac{\partial \phi}{\partial x} >^2 + \left(K - (\gamma+1)M_0^2\frac{\bar{\partial \phi}}{\partial x}\right) < \frac{\partial \phi}{\partial x} >^2 + < \frac{\partial \phi}{\partial y} >^2 = 0 \tag{5.65}$$

The equation is always of hyperbolic type. One family of characteristic surfaces through the origin comprises the plane $t = 0$ that corresponds to an infinite speed of propagation of the perturbations. The other family are planes that envelop a cone of equation

$$2Bxt + Ct^2 = (By)^2 \tag{5.66}$$

The cross section of the cone by a plane $t = const$ is a parabola having the x-axis as axis of symmetry, Fig. 5.10. The origin is inside the parabola when $M_0 < 1$, and outside when $M_0 > 1$. The tangents to the parabola from the origin represent the characteristic lines of the steady flow.

Next, the sources of material of this section are summarized.

First, the linear theory for high frequency transonic flow is discussed by Landhal [5]. The more recent, nonlinear small disturbance theory is discussed by Cole and Cook [6]. The numerical treatment and numerical results can be found in the articles of Ballhaus and Lomax [7], Bailey and Ballhaus [4], Goorjian [8] and van der Vooren

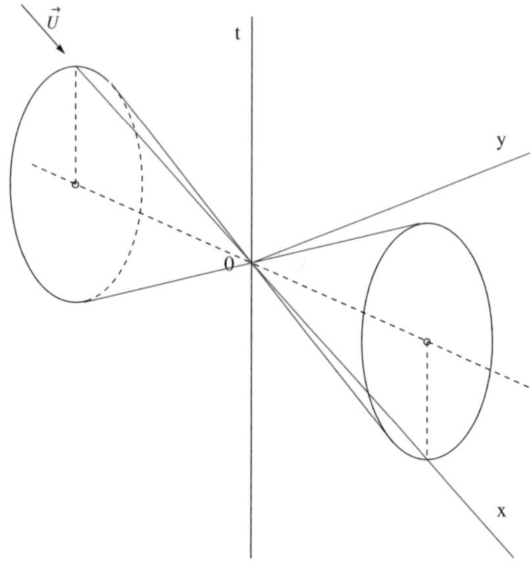

Fig. 5.10 Characteristic cone for the TSD equation

and Schippers [9]. Engquist and Majda covered the mathematics and numerics of absorbing boundary conditions in [3] and [10]. Applications of non-reflecting boundary conditions to unsteady transonic flow computations were reported by Kwak in Ref. [11]. Many others contributed to NASA codes for low frequency transonic flows and flutter analysis.

5.3 Summary of Chapter 5

In this chapter, unsteady incompressible and compressible flows are discussed within the framework of thin airfoil and small disturbance approximations.

The fundamental aspect of unsteady, inviscid flow is the shedding of vorticity from the trailing edge of the thin profile. A Kutta-Joukowski condition still applies at the trailing edge to render the solution unique. The steady linear models of subsonic compressible and supersonic potential flows require a cut across which the potential has a jump equal to the circulation around the profile, $\Gamma =< \phi >$. In steady flow, the cut could be placed anywhere to make the domain simply-connected. However, in unsteadys flow, the cut represents the inviscid "wake" or slip line, downstream from the trailing edge, along the x-axis, where the potential function is discontinuous and the jump in potential is a function of time and space. The unsteady Bernoulli's law (or generalized Bernoulli's law in compressible flow) provides the condition along the slip line by specifying the continuity of pressure.

The example of the plunging plate in incompressible flow is simulated numerically and the results are compared with the exact solution. Care must be exerted to include the suction force at the leading edge of the plate in order to obtain zero drag when the flow returns to steady-state. Noteworthy is the first instant in the simulation when the circulation in the plate is still close to zero and the vorticity distribution similar to that of an elliptic wing. The lift and drag are equal, resulting in a force perpendicular to the plate and their magnitude is inversely proportional to the time step.

Another example is the pitching NACA0012 at low speeds and low frequency.

Extension to the unsteady full potential equation is discussed. The unsteady transonic small disturbance equation (unsteady TSD) is derived and the magnitude of the coefficients of the time derivatives analyzed in relation to the reduced frequency. The low-frequency TSD is highlighted as it is of great importance to the study of transonic unsteady flows. Key papers are referenced for the reader interested in more in depth exposure.

5.4 Problems

5.4.1 Motion of a Vortex Pair Above Ground

Consider to counter-rotating vortices of strength $\pm\Gamma$, such as trailing the wings of an aircraft. Let the z-axis be the meridian plane and the x-axis be the ground, Fig. 5.11.

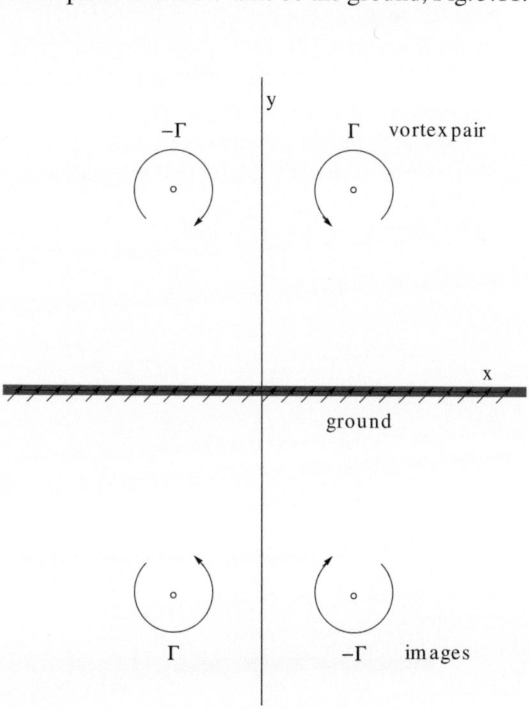

Fig. 5.11 Vortex pair

The two vortices are initially located at (a, b) and $(-a, b)$. Show, by using images of the vortices below the ground, that the trajectory of the vortices is described by

$$\frac{1}{x^2} + \frac{1}{z^2} = \frac{1}{a^2} + \frac{1}{b^2}$$

Show that the pair of vortices does not hit the ground, but gets as close as $z_{min} = c = ab/\sqrt{a^2 + b^2}$.

Show also that, far from the ground, the vortex pair moves parallel to the z-axis with velocity $w = -\Gamma/(4\pi c)$.

5.4.2 Falling Plate over Flat Surface

Consider the 2-D unsteady motion of a plate of width $2L$ falling on a flat surface a very small distance h apart where $h/L \ll 1$. The gap is occupied by air. The air viscosity is neglected and the flow is assumed incompressible, see Fig. 5.12.

Apply conservation of mass theorem to the control volume and show that the flow field corresponds to a stagnation point flow with a time dependent multiplication constant

$$u = K(t)\,x, \quad w = -K(t)\,z$$

Find the relationship between $K(t)$ and $h(t)$.

Find the pressure in the gap by integrating the x- and z-momentum equations, and find the pressure on the lower side of the plate by enforcing the condition that the pressure should be continuous and equal to p_{atm} at the edge $x = L$, $z = h(t)$.

Let M' be the mass per unit span of the plate and assume that $M' \ll \rho L^3/h(0)$, where ρ is the air density. Show that the plate motion is described by

$$\frac{h(t)}{h(0)} = \frac{1}{\cosh \sqrt{C}t}$$

where

$$C = \frac{3}{2}\frac{M'g}{\rho L^3}$$

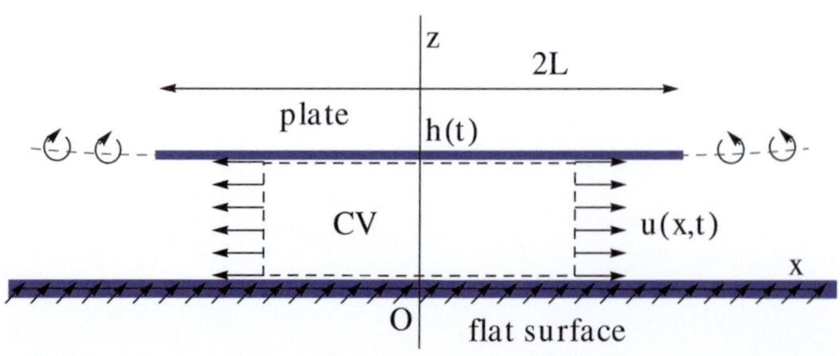

Fig. 5.12 Falling plate

References

1. Chattot, J.-J.: Helicoidal model for steady and unsteady flows. Comput. Fluids **35**(7), 733–741 (2006)
2. Wood, M.E.: Results from Oscillatory Pitch Tests on the NACA0012 Section. ARA Memo 220. Aircraft Research Association. Bedford (1979)
3. Engquist, B., Majda, A.: Numerical radiation boundary conditions for unsteady transonic flow. J. Comput. Phys. (1981)
4. Bailey, F.R., Ballhaus, W.F.: Relaxation methods for transonic flow about wing-cylinder combinations and lifting swept wings. Lecture Notes in Physics, vol. **19**, pp. 2–9. Springer (1972)
5. Landhal, M.: Unsteady Transonic Flow. Pergamon Press, New York (1961)
6. Cole, J.D., Cook, P.: Transonic Aerodynamics. North Holland, Amsterdam, The Netherlands (1986)
7. Ballhaus, W. F., Lomax, H.: The numerical simulation of low frequency unsteady transonic flow fields. Lecture Notes in Physics, vol. **35**, pp. 57–63. Springer (1975)
8. Goorjian, P.M.: Computations of Unsteady Transonic Flows. Advances in Computational Transonics. Pineridge Press, Swansea (1985)
9. van der Vooren, J., Schippers, H.: Implicit finite-difference methods for the calculation of unsteady transonic potential flow about 2d airfoils—a discussion. Advances in Computational Transonics. Pineridge Press, Swansea (1985)
10. Engquist, B., Majda, A.: Absorbing boundary conditions for the numerical simulation of waves. Math. Comput. **13**(139), 629–651 (1977)
11. Kwak, D.: Non-reflecting far field boundary conditions for unsteady transonic flow computations. AIAA paper 80-1393 (1980)

Chapter 6
Flow Past Large and Moderate Aspect Ratio Wings

In this chapter we will study 3-D steady flows past finite wings of large and moderate aspect ratios. The wing geometrical parameters have been defined in Chap. 1 with wing aspect ratio given by $AR = b^2/S$, where b is the wing span from tip to tip and S is the projected surface area of the wing onto the x, y plane, the x-axis being aligned and in the direction of the incoming flow in the plane of symmetry of the wing, the y-axis being along the right wing and the z-axis completing the direct orthonormal coordinate system. For a rectangular wing, this reduces to $AR = b/c$, where c is the constant chord. The assumption of a large aspect ratio wing will contribute to a major simplification which consists in neglecting geometric influence in the y-direction compared to the other directions, in other words

$$\frac{\partial}{\partial y} \ll \frac{\partial}{\partial x}, \frac{\partial}{\partial z} \tag{6.1}$$

which corresponds to the so-called strip theory. Large aspect ratio lifting elements make for the wings of commercial aircraft, glider, helicopter and wind turbine blades. Such a fundamental geometrical aspect has ramifications not only for aerodynamics, but also for structures with the beam theory. Aspect ratio varies widely:

- $AR \simeq 35$ for high performance sailplanes
- $AR = 7 - 2$ for commercial airplanes
- $AR = 4 - 5$ for transport jets
- $AR \simeq 2$ for supersonic jets

Note that $AR \geq 7$ will be considered high aspect ratio and $AR \leq 4$ low aspect ratio (Fig. 6.1).

J.J. Chattot and M.M. Hafez, *Theoretical and Applied Aerodynamics*,
DOI 10.1007/978-94-017-9825-9_6

Fig. 6.1 High aspect ratio wing. Pathfinder (from http://www.nasa.gov/vision/earth/technologies/solarFarm.html)

6.1 Wing Geometric Parameters

The wing planform geometry can be described with several parameters. Unswept wings are shown in Fig. 6.2. The quarter chord $c_{\frac{1}{4}}$ is parallel to the y-axis. In the case of the trapezoidal wing, the root chord, c_r or c_0, and the tip chord, c_t, are two parameters that define the planform geometry.

The quarter-chord sweep angle, $\Lambda_{\frac{c}{4}}$ is another geometric parameter for swept wings such as the rectangular swept wing and the delta wing, Fig. 6.3.

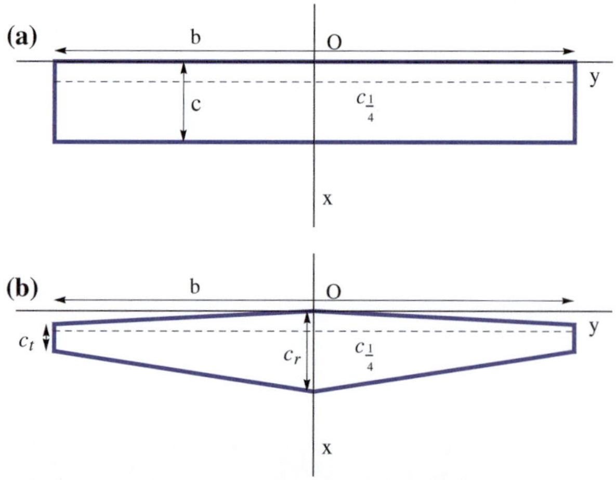

Fig. 6.2 Unswept wings: **a** rectangular wing; **b** trapezoidal wing

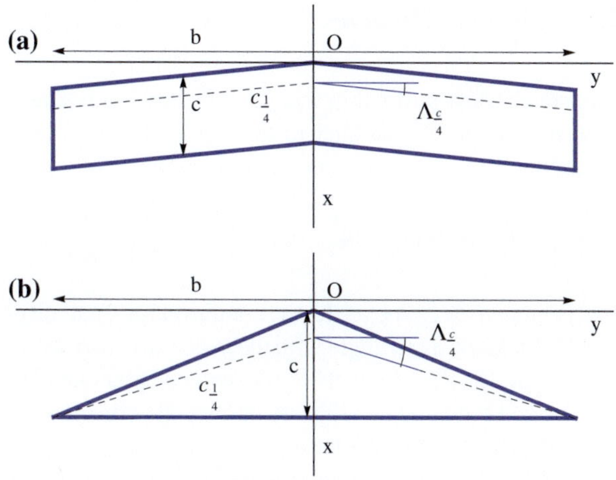

Fig. 6.3 **a** Swept rectangular wing; **b** delta wing

The delta wing is usually the preferred design for low aspect ratio supersonic aircraft wings.

One defines the *mean aerodynamic chord* as

$$\overline{c} = \frac{1}{S}\int_{-\frac{b}{2}}^{\frac{b}{2}} c^2(y)dy = \frac{1}{S}\int_{-\frac{b}{2}}^{\frac{b}{2}} c(y)dS(y) \tag{6.2}$$

The mean aerodynamic chord plays a role in the wing pitching moment.

When facing the wing and the fuselage, the angle of the wing with the horizontal plane is called the *dihedral angle*. The dihedral angle can be positive or negative, see Fig. 6.4. A positive dihedral angle introduces rolling stability, known as *dihedral effect*, but this and other unsymmetrical flow conditions such as yaw, are outside the scope of this book.

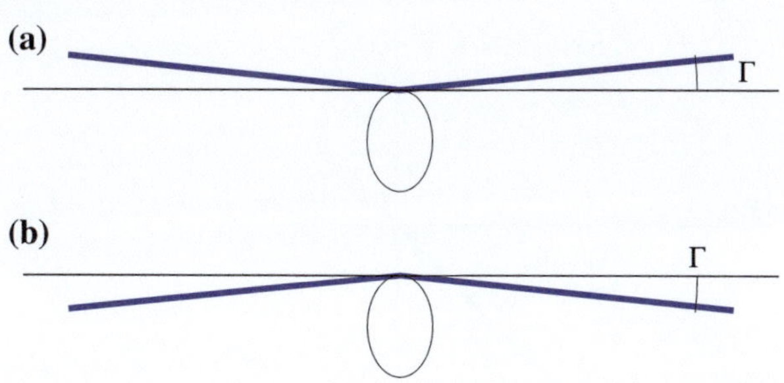

Fig. 6.4 Dihedral angles: **a** $\Gamma > 0$; **b** $\Gamma < 0$

6.2 Small Disturbance Theories

Assume inviscid, steady flow past a thin wing that disturbs only slightly the incoming uniform flow of velocity U, parallel to the x-axis. Let (u, v, w) be the small perturbation components. The velocity vector is given by

$$\mathbf{V} = (U + u)\,\mathbf{i} + v\mathbf{j} + w\mathbf{k}, \quad s.t. \quad \left|\frac{u}{U}\right|, \left|\frac{v}{U}\right|, \left|\frac{w}{U}\right| \ll 1 \qquad (6.3)$$

Since the flow is tangent to the obstacle, the condition for small disturbance is that the wing or the body makes a small angle with the x-axis. In other words, the obstacle must be close to a cylindrical surface, the *base surface*, generated by straight lines parallel to the x-axis and touching an arbitrary curve called the *base line*. In the case when the curve reduces to a point, the base surface degenerates to a line parallel to **Ox** called the *axis of the body*. In the first case, the obstacle is called a *thin wing*. In the second case, the obstacle is called a *slender body*, Fig. 6.5. Slender bodies and low aspect ratio wings are the object of Chap. 7, where slender body theory will be discussed. For the rest of this chapter, we will develop the theory for thin wings of large and moderate aspect ratios.

Note that the base line can be curved. For example, it can be an arc of a circle in the (y, z) plane, centered on the z-axis and tangent to the y-axis at the origin. If a wing geometry is drawn on such a cylindrical surface, an infinitely thin wing will be defined and the wing will have curved up tips, but as such will not disturb the incoming flow.

Thickness, camber and incidence can be added to the thin wing geometry as defined above, and the same restrictions for thin airfoils apply, i.e.

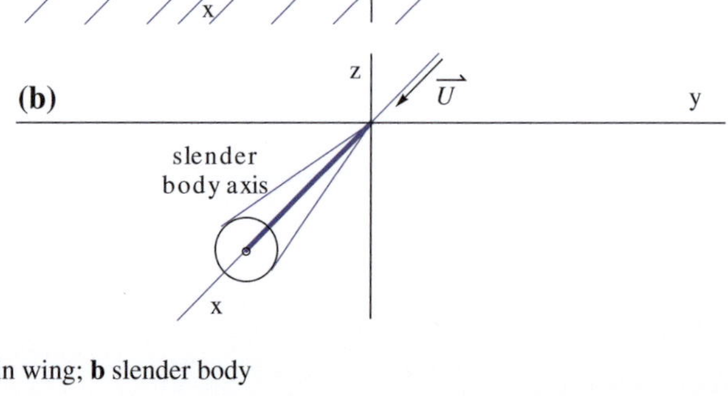

Fig. 6.5 **a** Thin wing; **b** slender body

$$\frac{e}{c}, \left|\frac{d}{c}\right|, |\alpha| \ll 1 \tag{6.4}$$

The addition of thickness, camber or incidence is expected to create a small perturbation, since in the limit as $\frac{e}{c}, \left|\frac{d}{c}\right|, |\alpha| \to 0$ the perturbation must vanish.

6.3 Flow Past Thin Wings

The above assumptions result in the existence of a perturbation potential $\phi(x, y, z)$ that governs the disturbance flow field $(u, v, w) = \nabla\phi$ and verifies

$$\left(1 - M_0^2\right)\frac{\partial^2\phi}{\partial x^2} + \frac{\partial^2\phi}{\partial y^2} + \frac{\partial^2\phi}{\partial z^2} = 0 \tag{6.5}$$

The full velocity is $\mathbf{V} = U\mathbf{i} + \nabla\phi$. This linear partial differential equation is not valid near Mach one, as was seen in 2-D. Unless otherwise stated, we will consider planar wings, i.e. wings whose base surface is in the (x, y) plane. Such a wing can be described by the equation $z = f^{\pm}(x, y)$ where the plus sign corresponds to the upper surface and the minus sign to the lower surface. $f^{\pm}$ is small and of order e/c, d/c. Setting the wing at small incidence α amounts to adding to $f^{\pm}$ the shearing term $\alpha(c_r - x)$, as was done in 2-D.

The tangency condition requires that the flow be tangent to the thin wing, $\mathbf{V}.\mathbf{n}|_{wing} = 0$. A vector, normal to the wing, has components

$$\nabla\left(f^{\pm}(x, y) + \alpha(c_r - x) - z\right) = \left(\frac{\partial f^{\pm}}{\partial x} - \alpha, \frac{\partial f^{\pm}}{\partial y}, -1\right) \tag{6.6}$$

Hence the tangency condition reads

$$\left\{(U + u)\left(\frac{\partial f^{\pm}}{\partial x} - \alpha\right) + v\frac{\partial f^{\pm}}{\partial y} - w\right\}_{wing} = 0 \tag{6.7}$$

The transfer of the tangency condition from the actual wing surface to the nearby base surface yields, to first-order accuracy, the following result

$$w^{\pm}(x, y, 0^{\pm}) = U\left(\frac{\partial f^{\pm}}{\partial x} - \alpha\right) \tag{6.8}$$

The Bernoulli equation for a small perturbation yields the same result for pressure as was derived in 2-D, i.e. $p = p_\infty - \rho_\infty U u$, hence the pressure coefficient is

$$C_p = -2\frac{u}{U} \tag{6.9}$$

6.4 Fundamental Feature of the Flow Past Finite Wings: The Vortex Sheet

The small disturbance potential equation, governing the flow past thin wings and slender bodies, admits jump conditions which represent discontinuities in the flow. Shock waves in linearized supersonic flow are such discontinuities and correspond to jump of velocity, density and pressure across characteristic surfaces. In incompressible or low speed flow, shock waves do not occur. However, at all flow regimes, discontinuities of a different nature are admissible that are called vortex sheets. A vortex sheet occurs at the sharp trailing edge of a wing and trails behind it to downstream infinity. To simplify the mathematical treatment, the vortex sheet rolling up at the edges into tip vortices is neglected. This will not affect the vorticity content of the vortex sheet, but only modify slightly its distribution in space. It has been found that this effect is of second or higher order, Fig. 6.6.

The study of the jump conditions associated with the perturbation potential equation provides insight into the type of discontinuities that can be present with this flow model. First, the second order PDE is transformed into an equivalent first order system in (u, v, w) as

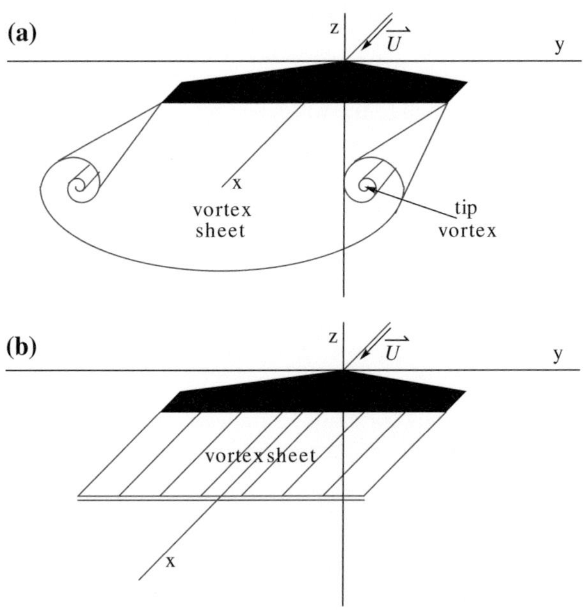

Fig. 6.6 Vortex sheet behind a finite wing: **a** with tip roll up; **b** simplified for mathematical treatment

$$\begin{cases} \left(1 - M_0^2\right) \frac{\partial u}{\partial x} + \frac{\partial v}{\partial y} + \frac{\partial w}{\partial z} = 0 \\ \frac{\partial w}{\partial y} - \frac{\partial v}{\partial z} = 0 \\ -\frac{\partial w}{\partial x} + \frac{\partial u}{\partial z} = 0 \end{cases} \tag{6.10}$$

Here we have used the governing equation and two components of the irrotationality condition. The third component is a consequence of the other two. The jump conditions for this system read

$$\begin{cases} \left(1 - M_0^2\right) < u > n_x + < v > n_y + < w > n_z = 0 \\ < w > n_y - < v > n_z = 0 \\ - < w > n_x + < u > n_z = 0 \end{cases} \tag{6.11}$$

This is a homogeneous system for the components (n_x, n_y, n_z) of the normal vector to the jump surface. For a non-trivial solution, the determinant must vanish, i.e.

$$\left\{\left(1 - M_0^2\right) < u >^2 + < v >^2 + < w >^2\right\} < w >= 0 \tag{6.12}$$

Two cases are possible:

(i) $< w > \neq 0$. The equation reduces to $\left(1 - M_0^2\right) < u >^2 + < v >^2 + < w >^2 = 0$. A non trivial solution can only be found if $1 - M_0^2 < 0$, that is if the flow is supersonic. The surfaces that admit this jump condition are shock waves that coincide with characteristic surfaces of the PDE. Note that Prandtl-Meyer expansions will be represented by "expansion shocks" in this linear theory. The flow crosses the shock surface which is not a stream surface since $< w > \neq 0$. In general $< v > \neq 0$ (except in a plane of symmetry). The component u must have a jump for a non-trivial solution, i.e. $< u > \neq 0$, which indicates that the pressure is discontinuous.

(ii) $< w >= 0$. The jump condition is satisfied. In general $< v > \neq 0$ (except in a plane of symmetry). The fluid is tangent to the surface of discontinuity since $< w >= 0$. It is called a vortex sheet. The vortex sheet is a stream surface, not a material surface, therefore it cannot withstand a pressure difference across it. This implies that $< u >= 0$. In consequence, the only velocity component that has a jump at a vortex sheet is the v-component.

6.5 Two-Dimensional and Three-Dimensional Vorticity Representations

The vorticity distributions in 2-D and 3-D have similarities and differences that can help explain the flow features and make connections between the two flow situations. First we look at the vortex filaments in a 2-D flow past an infinite wing and of the simplified 3-D model for a finite thin wing. In 2-D the vortex filaments are

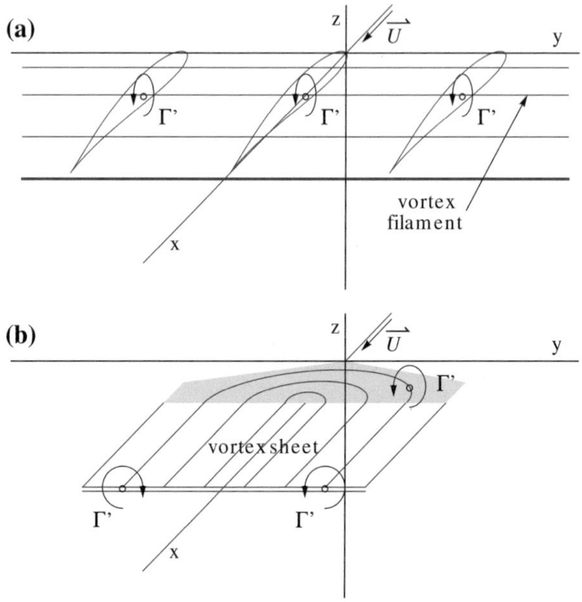

Fig. 6.7 Vortex filaments: **a** straight filaments in 2-D; **b** horseshoe filaments in 3-D

perpendicular to the plane of the profile and they carry a constant vorticity $\Gamma'(x)$. In 3-D, the vortex filaments have the shape of a horseshoe: they come from infinity downstream, enter the wing at the trailing edge, bend into the wing and exit at the trailing edge to trail to downstream infinity. The vorticity Γ' is constant along the filament. The part of the vortex filament inside the wing (or inside the profile) corresponds to the *bound vorticity*. Outside the wing, the filaments are parallel to the x-axis and the vorticity they carry is called *trailed vorticity*, all the way to the plane at downstream infinity called the Trefftz plane, Fig. 6.7.

The bound vorticity inside a profile and the trailed vorticity in the vortex sheet give rise to a discontinuity in the velocity field. Consider for example a thin cambered plate in 2-D and the trace of the vortex sheet in the Trefftz plane in 3-D, Fig. 6.8. In 2-D, the vorticity along the x-axis is responsible for the jump in u across it, with $u(x, 0^+) = \Gamma'(x)/2$ and $u(x, 0^-) = -\Gamma'(x)/2$. In the Trefftz plane, the situation is identical to that inside the profile in 2-D, with an infinite straight filament responsible for the jump in v across the vortex sheet, where $v(\infty, y, 0^+) = \Gamma'(y)/2$ and $v(\infty, y, 0^-) = -\Gamma'(y)/2$.

Note that, in general, $\Gamma'(\pm b/2) = \mp\infty$, that is the flow at the edge of the vortex sheet behaves as at the leading edge of a thin plate. The velocity components v and w are infinite. Note also that the flow satisfies a Kutta-Joukowski condition at the trailing edge in 2-D and 3-D.

The circulation is the integral of the bound vorticity inside a profile according to $d\Gamma = \Gamma' dx$. At the trailing edge the value $\Gamma(c)$ obtained corresponds to the total vorticity inside that profile. Downstream of the trailing edge the circulation is a

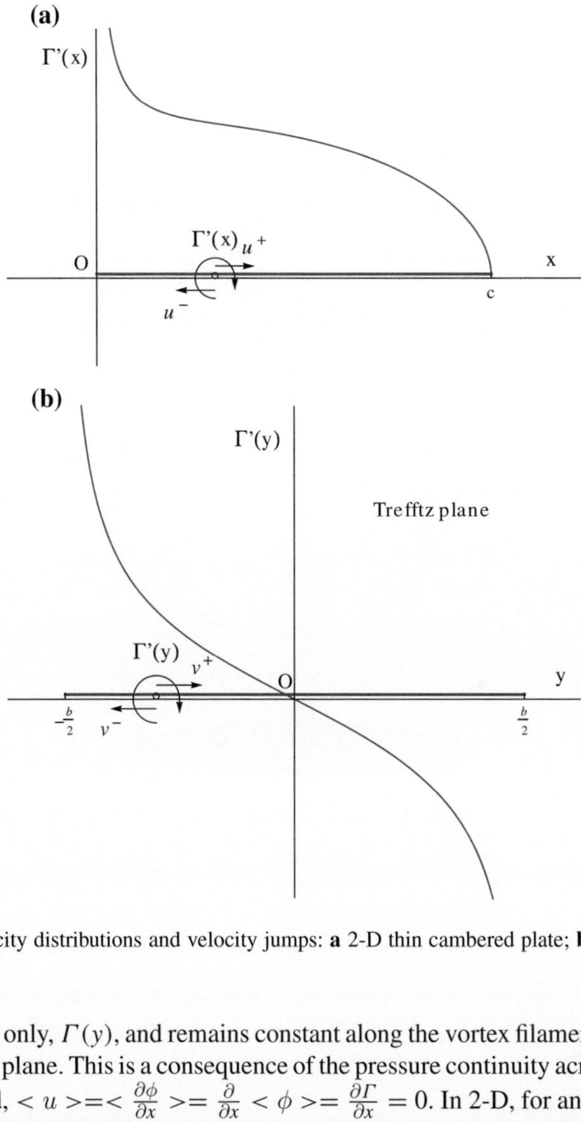

Fig. 6.8 Vorticity distributions and velocity jumps: **a** 2-D thin cambered plate; **b** vortex sheet in Trefftz plane

function of y only, $\Gamma(y)$, and remains constant along the vortex filaments all the way to the Trefftz plane. This is a consequence of the pressure continuity across the vortex sheet. Indeed, $< u >=< \frac{\partial \phi}{\partial x} >= \frac{\partial}{\partial x} < \phi >= \frac{\partial \Gamma}{\partial x} = 0$. In 2-D, for an infinite wing, the circulation at the trailing edge, $\Gamma(y) \equiv \Gamma(c)$ is the same for all profiles, hence the velocity component v is continuous across the $z = 0$ plane downstream of the trailing edge. There is no trailed vorticity and no vortex sheet. Two such circulation distributions are shown in Fig. 6.9.

The velocity components at the trailing edge, above and below the vortex sheet of a finite wing, are depicted in Fig. 6.10. The components are continuous across the vortex sheet, except for the y-component. The same result holds at any point on the vortex sheet.

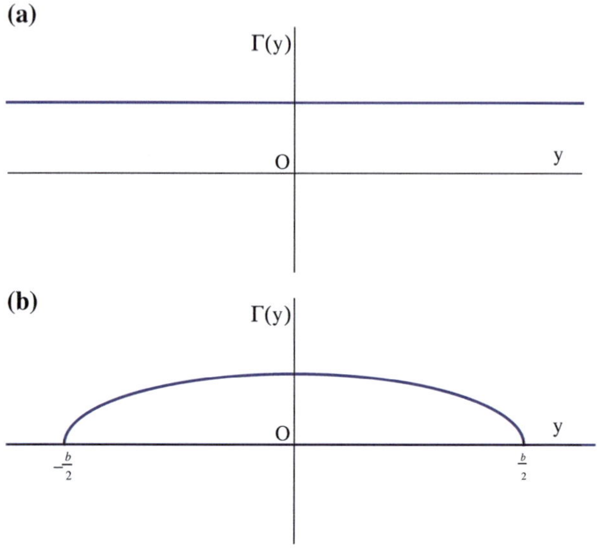

Fig. 6.9 Circulation distribution behind a wing: **a** infinite wing; **b** finite wing

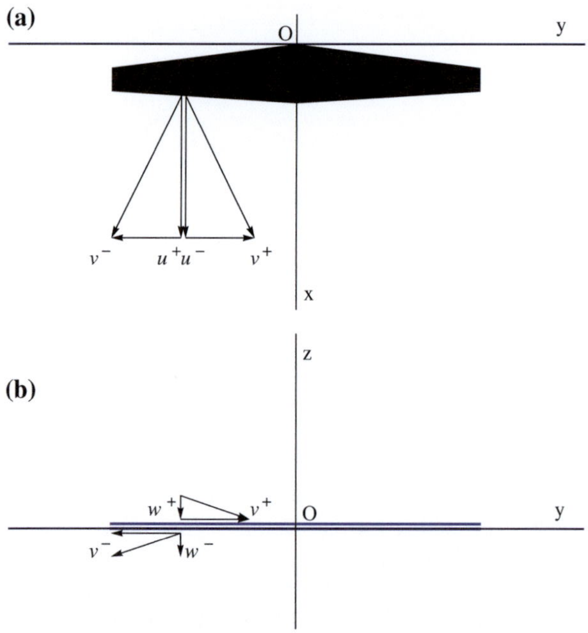

Fig. 6.10 Velocity vector jump at trailing edge: **a** in wing plane; **b** cross flow plane

6.6 Forces

The momentum theorem applied to a large control volume surrounding the wing and bounded by a surface Σ, yields the resulting force

$$\mathbf{F} = -\int_{\Sigma} (\rho \mathbf{V}(\mathbf{V}.\mathbf{n}) + p\mathbf{n})\, d\Sigma \tag{6.13}$$

When the flow on the wing does not present singularities, such as a leading edge singularity, thanks to an adapted leading edge, the surface Σ can be shrunk to the surface of the wing and the force reduces to

$$\mathbf{F} = -\int_{wing} p\mathbf{n}dA \tag{6.14}$$

However, the flow is in general singular along the leading edge of the wing, even in supersonic flow if the leading edge sweep is large enough, and the integration of pressure will miss the contribution due to that singularity, the suction force. Hence it is best to integrate on a surface that stands away from the wing. Typically one uses a cylinder of large radius R, denoted Σ_R, and two planes perpendicular to the x-axis to close the control volume, one far upstream where the flow is undisturbed, Σ_0, and one far downstream treated as the Trefftz plane, Σ_T, Fig. 6.11. The control volume surface is $\Sigma = \Sigma_R \cup \Sigma_0 \cup \Sigma_T$. On Σ_0, the flow is uniform and $\mathbf{V} = \{U, 0, 0\}$. Perturbations may be present on Σ_R due to shock waves crossing the boundary in supersonic flow regime and u, v, w have jumps along the shock trace. In Σ_T, far downstream, the flow will be perturbed in the vicinity of the vortex sheet. The pressure has returned to p_∞, i.e. $u = 0$, but v, w are different from zero, with v being discontinuous at the sheet and v, w tending to zero far from the x-axis.

The expression for $\mathbf{F}$ can be simplified by introducing a small perturbation assumption and expanding the dependent variables ρ, $\mathbf{V}$ and p with respect to a small parameter ϵ, where ϵ can represent relative thickness, camber or incidence, as

$$\begin{cases} \rho = \rho_\infty + \epsilon\rho_1(x, y, z) + \epsilon^2\rho_2(x, y, z) + \cdots \\ \mathbf{V} = U\mathbf{i} + \epsilon\mathbf{V}_1 + \epsilon^2\mathbf{V}_2 + \cdots = \\ \{U + \epsilon u_1(x, y, z) + \epsilon^2 u_2(x, y, z) + \cdots ; \\ \epsilon v_1(x, y, z) + \epsilon^2 v_2(x, y, z) + \cdots ; \\ \epsilon w_1(x, y, z) + \epsilon^2 w_2(x, y, z) + \cdots \} \\ p = p_\infty + \rho_\infty U^2 \left(\epsilon p_1(x, y, z) + \epsilon^2 p_2(x, y, z) + \cdots\right) \end{cases} \tag{6.15}$$

It has been established by Viviand [1] that, to first order, the following relations hold

$$\begin{cases} \rho_1 = -M_\infty^2 u_1 \\ \mathbf{V}_1 = \nabla\phi_1, \quad \mathbf{V}_2 = \nabla\phi_2 \\ p_1 = -u_1 \end{cases} \tag{6.16}$$

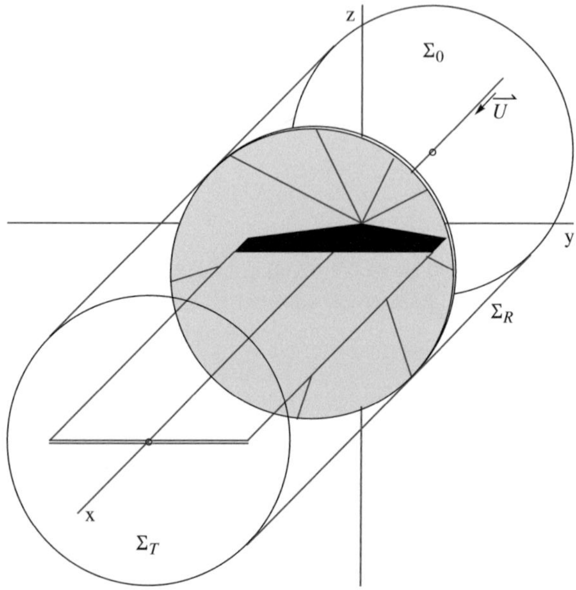

Fig. 6.11 Control volume for the evaluation of forces. The shock surface is represented in *grey*

which indicate that the flow remains irrotational at the first two orders, even in presence of shock waves (assumed to be weak). Rotation and entropy appear at third order. To find the forces, the expansions need to be carried to second order, but the result can be expressed in terms of the first order perturbations only.

6.6.1 Lift

Viviand showed, after lengthy calculations, that the lift coefficient is given by

$$A_{ref}C_L = \frac{2}{U}\int_{-\frac{b}{2}}^{\frac{b}{2}} \left(\phi^+ - \phi^-\right) dy = \frac{2}{U}\int_{-\frac{b}{2}}^{\frac{b}{2}} \Gamma(y)dy \tag{6.17}$$

Here we substituted ϕ for $\epsilon\phi_1$. This result is consistent with the Kutta-Joukowski lift theorem. Indeed, if one considers a section dy of the wing with bound vorticity $\Gamma(y)$, the element of lift is

$$dL = \rho_\infty U\Gamma(y)dy \tag{6.18}$$

and is perpendicular to the incoming flow. The total lift will be

$$L = \rho_\infty U\int_{-\frac{b}{2}}^{\frac{b}{2}} \Gamma(y)dy \tag{6.19}$$

and the lift coefficient based on the reference area A_{ref} given by the above formula.

6.6.2 Drag

The drag is given by Viviand as

$$A_{ref}C_D = \int_{\Sigma_T}\left[\left(\frac{v}{U}\right)^2 + \left(\frac{w}{U}\right)^2\right]d\Sigma_T - 2\int_{\Sigma_R}\frac{u}{U}\left(\frac{v}{U}n_y + \frac{w}{U}n_z\right)d\Sigma_R$$

$$= A_{ref}C_{Di} + A_{ref}C_{Dw} \tag{6.20}$$

where $\mathbf{n} = \{0; n_y; n_z\}$ is the normal unit vector to Σ_R. The two terms are interpreted, the former as the *induced drag* due to the vortex sheet, the latter as the *wave drag* due to shock waves in supersonic flow. The induced drag refers to the drag induced by lift or vortex drag. If the circulation is zero, there will be no vortex sheet and the first term will vanish.

It is also possible to derive the formula for induced drag from the steady energy equation for small disturbance flow. Consider the small stream tube entering through the plane Σ_0 and exiting at the Trefftz plane, Fig. 6.12. The steady energy equation applied to all such small stream tubes yields

$$\int_{\Sigma_0}\left(\hat{u} + \frac{p}{\rho} + \frac{V^2}{2} + gz\right)d\dot{m} = \int_{\Sigma_T}\left(\hat{u} + \frac{p}{\rho} + \frac{V^2}{2} + gz\right)d\dot{m} - \dot{W}_s \tag{6.21}$$

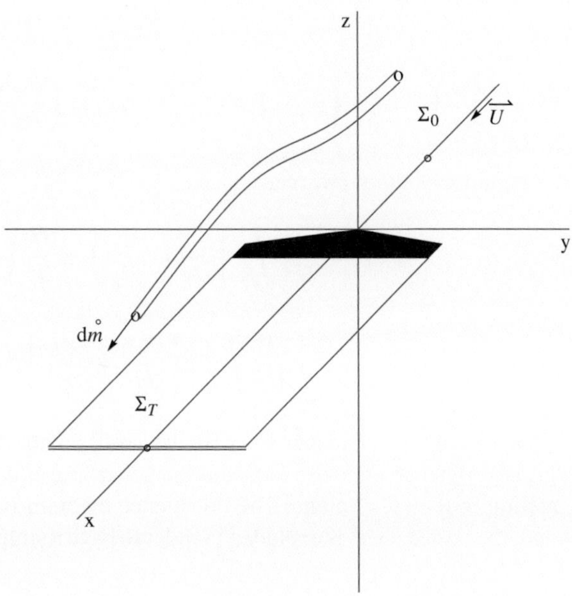

Fig. 6.12 Elementary stream tube used in application of the steady energy equation

The internal energy term $\hat{u} = C_v T$, and the pressure/density term cancel out as the flow returns to its free stream temperature, pressure and density. The gravity term is neglected. $\dot{W}_s$ represents the shaft work done to the fluid. Upon simplification, this reads

$$\int_{\Sigma_0} \left(\frac{U^2}{2}\right) d\dot{m} = \int_{\Sigma_T} \left(\frac{U^2 + v^2 + w^2}{2}\right) d\dot{m} - \dot{W}_s, \quad \Rightarrow$$

$$\dot{W}_s = \rho_\infty U \int_{\Sigma_T} \left(\frac{v^2 + w^2}{2}\right) d\Sigma_T \geq 0 \tag{6.22}$$

One can see that the kinetic energy leaving the Trefftz plane is larger than the kinetic energy entering through the Σ_0 plane. Work has been done to the fluid. The work is due to the thrust that is needed to balance the induced drag to maintain the steady motion of the wing in the frame where the fluid is at rest in front of the wing. The induced drag is present, even in inviscid flow. It can be written

$$\dot{W}_s = -\mathbf{T}.\mathbf{U} = \mathbf{D}.\mathbf{U} = \frac{1}{2}\rho_\infty U^3 A_{ref} C_{Di} \tag{6.23}$$

Upon substitution of the definition of the induced drag coefficient, one finds

$$A_{ref} C_{Di} = \int_{\Sigma_T} \left[\left(\frac{v}{U}\right)^2 + \left(\frac{w}{U}\right)^2\right] d\Sigma_T \tag{6.24}$$

The surface integral can be transformed into a contour integral along the trace of the vortex sheet. Consider the identities

$$v^2 = \left(\frac{\partial \phi}{\partial y}\right)^2 = \frac{\partial}{\partial y}\left(\phi \frac{\partial \phi}{\partial y}\right) - \phi \frac{\partial^2 \phi}{\partial y^2}, \quad w^2 = \left(\frac{\partial \phi}{\partial z}\right)^2 = \frac{\partial}{\partial z}\left(\phi \frac{\partial \phi}{\partial z}\right) - \phi \frac{\partial^2 \phi}{\partial z^2} \tag{6.25}$$

The induced drag now reads

$$A_{ref} C_{Di} = \frac{1}{U^2} \int_{\Sigma_T} \left[\frac{\partial}{\partial y}\left(\phi \frac{\partial \phi}{\partial y}\right) + \frac{\partial}{\partial z}\left(\phi \frac{\partial \phi}{\partial z}\right)\right] dydz$$

$$- \frac{1}{U^2} \int_{\Sigma_T} \phi \left[\frac{\partial^2 \phi}{\partial y^2} + \frac{\partial^2 \phi}{\partial z^2}\right] dydz \tag{6.26}$$

The last integral vanishes since, in the Trefftz plane, the flow is two-dimensional with $u = 0$ hence $\partial^2 \phi / \partial x^2 = 0$ and the governing equation reduces to the Laplace equation in the (y, z)-plane. The divergence theorem is applied to the first integral where the vortex sheet is excluded in order to have a simply connected domain with a

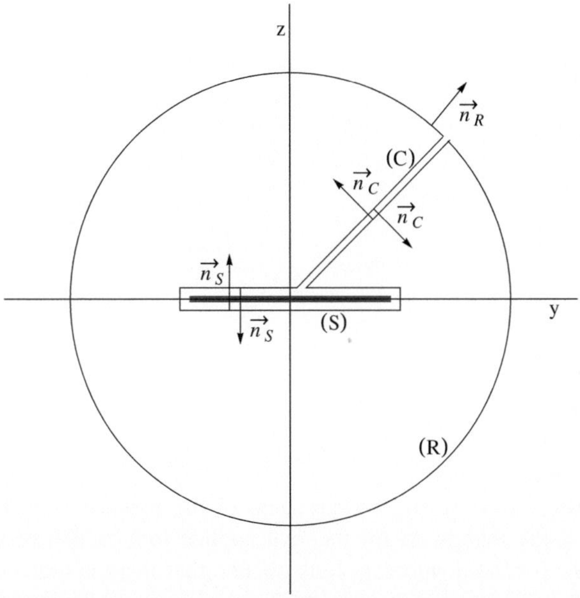

Fig. 6.13 Control volume for application of the divergence theorem to the drag integral

contour made of a large circle (R) of radius R in the far field, a contour (S) surrounding the sheet and a cut (C) to connect them, as shown in Fig. 6.13.

The integral now reads

$$A_{ref}C_{Di} = \frac{1}{U^2}\int_{R+S+C} \phi \left\{\frac{\partial \phi}{\partial y}; \frac{\partial \phi}{\partial z}\right\}.\mathbf{n}dl \tag{6.27}$$

The contribution on (R) goes to zero as $R \to \infty$ since ϕ and $\nabla\phi$ vanish far away from the vortex sheet. The integral on (C) is zero from continuity of the integrand and change of sign of the normal vectors. There remains

$$A_{ref}C_{Di} = \frac{1}{U^2}\int_{S} \phi \left\{\frac{\partial \phi}{\partial y}; \frac{\partial \phi}{\partial z}\right\}.\mathbf{n}dl = \frac{1}{U^2}\int_{-\frac{b}{2}}^{\frac{b}{2}} \left(\phi^-\frac{\partial \phi^-}{\partial z} - \phi^+\frac{\partial \phi^+}{\partial z}\right)dy \tag{6.28}$$

where we have accounted for $n_y = 0, n_z^+ = -1$ and $n_z^- = 1$, and replaced the contour (S) by the segment $[-\frac{b}{2}, \frac{b}{2}]$. We have seen that the w-component is continuous across the vortex sheet, and the jump of the potential is the circulation, i.e. $\Gamma(y) = \phi^+ - \phi^- =< \phi >$. The result follows

$$A_{ref}C_{Di} = -\frac{1}{U^2}\int_{-\frac{b}{2}}^{\frac{b}{2}} \Gamma(y)w_T(y)dy \tag{6.29}$$

where w_T stands for the vertical component of the induced velocity in the Trefftz plane. It will be seen that the vortex sheet trailing a thin wing with positive lift induces a negative component w above and below the sheet, called the *downwash*. The downwash is responsible for the induced drag. According to the 2-D result of thin airfoil theory in incompressible flow, the contribution to the downwash at a point y of the vortex sheet due to an infinite vortex line of intensity $d\Gamma = \Gamma' d\eta$ located at η is given by

$$dw_T = -\frac{1}{2\pi}\frac{\Gamma'(\eta)d\eta}{y-\eta} \tag{6.30}$$

and the resulting w_T reads

$$w_T(y) = -\frac{1}{2\pi}\int_{-\frac{b}{2}}^{\frac{b}{2}} \frac{\Gamma'(\eta)}{y-\eta}d\eta \tag{6.31}$$

In incompressible flow, one interpretation of the induced drag is obtained by revisiting the Kutta-Joukowski lift theorem applied to a section of the wing, but accounting now for the downwash. Lets assume that there is a downwash $w_w(y)$ induced by the vortex sheet in the neighborhood of the wing. The bound vortex sees an incoming flow that has been deflected down, as shown in Fig. 6.14.

The angle between the undisturbed velocity vector $\mathbf{U}$ and the velocity vector $\mathbf{q}$ seen by the profile is called *induced incidence*. The induced incidence $\alpha_i(y) = \tan^{-1}(w_w(y)/U) \simeq w_w(y)/U$ is negative, in general, for positive lift. The lift contribution is unchanged by the induced incidence since $\cos\alpha_i \simeq 1$. The induced element of drag is given by

$$dD_i = -dF\sin\alpha_i \simeq -\rho_\infty U\Gamma(y)\alpha_i dy = -\rho_\infty \Gamma(y) w_w(y)dy \tag{6.32}$$

The induced drag coefficient therefore becomes

$$A_{ref}C_{Di} = -\frac{2}{U^2}\int_{-\frac{b}{2}}^{\frac{b}{2}} \Gamma(y)w_w(y)dy \tag{6.33}$$

This result is consistent with that derived earlier, because at the wing, application of the Biot-Savart formula in incompressible flow shows that the induced velocity is

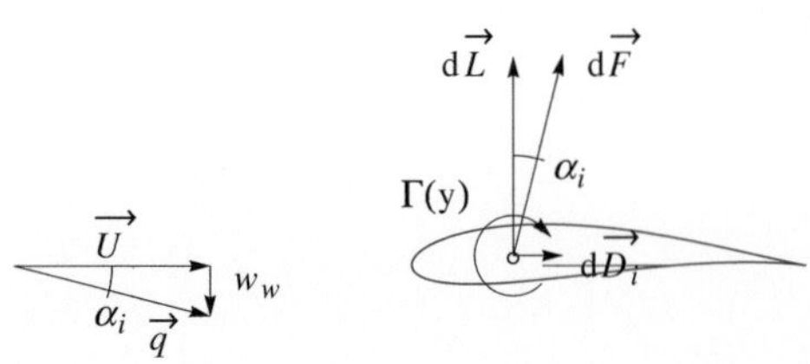

Fig. 6.14 Interpretation of induced drag in incompressible flow

half that in the Trefftz plane, since the contribution of a semi-infinite line vortex is half that of an infinite one, i.e.

$$w_w(y) = -\frac{1}{4\pi}\int_{-\frac{b}{2}}^{\frac{b}{2}} \frac{\Gamma'(\eta)}{y-\eta}d\eta = \frac{1}{2}w_T(y) \tag{6.34}$$

The formula for the induced drag can be manipulated upon substituting the downwash as

$$\begin{aligned} A_{ref}C_{Di} &= \frac{1}{2\pi U^2}\int_{-\frac{b}{2}}^{\frac{b}{2}}\int_{-\frac{b}{2}}^{\frac{b}{2}} \Gamma(y)\frac{\Gamma'(\eta)}{y-\eta}d\eta dy \\ &= \frac{1}{2\pi U^2}\int_{-\frac{b}{2}}^{\frac{b}{2}} \Gamma'(\eta)\left\{\int_{-\frac{b}{2}}^{\frac{b}{2}} \frac{\Gamma(y)}{y-\eta}dy\right\}d\eta \end{aligned} \tag{6.35}$$

Using integration by parts for the integral in the bracket, yields

$$A_{ref}C_{Di} = \frac{1}{2\pi U^2}\int_{-\frac{b}{2}}^{\frac{b}{2}} \Gamma'(\eta)\left\{\left[\Gamma(y)\ln|y-\eta|\right]_{-\frac{b}{2}}^{\frac{b}{2}} - \int_{-\frac{b}{2}}^{\frac{b}{2}} \Gamma'(y)\ln|y-\eta|dy\right\}d\eta \tag{6.36}$$

The integrated term in the bracket is zero since $\Gamma\left(\pm\frac{b}{2}\right) = 0$, thus we are left with a symmetrical expression for the induced drag that reads

$$A_{ref}C_{Di} = -\frac{1}{2\pi U^2}\int_{-\frac{b}{2}}^{\frac{b}{2}}\int_{-\frac{b}{2}}^{\frac{b}{2}} \Gamma'(y)\Gamma'(\eta)\ln|y-\eta|dyd\eta \tag{6.37}$$

Interestingly, in slender body theory, a similar formula can be derived for the wave drag. Indeed, away from the body, the disturbance potential behaves as if the body is a body of revolution and the leading term in the expansion of the potential solution corresponds to a distribution of sources along the axis that depend on $S'(x)$, where $S(x)$ represents the cross-section area of the body of length l. If furthermore the body satisfies $S'(0) = S'(l) = 0$, that is if the body is finite or connects to a cylindrical part, the wave drag reads

$$A_{ref}C_{Dw} = -\frac{1}{2\pi}\int_0^l\int_0^l S''(x)S''(\xi)\ln|x-\xi|dxd\xi \tag{6.38}$$

This analogy can be used to find the body of revolution with minimum wave drag in supersonic flow from the finite wing with minimum induced drag in incompressible flow.

6.7 Prandtl Lifting Line Theory (Incompressible Flow)

In incompressible flow, for large aspect ratio wings, i.e. $AR \geq 7$, Prandtl [2] imagined the following model for the flow, shown in Fig. 6.15.

The flow is described as a lifting line and a semi-infinite vortex sheet, made of vortex filaments which run from $x = -\infty$ to $x = 0$, turn abruptly along the y-axis and turn again to leave toward the Trefftz plane. The wing of Fig. 6.7b has been collapsed along the y-axis, between $-b/2$ and $b/2$. The circulation $\Gamma(y)$ fully describes the vorticity content of the flow, i.e. the bound vorticity inside the wing, as well as the trailed vorticity, $\Gamma'(y)$, of the vortex sheet. The circulation $\Gamma(y)$ is found by matching the local flow incidence with the profile characteristics at a given wing cross-section y, as if in isolation or part of an infinite wing. This is the so-called strip theory, which has proved useful for slender lifting elements such as large aspect ratio wings, helicopter and wind turbine blades. Mathematically, this corresponds to neglecting the derivatives along the span in comparison to the other derivatives, in other words, $\partial/\partial y \ll \partial/\partial x, \partial/\partial z$.

On a more historical note, Prandtl initially thought of representing the flow vorticity with a single horseshoe vortex of finite intensity corresponding approximately to the maximum circulation in the wing. However, he soon realized that the induced drag of a finite vortex is infinite because the azimuthal velocity component varies as $1/r$, where r is the distance from the vortex center, so that $v^2 + w^2 \propto 1/r^2$, which in the Trefftz plane integrates as $\ln r$, a diverging integral over the vortex, yielding an infinite drag.

Lets assume that the profiles that equip the wing are characterized by their 2-D lift coefficients $C_l(\alpha)$, known from thin airfoil theory, say $C_l(\alpha) = 2\pi(\alpha - \alpha_0)$, where $\alpha_0(y) = -2d(y)/c(y)$ represents the incidence of zero lift and $d(y)/c(y)$ is the relative camber. The setting angle or twist of the wing, $t(y)$, is an additional geometric parameter. Note, that, without restriction, one can set $t(0) = 0$ for a symmetric wing. Negative values of $t(y)$ or nose down toward the wing tips is called *washout*, whereas positive values or nose up is called *washin*. Figure 6.16 depicts the local working condition of the profile in the $y = const.$ plane.

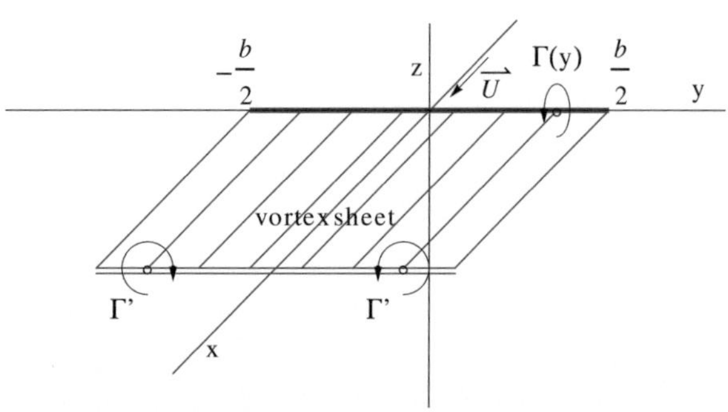

Fig. 6.15 Prandtl model for large aspect ratio wing

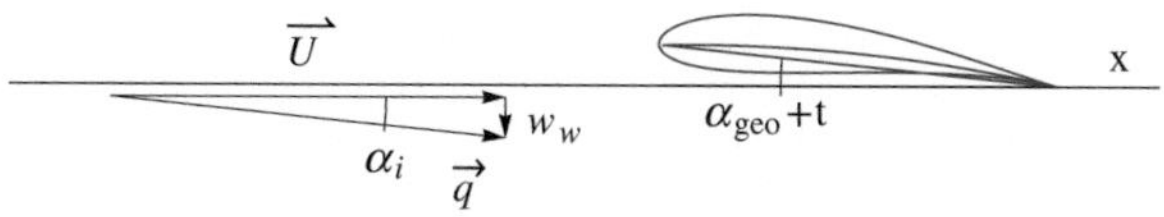

Fig. 6.16 Matching of local incidence and local lift coefficient

The profile setting angle with respect to the x-axis is given by $\alpha_{geo} + t(y)$, where α_{geo} corresponds to the geometric incidence given to the wing. The effective incidence of the profile is $\alpha_{eff} = \alpha_{geo} + t + \alpha_i$, where we have accounted for the induced incidence. The local lift coefficient is therefore

$$C_l(y) = 2\pi\left(\alpha_{eff}(y) + 2\frac{d(y)}{c(y)}\right) = 2\pi\left(\alpha_{geo} + t(y) + 2\frac{d(y)}{c(y)} + \alpha_i(y)\right)$$

$$= 2\pi\left(\alpha_{geo} - \alpha_0(y) + \alpha_i(y)\right) \tag{6.39}$$

For the wing with twist, $\alpha_0 = -(t + 2d/c)$. The relation between the circulation and the local lift coefficient, as seen before, is

$$C_l(y) = \frac{2\Gamma(y)}{Uc(y)} \tag{6.40}$$

Here, we have replaced q with U since $\alpha_i = w_w/U$ is small. Substitution of the local lift coefficient in terms of the circulation and of the formula for the downwash at the lifting line, yields the integro-differential equation of Prandtl

$$\Gamma(y) = \pi U c(y)\left\{\alpha_{geo} - \alpha_0(y) - \frac{1}{4\pi U}\int_{-\frac{b}{2}}^{\frac{b}{2}} \frac{\Gamma'(\eta)}{y-\eta}d\eta\right\} \tag{6.41}$$

Two types of problems are commonly considered: the analysis and the design problems. In the analysis problem, the wing geometry is given, that is $c(y)$ and $\alpha_0(y)$ are known. The objective is to find the circulation Γ, the lift and the induced drag for different values of the geometric incidence α_{geo}. In general, this problem does not have an analytical solution and can only be solved numerically. This is particularly true if the planform has kinks as in the case of the trapezoidal wing. The numerical approach will be described later.

As a designer, one can study the properties of a wing having a given loading, $\Gamma(y)$, and relate the loading with the induced drag in order to obtain the minimum drag for a given lift. It will remain to find how to build such a wing. In this inverse approach, as was done in thin airfoil theory, the circulation is represented as a Fourier series of sines, after a change of variables from y to t

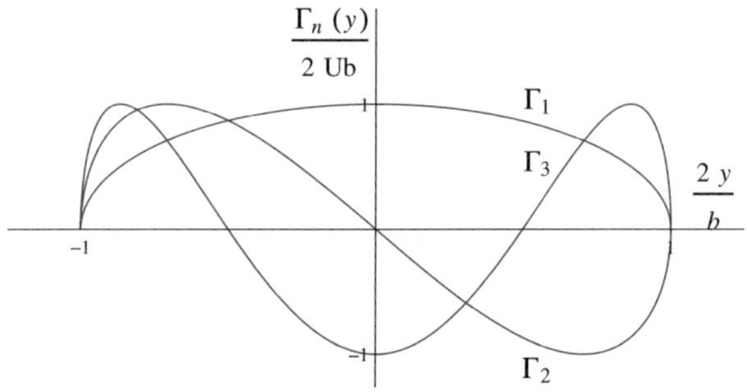

Fig. 6.17 First three modes in Fourier expansion of the circulation

$$\begin{cases} \Gamma(t) = 2Ub\sum_{n=1}^{\infty} A_n \sin nt \\ y(t) = -\frac{b}{2}\cos t \end{cases} \tag{6.42}$$

Note that, unlike in thin airfoil theory, the series contains only regular terms that satisfy the boundary conditions at the wing tips $\Gamma(0) = \Gamma(\pi) = 0$. Note also that the development is for Γ not Γ'. The first three modes $\Gamma_n(y)$ are shown in Fig. 6.17.

It can be seen that the even modes are antisymmetrical and will induce a rolling motion to the wing. In symmetrical flight, only the odd modes are present in the solution.

Substitution of the Fourier series in the formula for the lift coefficient along with the change of variable yields

$$A_{ref}C_L = 4b\sum_{n=1}^{\infty} A_n \int_0^{\pi} \sin nt \frac{b}{2} \sin t dt = 2b^2 A_1 \int_0^{\pi} \sin^2 t dt = \pi b^2 A_1 \tag{6.43}$$

where we have exchanged integration and summation, and used the orthogonality of the Fourier modes. If we choose as reference area the wing area, i.e. $A_{ref} = S$, then the result is

$$C_L = \pi AR A_1 \tag{6.44}$$

The lift depends on the first mode only.

The induced drag is now considered. First, one evaluates the downwash in the Trefftz plane by substituting the Fourier expansion in the formula for w_T to get with $\eta = -\frac{b}{2}\cos\theta$

$$w_T[y(t)] = -\frac{Ub}{\pi}\sum_{n=1}^{\infty} nA_n \int_0^{\pi} \frac{\cos n\theta}{\frac{b}{2}(\cos\theta - \cos t)} d\theta = -2U\sum_{n=1}^{\infty} nA_n \frac{\sin nt}{\sin t} \tag{6.45}$$

where again, integration and summation are switched, and the result, taken as principal value integral, can be found in books of integrals.

The expression for w_T can now be inserted in the formula for the induced drag as

$$A_{ref}C_{Di} = 4b \sum_{m=1}^{\infty} A_m \sum_{n=1}^{\infty} nA_n \int_0^{\pi} \{\sin mt\} \left\{ \frac{\sin nt}{\sin t} \right\} \frac{b}{2} \sin t \, dt$$

$$= 2b^2 \sum_{n=1}^{\infty} nA_n^2 \int_0^{\pi} \sin^2 nt \, dt = \pi b^2 \sum_{n=1}^{\infty} nA_n^2 \tag{6.46}$$

where we have use orthogonality of the Fourier modes, term by term, from the product of the two series. Let $A_{ref} = S$, then the result reads

$$C_{Di} = \pi AR \sum_{n=1}^{\infty} nA_n^2 \tag{6.47}$$

This result proves that the induced drag is zero only if all the coefficients are zero, that is if $\Gamma(y) = 0$. This is the case with a flat plate wing at zero incidence. For a given, non-zero lift, $A_1 \neq 0$, the minimum drag is obtained when $A_2 = A_3 = \cdots = A_n = \cdots = 0,\ n \geq 2$. Such an ideal wing has a circulation distribution given by

$$\Gamma[y(t)] = 2UbA_1 \sin t = 2UbA_1 \sqrt{1 - \left(\frac{2y}{b}\right)^2} \tag{6.48}$$

The ideal or optimum wing has an elliptic distribution of circulation or elliptic loading.

The corresponding downwash is constant

$$w_T(y) = 2w_w(y) = -2UA_1 = const. \tag{6.49}$$

See Fig. 6.18. The induced incidence is also constant and equal to

$$\alpha_i(y) = -A_1 = const. \tag{6.50}$$

In summary, the ideal wing has $C_L = \pi AR\, A_1$, $C_{Di} = \pi AR\, A_1^2$. Elimination of A_1 provides an important relationship between lift and induced drag coefficients as

$$C_{Di} = \frac{C_L^2}{\pi AR} \tag{6.51}$$

which demonstrates the benefit of a large aspect ratio wing over a low aspect ratio one, having the same area S: for the same lift, the induced drag is inversely proportional to the aspect ratio.

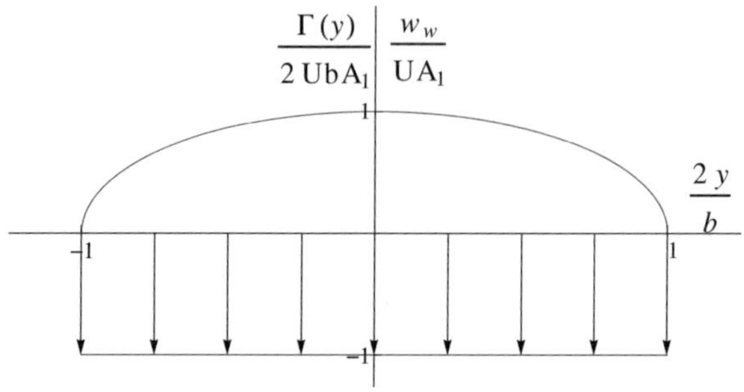

Fig. 6.18 Ideal wing loading and downwash

6.7.1 Induced Drag for Non-ideal Wings

For non-ideal wings, let δ be the relative drag increase compared to the ideal wing, i.e.

$$\delta = \sum_{n=2}^{\infty} n \left(\frac{A_n}{A_1}\right)^2 \tag{6.52}$$

With this notation, the induced drag reads

$$C_{Di} = \pi AR A_1^2 (1+\delta) = (C_{Di})_{elliptic} (1+\delta) = \frac{C_L^2}{\pi AR}(1+\delta) \tag{6.53}$$

The wing efficiency parameter is defined as $e = 1/(1+\delta)$ so that, in terms of e the induced drag becomes

$$C_{Di} = \frac{C_L^2}{\pi e AR} \tag{6.54}$$

6.7.2 Wing Lift Curve and Drag Polar

The relation $C_L(\alpha)$ is now derived for a non-ideal wing. Let α be the geometric incidence for short. Starting from the formula for the lift coefficient in which the circulation has been replaced by the right-hand-side of Prandtl integro-differential equation, one obtains

$$A_{ref}C_L = \frac{2}{U}\int_{-\frac{b}{2}}^{\frac{b}{2}} \Gamma(y)dy = 2\pi \int_{-\frac{b}{2}}^{\frac{b}{2}} \left\{\alpha - \alpha_0(y) + \frac{w_w(y)}{U}\right\} c(y)dy \tag{6.55}$$

where

$$\frac{w_w[y(t)]}{U} = -A_1\left(1 + \sum_{n=2}^{\infty} \frac{nA_n}{A_1}\frac{\sin nt}{\sin t}\right) \tag{6.56}$$

Now, define averaging quantities with the wing area S such as

$$\bar{\alpha}_0 = \frac{1}{S}\int_{-\frac{b}{2}}^{\frac{b}{2}} \alpha_0(y)c(y)dy, \quad \frac{1}{S}\int_{-\frac{b}{2}}^{\frac{b}{2}} \frac{w_w(y)}{U}c(y)dy = -A_1(1+\tau) \tag{6.57}$$

The value of τ ranges between 0.05 and 0.25 according to Glauert [3], for non-ideal wings. Often τ will be neglected. Upon elimination of A_1 this leads to

$$C_L = 2\pi\left\{\alpha - \bar{\alpha}_0 - A_1(1+\tau)\right\} = 2\pi\left\{\alpha - \bar{\alpha}_0 - \frac{C_L}{\pi AR}(1+\tau)\right\} \tag{6.58}$$

Finally, solving for C_L

$$C_L(\alpha) = \frac{2\pi}{1 + \frac{2(1+\tau)}{AR}}(\alpha - \bar{\alpha}_0) \tag{6.59}$$

This result is remarkable in that it gives the lift slope for a finite wing in terms of the aspect ratio

$$\frac{dC_L}{d\alpha} = \frac{2\pi}{1 + \frac{2(1+\tau)}{AR}} \simeq \frac{2\pi}{1 + \frac{2}{AR}} \tag{6.60}$$

In the limit of very large aspect ratio, the 2-D result of thin airfoil theory is recovered. Furthermore, $\bar{\alpha}_0$ can be interpreted as the incidence of zero lift for the finite wing. Note that, as was the case for thin airfoil theory, this linear model does not provide a maximum lift coefficient as the full potential equation would. The lift curve is a straight line, but good agreement with real flow lift coefficient will be obtained for small angles of incidence. If a viscous drag coefficient is added to the induced drag, using, say, a flat plate formula based on the wing Reynolds number and the wetted area, an approximation to the wing polar can be constructed as

$$C_D(\alpha) = C_{D0}(R_e) + \frac{C_L^2(\alpha)}{\pi e AR} \tag{6.61}$$

Assuming a constant Reynolds number, the polars are parabolas. The wing lift curves and polars are shown for different aspect ratios in Fig. 6.19.

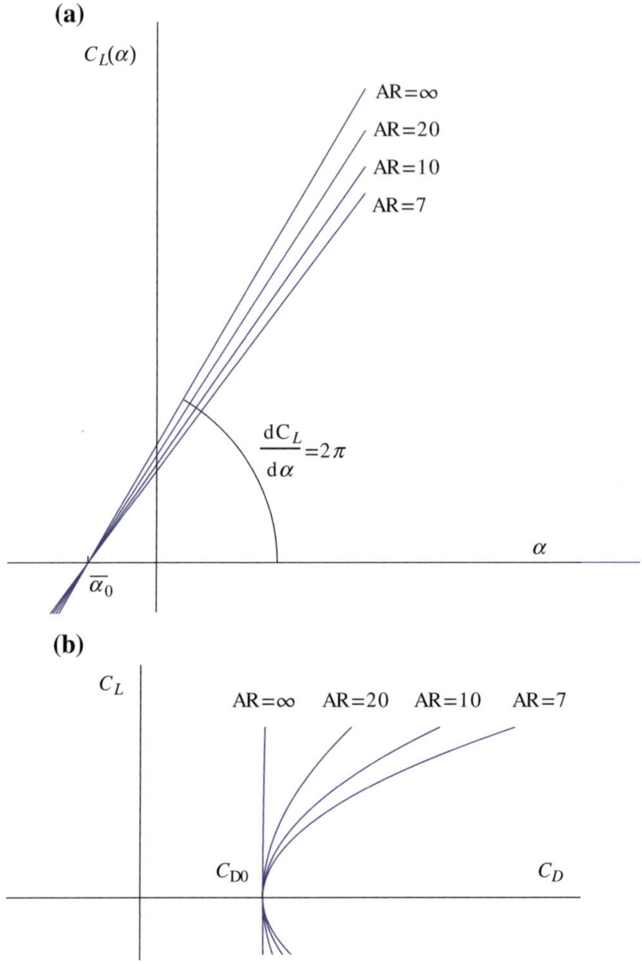

Fig. 6.19 Global coefficients: **a** lift coefficient; **b** wing polar

6.7.3 Design of an Ideal Wing

The ideal wing is a wing that has an elliptic circulation or loading.

$$\Gamma(y) = \Gamma_0\sqrt{1 - \left(\frac{2y}{b}\right)^2} \tag{6.62}$$

where Γ_0 is the root circulation to be selected. The question at hand is the following: How to design the wing so that the elliptic circulation is obtained at a given C_L? As we shall see, in inviscid flow, there is an infinite number of wings that will produce the target circulation. They differ by chord, camber and twist. Consider the Prandtl integro-differential equation, specialized to the case of ideal wing loading

$$\Gamma(y) = \pi U c(y) \{\alpha - \alpha_0(y) + \alpha_i\} \tag{6.63}$$

where α is the geometric incidence and α_i the constant induced incidence corresponding to elliptic loading $\Gamma(y) = \Gamma_0\sqrt{1-(2y/b)^2}$. One recalls that $\alpha_0(y) = -\left(t(y) + 2d(y)/c(y)\right)$.

The simplest design is one in which the twist is zero, $t(y) = 0$, and the relative camber is constant, hence $\alpha_0(y) = const.$ In this case, the bracket term is constant and the design is obtained with an elliptic chord distribution $c(y) = c_0\sqrt{1-(2y/b)^2}$. The root chord, aspect ratio and geometric incidence, can be found to be

$$c_0 = \frac{2\Gamma_0}{UC_L}, \quad AR = \frac{4b}{\pi c_0}, \quad \alpha = \alpha_0 + \left(1 + \frac{2}{AR}\right)\frac{C_L}{2\pi} \tag{6.64}$$

Here we have made use of the area of an ellipse, $S = \frac{\pi}{4} b c_0$. However, such a wing will develop an elliptic distribution of circulation at all incidences since it is easy to see that at a different incidence β

$$\Gamma[y(t)] = \pi U c_0 \sin t \left\{\beta - \alpha_0 - \sum_{n=1}^{\infty} nA_n \frac{\sin nt}{\sin t}\right\} = 2Ub\sum_{n=1}^{\infty} A_n \sin nt \tag{6.65}$$

where we have used the Fourier expansion of the circulation and the identity $\sin t = \sqrt{1-(2y/b)^2}$. The two Fourier series will be equal if only if

$$A_1 = \frac{2\pi(\beta - \alpha_0)}{\pi AR\left(1 + \frac{2}{AR}\right)}, \quad A_2 = A_3 = \cdots = A_n = 0, \; n \geq 2 \tag{6.66}$$

proving that the circulation is elliptic at the new lift coefficient

$$C_L(\beta) = \pi AR\, A_1 = \frac{2\pi}{\left(1 + \frac{2}{AR}\right)}(\beta - \alpha_0) \tag{6.67}$$

Such a planform corresponds to the WWII Spitfire airplane unswept wing shown in Fig. 6.20.

Another design, which is attractive for ease of construction of small remote control airplanes, consists of a constant chord wing. The integro-differential equation of Prandtl now reads

$$\Gamma(y) = \pi U c \{\alpha - \alpha_0(y) + \alpha_i\} \tag{6.68}$$

where c is the wing chord and α_i the constant induced incidence corresponding to elliptic loading. Solving for $\alpha_0(y)$ and letting $AR = b/c$ yields

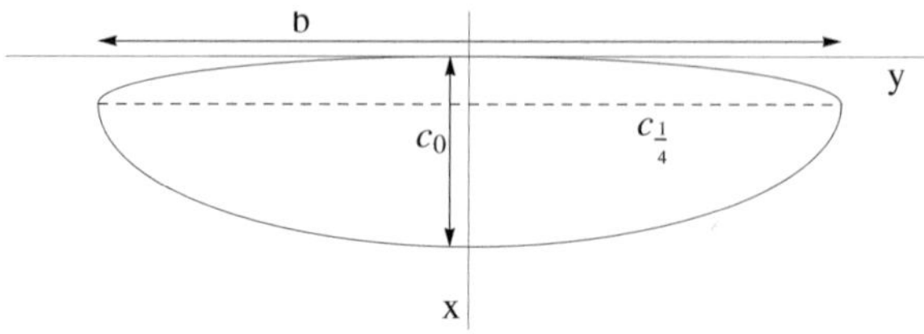

Fig. 6.20 Elliptic wing with unswept quarter-chord (*top*); Supermarine Spitfire (*bottom*) (from https://en.wikipedia.org/wiki/File:Spitfire_mk2a_p7350_arp.jpg Author: Arpingstone)

$$\alpha_0(y) = \alpha - \frac{C_L}{\pi AR}\left\{1 + \frac{2AR}{\pi}\sqrt{1 - \left(\frac{2y}{b}\right)^2}\right\} \tag{6.69}$$

For simplicity, assume that the relative camber is constant and that the twist is a function of y. One finds

$$t(y) = \frac{C_L}{\pi AR}\left\{1 + \frac{2AR}{\pi}\sqrt{1 - \left(\frac{2y}{b}\right)^2}\right\} - \alpha - \frac{2d}{c} \tag{6.70}$$

But, since without restriction, one can choose $t(0) = 0$, this determines the angle of incidence at design C_L

$$\alpha = \frac{C_L}{\pi AR}\left\{1 + \frac{2AR}{\pi}\right\} - \frac{2d}{c} \tag{6.71}$$

the twist distribution reduces to

$$t(y) = \frac{2C_L}{\pi^2}\left\{\sqrt{1-\left(\frac{2y}{b}\right)^2} - 1\right\} \tag{6.72}$$

The rectangular wing requires an elliptic washout with a tip twist of $-2C_L/\pi^2$.

If this twisted rectangular wing operates at an incidence β, different from the design incidence α, the circulation is no longer elliptic and is given by

$$\Gamma[y(\theta)] = \pi U c\left\{\beta - \alpha + \frac{2C_L}{\pi^2}\sin\theta - \sum_{n=1}^{\infty} nA_n \frac{\sin n\theta}{\sin\theta}\right\} = 2Ub\sum_{n=1}^{\infty} A_n \sin n\theta \tag{6.73}$$

A_1 is equal to the new value $C_L(\beta)/(\pi AR)$, unknown at this stage, and the other coefficients need to be calculated using the orthogonality property. For example, multiplying both hand-sides by $\sin\theta$ and integrating on $[0, \pi]$ gives a first relation between the Fourier coefficients (note that we only consider the odd modes for reason of symmetry)

$$\sum_{p=1}^{\infty} A_{2p+1} = \beta - \alpha \tag{6.74}$$

It seems likely that the solution will involve the infinite series of coefficients. A numerical solution is the best way to obtain the value of $C_L(\beta)$.

Clearly, other designs are possible with different distributions of twist, camber and chord.

6.7.4 Local and Global Lift Coefficients

As seen earlier, the local lift coefficient is related to the local circulation by

$$C_l(y) = \frac{2\Gamma(y)}{Uc(y)} \tag{6.75}$$

Substitution of the circulation in the formula for the global lift coefficient and choosing $A_{ref} = S$ yields

$$C_L = \frac{1}{S}\int_{-\frac{b}{2}}^{\frac{b}{2}} C_l(y)c(y)dy \tag{6.76}$$

The global lift coefficient is the area weighted average of the local lift coefficient. In particular, if $C_l(y) = const.$ the global lift is equal to the local lift coefficient, $C_L = C_l$. This is the case for the ideal wing with elliptic planform. For the ideal wing with rectangular planform, the local lift coefficient is elliptic and given by

$$C_l(y) = \frac{4C_L}{\pi}\sqrt{1-\left(\frac{2y}{b}\right)^2} \tag{6.77}$$

These distributions are shown in Fig. 6.21 for the two wings having the same C_L.

Here we have assumed that the wings are equipped with the same profile. If, at the operating Reynolds number, the profile lift curve has a maximum lift coefficient $C_{l,max}$, one can speculate that the elliptical wing will see stall occurring at once from tip to tip when the effective angle of incidence will reach the incidence of maximum C_l. On the other hand, the rectangular wing will stall first in the root section, where the maximum local lift coefficient is found. From the point of view of airplane design, it is desirable to have the wing stall in the root section first, because tip stall will send the plane into a spiral motion from which it is more difficult to escape. The delta wing of Fig. 6.3b is at higher risk of tip stall.

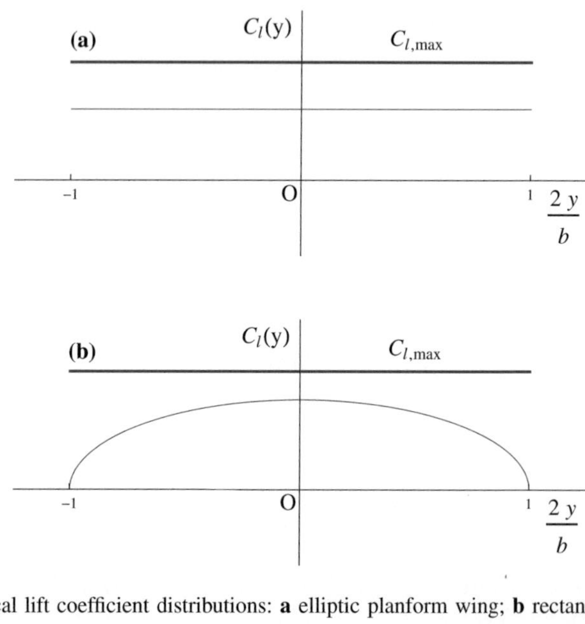

Fig. 6.21 Local lift coefficient distributions: **a** elliptic planform wing; **b** rectangular wing with washout

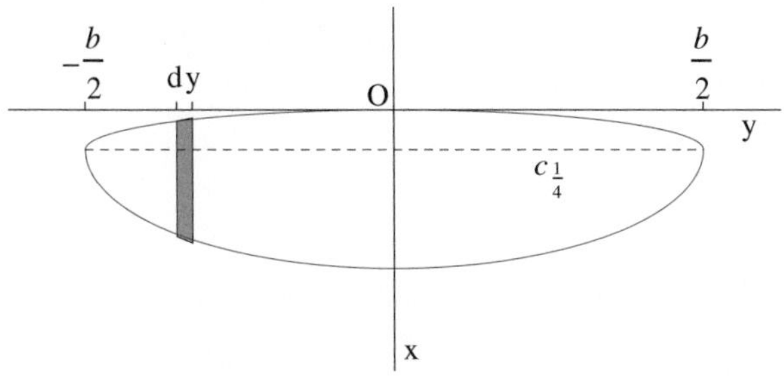

Fig. 6.22 Elementary contribution to pitching moment

6.7.5 Pitching Moment

Consider a thin wing with unswept quarter-chord as shown in Fig. 6.22.

A small section of the wing of thickness dy contributes to the lift force according to

$$dL = \rho U \Gamma(y)dy = \frac{1}{2}\rho U^2 C_l(y)c(y)dy \tag{6.78}$$

and to the pitching moment about the y-axis as

$$dM_{,o} = dM_{,a.c.} - x_{a.c.}dL \tag{6.79}$$

The term $dM_{,a.c.}$ is the contribution to the pitching moment at the aerodynamic center of the profile and is given by

$$dM_{,a.c.} = \frac{1}{2}\rho U^2 c^2(y) C_{m,a.c.}(y)dy \tag{6.80}$$

In thin airfoil theory the aerodynamic center pitching moment coefficient depends on the relative camber, as $C_{m,a.c.} = -\pi d(y)/c(y)$. The negative sign is for a nose down moment. Combining the two results and integrating along the span yields

$$M_{,o} = \frac{1}{2}\rho U^2 \int_{-\frac{b}{2}}^{\frac{b}{2}} C_{m,a.c.}(y)c^2(y)dy - \frac{1}{2}\rho U^2 x_{a.c.} \int_{-\frac{b}{2}}^{\frac{b}{2}} C_l(y)c(y)dy \tag{6.81}$$

Assuming for simplicity that the profile relative camber is constant along the span, the moment coefficient at the aerodynamic center is constant and the integration can be performed as

$$M_{,o} = \frac{1}{2}\rho U^2 S\bar{c} C_{m,a.c.} - \frac{1}{2}\rho U^2 x_{a.c.} S C_L \tag{6.82}$$

Here, we have introduced the average aerodynamic chord $\bar{c}$. It is used, with the wing area S, to define the pitching moment coefficient

$$C_{M,o} = C_{M,a.c.} - \frac{x_{a.c.}}{\bar{c}} C_L \tag{6.83}$$

The wing aerodynamic center pitching moment coefficient is defined more generally as

$$C_{M,a.c.} = \frac{1}{S\bar{c}} \int_{-\frac{b}{2}}^{\frac{b}{2}} C_{m,a.c.}(y) c^2(y) dy \tag{6.84}$$

6.7.6 Example of Wing Loading with Upwash

When considering Fig. 6.16, one is tempted to design a wing loading with "upwash" on a part of the wing, to locally create a "thrust" or "negative drag" to mitigate the induced drag on the rest of the wing. This can be achieved by combining some odd modes of the Fourier series for the circulation. As seen previously, the downwash in the Trefftz plane is given by

$$w_T(t) = -2U \sum_{n=1}^{\infty} n A_n \frac{\sin nt}{\sin t} \tag{6.85}$$

For mode 1 this is a constant downwash

$$w_{1T} = -2UA_1 \tag{6.86}$$

For mode 3 this is a parabolic downwash

$$w_{3T} = -6UA_3 \frac{\sin 3t}{\sin t} = -6UA_3 \left\{ 4 \left(\frac{2y}{b} \right)^2 - 1 \right\} \tag{6.87}$$

A combination for which $A_3 = -A_1/6$, with all the other modes suppressed, gives the results for the circulation and the downwash of Fig. 6.23.

The thin lines are for the elliptic loading and constant downwash. The thicker lines show the combined 1 and 3 modes. The last 10 % of the wing span sees an upwash and a thrust. Nevertheless, since it was proved that the elliptic loading is the optimum solution, the induced drag of the wing is increased according to the formula

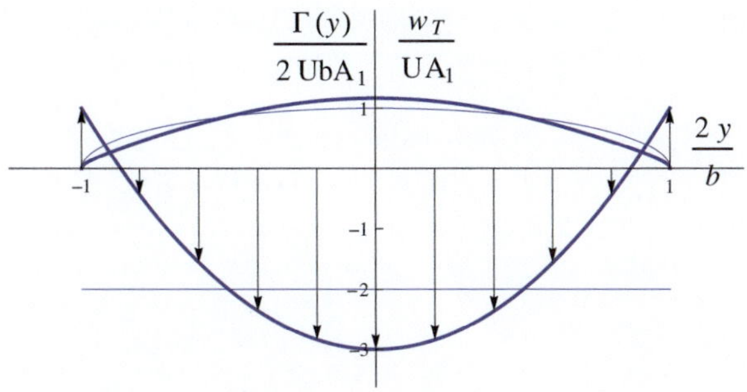

Fig. 6.23 Wing with upwash near the tips

$$C_{Di} = (C_{Di})_{elliptic} \left\{ 1 + 3 \left(\frac{A_3}{A_1} \right)^2 \right\} = \frac{13}{12} (C_{Di})_{elliptic} \tag{6.88}$$

a drag increase of 8 %.

6.7.7 Extension of the Theory to Non-straight Lifting Lines

Prandtl lifting line theory has been used to analyze wing/winglet combinations and, in inverse mode, to help design wings and winglets for practical low-speed aircraft applications. The seminal work of Munk [4], a student of Prandtl in Gottingen, paved the way to understanding the contributions of these small elements placed at 90° to the main wing, which clearly do not contribute to lift, but are shown to improve the wing efficiency by decreasing the induced drag of the main wing. In Fig. 6.24, winglets are equipped with tufts to analyze the flow field during flight.

For large aspect ratio wings, the theory can be extended to lifting lines having kinks as in the case of winglets, or having a continuous curvature that turns its direction ±90°, for so-called "blended winglets", or also being turned in the flow or against the flow direction for backward or forward sweep, as well as combinations of those. All cases have been considered by Munk, for single lifting line and multiple lifting lines. The vortex sheet takes the shape of the base surface, touching the lifting line, considered the base line. The rolling-up of the sheet edges is again neglected in the mathematical model.

JJC [5] has developed an optimization code that calculates the optimum circulation distribution along a lifting line that has a kink, for the design of winglets.

In order to find the wing/winglet combination that has the minimum induced drag for a given lift, the following objective function $F(\Gamma) = C_{Di} + \lambda C_L$ is defined. C_{Di} is the induced drag coefficient. λ is the Lagrange multiplier that governs the target lift. C_L is the lift coefficient and is added to the objective function as a means of enforcing the lift constraint. Let a be the dimensionless winglet height normalized

Fig. 6.24 Research on winglet from http://www.nasa.gov/centers/dryden/about/Organizations/Technology/Facts/TF-2004-15-DFRC.html

with the semi-span, $b/2$. The lift and induced drag coefficients are

$$C_L = \frac{AR}{2}\int_{-1-a}^{1+a} \Gamma(\sigma)\frac{dy}{d\sigma}d\sigma \tag{6.89}$$

$$C_{Di} = -\frac{AR}{2}\int_{-1-a}^{1+a} \Gamma(\sigma)q_n(\sigma)d\sigma \tag{6.90}$$

where σ is the curvilinear abscissa along the dihedral shape, and q_n is the induced velocity component perpendicular to the dihedral line in the $x = 0$ plane, with $q_n = w$ the usual downwash along the main part of the wing, $q_n = v$ on the left winglet and $q_n = -v$ on the right winglet. The "normalwash", q_n, depends linearly on Γ as

$$q_n(s) = -\frac{1}{4\pi}\int_{-1-a}^{1+a} \Gamma'(\sigma)\frac{(y(s)-y(\sigma))\frac{dy}{ds} + (z(s)-z(\sigma))\frac{dz}{ds}}{(y(s)-y(\sigma))^2 + (z(s)-z(\sigma))^2}d\sigma \tag{6.91}$$

All quantities have been made dimensionless with the half span, $b/2$, and the incoming flow velocity aligned with the x-axis, U. The area of reference is the projected wing area, S. The minimization equation is a function of Γ only, in this inviscid model, and is obtained by taking the Frechet derivative of the objective function, which reads

$$\frac{\partial F}{\partial \Gamma}(\delta\Gamma) = \frac{\partial C_{Di}}{\partial \Gamma}(\delta\Gamma) + \lambda\frac{\partial C_L}{\partial \Gamma}(\delta\Gamma) = 0, \ \forall\, \delta\Gamma \tag{6.92}$$

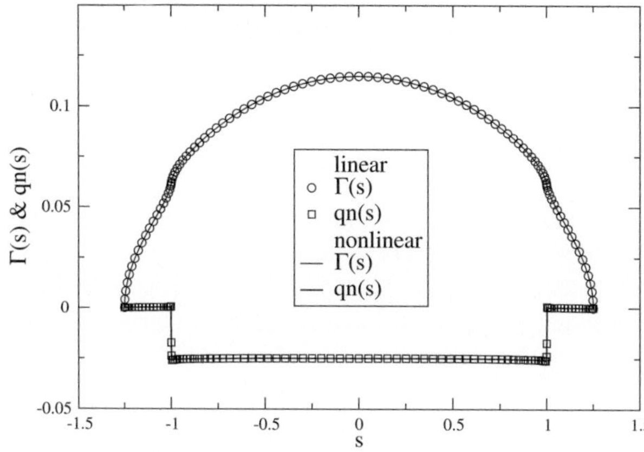

Fig. 6.25 Circulation and normalwash distributions for the optimum wing with 25 % winglet (analysis)

The boundary conditions are $\Gamma(-1-a) = \Gamma(1+a) = 0$. The Frechet derivative can be defined as

$$\lim_{\theta \to 0} \frac{d}{d\theta} \{F(\Gamma + \theta\delta\Gamma)\} \tag{6.93}$$

and is termed "the derivative of F w.r.t. Γ in the direction $\delta\Gamma$". The result is linear in $\delta\Gamma$.

Since C_{Di} is homogeneous of degree two in Γ and C_L of degree one, the minimization equation is linear non-homogeneous in Γ when $\lambda \neq 0$. The solution is obtained in two steps: first the Lagrange multiplier is set to an arbitrary non-zero value, say $\lambda = 1$ and the corresponding circulation and lift C_L are calculated. Then, the final answer is obtained by multiplying the circulation by $\kappa = C_{Ltarget}/C_L$.

As a test case, a wing with 25 % vertical winglets is designed and then analyzed for a lift coefficient $C_{Ltarget} = 1$. The results for the circulation and "normalwash" are shown as solid symbols in Fig. 6.25, for a selected geometry (zero twist, constant relative camber) corresponding to the optimum circulation and linear or nonlinear lift curves. The computation uses a fine mesh of $jx = 101$ points, including the winglets, with a cosine distributions of points on each element. This is required because the normal induced velocity is singular at the wing/winglet junction for the optimization equation, and small mesh steps are needed there. It can be noted that the winglets do not contribute to the induced drag as the normal induced velocity q_n is zero at the winglets, which is in agreement with the theory [4]. The winglets are loaded, although much less than the wing. Their role is to redistribute the loads on the wing and decrease the induced drag by lowering the root circulation and by increasing it near the wing/winglet junction. The downwash on the wing is again constant, but of smaller absolute value compared to the elliptic loading. More details on the design of the geometry can be found in Chap. 11, Sect. 11.7.

6.7.8 Numerical Solution of Prandtl Integro-Differential Equation

6.7.8.1 Linear Model

Given the wing geometry, wing span, chord, twist distributions and linear airfoil profile characteristics, $C_l(\alpha) = 2\pi(\alpha + 2d/c)$, the integro-differential equation of Prandtl allows to calculate the circulation $\Gamma(y)$ and subsequently the coefficients of lift C_L and induced drag C_{Di}, at a given geometric incidence α, in incompressible, inviscid flow. The equation reads

$$\Gamma(y) = \pi U c(y) \left\{ \alpha - \alpha_0(y) - \frac{1}{4\pi U} \int_{-\frac{b}{2}}^{\frac{b}{2}} \frac{\Gamma'(\eta)}{y - \eta} d\eta \right\} \tag{6.94}$$

The span is discretized using a cosine distribution of points which enhances the accuracy

$$y_j = -\frac{b}{2} \cos t_j, \quad t_j = (j-1)\Delta t = \frac{j-1}{jx-1}\pi, \quad 0 \leq t_j \leq \pi, \quad j = 1, \ldots, jx \tag{6.95}$$

where jx is the total number of control points along the span. In order to avoid the indefinite value of the integrand at $\eta = y$, the vortices of intensity $\delta\Gamma(\eta)$ are located between the control points according to

$$\eta_k = -\frac{b}{2} \cos \theta_k,$$

$$\theta_k = (k-1)\Delta\theta + \frac{\Delta\theta}{2} = \frac{2k-1}{2(jx-1)}\pi, \quad 0 < \theta_k < \pi, \quad k = 1, \ldots, jx-1 \tag{6.96}$$

The control points and vortex locations are shown in Fig. 6.26.

The geometry is discretized at the control points, i.e. the chord, c_j, the twist, t_j, and the camber, d_j. The principal value integral for w_{wj} is evaluated as

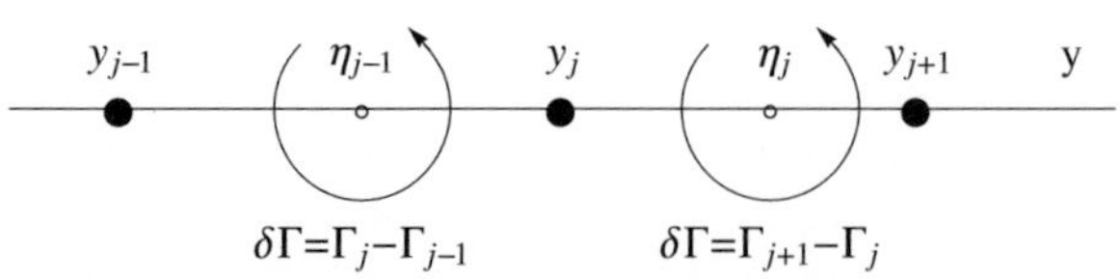

Fig. 6.26 Control points and vortices in the discrete representation of the lifting line

$$w_{wj} = -\frac{1}{4\pi} \sum_{k=1}^{jx-1} \frac{\Gamma_{k+1} - \Gamma_k}{y_j - \eta_k} \tag{6.97}$$

Substitution in the equation yields for the inner control points

$$\Gamma_j = \pi U c_j \left\{ \alpha + t_j + 2\frac{d_j}{c_j} - \frac{1}{4\pi U} \sum_{k=1}^{jx-1} \frac{\Gamma_{k+1} - \Gamma_k}{y_j - \eta_k} \right\}, \quad j = 2, \ldots, jx - 1 \tag{6.98}$$

At the wing tips, the boundary conditions are $\Gamma_1 = \Gamma_{jx} = 0$.

This is a linear system for the Γ_j's. The matrix is a full matrix with a strong term on the main diagonal which makes it amenable to point over-relaxation. Let $d\Gamma_j = \Gamma_j^{n+1} - \Gamma_j^n$, where the upper index represents the iteration level. The domain is initialized with zero circulation by letting $\Gamma_j^0 = 0, \ j = 1, \ldots, jx$. The domain is swept in increasing values of $j, \ j = 2, \ldots jx - 1$, and $d\Gamma_j$ is calculated and the circulation updated immediately as $\Gamma_j^{n+1} = \Gamma_j^n + \Delta\Gamma_j$ before moving to the next point. The contributions to $\Delta\Gamma_j$ come from the left- and right-hand-sides upon introducing $\Delta\Gamma_j$ in the equation

$$\Gamma_j^{n+1} = \Gamma_j^n + \Delta\Gamma_j = \pi U c_j \left\{ \alpha + t_j + 2\frac{d_j}{c_j} \right.$$

$$\left. - \frac{1}{4\pi U} \sum_{k=1}^{jx-1} \frac{\Gamma_{k+1}^{n,n+1} - \Gamma_k^{n,n+1}}{y_j - \eta_k} - \frac{1}{4\pi U} \left(\frac{\Delta\Gamma_j}{y_j - \eta_{j-1}} - \frac{\Delta\Gamma_j}{y_j - \eta_j} \right) \right\} \tag{6.99}$$

Regrouping the changes in circulation to the left-hand-side gives

$$\left\{ 1 + \frac{c_j}{4} \left(\frac{1}{y_j - \eta_{j-1}} - \frac{1}{y_j - \eta_j} \right) \right\} \frac{\Delta\Gamma_j}{\omega} = \pi U c_j \left\{ \alpha + t_j + 2\frac{d_j}{c_j} + \frac{w_{wj}^{n,n+1}}{U} \right\} - \Gamma_j^n \tag{6.100}$$

Here, a relaxation factor ω has been introduced, which can be set between 1 and 2, and contributes to accelerating the convergence. Typically $\omega = 1.5$ has been used with the linear model above. The upper index $(n, n+1)$ term is updated with each j-equation, but some values of the circulation Γ_k are new $(n+1)$ and some are old (n), depending on whether $k < j \ or \ k \geq j$. Once converged, the circulation distribution provides the lift and induced drag coefficients as

$$SC_L = \frac{2}{U} \sum_{j=2}^{jx-1} \Gamma_j (\eta_j - \eta_{j-1}) \tag{6.101}$$

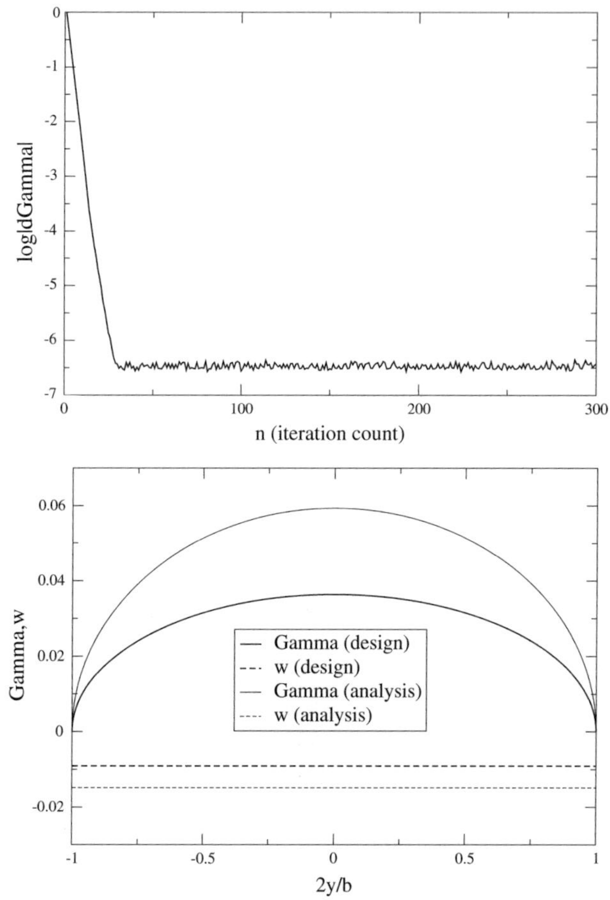

Fig. 6.27 Convergence and results for the elliptic planform wing

$$SC_{Di} = -\frac{2}{U^2} \sum_{j=2}^{jx-1} \Gamma_j w_{wj} (\eta_j - \eta_{j-1}) \qquad (6.102)$$

The design of an ideal wing with elliptic planform of span $b = 2.1$ m and root chord $c_0 = 0.382$ m, is carried out for a design lift coefficient $C_L = 0.2$. The profiles are parabolic with relative camber $d/c = 0.0159$. The design incidence is found to be $\alpha = 0.521°$. The numerical simulation is performed with $jx = 101$ points. The same wing is also analyzed at $\beta = 2°$. The results for the convergence, the circulation and the downwash are shown in Fig. 6.27.

Note that the circulation remains elliptic and the downwash constant at off design condition. The design condition has been selected so that the wing will have an

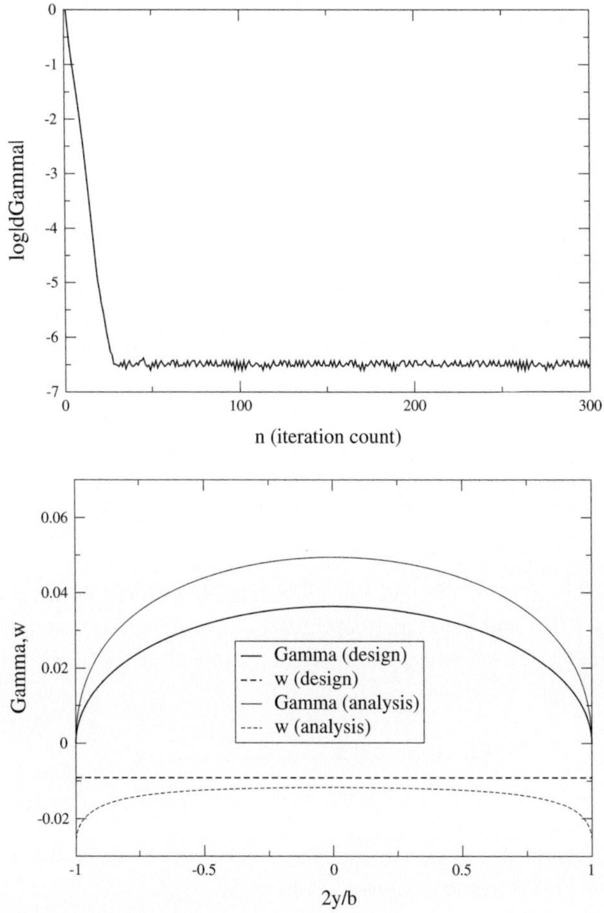

Fig. 6.28 Convergence and results for the rectangular wing

adapted leading edge. Indeed, at $\alpha = 0.521°$ and $C_L = 0.2$, this wing of aspect ratio $AR = 7$ sees an induced incidence $\alpha_i = -0.521°$, hence the effective incidence is $\alpha_{eff} = 0°$ which is the incidence of adaptation for a parabolic profile. Each profile operates as in Fig. 6.16 (here $t(y) = 0$), with an incoming velocity parallel to the profile chord.

The rectangular wing with span $b = 2.1$ m, chord $c = 0.3$ m of same aspect ratio as the previous one, is equipped with the same parabolic profile of relative camber $d/c = 0.0159$. The wing has a $t(\pm\frac{b}{2}) = -0.0405$ rad elliptical washout. At design $C_L = 0.2$, the geometric incidence is $\alpha = 1.019°$. The convergence history and the distributions of circulation and downwash are found in Fig. 6.28.

Although the analysis at off-design incidence $\beta = 2°$, shows a circulation that looks elliptic, the downwash indicates that this is not the case.

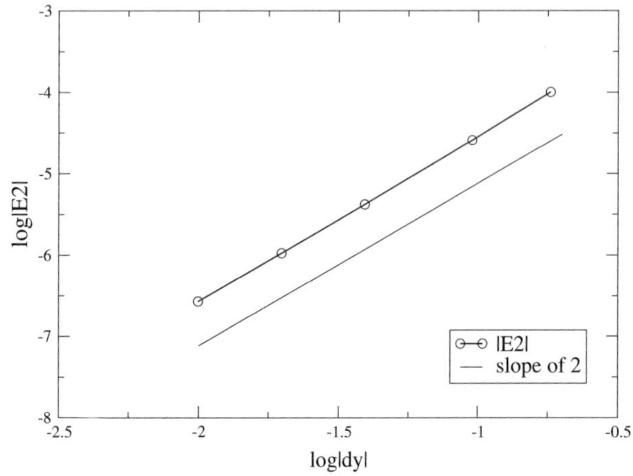

Fig. 6.29 Study of the scheme accuracy with mesh refinement

The method is second order accurate. The number of mesh points is varied from $jx = 11$ to $jx = 201$ and the numerical result compared with the analytical solution for the elliptical planform wing. The L_2 error norm is used

$$|E_2| = \left[\frac{\sum_{j=1}^{jx} \left(g_j - g_{exact}(y_j) \right)^2}{jx} \right]^{\frac{1}{2}} \tag{6.103}$$

and $\log |E_2|$ is plotted versus $\log \bar{\Delta y}$, where the average mesh size is defined as $\bar{\Delta y} = 2/(jx - 1)$. The result is shown in Fig. 6.29.

6.7.8.2 Nonlinear Lift Curve and Viscous Correction

A viscous correction can be added to the drag evaluation given by the lifting line method, by using a nonlinear viscous polar for the profiles. The viscous polar, from experiments or from the two-dimensional solution of the Navier-Stokes equations, typically displays a maximum lift coefficient at the incidence of stall α_{Clmax}. The numerical solution of the lifting line problem becomes highly nonlinear in the vicinity of α_{Clmax}, due to the existence of two or more incidence values for a given lift coefficient $C_l < C_{lmax}$. The method proposed by JJC [6], consists in adding an artificial viscosity term to the governing equation, whose role is to select the correct solution of the nonlinear problem. Lets consider the inviscid flow problem in which the lift curve presents a maximum at some value of incidence.

The governing equation for a nonlinear lift curve $C_l(\alpha)$ reads

$$\Gamma(y) = \frac{1}{2}Uc(y)Cl\left[\alpha_{eff} - \alpha_0(y)\right] = \frac{1}{2}Uc(y)Cl\left[\alpha - \alpha_0(y) + \arctan\left(\frac{w_w(y)}{U}\right)\right] \tag{6.104}$$

Newton's method is used to linearize locally, at control point j, the change in circulation

$$\Gamma_j + \Delta\Gamma_j = \frac{1}{2}Uc_j\left(C_{lj} + \frac{dC_{lj}}{d\alpha}\Delta\alpha_{effj}\right) \tag{6.105}$$

where the change of effective incidence is expressed using the discrete formula for the downwash

$$\Delta\alpha_{effj} = \frac{\frac{\Delta w_{wj}}{U}}{1 + \left(\frac{w_{wj}}{U}\right)^2} \tag{6.106}$$

and

$$\Delta w_{wj} = -\frac{1}{4\pi}\left(\frac{1}{y_j - \eta_{j-1}} - \frac{1}{y_j - \eta_j}\right)\Delta\Gamma_j = a_j\Delta\Gamma_j \tag{6.107}$$

Finally, the equation assumes the following iterative form

$$\left(1 - \frac{1}{2}c_j\frac{dC_{lj}}{d\alpha}\frac{a_j}{1 + \left(\frac{w_{wj}}{U}\right)^2}\right)\frac{\Delta\Gamma_j}{\omega} = \frac{1}{2}Uc_jC_{lj} - \Gamma_j^n \tag{6.108}$$

where ω is the relaxation factor and $\Delta\Gamma_j = \Gamma_j^{n+1} - \Gamma_j^n$. Note that $a_j < 0$, so that, the second term in the bracket in the l.h.s. is positive as long as the lift slope is positive, and the solution can be over-relaxed. When stall has occurred on part or the whole wing, the term in the bracket changes sign and the algorithm for the circulation will no longer converge. The equation is modified with the addition of an artificial viscosity term in the r.h.s., which also contributes to the coefficient of $\Delta\Gamma$, to counterbalance the negative coefficient. The iterative algorithm now reads

$$\left(1 - \frac{1}{2}c_j\frac{dC_{lj}}{d\alpha}\frac{a_j}{1 + \left(\frac{w_{wj}}{U}\right)^2} + 2\mu_j\right)\frac{\Delta\Gamma_j}{\omega}$$

$$= \frac{1}{2}Uc_jC_{lj} - \Gamma_j^n + \mu_j\left(\Gamma_{j+1}^n - 2\Gamma_j^n + \Gamma_{j-1}^{n+1}\right) \tag{6.109}$$

where the artificial viscosity coefficient μ_j is given by

$$\mu_j \geq \sup\left(\frac{1}{4}c_j\frac{dC_{lj}}{d\alpha}\frac{a_j}{1+\left(\frac{w_{wj}}{U}\right)^2}\,;\,0\right) \tag{6.110}$$

Under-relaxation is necessary ($\omega < 1$) when separation occurs.

A simple test case, for which an exact solution of Prandtl lifting line theory can be obtained, corresponds to the elliptic loading. Consider a wing of elliptic planform and of aspect ratio $AR = 7$, that has no twist and no camber. The two-dimensional lift curve is given analytically by

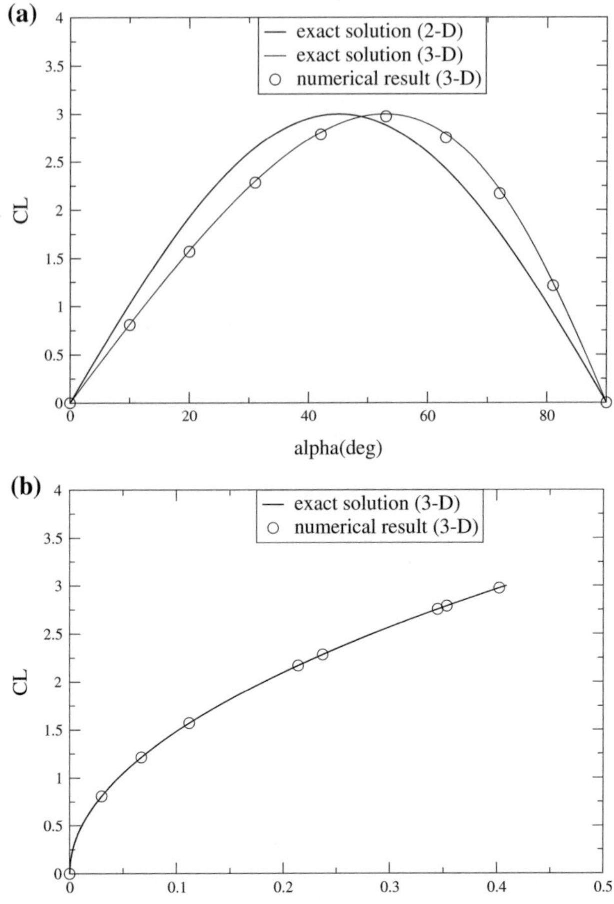

Fig. 6.30 Result of analytical test: **a** lift curves; **b** wing polar

$$C_l(\alpha) = C_{lmax} \sin 2\alpha \tag{6.111}$$

The maximum lift is obtained at $\alpha = \pi/4$. Only the first Fourier mode is present in the solution, and the downwash is constant $w_w/U = -A_1$. The governing equation reduces to

$$C_l(y) = \frac{2\Gamma(y)}{Uc(y)} = C_L = \pi AR\, A_1 = C_{lmax} \sin\left(2\alpha + 2\arctan\frac{w_w}{U}\right), \quad \Rightarrow$$

$$A_1 = \frac{C_{lmax}}{\pi AR} \sin(2\alpha - 2\arctan A_1) \tag{6.112}$$

This last equation is solved iteratively, using Newton's method to give $A_1 = A_1(\alpha)$. Hence the wing lift is obtained as $C_L(\alpha) = \pi AR\, A_1(\alpha)$. Here we have chosen $C_{lmax} = 3$. The inviscid nonlinear lift curves and wing polar are shown in Fig. 6.30. The numerical solution matches the exact solution very well.

6.8 Incompressible Flow Over Moderate Aspect Ratio Wings: The Vortex Lattice Method

In incompressible, inviscid flow, for aspect ratio wings which are less than 7, the Prandtl lifting line theory is extended to a lifting surface theory, the vortex lattice method. For the lifting problem, consider an infinitely thin wing of equation $z = f(x, y)$ at incidence α. The wing surface is covered with a mesh that conforms to the wing projected planform in the (x, y) plane. At the mesh node one defines the circulation $\Gamma_{i,j} = \Gamma(x_{i,j}, y_{i,j})$. Along the edges, the circulation is either zero, at leading edge and tips, or corresponds to the total bound vorticity for that cross-section inside the wing $\Gamma(y) = \Gamma[x_{te}(y), y]$, at the trailing edge. The distribution $\Gamma(y)$ is then carried, without alteration, along the vortex filaments to the Trefftz plane as trailed vorticity. To avoid difficulties with the singular integral, the vortices are placed between the control points as shown in Fig. 6.31.

The Biot-Savart law is used to calculate the components of the induced velocity at the interior control points due to the bound and trailed vorticity. For the bound vorticity, consider one of the two types of small, oriented elements of vorticity $\mathbf{dl}_{i,j} = \{dl_x; dl_y; 0\} = \mathbf{dl}_{i-\frac{1}{2},j}$ between control points $(i-1, j)$ and (i, j) or $\mathbf{dl}_{i,j} = \mathbf{dl}_{i,j+\frac{1}{2}}$ between points (i, j) and $(i, j+1)$. The induced velocity at point $M_k = (x_k, y_k, 0)$ of the wing is given by

$$d\mathbf{V}_{i-\frac{1}{2},j,k} = \frac{d\Gamma_{i,j}}{4\pi} \frac{\mathbf{r}_{i,j,k} \wedge \mathbf{dl}_{i,j}}{\left|\mathbf{r}_{i,j,k}\right|^3} \tag{6.113}$$

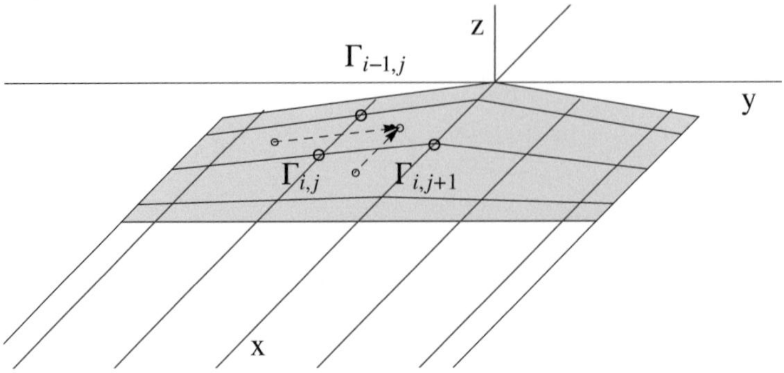

Fig. 6.31 Discrete representation of the vorticity with the lifting surface method

where

$$d\Gamma_{i,j} = \Gamma_{i,j} - \Gamma_{i-1,j} \; or \; d\Gamma_{i,j} = \Gamma_{i,j+1} - \Gamma_{i,j} \tag{6.114}$$

and

$$\mathbf{r}_{i,j,k} = (r_x; r_y; r_z) = \left(x_k - x_{i-\frac{1}{2},j}; y_k - y_{i-\frac{1}{2},j}; 0\right) \tag{6.115}$$

or

$$\mathbf{r}_{i,j,k} = \left(x_k - x_{i,j+\frac{1}{2}}; y_k - y_{i,j+\frac{1}{2}}; 0\right) \tag{6.116}$$

The contribution to the z-component reads

$$dw_{i,j,k} = \frac{d\Gamma_{i,j}}{4\pi} \frac{r_x dl_y - r_y dl_x}{\left| r_x^2 + r_y^2 \right|^{\frac{3}{2}}} \tag{6.117}$$

The semi-infinite vortex filaments $\mathbf{l}_j$ also contribute with the trailed vorticity to the z-component at all interior points as $dw_{j,k}$. The tangency condition is enforced at the interior control points k

$$w_k = \sum_{i,j}^{bound} dw_{i,j,k} + \sum_{j}^{trailed} dw_{j,k} = U \left\{ \frac{\partial f}{\partial x}(x_k, y_k) - \alpha \right\} \tag{6.118}$$

The Kutta-Joukowski condition is satisfied at the trailing edge by applying a parabolic extrapolation of the circulation that satisfies $\frac{\partial \Gamma}{\partial x} = 0$, as was done in 2-D. Accounting for the boundary conditions mentioned above, there are as many equations as there are unknowns $\Gamma_{i,j}$. The linear system can be solved by relaxation. The result is the circulation at the trailing edge, $\Gamma(y)$, that determine the lift and induced drag of the thin wing, using the general formula derived earlier. More details on the vortex lattice method can be found in Chap. 10 on wind turbine simulation with the vortex method.

6.9 Compressible Flow Over Moderate Aspect Ratio Wings

In this section, small disturbance theories for both subsonic and supersonic wings are discussed. As in two-dimensional cases, the symmetric and the lifting problems are treated separately, thanks to the linearity of the governing equations and boundary conditions, for thin wings with small relative camber and at small angles of attack.

The linearized boundary condition on the plane $z = 0$ are

$$\begin{cases} w(x, y, 0^+) = \frac{\partial \phi}{\partial z}(x, y, 0^+) = U\left(\frac{\partial d}{\partial x}(x, y) + \frac{1}{2}\frac{\partial e}{\partial x}(x, y) - \alpha\right) \\ w(x, y, 0^-) = \frac{\partial \phi}{\partial z}(x, y, 0^-) = U\left(\frac{\partial d}{\partial x}(x, y) - \frac{1}{2}\frac{\partial e}{\partial x}(x, y) - \alpha\right) \end{cases} \tag{6.119}$$

where $d(x, y)$ and $e(x, y)$ represent the camber and thickness of the thin wing, respectively.

Notice that in the symmetric problem involving the wing thickness distribution only, the potential is symmetric with $\phi(x, y, -z) = \phi(x, y, z)$, whereas for the lifting problem the potential is antisymmetric with $\phi(x, y, -z) = -\phi(x, y, z)$. As in two-dimensional thin airfoil theory, across the wing surface, u is continuous and w jumps for the symmetric problem and u jumps and w is continuous for the lifting problem. The solution will be obtained as superposition of sources and their derivatives in three-dimensional flow, such that the boundary conditions are satisfied.

6.9.1 Symmetric Problem in Subsonic Flow

The fundamental solution is a three dimensional source/sink in subsonic flow. It is related to that in incompressible flow via the Prandtl-Glauert transformation. Hence the perturbation potential at any point in the field is obtained from the surface integral over the wing planform given by (see Ashley and Landhal [7])

$$\phi(x, y, z) = -\frac{1}{4\pi}\int\int_S \frac{q(\xi, \eta)}{\sqrt{(x-\xi)^2 + \beta^2(y-\eta)^2 + \beta^2 z^2}} d\xi d\eta \tag{6.120}$$

where q is the density of source/sink and $\beta = \sqrt{1 - M_\infty^2}$.

From the boundary condition for the symmetric problem, one obtains

$$w(x, y, 0^+) = \frac{\partial \phi}{\partial z}(x, y, 0^+) = \frac{U}{2}\frac{\partial e}{\partial x}(x, y) \tag{6.121}$$

Calculating $w(x, y, z)$

$$w(x, y, z) = \frac{\partial \phi}{\partial z}(x, y, z) = \frac{\beta^2}{4\pi}\int\int_S \frac{z\, q(\xi, \eta)}{\left[(x-\xi)^2 + \beta^2(y-\eta)^2 + \beta^2 z^2\right]^{\frac{3}{2}}} d\xi d\eta \tag{6.122}$$

As $z \to 0^+,\ z > 0$, the integrands vanishes almost everywhere, except at $(\xi, \eta) = (x, y)$, where it is indeterminate of the form 0/0. It is possible to find the limit with the change of variables from (ξ, η) to (t, θ) as

$$\begin{cases} \xi = x + \beta^2 zt\cos\theta \\ \eta = y + \beta zt\sin\theta \end{cases},\quad 0 \le \beta zt \le \epsilon,\quad 0 \le \theta \le 2\pi \tag{6.123}$$

where ϵ is a small positive number. The condition $0 \le \beta zt \le \epsilon$ amounts to considering only the area within a small ellipse centered at (x, y) as shown in Fig. 6.32.

The integral reduces to

$$w_\epsilon(x, y, z) = \frac{\beta^2}{4\pi}\int_0^{2\pi}\int_0^{\frac{\epsilon}{\beta z}} \frac{q(x + \beta^2 zt\cos\theta, y + \beta zt\sin\theta)}{\left[1 + \beta^2 t^2\right]^{\frac{3}{2}}} tdtd\theta \tag{6.124}$$

Assuming $q(\xi, \eta)$ to be a continuous and differentiable function, for sufficiently small ϵ, it can be expanded as

$$q(x + \beta^2 zt\cos\theta, y + \beta zt\sin\theta) = q(x, y) + O(\epsilon) \tag{6.125}$$

the substitution of which gives

$$\begin{aligned} w_\epsilon(x, y, z) &= \frac{\beta^2}{4\pi} q(x, y)\int_0^{2\pi}\int_0^{\frac{\epsilon}{\beta z}} \frac{tdtd\theta}{\left[1 + \beta^2 t^2\right]^{\frac{3}{2}}} + O(\epsilon) \\ &= -\frac{1}{2} q(x, y)\left[\frac{1}{\sqrt{1 + \beta^2 t^2}}\right]_0^{\frac{\epsilon}{\beta z}} + O(\epsilon) \end{aligned} \tag{6.126}$$

as $z \to 0^+$ the integral converges to $\frac{1}{2}q(x, y) + O(\epsilon),\ \forall\epsilon$, which, since ϵ can be chosen as small as one wishes, gives the result

$$w(x, y, 0^+) = \frac{\partial \phi}{\partial z}(x, y, 0^+) = \frac{1}{2}q(x, y) = \frac{U}{2}\frac{\partial e}{\partial x}(x, y) \tag{6.127}$$

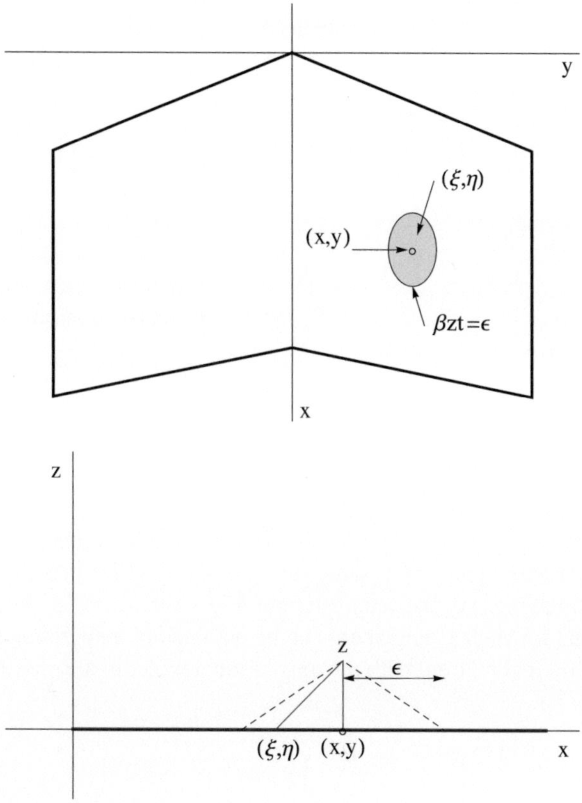

Fig. 6.32 Integration domain for singular integral

This determines the source/sink density $q(x, y) = U\frac{\partial e}{\partial x}(x, y)$. The potential now reads

$$\phi(x, y, z) = -\frac{U}{4\pi}\int\int_S \frac{1}{\sqrt{(x-\xi)^2 + \beta^2(y-\eta)^2 + \beta^2 z^2}} \frac{\partial e}{\partial x}(\xi, \eta) d\xi d\eta \quad (6.128)$$

One can see that, by symmetry, the boundary condition on the lower surface is satisfied.

The u-component of the perturbation is now found to be

$$u(x, y, z) = \frac{U}{4\pi}\int\int_S \frac{(x-\xi)}{\left[(x-\xi)^2 + \beta^2(y-\eta)^2 + \beta^2 z^2\right]^{\frac{3}{2}}} \frac{\partial e}{\partial x}(\xi, \eta) d\xi d\eta \quad (6.129)$$

The surface pressure coefficient is therefore

$$C_p = -2\frac{u(x,y,0)}{U} = -\frac{1}{2\pi}\int\int_S \frac{(x-\xi)}{\left[(x-\xi)^2+\beta^2(y-\eta)^2\right]^{\frac{3}{2}}}\frac{\partial e}{\partial x}(\xi,\eta)d\xi d\eta \tag{6.130}$$

It can be shown that the 2-D result is recovered for an infinite wing with $e = e(x)$ only.

The symmetric problems, like the 2-D and axisymmetric cases, are simply evaluations of singular integrals. In general, numerical approximations can be readily used in an effective way.

6.9.2 Lifting Problem in Subsonic Flow

The fundamental solution for this problem is the horseshoe vortex. It consists of two infinitely long vortex filaments of same strength but opposite signs, joined together by an infinitesimal piece of the same strength, Fig. 6.33.

The potential for such arrangement can be obtained by integrating a distribution of doublets in the x-direction, with the axis of the doublet in the z-direction

$$\phi(x,y,z) = \frac{\Gamma}{4\pi}\int_x^\infty \frac{\beta z}{\left[\xi^2+\beta^2y^2+\beta^2z^2\right]^{\frac{3}{2}}}d\xi$$

$$= \frac{\Gamma}{4\pi}\frac{\beta z}{\beta^2y^2+\beta^2z^2}\left(1+\frac{x}{\sqrt{x^2+\beta^2y^2+\beta^2z^2}}\right) \tag{6.131}$$

The downwash can be obtained by differentiation w.r.t. z

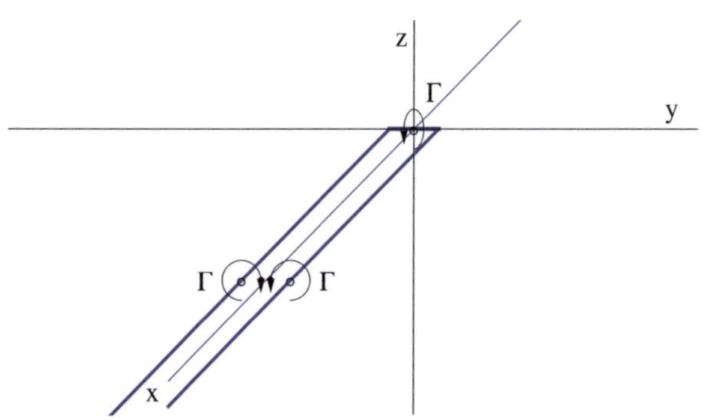

Fig. 6.33 Elementary horseshoe vortex of finite strength Γ

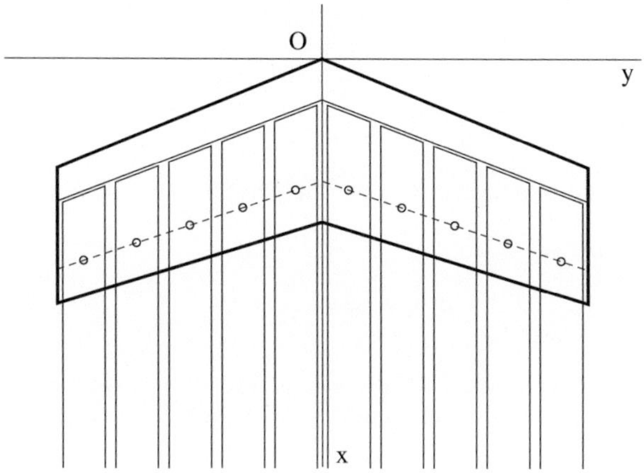

Fig. 6.34 Lifting surface method with a single row of panels

$$w(x, y, 0) = \frac{\partial \phi}{\partial z}(x, y, 0) \tag{6.132}$$

Alternatively, the latter can be obtained directly from Biot-Savart law, as we did before, (see for example Batchelor [8] or Moran [9]).

6.9.3 Extended Lifting Line Theory

For compressible flows, swept wings are more efficient, since what counts is the velocity component normal to the leading edge. In this case, the lifting line must be modified to take into consideration the sweep angle. A simplified vortex lattice method was used successfully by Weissinger [10] with one row of panels along the loci of the quarter-chord line, as shown in Fig. 6.34. The control points, where the tangency condition is enforced, are located at the three-quarter chord line. Note that, with a single row of panels, one can only represent a flat wing, with zero camber.

6.10 Supersonic Flow Over Moderate Aspect Ratio Wings

6.10.1 Symmetric Problem

The fundamental solution is a supersonic source and the solution of the linearized small disturbance equation is given by [7]

$$\phi(x, y, z) = -\frac{1}{2\pi} \int \int_{\Sigma} \frac{q(\xi, \eta)}{\sqrt{(x-\xi)^2 - \beta^2[(y-\eta)^2 + z^2]}} d\xi d\eta \tag{6.133}$$

where $\beta = \sqrt{M_0^2 - 1}$.

The source strength is determined by imposing the surface boundary condition. Hence

$$q(x, y) = U \frac{\partial e}{\partial x}(x, y) \tag{6.134}$$

The domain of integration Σ is the part of the (x, y)-plane intercepted by the upstream Mach cone from the field point (x, y, z).

The pressure coefficient is, as before

$$C_p = -\frac{2}{U} \frac{\partial \phi}{\partial x} \tag{6.135}$$

It can be shown that, for an infinite wing, the surface pressure coefficient, according to the above formula, is

$$C_p = \frac{2}{\beta} \frac{\partial f}{\partial x} = \frac{1}{\beta} \frac{\partial e}{\partial x} \tag{6.136}$$

which is consistent with the thin airfoil theory in Chap. 3.

6.10.2 *Lifting Problem*

We will consider first planforms with supersonic leading and trailing edges. For such wings, there is no interaction between the flows on the upper and lower surfaces, see Fig. 6.35.

Jones and Cohen analyzed uncambered delta wings of sweep angles less than the sweep angle of the Mach lines. It is shown that ΔC_p is constant along rays through the wing vertex, i.e. it is conical flow. The lift coefficient is independent of the sweep and it is the same as the two-dimensional result

$$C_L = \frac{4\alpha}{\beta} \tag{6.137}$$

Other planforms can be also easily analyzed. For example non rectangular wings with supersonic edges, see Fig. 6.36.

Using superposition principle, the same results are readily obtained, as depicted in Fig. 6.37, see Ref. [1].

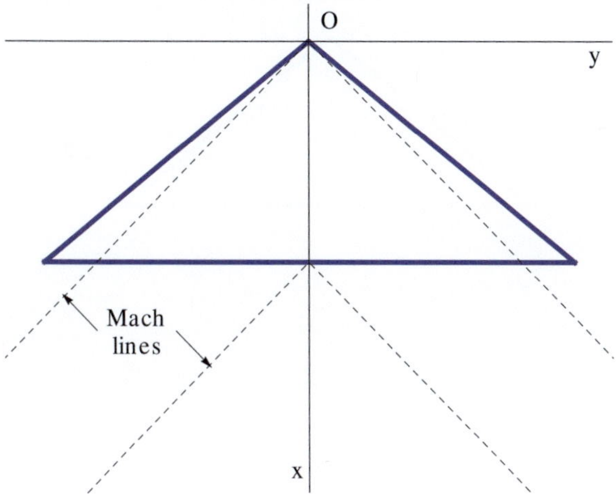

Fig. 6.35 Wing with supersonic leading and trailing edges

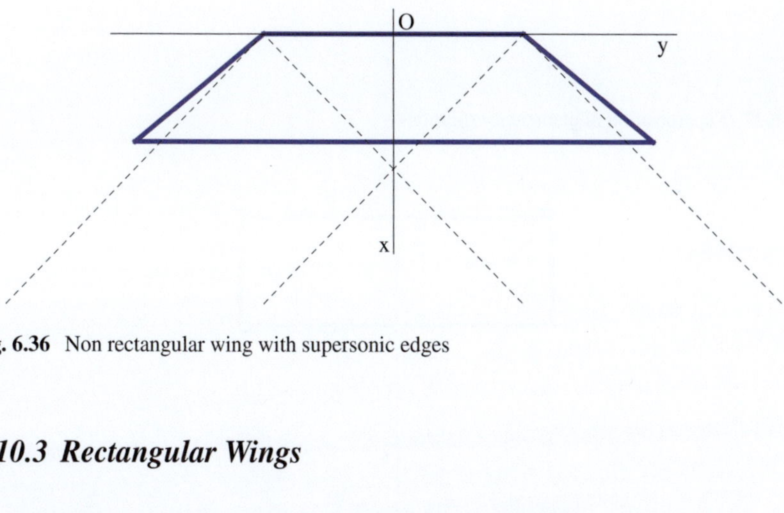

Fig. 6.36 Non rectangular wing with supersonic edges

6.10.3 Rectangular Wings

Consider supersonic flow over a rectangular wing, Fig. 6.38.

The pressure is constant along rays from the tips of the leading edge. From symmetry of pressure in the tip region, it may be shown that the average pressure coefficient is half the two-dimensional pressure coefficient. One finds the three-dimensional lift coefficient C_L from (see Ref. [11])

$$\Delta C_L = C_{L,2D} - C_L = \frac{1}{2}C_{L,2D}\frac{\frac{1}{2}cd + \frac{1}{2}cd}{bc} = \frac{1}{2}\frac{C_{L,2D}}{\frac{b}{d}} \tag{6.138}$$

Notice that $b/c = AR$ and $c/d = \sqrt{M_0^2 - 1}$, hence

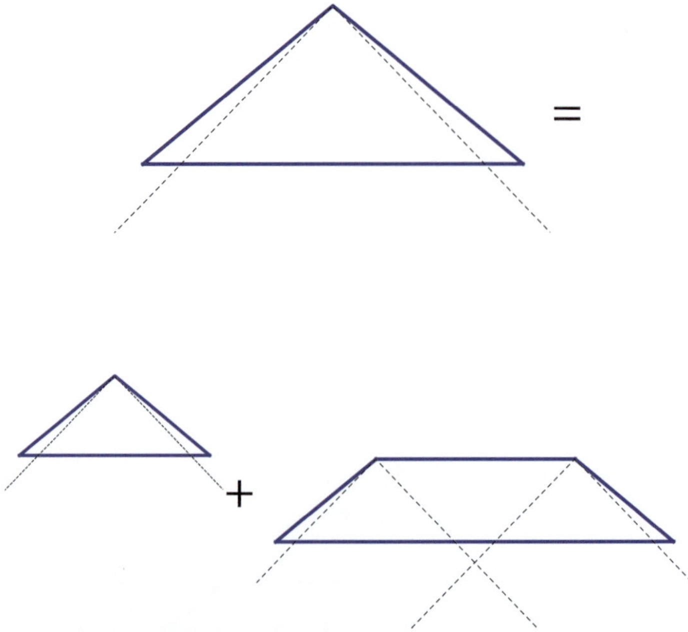

Fig. 6.37 Superposition of elementary solutions

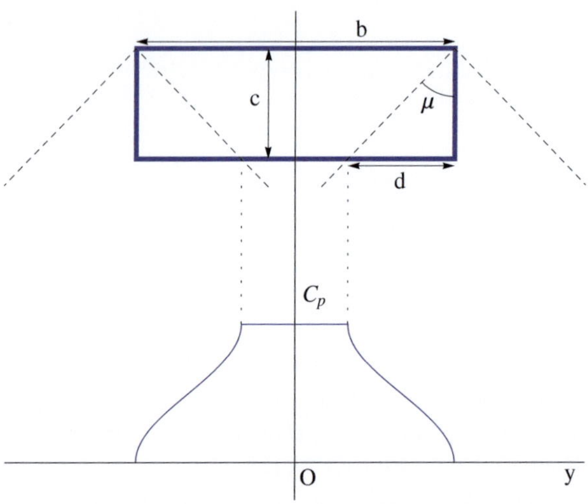

Fig. 6.38 Pressure distribution at the trailing edge (Elements of Gas Dynamics 2001 edition, by Liepmann and Roshko, with kind permission of Dover publication)

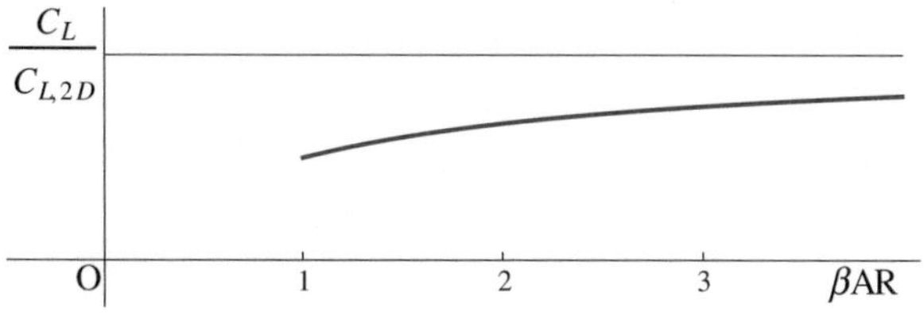

Fig. 6.39 Evolution of loss of lift coefficient with Mach number

$$\frac{C_L}{C_{L,2D}} = 1 - \frac{1}{2AR\sqrt{M_0^2 - 1}} \tag{6.139}$$

The loss of lift due to tip effect is plotted in Fig. 6.39.

Wings with subsonic edges are more complicated, due to the singularity of thin airfoil theory there.

6.11 General Wings in Subsonic and Supersonic Flows

To avoid the limitations of thin airfoil theories, the flow must be tangent to the body surface and the singularities should be distributed on the body surface rather than on the axis for a slender body, or on the base surface for a thin wing, for three-dimensional cases. In the panel methods, the body surface is divided into panels and a combination of sources and doublets (with their axis normal to the surface) are used to calculate the flow field, based on approximations of integral equations. In the following, the details for subsonic and supersonic flows are described.

6.11.1 Wings in Subsonic Flows

Panel methods are based on Green's identities. The derivation starts with Gauss' theorem (see Ref. [7])

$$\iiint_{\vartheta} \nabla.\mathbf{A}\, d\vartheta = -\iint_{S+\Sigma} \mathbf{A}.\mathbf{n}\, dS \tag{6.140}$$

where S is the surface of the obstacle and Σ is the outer boundary. The volume of fluid in between is denoted by ϑ (the minus sign results from taking the unit normal outward from the solid surface).

Let ϕ and ψ be two continuous functions with continuous first and second derivatives in the volume ϑ, and let

$$\mathbf{A} = \phi \nabla \psi \tag{6.141}$$

Substituting in Gauss' theorem yields the first Green's identity:

$$\int\int_{S+\Sigma} \phi \frac{\partial \psi}{\partial n} dS = -\int\int\int_{\vartheta} \left(\nabla \phi . \nabla \psi + \phi \nabla^2 \psi \right) d\vartheta \tag{6.142}$$

Interchanging ϕ and ψ gives a similar formula

$$\int\int_{S+\Sigma} \psi \frac{\partial \phi}{\partial n} dS = -\int\int\int_{\vartheta} \left(\nabla \psi . \nabla \phi + \psi \nabla^2 \phi \right) d\vartheta \tag{6.143}$$

Subtracting the two formulas yields the second Green's identity:

$$\int\int_{S+\Sigma} \left(\phi \frac{\partial \psi}{\partial n} - \psi \frac{\partial \phi}{\partial n} \right) dS = \int\int\int_{\vartheta} \left(\psi \nabla^2 \phi - \phi \nabla^2 \psi \right) d\vartheta \tag{6.144}$$

If both ϕ and ψ satisfy the Laplace equation, the right-hand-side vanishes and we have only surface integrals, in terms of ϕ and ψ and their derivatives normal to the surface.

Now, let ϕ be the potential of a flow over a wing and ψ be the potential due to a sink at a point P with coordinates (x_P, y_P, z_P). The potential ψ is given by

$$\psi = \frac{1}{4\pi} \frac{1}{r_P} \tag{6.145}$$

where $r_P = \sqrt{(x - x_P)^2 + (y - y_P)^2 + (z - z_P)^2}$.

To apply Green's identity, the point P must be excluded by centering a sphere σ around it with exterior normal, thus

$$\int\int_{S+\Sigma} \phi \frac{\partial \psi}{\partial n} dS + \int\int_{\sigma} \phi \frac{\partial \psi}{\partial n} d\sigma = \int\int_{S+\Sigma} \psi \frac{\partial \phi}{\partial n} dS + \int\int_{\sigma} \psi \frac{\partial \phi}{\partial n} d\sigma \tag{6.146}$$

Over the surface σ,

$$\lim_{\sigma \to 0} \int\int_{\sigma} \phi \frac{\partial \psi}{\partial n} d\sigma = -\phi_P \tag{6.147}$$

and

$$\lim_{\sigma \to 0} \int\int_{\sigma} \psi \frac{\partial \phi}{\partial n} d\sigma = 0 \tag{6.148}$$

hence,

$$\phi_P = \int\int_{S+\Sigma} \left(\frac{1}{4\pi r}\right) \frac{\partial \phi}{\partial n} dS - \int\int_{S+\Sigma} \phi \frac{\partial}{\partial n} \left(\frac{1}{4\pi r}\right) dS \tag{6.149}$$

In the above formula, the sink term $\left(\frac{1}{4\pi r}\right)$ and the doublet $\frac{\partial}{\partial n}\left(\frac{1}{4\pi r}\right)$ represent the Green's functions of the integral equation if the point P is chosen to lie on the surface S, while the integral on Σ in the far field yields the potential of the uniform flow.

The potential panel methods are based on this integral equation, where $\frac{\partial \phi}{\partial n} = 0$ on solid surfaces. After triangulation of the surface, the values of ϕ at the vertices are calculated from the algebraic equations approximating the integrals, assuming ϕ is linear over the triangles. other equivalent methods are available in literature.

The above formulation can be extended to subsonic flows via the Prandtl/Glauert transformation.

For lifting problems, special treatment of the vortex sheet is required, where the potential itself is discontinuous. In general, the vortex sheet is floating with the flow and the pressure is continuous across such a sheet. Away from the trailing edge, the sheet will roll up and the vortex core may be modeled in these calculations (the wake can be considered a contact discontinuity and associated jump conditions must be enforced).

Finally, For nonlinear subsonic flows, the governing equation can be rewritten as a Poisson's equation since

$$\nabla.(\rho\nabla\phi) = \rho\nabla^2\phi + \nabla\rho.\nabla\phi = 0 \tag{6.150}$$

hence

$$\nabla^2\phi = -\frac{1}{\rho}\nabla\rho.\nabla\phi = f \tag{6.151}$$

Iterative methods can be constructed to calculate these flows. The formulation must be extended since $\nabla^2\phi \neq 0$, with an extra term involving volume integrals of the non homogeneous term of the Poisson's equation. Panel methods become less attractive since three dimensional grids are needed and the convergence of the iterative methods become a problem as the local velocity becomes sonic.

Panel methods for transonic flows are available in literature, where the formulation is based on the unsteady flow equation and artificial viscosity is added in supersonic regions for numerical stability and to capture shock waves. The reader is referred to Refs. [12, 13], for example, for more details.

6.11.2 Wings in Supersonic Flows

The nature of supersonic flow is different from the subsonic one. The governing equations are hyperbolic, rather than elliptic, representing wave propagation. In two dimensional flows, the linearized small disturbance equation admits the d'Alembert solution (which is a special case of the method of characteristics). The corresponding solution of the three dimensional supersonic flow is based on Kirchoff's formula, used for example in acoustics, see Garrick [14]. It is given by

$$\phi_P = \frac{1}{4\pi}\int\int\left\{\frac{1}{r}\left(\frac{\partial\phi}{\partial n}\right)_{t-\frac{r}{a_\infty}} - \frac{\partial}{\partial n}\left(\frac{1}{r}\phi(t-\frac{r}{a_\infty})\right)\right\} dS$$

$$= \frac{1}{4\pi}\int\int G\left(t-\frac{r}{a_\infty}\right) dS \tag{6.152}$$

and ϕ is called "retarded potential" when t is replaced by $t-\frac{r}{a_\infty}$.
For steady supersonic flows, the linearized equation is given by

$$(1-M_\infty^2)\frac{\partial^2\phi}{\partial x^2}+\frac{\partial^2\phi}{\partial y^2}+\frac{\partial^2\phi}{\partial z^2}=0 \tag{6.153}$$

or

$$\beta^2\frac{\partial^2\phi}{\partial x^2}=\frac{\partial^2\phi}{\partial y^2}+\frac{\partial^2\phi}{\partial z^2} \tag{6.154}$$

where $\beta^2 = M_\infty^2 - 1,\ M_\infty > 1$.

Hence, the analogy between steady supersonic 3-D flow equation and the unsteady equation of acoustics in two space dimensions becomes clear.

The derivation of the main relation starts with Gauss' theorem

$$\int\int\int_\vartheta \nabla.\mathbf{A}d\vartheta = -\int\int_S \mathbf{A}.\mathbf{n}dS \tag{6.155}$$

Let

$$\mathbf{A} = \phi\left(\frac{\partial\psi}{\partial x}\mathbf{i}-\frac{\partial\psi}{\partial y}\mathbf{j}-\frac{\partial\psi}{\partial z}\mathbf{k}\right),\quad L(\phi)=\frac{\partial^2\phi}{\partial x^2}-\frac{\partial^2\phi}{\partial y^2}-\frac{\partial^2\phi}{\partial z^2},\quad (\beta^2=1) \tag{6.156}$$

then

$$\int\int_S \phi\frac{\partial\psi}{\partial\nu}dS = -\int\int\int_\vartheta\left(\frac{\partial\phi}{\partial x}\frac{\partial\psi}{\partial x}-\frac{\partial\phi}{\partial y}\frac{\partial\psi}{\partial y}-\frac{\partial\phi}{\partial z}\frac{\partial\psi}{\partial z}\right)d\vartheta - \int\int\int_\vartheta \phi L(\psi)d\vartheta \tag{6.157}$$

where $\boldsymbol{\nu}$ is the conormal to the surface $((\nu_1 = -n_1, \nu_2 = n_2, \nu_3 = n_3)$ and (n_1, n_2, n_3) are the direction cosines of the normal to the surface in general).

Interchanging ϕ and ψ and subtracting yields

$$\int\int_S \left(\phi\frac{\partial\psi}{\partial\nu} - \psi\frac{\partial\phi}{\partial\nu}\right) dS = -\int\int\int_\vartheta (\phi L(\psi) - \psi L(\phi))\, d\vartheta \tag{6.158}$$

if $L(\phi) = 0$ and $L(\psi) = 0$, we obtain

$$\int\int_S \left(\phi\frac{\partial\psi}{\partial\nu} - \psi\frac{\partial\phi}{\partial\nu}\right) dS = 0 \tag{6.159}$$

ψ is chosen to be a source in supersonic flow (with $\beta^2 \neq 1$)

$$\psi = -\frac{1}{2\pi}\frac{1}{\sqrt{(x-x_1)^2 - \beta^2(y-y_1)^2 - \beta^2(z-z_1)^2}} \tag{6.160}$$

and a doublet (with vertical axis) becomes

$$\psi = \frac{1}{2\pi}\frac{\beta^2(z-z_1)}{\left[(x-x_1)^2 - \beta^2(y-y_1)^2 - \beta^2(z-z_1)^2\right]^{\frac{3}{2}}} \tag{6.161}$$

In carrying the integration in the above formulas, the domain of influence and the domain of dependency must be taken into consideration.

Panel methods for supersonic flows are available, see Evvard [15] and Puckett and Stewart [16].

For nonlinear supersonic flows, $L(\phi) \neq 0$, again volume integrals are required and the problem is solved iteratively provided the local Mach number is always greater than one.

6.11.3 Wings in Transonic Flows

The governing equations in the transonic regime are nonlinear and of mixed type, to represent both locally subsonic and locally supersonic flow regions. The full potential equation is given by

$$\frac{\partial}{\partial x}\left(\rho\frac{\partial\phi}{\partial x}\right) + \frac{\partial}{\partial y}\left(\rho\frac{\partial\phi}{\partial y}\right) + \frac{\partial}{\partial z}\left(\rho\frac{\partial\phi}{\partial z}\right) = 0 \tag{6.162}$$

where ρ is obtained via Bernoulli's law in terms of $V^2 = (\nabla\phi)^2$. Together with the tangency boundary condition on a solid surface and imposing the right behavior of ϕ in the far field, the formulation is complete. Again, a special treatment is required to handle the wake, where ϕ is discontinuous.

The far field behavior, for subsonic incoming Mach number, is given by Klunker. Following Klunker, the integral equation for ϕ is analyzed to obtain the far field solution. The governing equation for $M_\infty < 1$ can be written in the form

$$\nabla^2 \varphi = f \tag{6.163}$$

where φ is the perturbation potential. Using

$$\psi = \frac{1}{4\pi} \frac{1}{\sqrt{(x-x_1)^2 + \beta^2 (y-y_1)^2 + \beta^2 (z-z_1)^2}} \tag{6.164}$$

Green's second identity leads to

$$\psi = \int\int\int_{\vartheta} \psi f d\vartheta + \int\int_{S} \left(\psi \frac{\partial \varphi}{\partial n} - \varphi \frac{\partial \psi}{\partial n} \right) dS \tag{6.165}$$

Asymptotically, ψ behaves as a doublet in 3-D and a horseshoe distribution over the wing, see Klunker [17].

$$\psi = \frac{1}{4\pi} \int\int_{W} \Gamma \frac{\beta z}{\beta^2 z^2 + \beta^2 (y-\eta)^2} \left(1 + \frac{x-\xi}{\sqrt{(x-\xi)^2 + \beta^2 (y-\eta)^2 + \beta^2 z^2}} \right) \beta d\xi d\eta$$
$$+ \frac{x}{4\pi \left[\left((x-\xi)^2 + \beta^2 (y-\eta)^2 + \beta^2 (z-\zeta)^2 \right) \right]^{\frac{3}{2}}} D \tag{6.166}$$

where $\Gamma = \left(\frac{\partial \psi}{\partial x}\right)_u - \left(\frac{\partial \psi}{\partial x}\right)_l$, and $D = \int\int\int_{\vartheta} \frac{\gamma+1}{2} M_\infty^2 \left(\frac{\partial \varphi}{\partial \xi}\right)^2 d\xi d\eta d\zeta$-Wing volume.

6.12 Transonic Lifting Line Theory

Cole and Cook [18], and Cheng and Meng [19], extended classical lifting line theory to unswept and swept wings in transonic flows.

For unswept wings in incompressible flows, Prandtl derived a single singular integral equation to account for the three-dimensional effects for large but finite aspect ratio wings. At each wing section, the two-dimensional theory can be used with an effective angle of attack, $\alpha_{eff} = \alpha - \alpha_i$, where

$$\alpha_i(y) = -\frac{1}{4\pi U} \int_{-\frac{b}{2}}^{\frac{b}{2}} \frac{\Gamma'(\eta)}{y-\eta} d\eta \tag{6.167}$$

The local two-dimensional theory can be replaced by numerical calculations, including viscous effects as discussed before this chapter.

To account for 3-D compressibility effects, an iterative method can be adopted as follows:

(i) First, the strip theory is used to calculate the circulation $\Gamma(y)$,
(ii) Secondly, the induced angle of attack is calculated numerically from the Prandtl singular integral equation, modified by Prandtl-Glauert's transformation for subsonic far field,
(iii) The local two-dimensional transonic flow calculations with effective angle of attack, at each spanwise location y, are performed independently,
(iv) The process is repeated until convergence.

According to Prandtl's theory, the downwash and the induced angle of attack are uniform across the span for an elliptic planform (without twist). In this case, thanks to similarity, only one section needs to be calculated.

In Hafez [20], numerical calculations confirm this result for both subsonic and transonic flows. The effective angle of attack is given by

$$\alpha_{eff} = \alpha_{2D} \frac{\beta AR}{\beta AR + 2}, \quad \beta = \sqrt{1 - M_0^2} \tag{6.168}$$

The corresponding formulas for the lift slope are

$$\frac{dC_L}{d\alpha} = 2\pi \frac{AR}{AR + 2} \quad (incompressible) \tag{6.169}$$

$$\frac{dC_L}{d\alpha} = \frac{2\pi}{\beta} \frac{\beta AR}{\beta AR + 2} \quad (subsonic) \tag{6.170}$$

$$\frac{dC_L}{d\alpha} = \left(\frac{dC_l}{d\alpha}\right)_{2D} \frac{\beta AR}{\beta AR + 2} \quad (transonic) \tag{6.171}$$

The following results of Ref. [20], confirm the validity of the above formula. The pressure distributions at the midspan for an elliptic planform ($AR = 12$) for $M_0 = 0.85$ and $\alpha = 2°$, are plotted for 2-D, corrected 2-D and 3-D calculations. The latter two sets are indistinguishable, see Fig. 6.40.

For the transonic case, the shock locus of 3-D flows for both upper and lower surfaces are plotted showing elliptic distributions, Fig. 6.41.

The three calculations are produced from NASA full potential codes.

The transonic small disturbance theory for 3-D is based on the von Karman-Guderley equation

$$\left((1 - M_0^2) - (\gamma + 1)M_0^2 \frac{\partial \phi}{\partial x}\right) \frac{\partial^2 \phi}{\partial x^2} + \frac{\partial^2 \phi}{\partial y^2} + \frac{\partial^2 \phi}{\partial z^2} = 0 \tag{6.172}$$

where ϕ is the dimensionless perturbation potential.

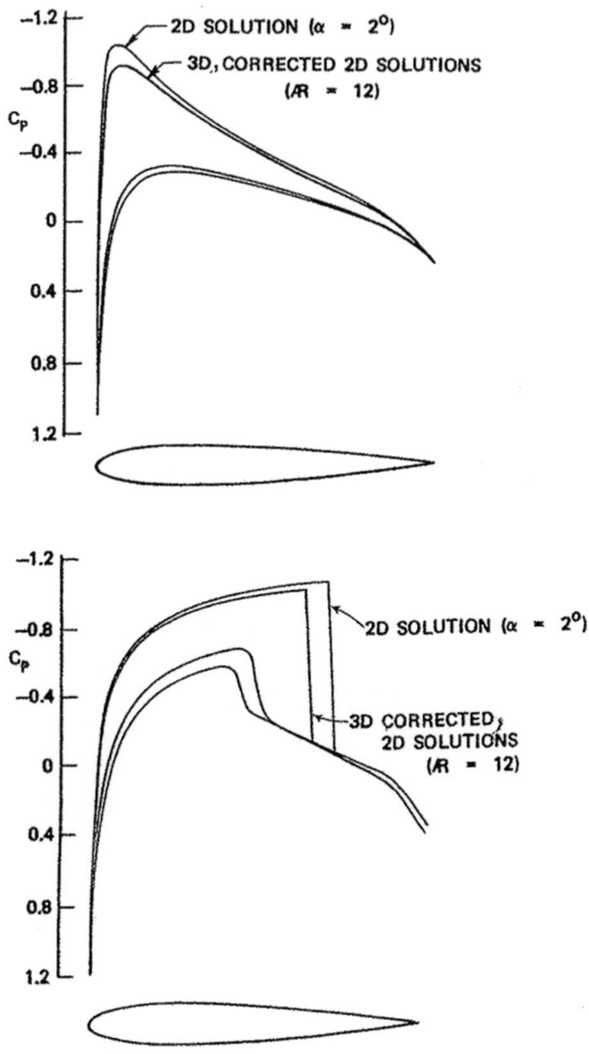

Fig. 6.40 *Top* Surface pressure distributions at mid-span, $M_0 = 0$; *Bottom* Surface pressure distributions at mid-span, $M_0 = 0.85$

The linearized boundary condition is

$$\frac{w(x, y, 0^{\pm})}{U} = \frac{\partial \phi}{\partial z}(x, y, 0^{\pm}) = \left(\frac{df^{\pm}}{dx}(x, y) - \alpha\right) \tag{6.173}$$

where $f^{\pm}(x, y)$ represents the upper and lower wing surface coordinate.

In the far-field, the potential vanishes.

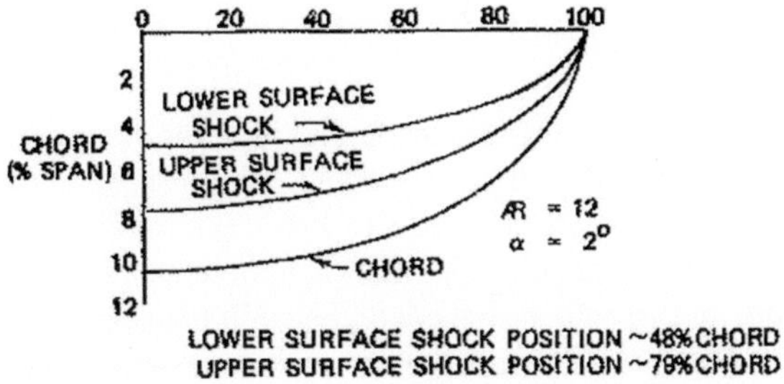

Fig. 6.41 The shock locus on elliptic wing at $M_0 = 0.85$ and $\alpha = 2°$

Across the wake there is a jump in potential to guaranty the continuity of pressure, ignoring the vortex sheet roll-up.

The lift is given by integration of surface pressure which reduces to the integration along the lifting line, of the circulation

$$A_{ref}C_L = \int_{-\frac{b}{2}}^{\frac{b}{2}} dy \int_{xle}^{xte} \left(C_p(x, y, 0^-) - C_p(x, y, 0^+)\right) dx = \frac{2}{U}\int_{-\frac{b}{2}}^{\frac{b}{2}} \Gamma(y)dy \tag{6.174}$$

As $x \to \infty$, the potential equation reduces in the Trefftz plane to

$$\frac{\partial^2\phi}{\partial y^2} + \frac{\partial^2\phi}{\partial z^2} = 0 \tag{6.175}$$

Across the vortex sheet, there is a jump in $v = U\frac{\partial\phi}{\partial y}$ corresponding to the vortex strength $\Gamma'(y) = U < v >$.

The drag is given by the integration in the Trefftz plane Σ_T, where $u = U\frac{\partial\phi}{\partial y} = 0$ and on the shock surfaces. Cole and Cook [18] showed that

$$A_{ref}C_D = \int_{\Sigma_T} \frac{1}{2}\left[\left(\frac{v}{U}\right)^2 + \left(\frac{w}{U}\right)^2\right] dydz - \frac{\gamma+1}{2}\int\int_{shocks}\left\langle\frac{u}{U}\right\rangle^3 dydz \tag{6.176}$$

The first term is a non-recoverable kinetic energy in the Trefftz plane representing induced drag. The second term is the wave drag evaluated in terms of "entropy" production at shock surfaces, which would have been produced if the nonisentropic shock jump conditions were enforced.

For given lift, elliptic loading is optimal, provided shocks are not strengthened.

Cheng and Meng [19] analyzed oblique wings using swept lifting line and analogy of 3-D formulation and unsteady two-dimensional flows.

For practical applications, the full potential equation, with exact boundary conditions, is solved numerically. Artificial viscosity is needed in the supersonic regions for numerical stability and to prevent the formation of expansion shocks as shown by Jameson [21], Caughey [22] Holst and Thomas [23] and Hafez et al. [24, 25]. The latter are admitted by the potential equation based on isentropic relation.

6.13 Summary of Chapter 6

In this chapter, inviscid incompressible and compressible flows past large and moderate aspect ratio wings are studied. Thin wings that only slightly disturb an otherwise uniform flow are defined. The three-dimensional potential flow equation governs the flow. The tangency condition is a mere extension of the two-dimensional one and the linearized pressure coefficient still reads $C_p = -2u/U$.

First the linear theories are considered and the admissible jump conditions investigated. It is shown that two types of discontinuities can occur, shock waves across which the pressure has a jump, i.e. $< C_p > \neq 0$ hence $< u > \neq 0$, and vortex sheets across which the pressure is continuous, i.e. $< C_p >= 0$ hence $< u >= 0$, and $< w >= 0$, but $< v > \neq 0$. In contrast with two-dimensional flow and its bound vorticity, three-dimensional flow combines bound vorticity in the finite wing and trailed vorticity in the vortex sheet. Forces and moments can be obtained from a momentum balance applied to a large control volume surrounding the finite wing. The formula for lift and pitching moment are consistent with the Kutta-Joukowski lift theorem applied to the bound vortex, whereas the drag exhibits two contributions, one corresponding to the wave drag, the other to the vortex drag or lift induced drag resulting from the trailed vortices.

For large aspect ratio wings, Prandtl lifting line theory is presented and the forces expressed as integrals along the span or lifting line. The matching between the local two-dimensional flow in a wing cross section and the three-dimensional flow superimposed by the vortex sheet, which determines the local downwash and induced incidence, results in the celebrated integro-differential equation of Prandtl. The general solution of the latter can be sought as an infinite Fourier series with unknown coefficients for the Fourier modes. Lift and drag are found to be related to the Fourier coefficients in a remarkable way, the lift depending on the first mode only and the induced drag on a summation with positive weights of the squares of the coefficients in the series. This leads to the ideal wing having minimum drag for a given lift and corresponding to the first mode only. The circulation is elliptical and the downwash constant. The geometry of the ideal wing is not unique. In fact, in inviscid flow, there is an infinite number of wings that can produce the ideal loading. They differ by chord and twist distributions. Examples are shown of an elliptic planform without twist and a rectangular planform with twist that satisfy the requirement. Extension to non-straight lifting lines is also discussed and the design and analysis of winglets presented. For wings of arbitrary planform, the numerical simulation is the most practical approach and illustrated with the two examples above of ideal loading. The simulation includes also viscous effects and nonlinear profile polars, with a penalization technique to capture separated flow regions on the wing.

The vortex lattice method is described for flow past moderate aspect ratio wings. It is a natural extension of the lifting line method that has close connection with the treatment of unsteady flow past wings and wind turbines (see Chap. 10).

Compressible flow over moderate aspect ratio wings based on small disturbance approximations, in both subsonic and supersonic regimes, are treated with linearized boundary conditions. General wings in subsonic flows are treated based on Green's theorems, while for general wings in supersonic flows, Kirchoff's formula is used. For wings in transonic flows, full potential equations should be solved numerically. For high aspect ratio wings, transonic lifting line theory is briefly discussed.

6.14 Problems

6.14.1 Analysis Problem

Consider a wing with elliptic planform, of span b and chord $c(y) = c_0\sqrt{1-(2y/b)^2}$. The wing has constant relative camber $d(y)/c(y) = d/c = const$ and has washout given by $t(y) = -4t_x(2y/b)^2$, where $-4t_x$ is the twist at the wing tips.

Calculate the coefficients $A_1,\ A_2, \ldots, A_n,\ n = 1, \infty$. Hint: Write Prandtl integro-differential equation, where the circulation in the l.h.s. is represented by the Fourier series, and the wing geometry is introduced in the r.h.s. Use the change of variables $y = -\frac{b}{2}\cos t,\ 0 \le t \le \pi$ and the identity $4\cos^2 t\ \sin t = \sin t + \sin 3t$.

From the equation for A_1 give the equation of the lift curve $C_L(\alpha)$. What is the incidence of zero lift, $\bar{\alpha}_0$?

Calculate the induced drag if $\alpha = 3t_x - 2d/c$.

6.14.2 Design Problem

Considering the lifting problem, design a rectangular wing of span b and chord c, ($AR = b/c$), with zero twist, and equipped with parabolic thin airfoils, such that the circulation is elliptic at the given lift coefficient $C_{L,des} = 0.5$. Furthermore, at the design $C_{L,des}$ the wing has an adapted leading edge.

Find the relative camber distribution $d(y)/c$ that will satisfy all the requirements. Where is the maximum camber located along the wing span? Calculate the value of the maximum camber.

Give the value of the geometric incidence at design, α, in terms of $C_{L,des}$ and AR.

Show that at a different incidence, $\beta \neq \alpha$, the circulation is no longer elliptic.

If viscous effects limit the maximum local lift coefficient to a value $C_{l,max}$, which is the same for all profiles, where will stall occur first on the wing?

References

1. Viviand, H.: Ailes et Corps Elancés en Théorie des petites Perturbations, Ecole Nationale Supérieure de L'Aéronautique et de l'Espace: Première Partie, Class Notes (1972)
2. Prandtl, L.: Applications of Modern Hydrodynamics to Aeronautics. NACA, report 116 (1922)
3. Glauert, H.: The Elements of Aerofoil and Airscrew Theory. Cambridge University Press, Cambridge (1926)
4. Munk, M.M.: The Minimum Induced Drag of Aerofoils. NACA, Report 121 (1921)
5. Chattot, J.-J.: Low speed design and analysis of wing/winglet combinations including viscous effects. J. Aircr. **43**(2), 386–389 (2006)
6. Chattot, J.-J.: Analysis and design of wings and wing/winglet combinations at low speeds. Comput. Fluid Dyn. J. **13**(3), 597–604 (2006)
7. Ashley, H., Landhal, H.: Aerodynamics of Wings and Bodies. Addison Wesley Longman, Reading (1965)
8. Batchelor, G.K.: An Introduction to Fluid Dynamics. Cambridge University Press, New York (1967)
9. Moran, J.: An Introduction to Theoretical and Computational Aerodynamics. Wiley, New York (1984)
10. Weissinger, J.: NASA TM 1120 (1947)
11. Liepmann, H.W., Roshko, A.: Elements of Gas Dynamics. Wiley, New York (1957)
12. Nixon, D., Hancock, G.J.: Integral equation methods—a reappraisal. Symposium Transsonicum II. Springer, New York (1976)
13. Tseng, K., Morino, L.: Nonlinear Green's function method for unsteady transonic flows. Transonic Aerodynamics. Progress in Aeronautics and Astronautics, vol. 81. AIAA, New York (1982)
14. Garrick, I.E.: Non steady wing characteristics. Aerodynamic Components of Aircraft at High Speed. Princeton University Press, Princeton (1957)
15. Evvard, J.C.: Use of Source Distributions for Evaluating Theoretical Aerodynamics of Thin Finite Wings at Supersonic Speeds. NACA, Report 951 (1950)
16. Puckett, A.E., Stewart, H.J.: Aerodynamic performance of delta wings at supersonic speeds. J. Aerosp. Sci. **14**, 567–578 (1947)
17. Klunker, E.B.: Contribution to Methods for Calculating the Flow About Thin Wings at Transonic Speeds—Analytic Expression for the Far Field. NACA TN D-6530 (1971)
18. Cole, J.D., Cook, P.: Transonic Aerodynamics. North Holland, Amsterdam (1986)
19. Cheng, H.K., Meng, S.Y.: The oblique wing as a lifting-line problem in transonic flow. J. Fluid Mech. **97**, 531–556 (1980)
20. Hafez, M.: Perturbation of transonic flow with shocks. Numerical and Physical Aspects of Aerodynamic Flows. Springer, New York (1982)
21. Jameson, A.: Transonic potential flow calculations using conservative form. In: Proceedings of 2nd AIAA CFD Conference (1975)
22. Caughey, D., Jameson, A.: Basic advances in the finite volume method for transonic potential flow calculations. Numerical and Physical Aspects of Aerodynamic Flows. Springer, New York (1982)
23. Holst, T., Thomas, S.D.: Numerical solution of transonic wing flow field. AIAA paper 82-0105 (1982)
24. Hafez, M., South, J., Murman, E.: Artificial compressibility methods for the numerical solution of the full potential equation. AIAA J. **16**, 573 (1978)
25. Hafez, M., Osher, S., Whitlow, W.: Improved finite difference schemes for transonic potential calculations. AIAA paper 84-0092 (1984)

Chapter 7
Axisymmetric Flows and Slender Body Theories

Axisymmetric flows are needed to analyze flow over bodies of revolution. Extensions to bodies of revolution with fins (missiles) or slender wings (low aspect ratio) are also possible, using perturbations of axisymmetric flows. In this chapter, it is natural to use cylindrical coordinates. After the derivation of the governing equations at different Mach number regimes (subsonic, supersonic, transonic), lift and drag are calculated for standard shapes (Fig. 7.1).

7.1 Governing Equations in Cylindrical Coordinates

Assuming steady, inviscid, adiabatic, irrotational and isentropic flow with uniform upstream conditions, the governing equations, in vector notation, are

$$\nabla.(\rho \mathbf{V}) = 0, \quad \nabla \times \mathbf{V} = \boldsymbol{\omega} = \mathbf{0} \tag{7.1}$$

In cylindrical coordinates, Fig. 7.2, the conservation of mass is

$$\frac{\partial}{\partial x}(\rho u_x) + \frac{1}{r}\frac{\partial}{\partial r}(\rho r u_r) + \frac{1}{r}\frac{\partial}{\partial \theta}(\rho u_\theta) = 0 \tag{7.2}$$

where (u_x, u_r, u_θ) are the velocity components in x, r and θ-directions, respectively.

The vorticity components are

$$\omega_x = \frac{1}{r}\frac{\partial}{\partial r}(r u_\theta) - \frac{1}{r}\frac{\partial}{\partial \theta}(u_r) \tag{7.3}$$

$$\omega_r = \frac{1}{r}\frac{\partial}{\partial \theta}(u_x) - \frac{\partial}{\partial x}(u_\theta) \tag{7.4}$$

J.J. Chattot and M.M. Hafez, *Theoretical and Applied Aerodynamics*,
DOI 10.1007/978-94-017-9825-9_7

Fig. 7.1 Slender wing-body configuration in supersonic flow (from http://www.aeronautics.nasa.gov/x_15/photos/images/26.jpg)

$$\omega_\theta = \frac{\partial}{\partial x}(u_r) - \frac{\partial}{\partial r}(u_x) \tag{7.5}$$

Notice that $\nabla.\boldsymbol{\omega} = 0$, since $\boldsymbol{\omega} = \nabla \times \mathbf{V}$. The condition is a consequence of the vector identity: the divergence of a curl vanishes.

The density can be calculated from Bernoulli's law, assuming isentropic conditions

$$\frac{p}{p_0} = \left(\frac{\rho}{\rho_0}\right)^\gamma \tag{7.6}$$

Fig. 7.2 Slender body of revolution and coordinate system

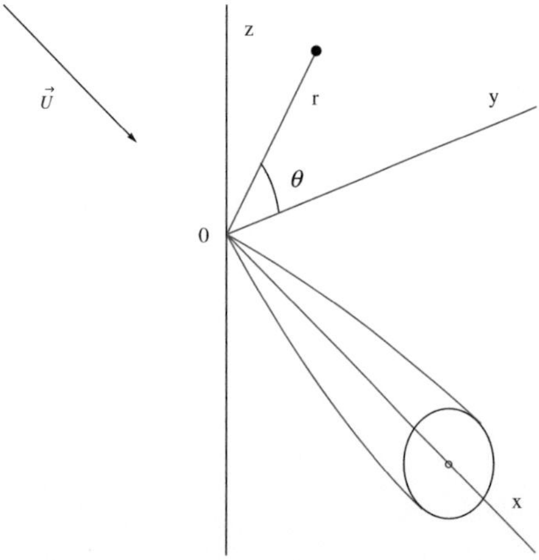

and

$$\frac{\gamma}{\gamma-1}\frac{p}{\rho}+\frac{V^2}{2}=H_0=const. \tag{7.7}$$

In the latter equation of isoenergetic flow, the potential energy is assumed constant (and does not appear in the total energy).

From the above two equations, one obtains

$$\frac{\rho}{\rho_0}=\left(1-\frac{\gamma-1}{2}M_0^2(V^2-1)\right)^{\frac{1}{\gamma-1}} \tag{7.8}$$

For the irrotational flow, the velocity can be described by the gradient of a potential function

$$\mathbf{V}=\nabla\Phi=\left(\frac{\partial\Phi}{\partial x},\frac{\partial\Phi}{\partial r},\frac{1}{r}\frac{\partial\Phi}{\partial\theta}\right) \tag{7.9}$$

The potential equation in cylindrical coordinates is given by

$$\frac{\partial}{\partial x}\left(\rho\frac{\partial\Phi}{\partial x}\right)+\frac{1}{r}\frac{\partial}{\partial r}\left(\rho r\frac{\partial\Phi}{\partial r}\right)+\frac{1}{r^2}\frac{\partial}{\partial\theta}\left(\rho\frac{\partial\Phi}{\partial\theta}\right)=0 \tag{7.10}$$

The tangency boundary condition on a solid surface is $\nabla\Phi.\mathbf{n}=0$, where $\mathbf{n}$ is the unit vector normal to the surface. Together with the far field condition, $\Phi\to 0$, as $x^2+r^2\to\infty$, the formulation is complete.

For a body of revolution at zero angle of attack, there is no dependency on θ, and the governing equation reduces to

$$\frac{\partial}{\partial x}\left(\rho\frac{\partial\Phi}{\partial x}\right)+\frac{1}{r}\frac{\partial}{\partial r}\left(\rho r\frac{\partial\Phi}{\partial r}\right)=0 \tag{7.11}$$

and the boundary condition on a solid surface is given by requiring

$$\frac{\partial\Phi}{\partial r}/\frac{\partial\Phi}{\partial x}=\frac{u_r}{u_x}=\frac{dR(x)}{dx},\quad at\ r=R(x) \tag{7.12}$$

or

$$r\frac{\partial\Phi}{\partial r}=\frac{\partial\Phi}{\partial x}R\frac{dR}{dx}=\frac{1}{2\pi}\frac{\partial\Phi}{\partial x}\frac{dS}{dx} \tag{7.13}$$

where $S(x)=\pi R^2(x)$, is the cross-sectional area of the body.

The pressure coefficient is

$$C_p=\frac{p-p_0}{\frac{1}{2}\rho V^2}=\frac{2}{\gamma M_0^2}\left(\left(\frac{\rho}{\rho_0}\right)^{\gamma}-1\right) \tag{7.14}$$

Alternatively, Stokes' stream function can be used, leading to

$$\frac{\partial}{\partial x}\left(\frac{1}{\rho}\frac{\partial \Psi}{\partial x}\right) + r\frac{\partial}{\partial r}\left(\frac{1}{r\rho}\frac{\partial \Psi}{\partial r}\right) = 0 \tag{7.15}$$

where $\psi = const.$ on the solid surface and

$$\rho u_x = \frac{1}{r}\frac{\partial \Psi}{\partial r}, \quad \rho u_r = -\frac{1}{r}\frac{\partial \Psi}{\partial x} \tag{7.16}$$

The Stokes stream function is defined by

$$\Psi - \Psi_0 = \int \{\rho r u_x dr - \rho r u_r dx\} \tag{7.17}$$

and the integral is taken along an arbitrary curve in an axial plane (x, r).

Both ϕ and ψ formulations have been used in numerical simulations of axisymmetric flows over bodies of revolution and inside nozzles.

Notice that as $M_0 \to 0$, $\rho/\rho_0 \to 1$ and, unlike the two-dimensional case, the potential and stream function equations are not the same.

7.2 Small Disturbance Theory

For slender bodies at zero angle of attack, the disturbances due to these bodies are small. Let $\mathbf{V} = \mathbf{U} + \mathbf{v}$, the perturbation potential and stream function become (see Ashley and Landhal [1])

$$\Phi = Ux + \phi, \quad \Psi = \frac{1}{2}\rho_0 U r^2 + \psi \tag{7.18}$$

Upon expanding the density, the linearized and nonlinear transonic small disturbance equations become

$$(1 - M_0^2)\frac{\partial^2 \phi}{\partial x^2} + \frac{1}{r}\frac{\partial}{\partial r}\left(r\frac{\partial \phi}{\partial r}\right) + \frac{1}{r^2}\frac{\partial^2 \phi}{\partial \theta^2} = 0 \tag{7.19}$$

and

$$\left(1 - M_0^2 - (\gamma + 1)M_0^2\frac{\partial \phi}{\partial x}\right)\frac{\partial^2 \phi}{\partial x^2} + \frac{1}{r}\frac{\partial}{\partial r}\left(r\frac{\partial \phi}{\partial r}\right) + \frac{1}{r^2}\frac{\partial^2 \phi}{\partial \theta^2} = 0 \tag{7.20}$$

respectively. Also, the pressure coefficient becomes

$$C_p = -\frac{2}{U}\frac{\partial \phi}{\partial x} - \frac{1}{U^2}\left[\left(\frac{\partial \phi}{\partial r}\right)^2 + \left(\frac{1}{r}\frac{\partial \phi}{\partial \theta}\right)^2\right] \tag{7.21}$$

Note that, for body of revolution, $\partial/\partial\theta = 0$, and the last term in the above equations vanishes.

For incompressible flows ($M_0 = 0$), the fundamental solution of the Laplace equation is

$$\phi_S = -\frac{1}{4\pi}\frac{1}{\sqrt{x^2 + r^2}} \tag{7.22}$$

For subsonic flows ($M_0 < 1$), using the Prandtl-Glauert transformation, the fundamental solution becomes

$$\phi_S = -\frac{1}{4\pi}\frac{1}{\sqrt{x^2 + \beta^2 r^2}}, \quad \beta^2 = 1 - M_0^2 \tag{7.23}$$

For a source distribution $q(x)$ per unit length along the x-axis, the perturbation potential at a point (x, r) reads

$$\phi = -\frac{1}{4\pi}\int_{-\infty}^{\infty}\frac{q(\xi)}{\sqrt{(x-\xi)^2 + \beta^2 r^2}}d\xi \tag{7.24}$$

The strength of the sources is determined through imposing the boundary condition

$$\lim_{r\to 0} r u_r = \lim_{r\to 0} r\frac{\partial \phi}{\partial r} = \frac{U}{2\pi}\frac{dS(x)}{dx} = U R(x)\frac{dR(x)}{dx} \tag{7.25}$$

where $S(x) = \pi R^2(x)$ represents the cross section of the body of revolution. Notice from mass balance, one obtains (see Fig. 7.3)

$$q(x)\Delta x = 2\pi R \Delta R\, U = U \Delta S \tag{7.26}$$

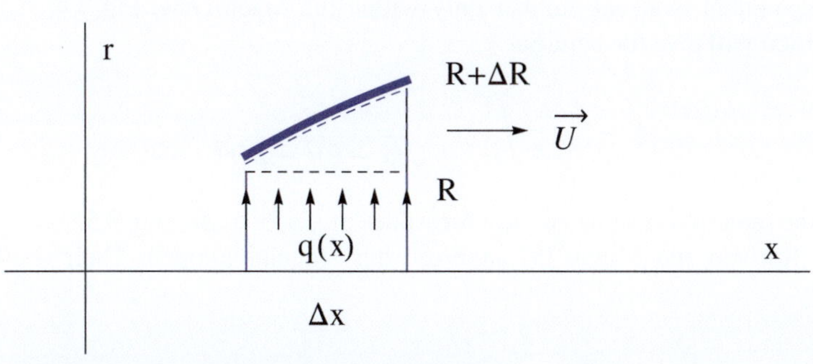

Fig. 7.3 Mass balance involving the source intensity

hence

$$q(x) = U\frac{dS(x)}{dx} \tag{7.27}$$

and

$$\phi = -\frac{U}{4\pi}\int_{-\infty}^{\infty}\frac{S'(\xi)}{\sqrt{(x-\xi)^2+\beta^2 r^2}}d\xi \tag{7.28}$$

Also, integrating the tangency condition in the proximity of the x-axis (small r)

$$r\frac{\partial\phi}{\partial r} = \frac{U}{2\pi}S'(x), \quad \Rightarrow \quad \phi \simeq \frac{U}{2\pi}S'(x)\ln r + g(x) \tag{7.29}$$

Oswatitsch and Keune [2] provided an approximation of the singular integral leading to

$$g(x) \simeq \frac{U}{4\pi}S'(x)\ln\left(\frac{\beta^2}{4x(l-x)}\right) - \frac{U}{4\pi}\int_0^l \frac{S'(\xi)-S'(x)}{|x-\xi|}d\xi \tag{7.30}$$

Note that the term $g(x)$ depends in general on M_0. For a body with a pointed nose and ending at a point or as a cylinder ($S'(l) = 0$)

$$g(x) \simeq \frac{U}{2\pi}S'(x)\ln\left(\frac{\beta}{2}\right) - \frac{U}{4\pi}\int_0^x S''(\xi)\ln(x-\xi)d\xi + \frac{U}{4\pi}\int_x^l S''(\xi)\ln(\xi-x)d\xi \tag{7.31}$$

For supersonic flow ($M_0 > 1$), the fundamental solution is different. The solution for a supersonic source is

$$\phi_S = -\frac{1}{2\pi}\frac{1}{\sqrt{x^2-\beta^2 r^2}}, \quad \beta^2 = M_0^2 - 1 \tag{7.32}$$

The potential is a real number only within the Mach cone, Fig. 7.4. A source distribution will give the solution

$$\phi = -\frac{1}{2\pi}\int_0^{x-\beta r}\frac{S'(\xi)}{\sqrt{(x-\xi)^2-\beta^2 r^2}}d\xi \tag{7.33}$$

where the integral is evaluated only for values of $\xi \le x - \beta r$, Fig. 7.5.

Von Karman and Moore [3] analyzed this singular integral. Their results are given by

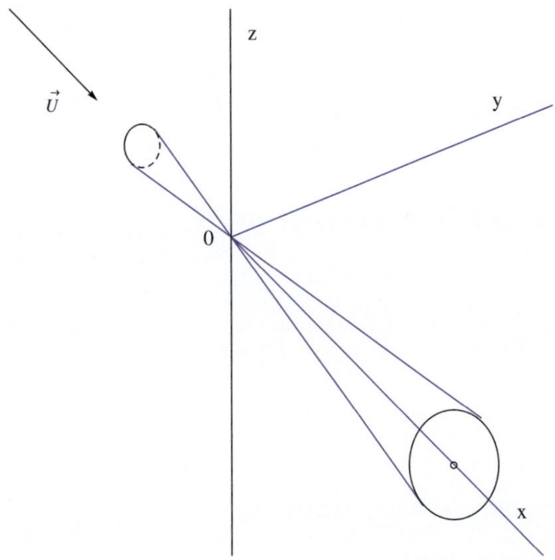

Fig. 7.4 Mach cone

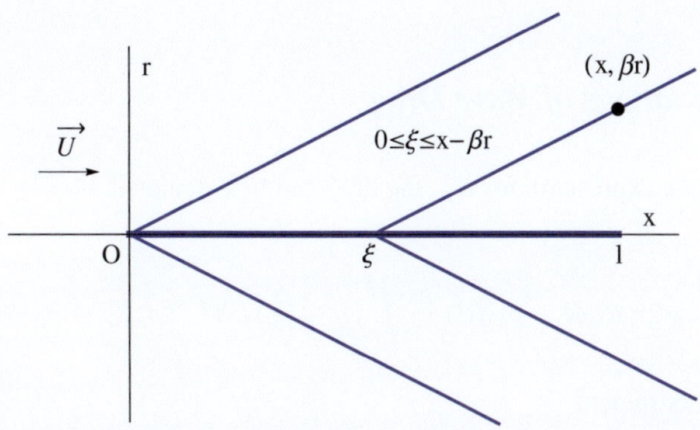

Fig. 7.5 Traces of Mach cones in meridian plane

$$\phi \simeq \frac{U}{2\pi} S'(x) \, \ln r + \frac{U}{2\pi} S'(x) \frac{\beta}{2x} - \frac{U}{2\pi} \int_0^x \frac{S'(\xi) - S'(x)}{x - \xi} d\xi \tag{7.34}$$

Thus, for supersonic flows

$$g(x) = \frac{U}{2\pi} S'(x) \, \ln \left(\frac{\beta}{2x} \right) - \frac{U}{2\pi} \int_0^x \frac{S'(\xi) - S'(x)}{x - \xi} d\xi \tag{7.35}$$

Integration of the last term by parts yields

$$g(x) = \frac{U}{2\pi} S'(x) \ln\left(\frac{\beta}{2}\right) - \frac{U}{2\pi} \int_0^x S''(\xi) \ln(x-\xi) d\xi \tag{7.36}$$

assuming $S'(0) = 0$.

The velocity components u_x and u_r are given by

$$u_x = \frac{\partial\phi}{\partial x} = \frac{U}{2\pi} S''(x) \ln\left(\frac{\beta r}{2}\right) - \frac{U}{2\pi}\frac{d}{dx} \int_0^x S''(\xi) \ln(x-\xi) d\xi \tag{7.37}$$

$$r u_r = r\frac{\partial\phi}{\partial r} = \frac{U}{2\pi}\frac{dS(x)}{dx} = U R(x)\frac{dR(x)}{dx} \tag{7.38}$$

The surface pressure coefficient is given by

$$C_p = \frac{1}{\pi} S''(x) \ln\left(\frac{2}{\beta r}\right) + \frac{1}{\pi}\frac{d}{dx} \int_0^x S''(\xi) \ln(x-\xi) d\xi - \left(\frac{dR}{dx}\right)^2 \tag{7.39}$$

7.2.1 Calculations of Wave Drag

Using the above expression for C_p, the drag can be calculated.

$$D_w = \int_0^l p\, 2\pi R dR - p_b S(l) = \int_0^l (p - p_0)\, dS - (p_{base} - p_0) S(l) \tag{7.40}$$

and the drag coefficient

$$A_{ref} C_{Dw} = \frac{D}{\frac{1}{2}\rho U^2} = \int_0^l C_p dS - C_{p,base} S_{base} \tag{7.41}$$

The last term, $C_{p,base}$, is the base pressure coefficient and depends on viscous characteristics of the flow. It will not be discussed here.

Assuming $S(l) = S'(l) = 0$, the drag coefficient is given by

$$A_{ref} C_{Dw} = \frac{1}{\pi} \int_0^l \left\{ \frac{d}{dx} \int_0^x S''(\xi) \ln(x-\xi) d\xi \right\} S'(x) dx \tag{7.42}$$

Integrating by parts, one obtains

$$A_{ref}C_{Dw}$$

$$= \frac{1}{\pi}\left\{\left[S'(x)\int_0^x S''(\xi)\ \ln(x-\xi)d\xi\right]_0^l - \int_0^l\int_0^x S''(x)\,S''(\xi)\ \ln(x-\xi)d\xi dx\right\} \tag{7.43}$$

The integrated term vanishes since $S'(0) = S'(l) = 0$. If the domain of integration is extended from the triangle, below the diagonal in the (ξ, x)-plane, to the square $[0, l] \times [0, l]$, the integral can be made to read

$$A_{ref}C_{Dw} = -\frac{1}{2\pi}\int_0^l\int_0^l S''(x)\,S''(\xi)\ \ln|x-\xi|d\xi dx \tag{7.44}$$

Notice that the above expression is independent of M_0. The analogy with the induced drag of a finite wing in incompressible flow has been mentioned in the previous chapter. The two formula are identical if one lets $\Gamma = US'$ and $[-b/2, b/2] = [0, l]$. Hence, the optimum elliptic loading corresponds to the optimum shape derived by Von Karman [4].

It can be shown that the drag for subsonic flows vanishes as the d'Alembert paradox indicates.

7.2.2 Optimum Shapes

To find the area distribution which gives the lowest wave drag under certain constraints, the Fourier series is used. Let

$$S'(x) = l\sum_{n=1}^{\infty} A_n \sin nt \tag{7.45}$$

where

$$x(t) = \frac{l}{2}(1-\cos t), \quad 0 \le t \le \pi \tag{7.46}$$

thus, the nose is located at $t = 0$ and the base at $t = \pi$.

First, rewrite the double integral as

$$A_{ref}C_{Dw} = -\frac{1}{2\pi}\int_0^l S''(x)\left\{\int_0^l S''(\xi)\ \ln|x-\xi|d\xi\right\}dx$$

$$= -\frac{1}{2\pi} \int_0^l S''(x) \left\{ \int_0^l \frac{S'(\xi)}{x-\xi} d\xi \right\} dx \tag{7.47}$$

where we have made use of $S'(0) = S'(l) = 0$.

Substituting this expression in the double integral of the wave drag and exchanging integration and summation gives

$$A_{ref} C_{Dw} = -\frac{l^2}{2\pi} \sum_{m=1}^{\infty} m A_m \sum_{n=1}^{\infty} n A_n \int_0^{\pi} \cos mt \left\{ \int_0^{\pi} \frac{\sin n\theta \, \sin\theta}{\cos\theta - \cos t} d\theta \right\} dt \tag{7.48}$$

The inner integral is to be taken as principal value and, as seen in Chap. 3, is equal to $-\pi \cos nt$. Orthogonality of the Fourier modes reduces the double summation to a single one as $m = n$

$$A_{ref} C_{Dw} = \frac{\pi l^2}{4} \sum_{n=1}^{\infty} n A_n^2 \tag{7.49}$$

The area distribution $S[x(t)]$ can be obtained via integration

$$S(t) = \frac{l^2}{4} \left\{ A_1 \left(t - \frac{\sin 2t}{2} \right) + \sum_{n=2}^{\infty} A_n \left[\frac{\sin(n-1)t}{n-1} - \frac{\sin(n+1)t}{n+1} \right] \right\} \tag{7.50}$$

Note that the first mode corresponds to an ogive blending into a cylinder at $x = l$, while all the other modes are for bodies pointed at both ends.

The total volume is thus

$$\vartheta = \frac{\pi l^3}{8} \left(A_1 + \frac{A_2}{2} \right) \tag{7.51}$$

For a given base area $S_1(l)$ for mode 1

$$A_1 = \frac{4}{\pi} \frac{S_1(l)}{l^2} \neq 0 \tag{7.52}$$

The minimum wave drag is given by the conditions $A_n = 0, \ n > 1$, hence

$$A_{ref} C_{Dw} = \frac{4}{\pi} \frac{S_1(l)^2}{l^2} \tag{7.53}$$

Choosing the base area as reference area

$$C_{Dw} = \frac{4}{\pi} \frac{S_1(l)}{l^2} \tag{7.54}$$

The wave drag for the optimum ogive is

$$D_w = \frac{2}{\pi}\rho_0 U^2 \frac{S_1(l)^2}{l^2} \tag{7.55}$$

The associated area distribution is given by

$$S_1(t) = \frac{S_1(l)}{\pi}\left(t - \frac{\sin 2t}{2}\right),\ 0 \le t \le \pi \tag{7.56}$$

This is the Von Karman ogive [4].

On the other hand, if the volume is given by a body pointed at both ends, then

$$A_1 = 0,\quad A_2 = \frac{16\vartheta}{\pi l^3},\quad A_3 = ... = A_n = 0,\ n \ge 3 \tag{7.57}$$

For minimum drag, all the other Fourier coefficients must vanish, leading to the Sears-Haaks body, where the area is

$$S_2(t) = \frac{l^2}{3} A_2 \sin^3 t,\ 0 \le t \le \pi \tag{7.58}$$

The maximum cross-section of the body is at $x = l/2$ $(t = \pi/2)$ and can be chosen as reference area, $A_{ref} = l^2 A_2/3$, to give

$$C_{Dw} = 24\frac{\vartheta}{l^3},\quad D_w = \frac{64}{\pi}\rho_0 U^2 \frac{\vartheta^2}{l^4} \tag{7.59}$$

The two distributions are plotted in Fig. 7.6. Note that only the front half of the Sears-Haaks [5] body is represented with maximum cross section located at $x = l$.

Interestingly, both geometries have a singular behavior at the nose, when considering the radius of the body of revolution with the same cross section. Let $R(x)$ be the radius such that $S(x) = \pi R(x)^2$. One finds the following results

$$as\ t \to 0,\quad x \propto t^2,\quad S \propto t^3,\quad R \propto t^{\frac{3}{2}},\quad \frac{dS}{dx} \propto t,$$

$$\frac{dR}{dx} \propto t^{-\frac{1}{2}},\quad \frac{d^2S}{dx^2} \propto t^{-1},\quad \frac{d^2R}{dx^2} \propto -t^{-\frac{5}{2}} \tag{7.60}$$

The nose is "blunt" with a vertical tangent and zero radius of curvature, Fig. 7.7.

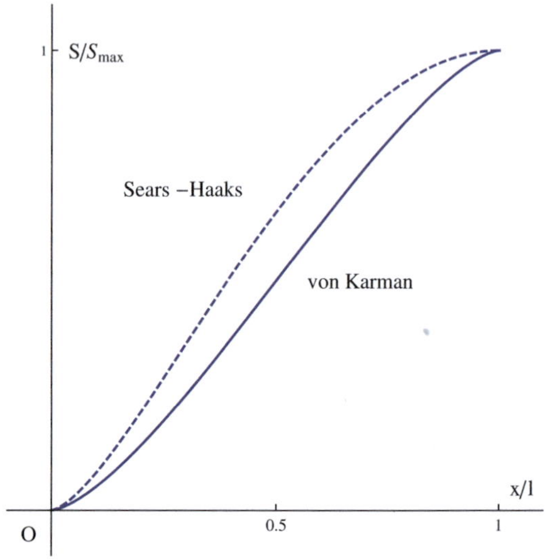

Fig. 7.6 Von Karman and half Sears-Haaks area distributions

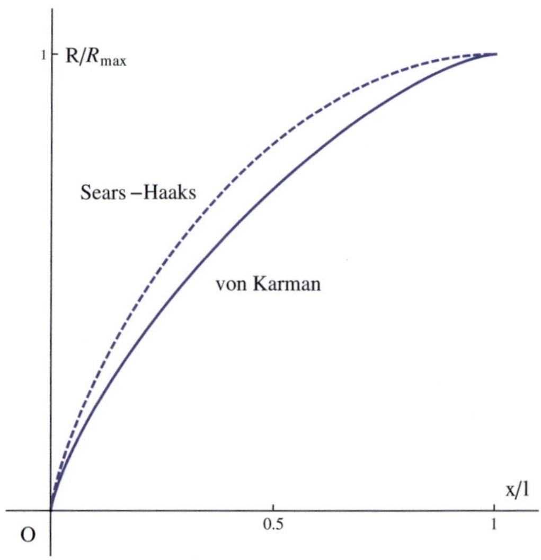

Fig. 7.7 Von Karman and half Sears-Haaks ogive geometries

7.3 Lift and Induced Drag of a Body of Revolution at Angle of Attack

The flow situation is depicted in Fig. 7.8. The perturbation potential ϕ is governed in this case by the three dimensional equation

$$(1 - M_0^2)\frac{\partial^2 \phi}{\partial x^2} + \frac{1}{r}\frac{\partial}{\partial r}\left(r\frac{\partial \phi}{\partial r}\right) + \frac{1}{r^2}\frac{\partial^2 \phi}{\partial \theta^2} = 0 \tag{7.61}$$

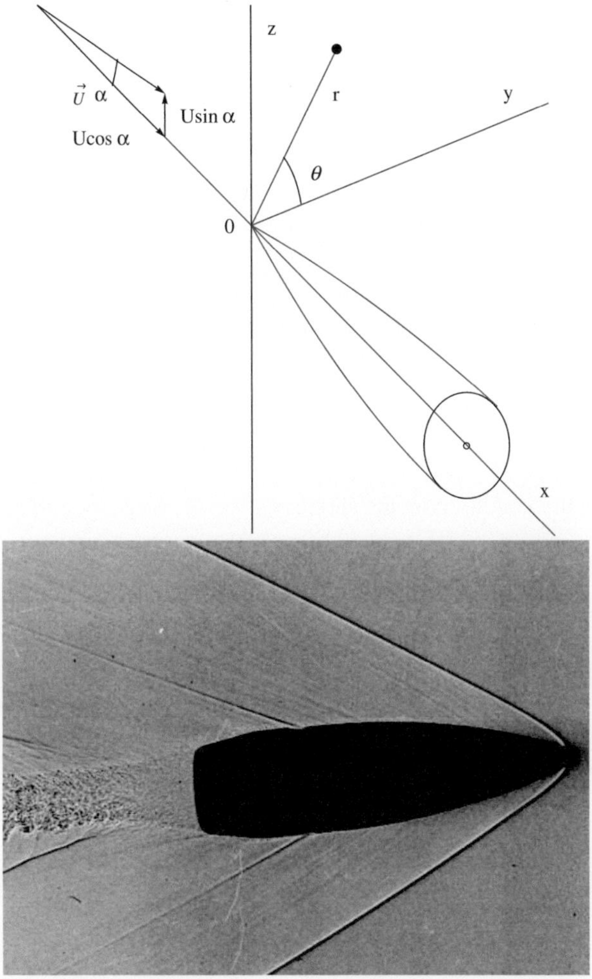

Fig. 7.8 Body of revolution at angle of attack: *Top*, coordinate system; *Bottom*, bullet at supersonic speed (from http://en.wikipedia.org)

The boundary condition is given by

$$\left(\frac{u_r}{u_x}\right)_{body} = \frac{dR}{dx} \tag{7.62}$$

In terms of the incoming flow and perturbation velocity, this becomes

$$\left(\frac{U \sin\alpha \sin\theta + \frac{\partial\phi}{\partial r}}{U \cos\alpha + \frac{\partial\phi}{\partial x}}\right)_{body} = \frac{dR}{dx} \tag{7.63}$$

Because of linearity, ϕ can be written as (see Liepmann and Roshko [6])

$$\phi(x, r, \theta) = \phi_0(x, r) + \phi_\alpha(x, r, \theta) \tag{7.64}$$

where ϕ_0 is governed by the equation for axisymmetric flow ($\frac{\partial}{\partial\theta} = 0$) with boundary condition ($\alpha = 0$)

$$\left(\frac{\frac{\partial \phi_0}{\partial r}}{U + \frac{\partial \phi_0}{\partial x}} \right)_{body} = \frac{dR}{dx} \tag{7.65}$$

and ϕ_α is solution of the full equation that can be modeled as a doublet distribution along the x-axis. The axis of the doublet is parallel to the z-axis to represent the cross flow, namely

$$\phi_\alpha(x, r, \theta) = \frac{\partial \phi_0}{\partial z} = \sin\theta \frac{\partial \phi_0}{\partial r} \tag{7.66}$$

with the boundary condition on the surface as

$$\left(\frac{dR}{dx}\frac{\partial \phi_\alpha}{\partial x} - \frac{\partial \phi_\alpha}{\partial r} \right)_{body} = U \sin\alpha \sin\theta \tag{7.67}$$

This last equation is obtained to second-order $O(\alpha^2)$ from the general boundary condition for ϕ, upon subtraction of the axisymmetric boundary condition for ϕ_0.

Notice that the first term in the left-hand-side is very small for slender configurations. In this case

$$\phi_\alpha = \frac{q(x)}{r} \sin\theta \tag{7.68}$$

The strength of the doublet $q(x)$ is proportional to the cross sectional area

$$\phi_\alpha = U \sin\alpha \sin\theta \frac{R^2(x)}{r} = \frac{U}{\pi} \sin\alpha \sin\theta \frac{S(x)}{r} \tag{7.69}$$

Therefore, the cross flow solution at any section is the same as the incompressible flow solution over a cylinder, independent of M_0.

The surface pressure coefficient is given by

$$C_p \simeq -\frac{2}{U\cos\alpha}\frac{\partial \phi}{\partial x} - \frac{1}{(U\cos\alpha)^2}\left[\left(\frac{\partial \phi}{\partial r}\right)^2 + \left(\frac{1}{r}\frac{\partial \phi}{\partial \theta}\right)^2 \right] \tag{7.70}$$

hence

$$C_{p,0} \simeq -\frac{2}{U\cos\alpha}\frac{\partial \phi}{\partial x} - \left(\frac{dR}{dx}\right)^2 \tag{7.71}$$

and

$$C_{p,\alpha} \simeq -\frac{2}{U\cos\alpha}\frac{\partial\phi_\alpha}{\partial x} = -4\alpha\frac{dR}{dx}\sin\theta \tag{7.72}$$

In the latter expression, α^2 terms are neglected.

The normal force in the z-direction can be calculated via integration

$$N = \frac{1}{2}\rho_0 U^2 \int_0^l \int_0^{2\pi} \left(-C_{p,\alpha}\sin\theta\right) R d\theta dx$$

$$= 4\alpha\frac{1}{2}\rho_0 U^2 \int_0^l \int_0^{2\pi} \sin^2\theta R\frac{dR}{dx} d\theta dx = 2\alpha\frac{1}{2}\rho_0 U^2 S(l) \tag{7.73}$$

Hence, the normal force coefficient, using the base area as reference, is given by

$$C_N = 2\alpha \tag{7.74}$$

Notice that, if the body is closed at both ends, $N = 0$.

The calculation of the drag is more delicate, since it is a small quantity (of order α^2). For slender bodies, $C_L \simeq C_N = 2\alpha$, independent of M_0, and the lift slope is $dC_L/d\alpha = 2$ for incompressible and compressible flows.

If one considers a cross section of the body at a given x location and let the circulation be defined on the body as

$$\Gamma(x,\theta) = \int_{-\theta}^{\theta} u_\theta R d\theta, \quad 0 \le \theta \le \pi \tag{7.75}$$

then

$$\Gamma(x,\theta) = \phi_\alpha(x,R,\theta) - \phi_\alpha(x,R,-\theta)$$

$$= 2\phi_\alpha(x,R,\theta) = 2U\sin\alpha R(x)\sin\theta = 2U\sin\alpha\sqrt{R^2(x) - y^2} \tag{7.76}$$

In term of the "span" variable y, this is an elliptic loading with induced drag $C_{Di} = \alpha_i C_L$.

It can be shown that $\alpha_i = \alpha/2$ and $C_{Di} = \alpha^2$.

According to Viviand [7], the induced drag coefficient is given by the same formula as for a finite wing

$$A_{ref} C_{Di} = \int_{\Sigma base} \left[\left(\frac{v}{U}\right)^2 + \left(\frac{w}{U}\right)^2\right] dydz \tag{7.77}$$

where the surface integral is performed on Σ_{base}, the part of the cross flow plane outside the base of the body and approximately perpendicular to the incoming flow, Fig. 7.9.

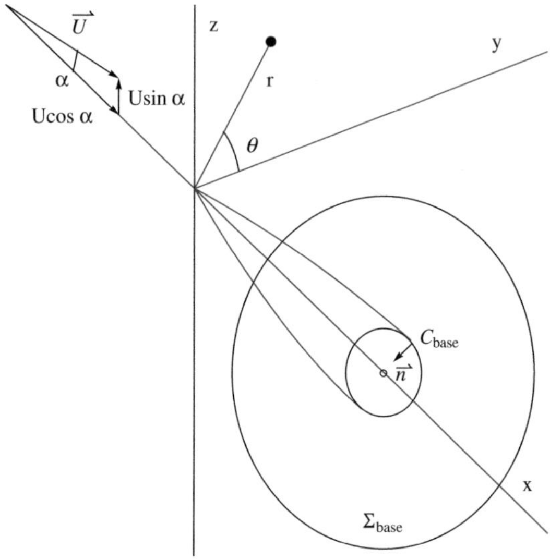

Fig. 7.9 Base plane for integration of induced drag

As seen in Chap. 6, this surface integral can be reduced to a line integral along the body base, C_{base}, as

$$A_{ref}C_{Di} = \frac{1}{U^2}\oint_{Cbase} \phi_\alpha\,(\nabla\phi_\alpha.\mathbf{n})\,dl = \frac{1}{U^2}\oint_{Cbase} \phi_\alpha \frac{\partial\phi_\alpha}{\partial n} dl \tag{7.78}$$

where **n** is the normal pointing into the body. Here we have

$$\frac{\partial\phi_\alpha}{\partial n} = -\frac{\partial\phi_\alpha}{\partial r} = U\sin\alpha\sin\theta \tag{7.79}$$

Hence

$$A_{ref}C_{Di} = \sin^2\alpha R^2(l)\oint_{Cbase} \sin^2\theta d\theta = S(l)\sin^2\alpha \simeq S(l)\alpha^2 \tag{7.80}$$

With the base as reference area, one finds

$$C_{Di} = \alpha^2 \tag{7.81}$$

The induced incidence is such that $C_{Di} = C_L\alpha_i$. But, as seen previously, $C_L = 2\alpha$, hence $C_{Di} = \alpha^2 = 2\alpha\alpha_i$, which proves that $\alpha_i = \alpha/2$.

For completeness, the pitching moment about the nose is obtained by integration as (derivation in Chap. 10)

$$A_{ref}L_{ref}C_{M,o} = 2\vartheta\alpha \tag{7.82}$$

7.4 Low Aspect Ratio Flat Wings

The cross flow over a flat plate is known from Chap. 2, Eq. (2.84), and with the current coordinate system and notation, the velocity components on the plate, in terms of the perturbation potential, are (see Katz and Plotkin [8], Duncan et al. [9])

$$\frac{\partial \phi}{\partial y} = \mp 2U \sin\alpha \frac{\frac{2y}{b}}{\sqrt{1-\left(\frac{2y}{b}\right)^2}} \simeq \mp 2U\alpha \frac{\frac{2y}{b}}{\sqrt{1-\left(\frac{2y}{b}\right)^2}}, \qquad \frac{\partial \phi}{\partial z} = 0 \tag{7.83}$$

where $b(x)$ is the wing span, the plus sign is for the lower surface, see Fig. 7.10.

Since $\partial u/\partial y = \partial v/\partial x$, one can find u by integration of $\partial u/\partial y$

$$\frac{\partial u}{\partial y} = \frac{\partial}{\partial x}\left\{\mp 2U\alpha \frac{\frac{2y}{b}}{\sqrt{1-\left(\frac{2y}{b}\right)^2}}\right\}, \quad \Rightarrow$$

$$u = \frac{\partial \phi}{\partial x} = \mp 2U\alpha \frac{\partial}{\partial x}\int_0^y \frac{\frac{2\eta}{b}}{\sqrt{1-\left(\frac{2\eta}{b}\right)^2}} d\eta = \pm U b\alpha \frac{\partial}{\partial x}\sqrt{1-\left(\frac{2y}{b}\right)^2} \tag{7.84}$$

From the Bernoulli equation, the pressure jump at a point $(x, y, 0)$ of the wing is given by $\Delta p = p(x, y, 0^-) - p(x, y, 0^+)$. This reduces to

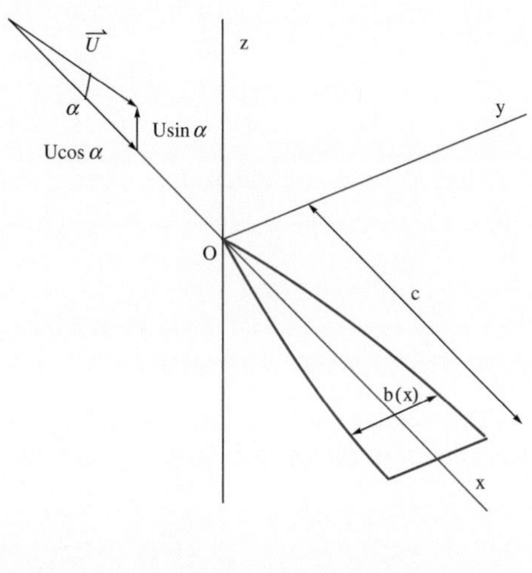

Fig. 7.10 Low aspect ratio flat wing

$$\Delta p = \frac{1}{2}\rho\left[(U+u(x,y,0^+))^2 - (U+u(x,y,0^-))^2\right] \simeq 2\rho U^2 b\alpha \frac{\partial}{\partial x}\sqrt{1-\left(\frac{2y}{b}\right)^2} \tag{7.85}$$

A section of the wing of length dx will contribute a lift force dL as

$$dL = dx\int_{-\frac{b}{2}}^{\frac{b}{2}} \Delta p\, dy = dx\, 2\rho U^2 b\alpha \frac{\partial}{\partial x}\int_0^{\pi} \sin t\, \frac{b}{2}\sin t\, dt \tag{7.86}$$

where we have made the change of variables $y = -\frac{b}{2}\cos t,\ 0 \le t \le \pi$. The result reads

$$dL = \frac{\pi}{2}\rho U^2 \alpha b \frac{db}{dx} dx \tag{7.87}$$

Upon integration in x, the partial lift force, from $x = 0$ to x becomes

$$L(x) = \frac{\pi}{4}\rho U^2 b^2 \alpha \tag{7.88}$$

For the whole wing, $L = \frac{\pi}{4}\rho U^2 b^2(c)\alpha$. Let A_{ref} be the wing projected area, then the lift coefficient can be written as

$$C_L = \frac{\pi AR}{2}\alpha \tag{7.89}$$

The v-component provides, upon integration, the perturbation potential

$$\phi(x,y,0^{\pm}) = \mp 2U\alpha \int_0^y \frac{\frac{2\eta}{b}}{\sqrt{1-\left(\frac{2\eta}{b}\right)^2}} d\eta = \pm 2Ub\alpha\sqrt{1-\left(\frac{2y}{b}\right)^2} + f(x) \tag{7.90}$$

Since ϕ is an odd function of z and ϕ is continuous at the wing edge, $f(x) = 0$. The jump of ϕ is the circulation $\Gamma(x, y)$, hence

$$\Gamma(x,y) = 4Ub\alpha\sqrt{1-\left(\frac{2y}{b}\right)^2} \tag{7.91}$$

The slender flat wing has elliptic loading. The induced drag is therefore given by the same formula as for a large aspect ratio wing, i.e.

$$C_{Di} = \frac{C_L^2}{\pi AR} = \frac{\pi AR}{4}\alpha^2 \tag{7.92}$$

The induced incidence α_i is such that $C_{Di} = -C_L\alpha_i$. One finds $\alpha_i = -\alpha/2$. The resulting force due to pressure is not perpendicular to the flat wing, as would be expected, because as in thin airfoil theory, the wing edge experiences a suction force due to the infinite velocity and vacuum pressure at the sharp edge. The force is in the $z = 0$ plane and its x-component is equal to $F_s = -L\alpha/2$. Note that the suction force does not exist if the leading edge is supersonic.

For triangular wings, the loading is constant along any straight line through the vertex (i.e. conical flow), the angle t with the x-axis being constant along such a line since $\sin t = \sqrt{1 - \left(\frac{2y}{b}\right)^2} = const.$

The pitching moment about the y-axis is given by

$$M_{,o} = -\int_0^c x dL = -\frac{\pi}{2}\rho U^2\alpha \int_0^c b\frac{db}{dx}x dx = -\frac{\pi}{4}\rho U^2\alpha \int_0^c \frac{db^2}{dx}x dx \quad (7.93)$$

For triangular wing with $b(x) = b\frac{x}{c}$, hence

$$M_{,o} = -\frac{\pi}{4}\rho U^2\frac{b^2}{c^2}\alpha \int_0^c \frac{dx^2}{dx}x dx = -\frac{\pi}{6}\rho U^2 b^2 c\alpha \quad (7.94)$$

The pitching moment is made dimensionless with the wing area and the chord as

$$C_{M,o} = -\frac{\pi AR}{3}\alpha \quad (7.95)$$

The center of pressure, as well as the aerodynamic center, for the flat plate wing are found to be located at

$$\frac{x_{c.p.}}{c} = -\frac{C_{M,o}}{C_L} = \frac{2}{3} \quad (7.96)$$

Finally, between the body of revolution and the flat wing, one can think of wings with elliptic cross sections. The flat plate cross section is the limiting case of the ellipse.

The order of magnitude for the lift coefficient in this section is consistent with the lifting line theory for elliptic loading of small aspect ratio wings, where C_L is proportional to $\pi AR\alpha$, and independent of M_0.

Following Weissinger [10], the low and high aspect ratio wing formulae are combined in a single expression for extended lifting line theory and after applying the Prandtl/Glauert rule, yielding

$$C_L = \frac{\frac{2\pi}{\beta}\beta AR}{\sqrt{\beta^2 AR^2 + 4} + 2} \quad (7.97)$$

7.5 Swept and Oblique Wings

For high speeds, swept wings are used since the velocity component normal to the leading edge determines the aerodynamic performance of the wing. R. T. Jones [11, 12] claims, based on theoretical and experimental results, that oblique wings are better, see Fig. 7.11.

Comparison with a delta wing is also shown in Fig. 7.12.

Jones used Kogan's theory [13] to determine minimum drag of elliptic wings. For details, see Ref. [11].

7.6 Wing-Body Combinations

A body of revolution, carrying plane wings of small aspect ratio can be analyzed using the results of the previous sections, via the Joukowski transformation.

Consider a mid-winged body of revolution as in Fig. 7.13.

For the calculation of the lift, we will follow Spreiter [14], who solved the cross flow problem by mapping the cylinder onto the Y-axis using

Fig. 7.11 Comparison of oblique and sweptback wings

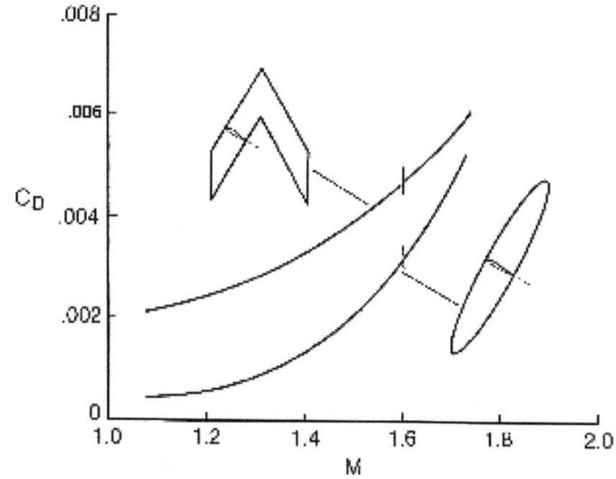

Fig. 7.12 Drag due to lift: oblique elliptic wing and delta wing, $M = \sqrt{2}$

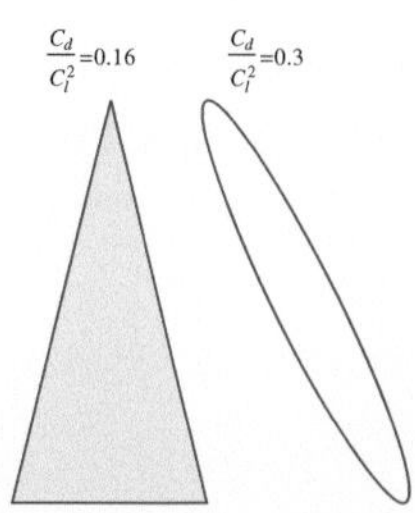

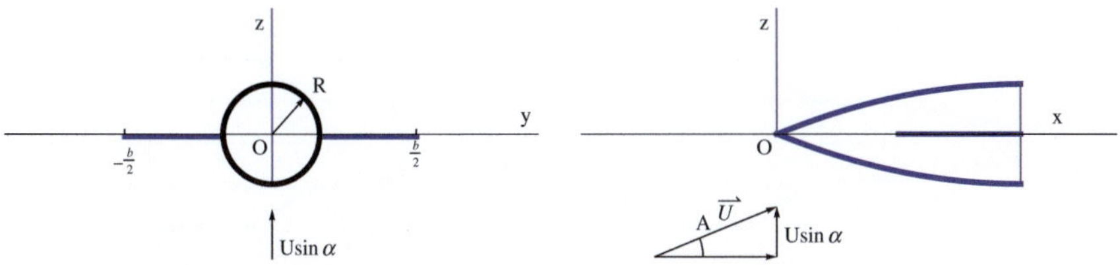

Fig. 7.13 Wing-body combination

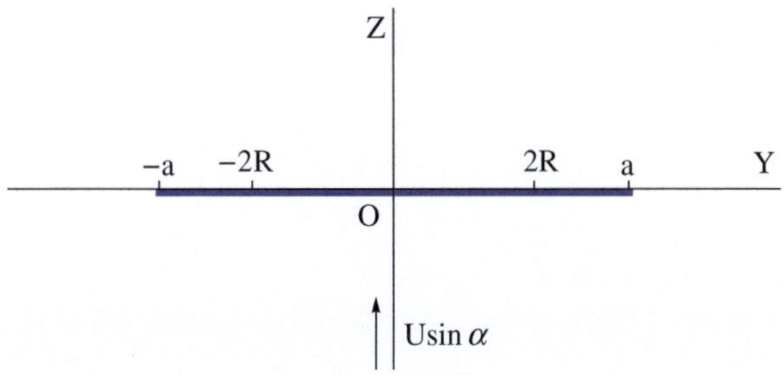

Fig. 7.14 Mapping of wing-body cross section onto the Y coordinate axis

$$Y = y\left(1 + \frac{R^2}{y^2 + z^2}\right), \quad Z = z\left(1 - \frac{R^2}{y^2 + z^2}\right) \tag{7.98}$$

The wings map also onto the Y-axis and the wing edges are located at $Y = \pm\frac{b}{2}\left(1 + \frac{4R^2}{b^2}\right) = \pm a$, Fig. 7.14.

The problem of the flow normal to a flat plate has been discussed before (the flow over a flat plate is related to a flow over a cylinder).

It should be mentioned that the two-dimensional cross flow solution is unrealistic, since the flow will separate at the plate edges. With a three-dimensional slender wing-body configuration, the axial flow keeps the cross flow from separating so the potential solution is a good approximation, for small angles of attack.

Spreiter [14] obtained the following result for the lift (see also Ashley and Landhal [1]):

$$L = \pi\rho U^2\left(\frac{b^2}{4} - R^2 + \frac{4R^4}{b^2}\right)\alpha \tag{7.99}$$

Special cases of the body alone or the wing alone are included in this formula.

To estimate the effect of the fuselage on total lift, i.e. body interference, consider the ratio

$$\frac{L_{W+B}}{L_W} = 1 - \left(\frac{2R}{b}\right)^2 + \left(\frac{2R}{b}\right)^4 \tag{7.100}$$

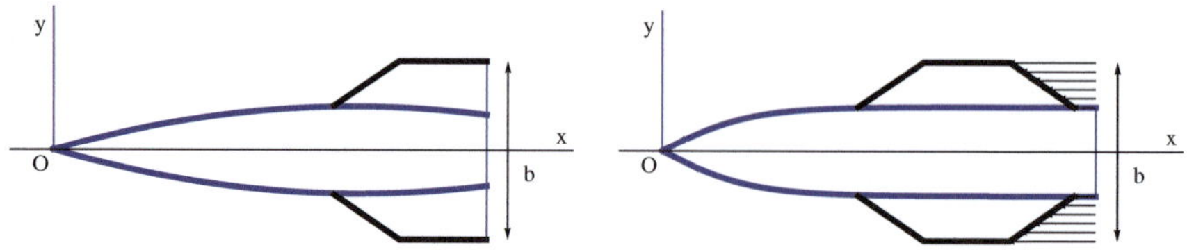

Fig. 7.15 Two wing-body configurations

The above ratio is less than one, since $b > 2R$.

The inviscid drag for supersonic flow is given by

$$D = D_0 + \frac{\pi}{2}\rho U^2 \frac{b^2}{4}\left[1 - \left(\frac{2R}{b}\right)^2 + \left(\frac{2R}{b}\right)^4\right]\alpha^2 \tag{7.101}$$

where D_0 is the wave drag at zero incidence.

The drag for subsonic flow reduces to induced drag since $D_0 = 0$, and in case of $R = 0$, it becomes

$$D = \frac{\pi}{2}\rho U^2 \frac{b^2}{4}\alpha^2 \tag{7.102}$$

The above formulas are valid for two types of configurations: a body of revolution having general cross section with wings whose edges are everywhere leading edges, and for a body of revolution with a uniform cross section downstream from the section of maximum wing span, as shown in the following sketches, Fig. 7.15.

For the case of an uncambered wing with swept-forward trailing edges, the lift on sections behind that of maximum span is zero in the present approximation and b is replaced by b_{max} in the above formulas.

In the case of swept-back trailing edges, as for an arrowhead wing, the analysis requires the solution of hyper integral equations (see Mangler [15]). See also Brandao [16].

For supersonic flows past bodies of revolution, when $S'(x)$ is discontinuous, the reader is referred to the work of Lighthill [17].

For ducted bodies of revolution with annular intakes, see Ward [18–20].

7.7 Slender Bodies with General Cross Sections

Consider a slender body, with a contour A at x location and contour B at $x + dx$, as shown in Fig. 7.16.

In the neighborhood of the body, the governing equation in terms of a small disturbance potential is the Laplace equation in the (y, z) cross plane

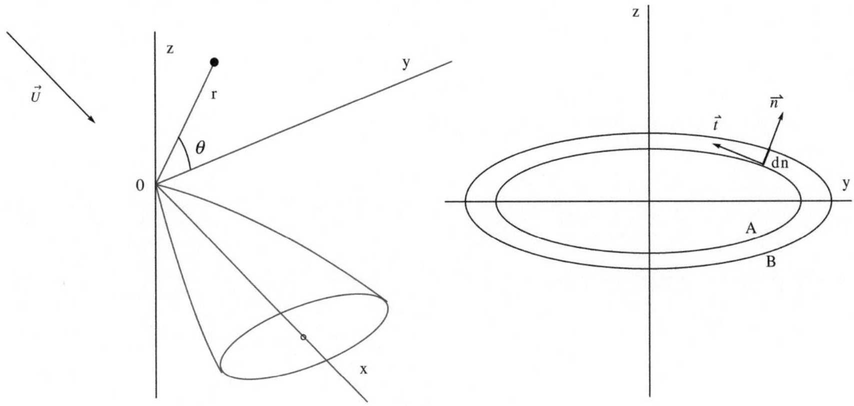

Fig. 7.16 Slender body with general cross section

$$\frac{\partial^2 \phi}{\partial y^2} + \frac{\partial^2 \phi}{\partial z^2} = 0 \tag{7.103}$$

The boundary condition requires the flow to be tangent to the surface

$$\frac{\partial \phi}{\partial n}|_{body} = U\frac{dn}{dx} \tag{7.104}$$

where n and t are in the normal and tangential directions of the contour of the cross section, respectively.

Applying Green's theorem in two-dimensions, yields a particular solution

$$\phi = \frac{1}{2\pi} \oint \left(\frac{\partial \phi}{\partial n} - \phi \frac{\partial}{\partial n} \right) \ln r \, dt \tag{7.105}$$

where $r = \sqrt{(y-\eta)^2 + (z-\zeta)^2}$.

For large r, ϕ can be approximated by

$$\phi = \frac{1}{2\pi} \ln r \oint \frac{\partial \phi}{\partial n} dt \tag{7.106}$$

In this approximation, $\frac{\partial}{\partial n} \ln r$ is neglected compared to $\ln r$ and $r \simeq \sqrt{y^2 + z^2}$. From the boundary condition

$$\oint \frac{\partial \phi}{\partial n} dt = U \oint \frac{dn}{dx} dt = U \frac{dS}{dx} \tag{7.107}$$

where $S(x)$ is the cross sectional area.

Hence

$$\phi \simeq \frac{U}{2\pi} S'(x) \ln r \tag{7.108}$$

The general solution is then

$$\phi \simeq \frac{U}{2\pi} S'(x) \ln r + g(x) \tag{7.109}$$

The pressure coefficient is

$$C_p = -\frac{2}{U}\frac{\partial \phi}{\partial x} - \frac{1}{U^2}\left[\left(\frac{\partial \phi}{\partial y}\right)^2 + \left(\frac{\partial \phi}{\partial z}\right)^2\right] \tag{7.110}$$

This is the same as the solution for an axisymmetric body having the same cross sectional area distribution as the actual body with a general cross section.

Oswatitsch and Keune [2], Ward [18–20] and Ashley, and Landhal [1] were among the first who derived the *equivalence rule* for transonic, subsonic and supersonic flows: "Far away from a general slender body, the flow becomes axisymmetric and equal to the flow around the equivalent body of revolution." "Near the slender body, the flow differs from that around the equivalent body of revolution by a two-dimensional constant-density cross flow part that makes the tangency condition at the body surface satisfied." (See also Harder and Klunker [21], Guderley [22] and Cole and Cook [23]).

One can solve the axisymmetric small disturbance equation for transonic, subsonic or supersonic flows over the equivalent body of revolution to provide the far field boundary condition for a two-dimensional cross flow calculation at each cross section.

It can be shown that the pressure drag of the general slender body is equal to the drag of the equivalent body of revolution if the body ends in a point ($S = 0$) or in a cylindrical portion parallel to the free stream ($S' = 0$ and $\frac{\partial \phi}{\partial n} = 0$). It is also true if the body ends with an axisymmetric portion.

Whitcomb [24] had verified these results experimentally, and the agreement was good in the transonic regime. Therefore, it is possible to reduce the wave drag of a wing-body combination by indenting the body such that the equivalent body of revolution has a smooth area distribution. This is the basis of "transonic area rule."

Whitcomb has also shown experimentally that at high angle of attack, the lift contributes to the cross sectional area distribution of the equivalent body.

Cheng and Hafez [25], and Cheng [26], developed a theoretical justification in the transonic regime due to the nonlinearity of the governing equation, therefore in a good design, the fuselage will have a reduced waist at the location of the wing. See Fig. 7.17 for several arrangements.

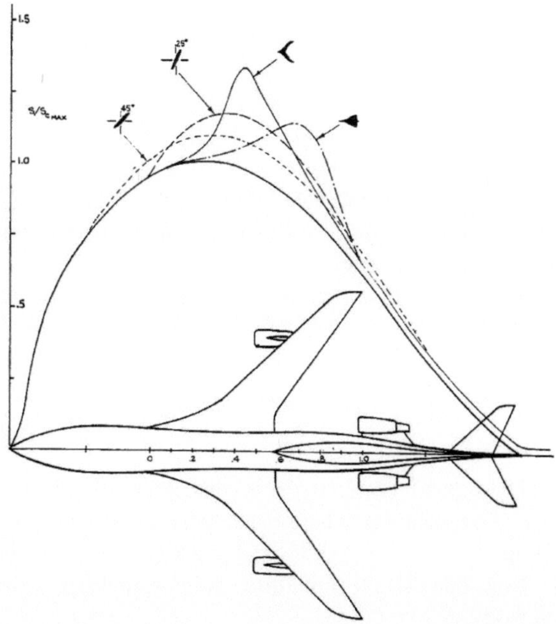

Fig. 7.17 Equivalent-body cross-section area at Mach 0.98 (from Chap. 23 of Numerical and Physical Aspects of Aerodynamic Flows, Ed. T. Cebeci, published by Springer Verlag, 1982, with kind permission of Springer)

Basic to their study are the parameters

$$\lambda = \frac{b}{l}, \quad \bar{\tau} = \frac{S_{max}}{bl}, \quad \bar{\alpha} = \frac{L_{max}}{\rho U^2 bl} \tag{7.111}$$

where b is the span of the slender wing ($\lambda = O(1)$), l a characteristic length, S_{max} is the maximum cross sectional area and L_{max} the maximum lift.

In the neighborhood of the body, the perturbation potential can be written in the form

$$\frac{\phi}{Ub} = \bar{\tau}\phi_1 + \bar{\alpha}\phi_2 + g(x) \tag{7.112}$$

where ϕ_1 and ϕ_2 are governed by two-dimensional Laplace's equations in y and z.

For larger r

$$\frac{\phi}{Ub} = \frac{1}{2\pi}\bar{\tau}\frac{dS}{dx}\ln r + \frac{1}{2\pi}\bar{\alpha}\, D(x)\frac{\sin\theta}{r} + g(x) \tag{7.113}$$

The above expression yields an axisymmetric flow as $r \to \infty$, confirming the area rule.

However, if $\bar{\tau}$ is much smaller than $\bar{\alpha}$, the cross flow solution breaks down before the source distribution prevails over the doublet distribution. in this case, nonlinear effects near the wing are important. The fuller equation governing the perturbation potential near the wing is

$$\frac{\partial^2 \phi}{\partial y^2} + \frac{\partial^2 \phi}{\partial z^2} = \left[(M_0^2 - 1) + (\gamma + 1) M_0^2 \frac{1}{U} \frac{\partial \phi}{\partial x} \right] \frac{\partial^2 \phi}{\partial x^2} \tag{7.114}$$

The nonlinear terms on the right hand side act as a source term in the cross flow equation (hence the doublet distribution contributes to the source distribution).

Away from the body, the three-dimensional equation must be solved to fully account for these nonlinear effects, and the solution is determined principally by the doublet and the source distributions from the solution of the cross flow equation, including the nonlinear corrections.

The doublet distribution is still proportional to $b^2(x)$, as in linear theory, however the effective source distribution consists of two parts

$$\frac{dS_e}{dx} = \frac{dS}{dx} + \sigma \frac{d}{dx} \left(\frac{dD}{dx} \right)^2 \tag{7.115}$$

The second term in the above expression is due to the nonlinear correction in the cross flow solution, where σ is a parameter representing the relative importance of the lift.

Now, one can solve the three dimensional small disturbance equation away from the body

$$\left[(1 - M_0^2) - (\gamma + 1) M_0^2 \frac{1}{U} \frac{\partial \phi}{\partial x} \right] \frac{\partial^2 \phi}{\partial x^2} + \frac{1}{r} \frac{\partial}{\partial r} \left(r \frac{\partial \phi}{\partial r} \right) + \frac{1}{r^2} \frac{\partial^2 \phi}{\partial \theta^2} = 0 \tag{7.116}$$

with

$$\phi \to 0 \quad as \quad x^2 + r^2 \to \infty \tag{7.117}$$

and

$$\frac{r}{Ub} \frac{\partial \phi}{\partial r} = \frac{1}{2\pi} \bar{\tau} \frac{dS_e}{dx} - \frac{1}{2\pi} \bar{\alpha}\, D(x) \frac{\sin \theta}{r} \quad as \quad r \to 0 \tag{7.118}$$

the solution of the above three dimensional problem provides the far field boundary conditions for the cross flow problems at each cross section, including the nonlinear effects as a forcing function and with the tangency condition at the solid surface. Notice iterations are needed to account for the forcing function.

If lift effects are not dominant, the three dimensional problem can be split into two axisymmetric ones where

$$\frac{\phi(x, r, \theta)}{Ub} = \frac{\phi_0(x, r)}{Ub} + \frac{1}{2\pi} \bar{\alpha} \sin \theta \frac{\phi_{12}(x, r)}{Ub} + g(x) \tag{7.119}$$

and ϕ_0 and ϕ_{12} are governed by

$$\left[(1-M_0^2)-(\gamma+1)M_0^2\frac{1}{U}\frac{\partial\phi}{\partial x}\right]\frac{\partial^2\phi_0}{\partial x^2}+\frac{1}{r}\frac{\partial}{\partial r}\left(r\frac{\partial\phi_0}{\partial r}\right)=0 \tag{7.120}$$

$$\left[(1-M_0^2)-(\gamma+1)M_0^2\frac{1}{U}\frac{\partial\phi_0}{\partial x}\right]\frac{\partial^2\phi_{12}}{\partial x^2}$$

$$+\frac{1}{r}\frac{\partial}{\partial r}\left(r\frac{\partial\phi_{12}}{\partial r}\right)-\frac{\phi_{12}}{r^2}-(\gamma+1)M_0^2\frac{1}{U}\frac{\partial^2\phi_0}{\partial x^2}\frac{\partial\phi_{12}}{\partial x}=0 \tag{7.121}$$

Both $\phi_0 \to 0$, and $\phi_{12} \to 0$ as $r \to \infty$, and as approaching the axis, we have

$$\frac{r}{Ub}\frac{\partial\phi_0}{\partial r}=\frac{1}{2\pi}\bar{\tau}\frac{dS_e}{dx},\quad \frac{r^2}{Ub}\frac{\partial\phi_{12}}{\partial r}=-D(x) \tag{7.122}$$

Special treatment is required to enforce the perturbed shock jump condition for ϕ_{12} problem.

For more details, the reader is referred to Cheng and Hafez [25], particularly for the estimate of the size of the region in the neighborhood of the body, which is obtained via asymptotic analysis.

7.8 Supersonic Area Rule

Hayes [27] derived a formula for the wave drag of a wing-body configuration of arbitrary thickness distribution and experiencing both lift and side forces.

For a thin non lifting wing, the source distribution is related to the geometry via (see Ashley and Landhal [1])

$$q(x, y) = \tau\frac{\partial g}{\partial x} \tag{7.123}$$

Consider the Mach planes passing through a point x_i along the x-axis and tangent to the Mach cone with apex at x_i, see Fig. 7.18. Let θ be the angle of the normal to the plane with the y-axis. The equation of such a plane is $(P_\theta):\ x - x_i + \beta y\cos\theta + \beta z\sin\theta$. The intersection of two parallel neighboring Mach planes with the (x, y)-plane is shown in Fig. 7.19. The intersection with the slender wing is projected on a plane normal to the axis as $S_w(x, \theta)$, also shown in the figure. Then upon integrating in y and lumping the source along the x-axis, on finds

$$Q(x_i, \theta) = \tau\int\frac{\partial g(x_i+\beta y\cos\theta, y)}{\partial x}dy$$

$$=\frac{\partial}{\partial x}\left\{\tau\int g(x_i+\beta y\cos\theta, y)dy\right\}=S'_w(x,\theta) \tag{7.124}$$

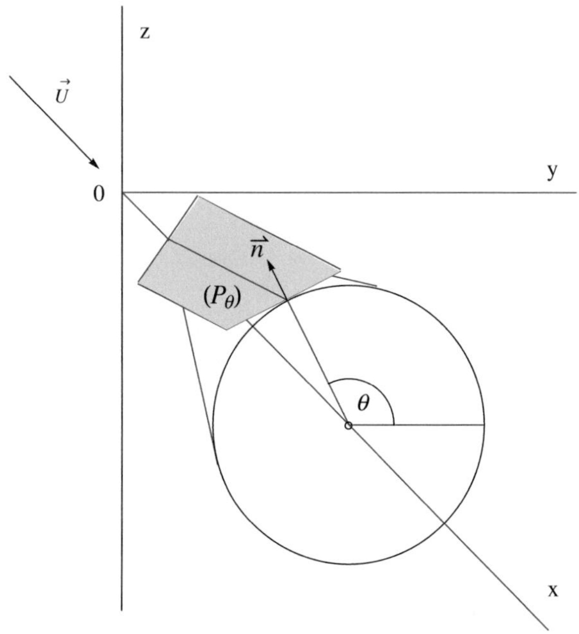

Fig. 7.18 Mach cone and Mach plane

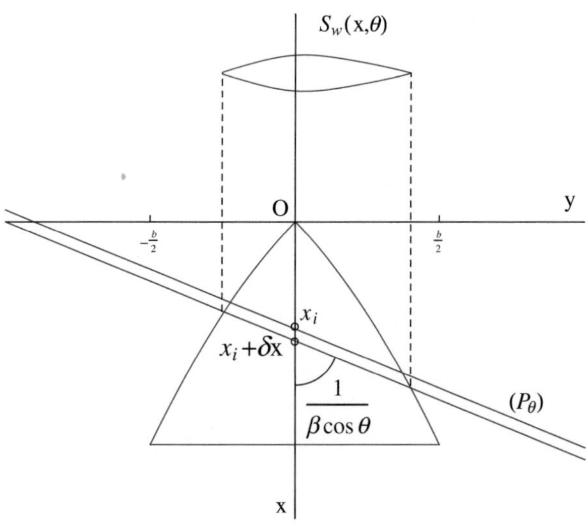

Fig. 7.19 Section of wing by characteristic plane (P_θ)

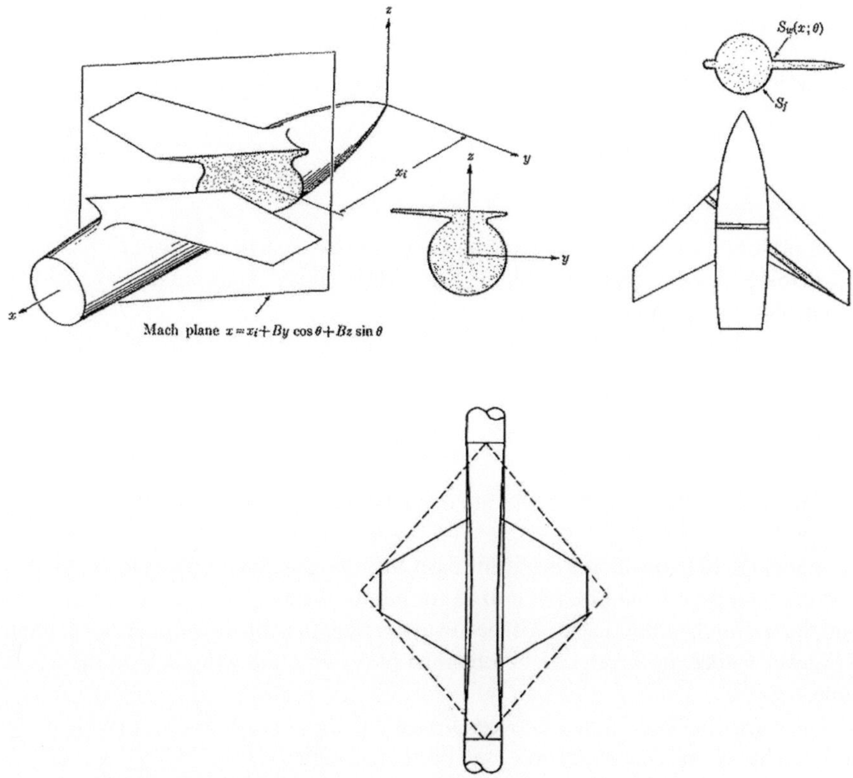

Fig. 7.20 Supersonic area rule methodology (from Aerodynamics of Wings and Body, by H. Ashley and M. Landhal, with kind permission of Dover Publication, 1985 edition)

For a body of revolution

$$Q(x_i) \simeq S'_f(x_i) \tag{7.125}$$

since, for a slender body, the area of the oblique cut will not differ much from its projection.

For a winged body of revolution, the area distribution of the equivalent body is

$$S(x,\theta) = S_f(x) + S_w(x,\theta) \tag{7.126}$$

Following Lomax and Heaslet [28], Fig. 7.20, the effect of wing-body interference can be represented through

$$S'_e(x) = S'_f(x) + A_0(x) \tag{7.127}$$

where

$$A_0(x) = \frac{1}{2\pi}\int_0^{2\pi} S'_w(x,\theta)d\theta \tag{7.128}$$

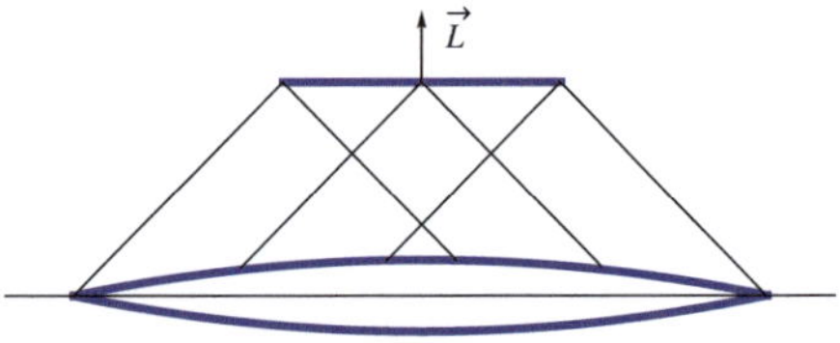

Fig. 7.21 Flat plate lift induced by a body of revolution in supersonic flow

Minimum wave drag (for non lifting configuration) can be achieved by choosing $S_e'(x)$ to be an optimum shape, i.e. the Sears-Haaks body. Thus the fuselage will be indented in the region of the wing.

If the body is approximated by an infinite cylinder, the optimum body indentation is then given by

$$\Delta S_f' + A_0(x) = 0 \tag{7.129}$$

The above supersonic area rule includes the transonic area rule, in the limit of $M_0 \to 1$.

Jones [11, 12], considered the lifting case, including the vortex drag and concluded that the total drag would be minimized for elliptic loading. The three-dimensional drag for a straight elliptic wing will be always greater than the two-dimensional drag.

Lower values, however, can be obtained by yawing the wing behind the Mach cone.

The wave drag interference between lift and thickness is zero for a mid wing body. A favorable interaction is possible (see Ferri et al. [29]).

For a flat plate at zero angle of attack located above the body, a lift will be generated without drag, and the drag of the body will be lower due to the wave reflection from the wing impinging on the rear part of the body (Ashley and Landhal [1]), Fig. 7.21.

7.9 Conical Flows

7.9.1 Method of Taylor and Maccoll

Consider supersonic flow over a cone at zero incidence, with a conical shock attached to the tip of the cone. The shock is of uniform strength and the entropy jump across the shock is also uniform, hence the flow between the shock and the cone is isentropic and irrotational. Unlike the two-dimensional wedge case, the corresponding governing equations are complicated. In the following, numerical integration due to Taylor and Maccoll [30] will be discussed.

Assuming that all the fluid properties are constant on cones with common vertex at the tip of the cone, and using spherical coordinates, the equation of continuity, the definition of vorticity and Bernoulli's law are given below (Shapiro [31]). Conservation of mass from control volume analysis, Fig. 7.22, reads

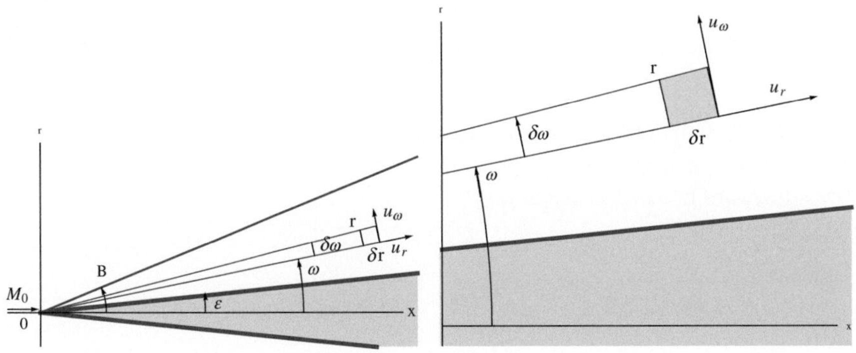

Fig. 7.22 Conical flow in meridian plane and control volume

$$\frac{\partial}{\partial r}\left(2\pi\rho u_r r^2 \sin\omega\right) + \frac{\partial}{\partial\omega}\left(2\pi\rho u_\omega r \sin\omega\right) = 0 \tag{7.130}$$

Since $\partial/\partial r = 0$, then $\partial/\partial\omega = d/d\omega$ and the above equation becomes

$$2\rho u_r + \rho u_\omega \cot\omega + \rho\frac{du_\omega}{d\omega} + u_\omega\frac{d\rho}{d\omega} = 0 \tag{7.131}$$

Irrotationality, Fig. 7.22, yields

$$u_\omega = \frac{du_r}{d\omega} \tag{7.132}$$

The generalized Bernoulli's law is

$$\frac{a^2}{\gamma-1} + \frac{V^2}{2} = \frac{V_{max}^2}{2} \tag{7.133}$$

Furthermore

$$-\frac{dp}{\rho} = u_r du_r + u_\omega du_\omega = -a^2\frac{d\rho}{\rho} \tag{7.134}$$

After substituting in the continuity equation and rearrangement, one obtains a second order nonlinear ordinary differential equation for u_r in terms of ω

$$-\left[\frac{\gamma+1}{2}\left(\frac{du_r}{d\omega}\right)^2 - \frac{\gamma-1}{2}\left(V_{max}^2 - u_r^2\right)\right]\frac{d^2u_r}{d\omega^2} - \frac{\gamma-1}{2}\cot\omega\left(\frac{du_r}{d\omega}\right)^3$$

$$- ku_r \left(\frac{du_r}{d\omega}\right)^2 + \frac{\gamma - 1}{2}\left(V_{max}^2 - u_r^2\right)\cot\omega\,\frac{du_r}{d\omega} + (\gamma - 1)\left(V_{max}^2 - u_r^2\right)u_r = 0 \tag{7.135}$$

The integration of the above equation was first done by Busemann, using a graphical construction in the hodograph plane. Later, Taylor and Maccoll [30] used a straightforward numerical integration.

(i) Begin with

$$u_r = (u_r)_{cone}, \quad u_\omega = \left(\frac{du_r}{d\omega}\right)_{cone} = 0 \quad at \quad (\omega)_{cone} = \epsilon \tag{7.136}$$

(ii) Integrate the nonlinear O.D.E. stepwise, using small steps in ω, by replacing the derivatives by finite differences.

(iii) After finding u_r/V_{max} for each value of ω, calculate u_ω/V_{max} using the irrotationality condition in discrete form.

(iv) The final step is to find the shock angle β and the free stream velocity U/V_{max}. For each value of ω during the integration, the downstream Mach number M of a shock with an angle $\beta = \omega$ and turning angle θ are compared with the value of M obtained from the integration process. If the two numbers agree, the limit of integration has been reached.

The incoming Mach number M_0 can be found from the shock tables. The shock relations together with the isentropic relations for the conical flow region provide all the necessary information.

As in the wedge case, one may construct a shock polar in the hodograph plane.

For each value of M_0 there exists a maximum value of the cone semi-angle ϵ for which there is a solution to the conical flow equations, and vice-versa, for each value of ϵ, there is a minimum M_0. Also, for a given M_0 and for $\epsilon < \epsilon_{max}$, there are two solutions, with strong and weak shocks. Since the flow behind the strong shock is always subsonic, which cannot be conical, the strong shock is not possible and should be eliminated. On the other hand, the weak shock solution is possible with finite cones in supersonic flow because cutting off the downstream part of the cone cannot affect the upstream flow.

Notice for equal cone and wedge angles, the surface pressure rise and shock angle are smaller for cones due to the three dimensional effects.

7.9.2 *Small Disturbance Approximations for Flows Over a Cone*

Von Karman and Moore [3] showed that the perturbation potential may be represented as an integral over a distribution of sources

$$\phi(x, r) = -\int_0^{x-\beta r} \frac{g(\xi)}{\sqrt{(x-\xi)^2 - \beta^2 r^2}} d\xi, \quad \xi \leq x - \beta r \tag{7.137}$$

where $\beta = \sqrt{M_0^2 - 1}$. Following the presentation of Liepmann and Roshko [6], a change of variable is recommended from ξ to σ

$$\xi = x - \beta r \cosh \sigma, \quad \Rightarrow \quad d\xi = -\beta r \sinh \sigma d\sigma = -\sqrt{(x-\xi)^2 - \beta^2 r^2} d\sigma \tag{7.138}$$

The integral and the limit of integration become

$$\phi(x, r) = -\int_0^{\cosh^{-1}(x/\beta r)} g(x - \beta r \cosh \sigma) d\sigma \tag{7.139}$$

Consider $g(\xi) = a\xi$, a linear distribution, then

$$\phi(x, r) = -ax \left[\cosh^{-1}\left(\frac{x}{\beta r}\right) - \sqrt{1 - \left(\frac{\beta r}{x}\right)} \right] \tag{7.140}$$

$$u = \frac{\partial \phi}{\partial x} = -a \cosh^{-1}\left(\frac{x}{\beta r}\right), \quad v = \frac{\partial \phi}{\partial y} = a\beta \sqrt{\left(\frac{x}{\beta r}\right)^2 - 1} \tag{7.141}$$

Notice, u and v are invariant along lines $x/r = const$. The solution can represent supersonic flow over a cone, where the vertex angle depends on a.

Applying the boundary condition

$$\left(\frac{v}{U + u}\right)_{cone} = \tan \delta \tag{7.142}$$

where δ is the cone half-angle, and on the cone surface

$$u = -a \cosh^{-1}\left(\frac{\cot \delta}{\beta}\right), \quad v = a\sqrt{\cot^2 \delta - \beta^2} \tag{7.143}$$

Hence

$$a = \frac{U \tan \delta}{\sqrt{\cot^2 \delta - \beta^2} + \tan \delta \cosh^{-1}\left(\frac{\cot \delta}{\beta}\right)} \tag{7.144}$$

For slender cones, with $\delta \ll 1$, and $\cot \delta \gg \beta$, then

$$\cosh^{-1}\left(\frac{\cot \delta}{\beta}\right) \simeq \ln\left(\frac{2}{\beta \delta}\right), \quad and \quad \tan \delta \cosh^{-1}\left(\frac{\cot \delta}{\beta}\right) \rightarrow 0 \quad as \quad \delta \rightarrow 0 \tag{7.145}$$

a has an asymptotic form

$$a \simeq \frac{U\delta}{\cot\delta} \propto U\delta^2 \tag{7.146}$$

Since r/x will be small, then

$$\phi(x, r) = -U\delta^2 x \left(\ln\left(\frac{2x}{\beta r}\right) - 1 \right) \tag{7.147}$$

and

$$\frac{u}{U} = -\delta^2 \ln\left(\frac{2x}{\beta r}\right), \quad \frac{v}{U} = \delta^2 \frac{x}{r} \tag{7.148}$$

On the surface of the cone $r/x = \delta$ and

$$\left(\frac{u}{U}\right)_{cone} = -\delta^2 \ln\left(\frac{2}{\beta\delta}\right), \quad \left(\frac{v}{U}\right)_{cone} = \epsilon \tag{7.149}$$

Hence

$$C_p = -2\frac{u}{U} - \left(\frac{v}{U}\right)^2 = \delta^2 \left(2\ln\left(\frac{2}{\beta\delta}\right) - 1\right) \tag{7.150}$$

The corresponding result for a wedge of nose angle 2δ is given by

$$C_p = 2\frac{\delta}{\beta} = 2\frac{\delta}{\sqrt{M_0^2 - 1}} \tag{7.151}$$

The pressure rise on the cone is much less than on the wedge, as shown in Fig. 7.23.

Notice the pressure is uniform over the surface, however the distribution of the pressure coefficient along a line $r = const.$ rises continuously downstream of the shock. Also, in the slender body approximation, there is no pressure jump at the nose.

The accuracy of the slender body approximation can be numerically evaluated by comparison to the numerical results of Maccoll and the tables of Kopal. A typical case is shown in Fig. 7.24 (see Liepmann and Roshko [6]).

7.9.3 Conical Flow Past a Delta Wing

In general, in a conical flow, all fluid properties are invariant along rays through a common vertex. An example of such flows is a triangular wing in supersonic flow, where disturbances are propagated only in the downstream Mach cone. Notice conical flows are not possible in subsonic flow regime.

Following Shapiro [31] and Ashley and Landhal [1], the governing perturbation potential equation is given by

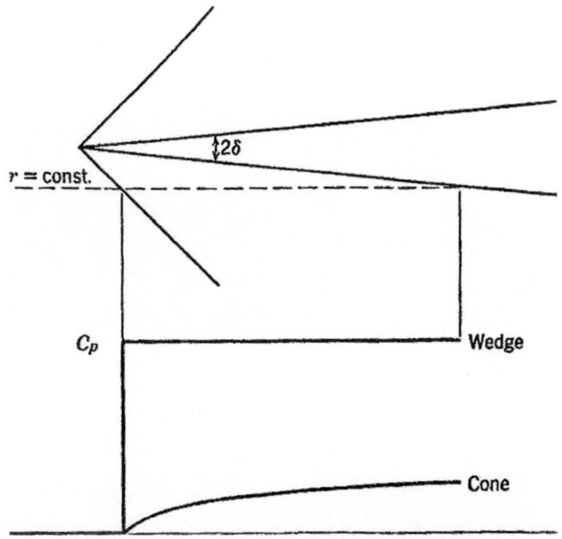

Fig. 7.23 Pressure distribution comparison between shock and body for a cone and a wedge with the small disturbance approximation (from Elements of Gas Dynamics, 2001 edition, by Liepmann and Roshko, with kind permission of Dover publication)

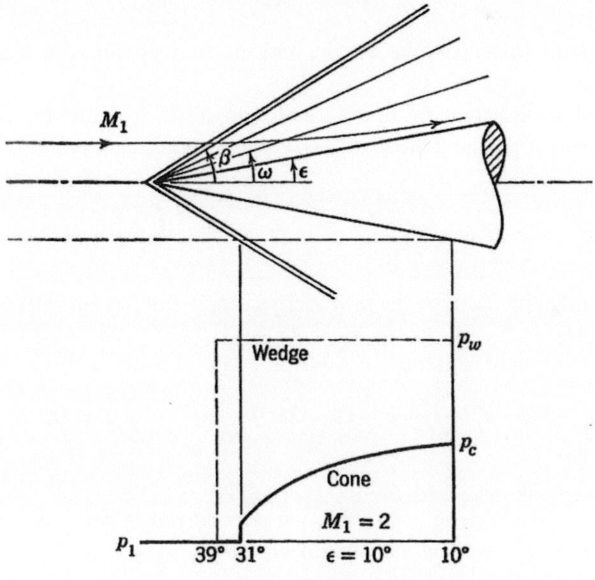

Fig. 7.24 Comparison between surface pressure distributions on a wedge and on a cone (from Elements of Gas Dynamics, 2001 edition, by Liepmann and Roshko, with kind permission of Dover publication)

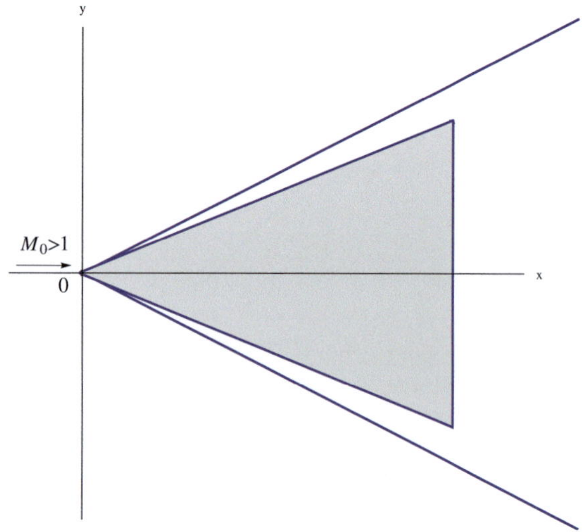

Fig. 7.25 Delta wing in supersonic flow

$$(1 - M_0^2)\frac{\partial^2 \phi}{\partial x^2} + \frac{\partial^2 \phi}{\partial y^2} + \frac{\partial^2 \phi}{\partial z^2} = 0 \tag{7.152}$$

where the origin is placed at the vertex, and the free stream is in the x-direction, Fig. 7.25.

The velocity components (u, v, w) are also governed by similar equations.

Following Hayes [27] and Stewart [32], conical coordinates are introduced where

$$\sigma = \frac{\beta}{x}\sqrt{y^2 + z^2}, \quad \theta = \tan^{-1}\left(\frac{y}{z}\right) \tag{7.153}$$

The governing equation for the u-velocity component becomes (see also Ashley and Landhal [1])

$$(1 - \sigma^2)\frac{\partial^2 u}{\partial \sigma^2} + \frac{1}{\sigma}(1 - 2\sigma^2)\frac{\partial u}{\partial \sigma} + \frac{1}{\sigma^2}\frac{\partial^2 u}{\partial \theta^2} = 0 \tag{7.154}$$

In the (η, ζ) plane. the Mach cone is represented by a unit circle and the wing by a slit along the η-axis of length $2\mu = \beta / \tan \Lambda$, where Λ is the sweep angle.

On the mach cone, u, v, w vanish and on the slit $w = -\alpha$. Using Chaplygin's transformation, where

$$\varrho = \frac{1 - \sqrt{1 - \sigma^2}}{\sigma}, \quad \Rightarrow \quad \frac{d\varrho}{d\sigma} = \frac{1}{\sqrt{1 - \sigma^2}}\frac{\varrho}{\sigma} \tag{7.155}$$

the above equation is reduced to the standard, two dimensional Laplace equation

$$\varrho^2 \frac{\partial^2 u}{\partial \varrho^2} + \varrho \frac{\partial u}{\partial \varrho} + \frac{\partial^2 u}{\partial \theta^2} = 0 \quad (7.156)$$

Let, $\eta = \varrho \cos\theta$, $\zeta = \varrho \sin\theta$, now the slit is between $\eta = -m'$ and $\eta = m'$, where

$$m' = \frac{1 - \sqrt{1 - m^2}}{m} \quad (7.157)$$

The velocity components are related through the irrotationality conditions, hence on the slit

$$\frac{\partial u}{\partial \zeta} = 0, \quad |\eta| < m' \quad (7.158)$$

On the Mach cone

$$u = 0, \quad on \quad \eta = 0 \quad (7.159)$$

From the incompressible flow solution for a flow normal to a plate, one expects that u has a square-root singularity at the leading edge. Indeed, for a delta wing with

$$u = C \frac{\beta}{\tan^2 \Lambda \sqrt{m^2 - \beta^2 y^2 / x^2}} \alpha \quad (7.160)$$

where C is a real constant (since the boundary conditions are homogeneous).

Obviously, $C \to 1$ as $m' \to 0$ to recover the slender body solution for a delta wing with vanishing span.

The determination of C (to satisfy the boundary condition $w = -\alpha$) is complicated. Referring to Ashley and Landhal [1]

$$C = \frac{1}{E(k)} \quad (7.161)$$

where $E(k)$ is the complete elliptic integral of second kind with modulus k. The pressure coefficient is given by

$$C_p = \frac{4C\beta}{\tan^2 \Lambda \sqrt{m^2 - \beta^2 y^2 / x^2}} \alpha \quad (7.162)$$

The lift coefficient is found to be

$$C_L = \frac{\pi AR\, C}{2} \alpha \quad (7.163)$$

The above results are consistent with slender body theory for $m = 1$ (or $k = 1$) and with supersonic edge theory for $m = 1$ (or $k = 0$).

The drag is given by

$$D = L\alpha \quad (7.164)$$

for supersonic edges. When the leading edge is subsonic, Jones and Cohen [11] accounted for leading edge suction, where

$$D = L\alpha - F_s \tag{7.165}$$

with their estimate of F_s

$$C_D = \frac{C_L^2}{\pi AR}\left(2E(k) - \sqrt{1-m^2}\right) \tag{7.166}$$

7.9.4 Rectangular Wings at Angle of Attack

Assuming small disturbances, the flow past a rectangular wing is two-dimensional except within the mach cones generated at the tips, where conical flow exists, Fig. 7.26.

For the region outside the Mach cone, a transformation similar to Chaplygin's produces two-dimensional wave equation. The solutions in the hyperbolic and elliptic regions must be joined such that the velocity components are continuous across the Mach cone.

More complicated analysis includes, Prandtl-Meyer's expansion, shock waves and formation of vortex flow.

7.9.5 Numerical Solution of Conical Euler Equations

The conservation laws in conical coordinates are given by (see Powell [33, 34])

$$\frac{r}{\kappa}\frac{\partial U}{\partial t} + \frac{r}{\kappa^2}\frac{\partial}{\partial r}(F + \eta G + \zeta H) + \frac{\partial}{\partial \eta}(G - \eta F) + \frac{\partial}{\partial \zeta}(H - \zeta F) + 2F = 0 \tag{7.167}$$

where

$$r = \sqrt{x^2 + y^2 + z^2}, \quad \eta = \frac{y}{x}, \quad \zeta = \frac{z}{x}, \quad \kappa = \sqrt{1 + \eta^2 + \zeta^2} \tag{7.168}$$

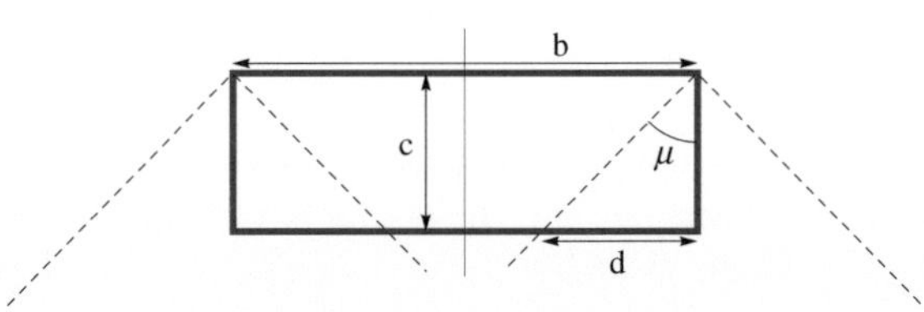

Fig. 7.26 Rectangular wing at angle of attack

U, F, G and H are the standard state vector and flux vectors of the Euler equations in Cartesian coordinates.

The conical velocities are given by

$$u_\kappa = u + \eta v + \zeta w \tag{7.169}$$

$$u_\eta = v - \eta u \tag{7.170}$$

$$u_\zeta = w - \zeta u \tag{7.171}$$

For conical flows, $\partial U/\partial r = 0$ and the boundary conditions must be consistent with conical flow assumption.

The governing equations become

$$\frac{\kappa}{r}\frac{\partial}{\partial t}\begin{bmatrix}\rho\\ \rho u\\ \rho v\\ \rho w\\ \rho E\end{bmatrix} + \frac{\partial}{\partial \eta}\begin{bmatrix}\rho u_\eta\\ \rho u u_\eta - \eta p\\ \rho u u_\eta + p\\ \rho w u_\eta\\ \rho H u_\eta\end{bmatrix} + \frac{\partial}{\partial \zeta}\begin{bmatrix}\rho u_\zeta\\ \rho u u_\zeta - \zeta p\\ \rho v u_\zeta\\ \rho w u_\zeta + p\\ \rho H u_\zeta\end{bmatrix} + 2\begin{bmatrix}\rho u\\ \rho u^2 + p\\ \rho u v\\ \rho u w\\ \rho u H\end{bmatrix} = 0 \tag{7.172}$$

where $H = E + p/\rho$.

Based on topological considerations, Powell indicated that the physically meaningful singularities of conical flows are nodes, saddles and spirals.

A cell vertex, finite volume, multi-stage Runge-Kutta scheme, with added artificial viscosity, was used to obtain numerical solutions over delta wings with grid generation allowing multiple levels of refinement. Typical results are shown in Fig. 7.27 (from Ref. [33, 34]).

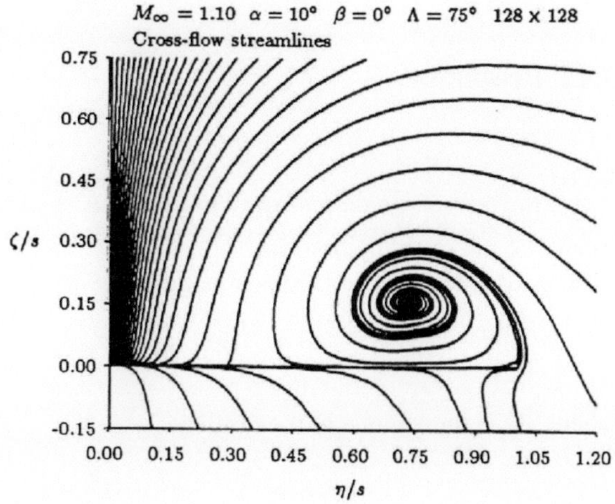

Fig. 7.27 Conical flow past a delta wing at incidence: cross-flow streamlines (with kind permission of the author, Prof. K. Powell)

The above study concluded that low pressure and density were seen in the vortex region.

Numerical parameters such as grid resolution and artificial viscosity did not affect the level of the total pressure loss, only the distribution of it. Modeling the core region remains however to be a problem.

In general, the conical flow calculation is less expensive than the full three-dimensional one. However, since the conical flow is supersonic everywhere, a space marching technique is an alternative. simply, artificial time is equivalent to a certain space marching and the latter is less restrictive.

Today, full three-dimensional Navier-Stokes solutions for high speed flows over delta wings are available, including simulation of important phenomena, for example vortex breakdown. CFD was an important tool in modifying the design of fighter jets where the vortex generated from the leading edge and its interaction with the tail was a serious problem.

7.10 Summary of Chapter 7

First, the governing equations are derived in cylindrical coordinates. The full potential equation and the small disturbance theory are followed including calculations of wave drag and optimum shapes (due to von Karman and Sears and Haaks) for bodies of revolution in supersonic flows. Slender wings of low aspect ratios, including swept and oblique wings, wing-body combinations and slender bodies with general cross-sections are also discussed.

The transonic area rule, involving lift, as well as the supersonic area rule are briefly covered.

Finally, conical flows are studied using the full potential equation and small disturbance approximation, as well as numerical solution of the conical Euler equations.

7.11 Problems

7.11.1

According to slender body theory, the fuselage does not contribute to lift, but contributes to the pitching moment.

Show that the fuselage moment is given by (see Fig. 7.28)

$$M_{,of} = \rho U^2 \vartheta_f \alpha$$

where, ϑ_f is the volume of the fuselage.

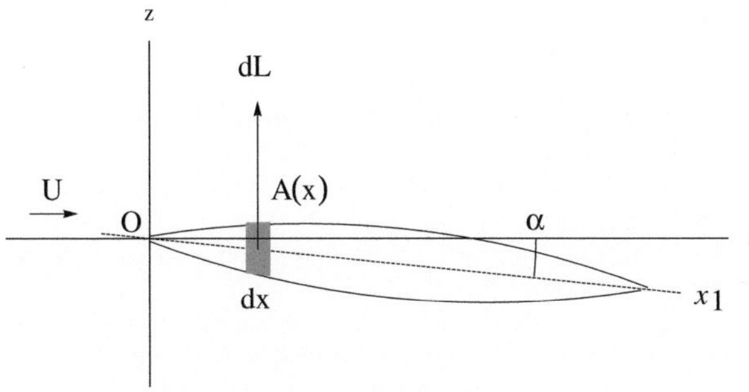

Fig. 7.28 Moment of a body of revolution at incidence

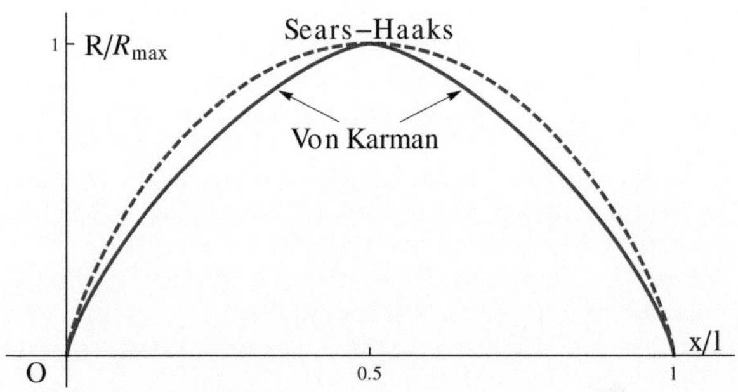

Fig. 7.29 Geometries of two pointed bodies

7.11.2

Consider two bodies pointed at both ends, the Sears-Haaks body and a body obtained by combining two Von Karman ogives (VK-KV), one facing forward and the other facing backward, see Fig. 7.29. The equation for the latter can be described as

$$\begin{cases} S'(x) = 2l\,A\sqrt{\frac{2x}{l}\left(1-\frac{2x}{l}\right)}, & 0 \le x \le \frac{l}{2} \\ S'(x) = -2l\,A\sqrt{\left(\frac{2x}{l}-1\right)\left(2-\frac{2x}{l}\right)}, & \frac{l}{2} \le x \le l \end{cases}$$

corresponding to two elliptic distributions. Calculate the wave drag of the VK-KV configuration and compare to the Sears-Haaks body for the same maximum cross section. Hint: one approach is to calculate the Fourier coefficients in terms of A with a computer algebra system such as Mathematica to find the drag as

$$A_{ref}C_{Dw} = \frac{\pi l^2}{4}\sum_{n=1}^{\infty} n A_n^2$$

In particular show that

$$A_1 = 0,\ A_2 = A,\ A_3 = 0,\ A_4 = -\frac{1}{4},\ \ldots$$

References

1. Ashley, H., Landhal, M.: Aerodynamics of Wings and Bodies. Addison Wesley Longman, Reading (1965)
2. Oswatitsch, K., Keune, F.: Ein Äquivalenzsatz für nicht-Angestellte Flügel Kleiner Spannweite in Schallnaher Strömung, Zeischrift Flugwiss., vol. 3(2), pp. 29–46. Also, The Flow around Bodies of Revolution at Mach Number 1, Proceedings Conference on High Speed Aeronautics, Polytechnic Institute of Brooklyn, New York (1995)
3. von Karman, Th., Moore, N.: Resistance of slender bodies moving with supersonic velocities, with special reference to projectiles. Trans. ASME **54**, 303–310 (1932)
4. von Karman, Th.: The problem of resistance in compressible fluids. In: Proceedings of the Fifth Volta Congress, Rome, Italy (1936)
5. Sears, W.R.: On projectiles of minimum drag. Q. Appl. Math. **4**(4), 361–366 (1947)
6. Liepmann, H., Roshko, A.: Elements of Gas Dynamics. Wiley, New York (1957)
7. Viviand, H.: Ailes et Corps Elancés en Théorie des Petites Perturbations, Ecole National Supérieure de l'Aéronautique et de l'Espace, Première Partie, Class Notes (1972)
8. Katz, J., Plotkin, A.: Low-Speed Aerodynamics, 2nd edn. Cambridge University Press, New York (2001)
9. Duncan, W.J., Thom, A.S., Young, A.D.: The Mechanics of Fluids. Arnold, London (1960)
10. Weissinger, J.: The Lift Distribution of Swept-Back Wings. NACA TM 1120 (1947)
11. Jones, R.T.: Collected Works of R.T. Jones. NASA Publication (1976)
12. Jones, R.T., Cohen, D.: Aerodynamic Components of Aircraft at High Speeds, Eds. Donovan, Lawrence, Princeton University Press, New Jersey (1990)
13. Kogan, N.M.: On bodies of minimum drag in a supersonic gas stream. Prikl. Mat. Mekhan **XXI**(2), 207–212 (1957)
14. Spreiter, J.R.: The Aerodynamic Forces on Slender Plane- and Cruciform-Wing and Body Combination. NACA Report 962 (1950)
15. Mangler, K.W.: Improper integrals in theoretical aerodynamics. British A.R.C., R&M No. 2424 (1951)
16. Brandao, M.P.: Improper integrals in theoretical aerodynamics. AIAA J. **25**(9), 1258–1260 (1987)
17. Lighthill, M.J.: Supersonic flow past slender bodies of revolution, the slope of whose meridian section is discontinuous. Q. J. Mech. Appl. Math. **1**, 90–102 (1947)
18. Ward, G.N.: Supersonic flow past slender pointed bodies. Q. J. Mech. Appl. Math. **2**, Part I, 75–97 (1949)
19. Ward, G.N.: On the minimum drag of thin lifting bodies in steady supersonic flows. British A.R.C., Report No. 18 (1949)
20. Ward, G.N.: Linearized Theory of Steady High-Speed Flow. Cambridge University Press, London (1955)
21. Harder, K.C., Klunker, E.B.: On Slender-Body Theory and the Area Rule at Transonic Speeds. NACA Report 1315 (1957)
22. Guderley, K.G.: Theory of Transonic Flow. Pergamon Press, London (1962)
23. Cole, J.D., Cook, P.: Transonic Aerodynamics. Elsevier Science Publishing Company, Amsterdam (1986)
24. Whitcomb, R.: A Study of the Zero-Lift Drag Rise Characteristics of Wing-Body Combinations Near the Speed of Sound. NACA Report 1273 (1956)
25. Cheng, H.K., Hafez, M.M.: Transonic equivalence rule: a nonlinear problem involving lift. J. Fluid Mech. **72**, 161–187 (1975)
26. Cheng, H.K.: Lift corrections to transonic equivalence rule: examples. AIAA J. **15**(3), 366 (1977)
27. Hayes, W.: Linearized supersonic flow. North American Aviation Report No., A.L. 222 (1947)
28. Lomax, H., Heaslet, M.: Recent developments in the theory of wing-body wave drag. J. Aeronaut. Sci. **23**(12), 1061–1074 (1956)

29. Ferri, A., Clarke, J., Ting, L.: Favorable interference in lifting systems in supersonic flow. J. Aeronaut. Sci. **24**(11), 791–804 (1957)
30. Taylor, G.I., Maccoll, J.W.: The air pressure on a cone moving at high speed. Proc. R. Soc. (A) **139**, 278 (1933)
31. Shapiro, A.: The Dynamics and Thermodynamics of Compressible Fluid Flow, vol. I and II. Arnold, New York (1953)
32. Stewart, A.J.: The lift of a delta wing at supersonic speeds. Q. Appl. Math. **4**, 246–254 (1946)
33. Powell, K.G.: Vortical Solutions of the Conical Euler Equations. Notes on Numerical Fluid Mechanics, vol. 28. Vieweg, Braunschweig (1990)
34. Powell, K.G., Murman, E.M.: Trajectory integration in vortical flows. AIAA J. **7**, 982–984 (1989)

Chapter 8
Viscous Fluid Flow and Laminar Boundary Layers

8.1 Incompressible 2-D Flows

Viscous stresses are important in studying aerodynamics and calculating the forces. So far, we have ignored the viscous stresses compared to the pressure and acceleration in the so called inviscid flow in the region outside the viscous layers (around the body and in the wake) (Fig. 8.1).

In the neighborhood of the solid surface, the flow slows down and the relative velocity is zero on the surface because of viscosity (no slip condition). The velocity profile $u(y)$ is shown in the sketch, Fig. 8.2. Note that in this chapter we will use a local coordinate system with x the curvilinear abscissa along the profile and y the normal distance to the wall, with unit vectors $(\mathbf{i}, \mathbf{j})$. Note also that the velocity components u and v represent the full velocity, not just the perturbations from the undisturbed flow.

The friction drag per unit span, D'_f is given by the shear stress τ integrated over the area:

$$\boldsymbol{\tau}_w = \mu \frac{\partial u}{\partial y}|_w \mathbf{i} \quad and \quad D'_f = \int_0^c \tau_{w\,\infty} dx \tag{8.1}$$

where μ is the viscosity coefficient and $\tau_{w\,\infty}$ is the component of the shear stress at the wall in the direction of the incoming flow.

In general μ is a function of temperature and slightly depends on pressure. It can be assumed constant for low and moderate speeds.

In our study, we will consider the flow to be laminar; transitional and turbulent flows are more complicated and little is known, in this regard, for practical applications. Therefore, the Reynolds number, $Re = \frac{\rho U c}{\mu}$ in our study is limited. Laminar separated flows will be discussed, and the effect of pressure gradient (retarding the flow) leading to stall (less lift and more drag) will be emphasized.

J.J. Chattot and M.M. Hafez, *Theoretical and Applied Aerodynamics*,
DOI 10.1007/978-94-017-9825-9_8

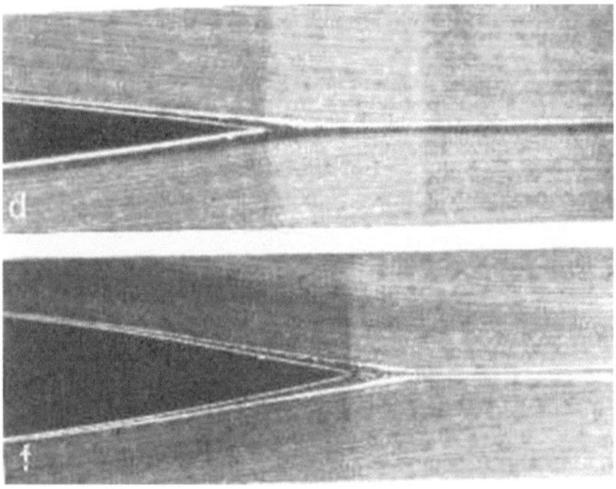

Fig. 8.1 Hydrodynamic visualization, courtesy H. Werlé ONERA—the French Aerospace Lab, ONERA RA 1987-3

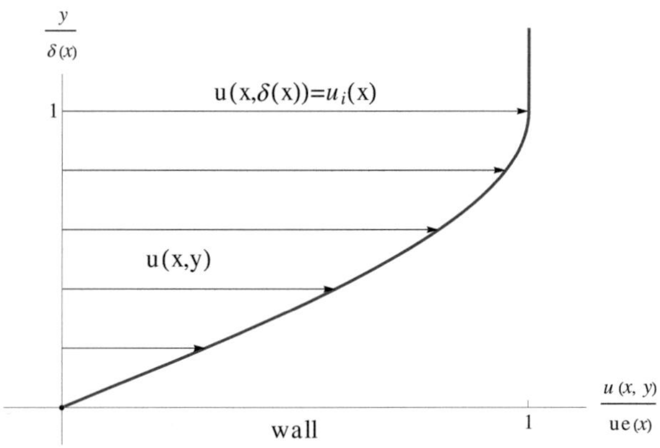

Fig. 8.2 Boundary layer u-velocity profile near a wall

8.1.1 Vorticity Versus Strain Rate in 2-D

Consider a rectangular particle of initial size $(\Delta x, \Delta y)$ near the wall $y = 0$. On the upper edge the velocity is $u = \Delta u$. Equivalently, this corresponds to moving with a particle located at $(0, 0)$ in a velocity field where $v = 0$. After a short time Δt the particle is sheared and takes the shape of a parallelogram. The strain rate is defined as $\dot{\epsilon}_{1,2} = \lim_{\Delta y \to 0} \frac{\Delta u}{\Delta y} = \frac{\partial u}{\partial y}$. Similarly, considering a velocity field with $u = 0$ and upon superposition with the previous flow field, one finds the following strain rate, Fig. 8.3

$$\dot{\epsilon}_{1,2} = \dot{\epsilon}_{2,1} = \frac{\partial u}{\partial y} + \frac{\partial v}{\partial x} \tag{8.2}$$

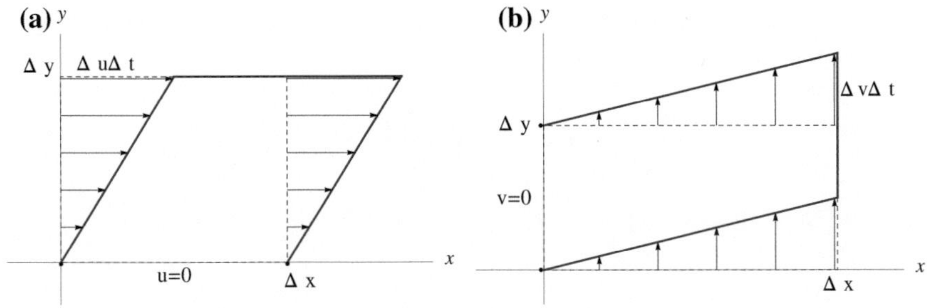

Fig. 8.3 Shear **a** in x-direction and **b** in y-direction

Further, we define the following strain rates

$$\dot{\epsilon}_{1,1} = 2\frac{\partial u}{\partial x} \quad and \quad \dot{\epsilon}_{2,2} = 2\frac{\partial v}{\partial y} \tag{8.3}$$

Vorticity is defined as

$$\omega = -\frac{\partial u}{\partial y} + \frac{\partial v}{\partial x} \tag{8.4}$$

and the angular velocity of the particle as

$$\Omega = \frac{1}{2}\omega = \frac{1}{2}\left(-\frac{\partial u}{\partial y} + \frac{\partial v}{\partial x}\right) \tag{8.5}$$

A general deformation reads:

$$\begin{cases} u(x + \Delta x, y + \Delta y, t) = u(x, y, t) + \Delta u(x, y, t) + \cdots \\ v(x + \Delta x, y + \Delta y, t) = v(x, y, t) + \Delta v(x, y, t) + \cdots \end{cases} \tag{8.6}$$

where

$$\begin{cases} \Delta u(x, y, t) = \frac{\partial u}{\partial x}\Delta x + \frac{\partial u}{\partial y}\Delta y + \cdots \\ \Delta v(x, y, t) = \frac{\partial v}{\partial x}\Delta x + \frac{\partial v}{\partial y}\Delta y + \cdots \end{cases} \tag{8.7}$$

or, in matrix form

$$\begin{pmatrix} \Delta u \\ \Delta v \end{pmatrix} = \begin{pmatrix} \frac{\partial u}{\partial x} & \frac{\partial u}{\partial y} \\ \frac{\partial v}{\partial x} & \frac{\partial v}{\partial y} \end{pmatrix} \begin{pmatrix} \Delta x \\ \Delta y \end{pmatrix} = A \begin{pmatrix} \Delta x \\ \Delta y \end{pmatrix} \tag{8.8}$$

Note: any square matrix A can be decomposed uniquely into the sum of a symmetric and antisymmetric matrices

$$A = \frac{1}{2}\left(A + A^T\right) + \frac{1}{2}\left(A - A^T\right) \tag{8.9}$$

where A^T is the transposed of A. Hence

$$\frac{1}{2}\left(A + A^T\right) = \begin{pmatrix} \frac{\partial u}{\partial x} & \frac{1}{2}(\frac{\partial u}{\partial y} + \frac{\partial v}{\partial x}) \\ \frac{1}{2}(\frac{\partial u}{\partial y} + \frac{\partial v}{\partial x}) & \frac{\partial v}{\partial y} \end{pmatrix} = \frac{1}{2}\begin{pmatrix} \dot{\epsilon}_{1,1} & \dot{\epsilon}_{1,2} \\ \dot{\epsilon}_{1,2} & \dot{\epsilon}_{2,2} \end{pmatrix} \tag{8.10}$$

$$\frac{1}{2}\left(A - A^T\right) = \begin{pmatrix} 0 & \frac{1}{2}(\frac{\partial u}{\partial y} - \frac{\partial v}{\partial x}) \\ -\frac{1}{2}(\frac{\partial u}{\partial y} - \frac{\partial v}{\partial x}) & 0 \end{pmatrix} = \frac{1}{2}\begin{pmatrix} 0 & -\omega \\ \omega & 0 \end{pmatrix} \tag{8.11}$$

Therefore

$$\left\{\begin{array}{l} \begin{pmatrix} u(x+\Delta x, y+\Delta y, t) \\ v(x+\Delta x, y+\Delta y, t) \end{pmatrix} \\ general\ 2-D\ motion \end{array}\right.$$

$$\left.\begin{array}{l} = \begin{pmatrix} u(x,y,t) \\ v(x,y,t) \end{pmatrix} + \begin{pmatrix} 0 & -\Omega \\ \Omega & 0 \end{pmatrix}\begin{pmatrix} \Delta x \\ \Delta y \end{pmatrix} + \frac{1}{2}\begin{pmatrix} \dot{\epsilon}_{1,1} & \dot{\epsilon}_{1,2} \\ \dot{\epsilon}_{1,2} & \dot{\epsilon}_{2,2} \end{pmatrix}\begin{pmatrix} \Delta x \\ \Delta y \end{pmatrix} \\ = \quad translation \quad + \quad rotation \quad + \quad strain\ rate \end{array}\right\} \tag{8.12}$$

8.1.2 Viscous Stresses in 2-D (Cartesian Coordinates)

To account for viscous stresses in the momentum equations, consider an element as in the sketch, Fig. 8.4.

The viscous forces in the x-direction are

$$\left(\sigma_{1,1}(x + \frac{\Delta x}{2}, y, t) - \sigma_{1,1}(x - \frac{\Delta x}{2}, y, t)\right)\Delta y$$

$$+ \left(\sigma_{2,1}(x, y + \frac{\Delta y}{2}, t) - \sigma_{2,1}(x, y - \frac{\Delta x}{2}, t)\right)\Delta x \tag{8.13}$$

Dividing by $\Delta x.\Delta y$ and taking the limit gives

$$\frac{\partial \sigma_{1,1}}{\partial x} + \frac{\partial \sigma_{2,1}}{\partial y} \tag{8.14}$$

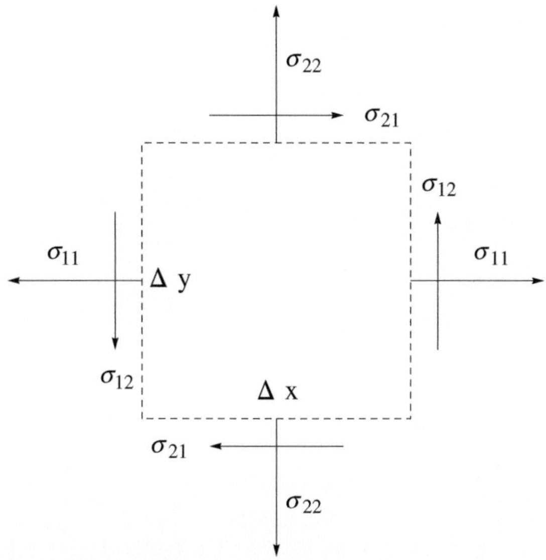

Fig. 8.4 Viscous stresses in 2-*D*

Similarly, the viscous forces in the y-direction is

$$\left(\sigma_{1,2}(x+\frac{\Delta x}{2},y,t)-\sigma_{1,2}(x-\frac{\Delta x}{2},y,t)\right)\Delta y$$

$$+\left(\sigma_{2,2}(x,y+\frac{\Delta y}{2},t)-\sigma_{2,2}(x,y-\frac{\Delta x}{2},t)\right)\Delta x \tag{8.15}$$

Again, dividing by $\Delta x.\Delta y$ and taking the limit gives

$$\frac{\partial\sigma_{1,2}}{\partial x}+\frac{\partial\sigma_{2,2}}{\partial y} \tag{8.16}$$

Note: taking moments about the centroid (x, y)

$$0=\sum M=\sigma_{1,2}\Delta x.\Delta y-\sigma_{2,1}\Delta x.\Delta y \tag{8.17}$$

hence $\sigma_{1,2}=\sigma_{2,1}$.

The momentum equations for viscous fluid flow become:

$$\begin{cases}\rho u\frac{\partial u}{\partial x}+\rho v\frac{\partial u}{\partial y}=-\frac{\partial p}{\partial x}+\frac{\partial\sigma_{1,1}}{\partial x}+\frac{\partial\sigma_{2,1}}{\partial y}\\ \rho u\frac{\partial v}{\partial x}+\rho v\frac{\partial v}{\partial y}=-\frac{\partial p}{\partial y}+\frac{\partial\sigma_{1,2}}{\partial x}+\frac{\partial\sigma_{2,2}}{\partial y}\end{cases} \tag{8.18}$$

8.1.3 Constitutive Relations

To find the viscous stresses in terms of velocity derivatives, we use the following relations:

$$\sigma_{1,1} = 2\mu\dot{\epsilon}_{1,1} = 2\mu\frac{\partial u}{\partial x} \tag{8.19}$$

$$\sigma_{2,2} = 2\mu\dot{\epsilon}_{2,2} = 2\mu\frac{\partial v}{\partial y} \tag{8.20}$$

$$\sigma_{1,2} = \sigma_{2,1} = \mu\dot{\epsilon}_{1,2} = \mu\dot{\epsilon}_{2,1} = \mu\left(\frac{\partial u}{\partial y} + \frac{\partial v}{\partial x}\right) \tag{8.21}$$

Hence

$$\frac{\partial\sigma_{1,1}}{\partial x} + \frac{\partial\sigma_{2,1}}{\partial y} = 2\mu\frac{\partial^2 u}{\partial x^2} + \mu\left(\frac{\partial^2 u}{\partial y^2} + \frac{\partial^2 v}{\partial x\partial y}\right) = \mu\left(\frac{\partial^2 u}{\partial x^2} + \frac{\partial^2 u}{\partial y^2}\right) \tag{8.22}$$

where we have made use of conservation of mass, i.e.

$$\frac{\partial}{\partial x}\left(\frac{\partial u}{\partial x} + \frac{\partial v}{\partial y}\right) = 0 \tag{8.23}$$

Similarly,

$$\frac{\partial\sigma_{1,2}}{\partial x} + \frac{\partial\sigma_{2,2}}{\partial y} = \mu\left(\frac{\partial^2 u}{\partial x\partial y} + \frac{\partial^2 v}{\partial x^2}\right) + 2\mu\frac{\partial^2 v}{\partial y^2} = \mu\left(\frac{\partial^2 v}{\partial x^2} + \frac{\partial^2 v}{\partial y^2}\right) \tag{8.24}$$

since

$$\frac{\partial}{\partial y}\left(\frac{\partial u}{\partial x} + \frac{\partial v}{\partial y}\right) = 0 \tag{8.25}$$

8.1.4 Navier-Stokes Equations for 2-D Incompressible Flows

Conservation of mass and the momentum equations read

$$\left\{\begin{aligned} \frac{\partial u}{\partial x} + \frac{\partial v}{\partial y} &= 0 \\ \rho u\frac{\partial u}{\partial x} + \rho v\frac{\partial u}{\partial y} &= -\frac{\partial p}{\partial x} + \mu\left(\frac{\partial^2 u}{\partial x^2} + \frac{\partial^2 u}{\partial y^2}\right) \\ \rho u\frac{\partial v}{\partial x} + \rho v\frac{\partial v}{\partial y} &= -\frac{\partial p}{\partial y} + \mu\left(\frac{\partial^2 v}{\partial x^2} + \frac{\partial^2 v}{\partial y^2}\right) \end{aligned}\right. \tag{8.26}$$

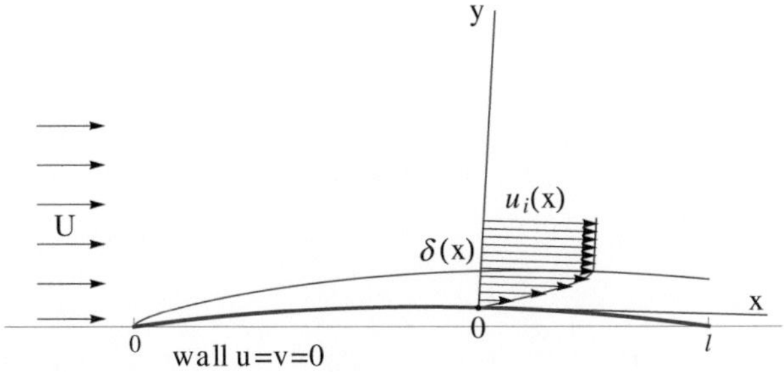

Fig. 8.5 Boundary layer development on a thin airfoil

The boundary conditions for the flow past a profile or a plate are the following (Fig. 8.5):

- on the solid surface, no penetration, no slip,
- in the far field, no disturbance.

8.1.5 Laminar Boundary Layer Theory (Prandtl 1904)

Let δ be the boundary layer thickness which represents the limit beyond which viscous effects are negligible. Let l be the characteristic length of the obstacle. We assume that $\delta \ll l$, see Fig. 8.6. Inside the boundary layer one can simplify the Navier-Stokes equations, using order of magnitude analysis which is based on the rigorous technique of asymptotic expansion. The u-component of velocity is of order U, the incoming flow velocity. Pressure is assumed to be of order $p = O\left(\rho U^2\right)$, in other words, it can be written

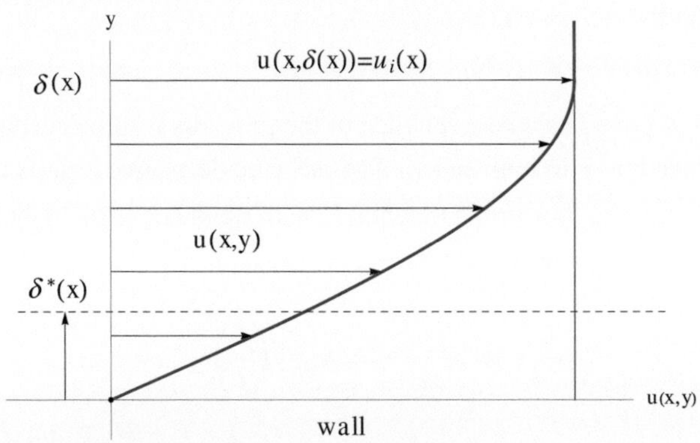

Fig. 8.6 Geometric interpretation of displacement thickness $\delta^*(x)$

$$p(x, y) = \rho U^2 \Pi \left(\frac{x}{l}, \frac{y}{\delta(x)} \right) \tag{8.27}$$

where Π is a dimensionless function of x/l and $y/\delta(x)$ only. The coefficient in front of the dimensionless function, ρU^2, is called a "gage". It gives the order of magnitude of the term. Furthermore, the partial derivatives are of order, respectively:

$$\frac{\partial}{\partial x} = O\left(\frac{1}{l}\right) \ll \frac{\partial}{\partial y} = O\left(\frac{1}{\delta}\right) \quad \Rightarrow \quad \frac{\partial^2}{\partial x^2} \ll \frac{\partial^2}{\partial y^2} \tag{8.28}$$

(a) From continuity, the two terms must balance, hence

$$\frac{v}{\delta} = O\left(\frac{U}{l}\right), \quad \Rightarrow \quad v = O\left(U\frac{\delta}{l}\right) \ll u \tag{8.29}$$

The flow is almost aligned with the body.

(b) From the x-momentum equation: the acceleration terms in the left-hand-side are both of order $O\left(\rho\frac{U^2}{l}\right)$ and must balance the viscous term $\mu\frac{\partial^2 u}{\partial y^2} = O\left(\mu\frac{U}{\delta^2}\right)$, hence

$$\delta^2 = O\left(l^2\frac{\mu}{\rho U l}\right) = l^2 O\left(\frac{1}{Re_l}\right) \quad \Rightarrow \quad \frac{\delta}{l} = O\left(\frac{1}{\sqrt{Re_l}}\right) \tag{8.30}$$

where the Reynolds number is defined as $Re_l = \frac{\rho U l}{\mu}$. In fact, this is true for any x sufficiently large, i.e.

$$\frac{\delta(x)}{x} = O\left(\frac{1}{\sqrt{Re_x}}\right) \tag{8.31}$$

where $Re_x = \frac{\rho U x}{\mu}$. Note that if there is a pressure gradient in the x-direction, its order will match the other terms, as $\frac{\partial p}{\partial x} = O\left(\rho\frac{U^2}{l}\right)$.

(c) Consider the y-momentum equation. First we multiply it by $\frac{\delta}{l}$: the acceleration terms in the left-hand-side are both now of order $O\left(\rho\frac{U^2\delta^2}{l^3}\right)$ and balance the viscous term of order $O\left(\mu\frac{U}{l^2}\right)$ as a consequence of the previous result (indeed, one term u has been replaced by v in both sides). The pressure derivative term is now of order $O\left(\rho\frac{U^2}{l}\right)$. Comparing this term's magnitude with the other terms indicates that this term dominates the equation since

$$\rho\frac{U^2}{l} \gg \rho\frac{U^2\delta^2}{l^3} = \rho\frac{U^2}{l}\left(\frac{\delta}{l}\right)^2 \quad \Rightarrow$$

$$\frac{\partial p}{\partial y} \gg \rho u\frac{\partial v}{\partial x} + \rho v\frac{\partial v}{\partial y} \; or \; \mu\frac{\partial^2 v}{\partial y^2} \tag{8.32}$$

As a result, to first order, the y-momentum equation reduces to one term and the pressure gradient in the y-direction must vanish

$$\frac{\partial p}{\partial y} = 0 \quad \Rightarrow \quad p = p(x) = p_i(x) \tag{8.33}$$

where the subscript i stands for "inviscid", i.e. in the inviscid flow. Let $u_i(x)$ be the inviscid flow velocity at the edge of the boundary layer. From Bernoulli's equation

$$p_\infty + \frac{1}{2}\rho U^2 = p_i(x) + \frac{1}{2}\rho\left(u_i^2(x) + v_i^2(x)\right) \simeq p_i(x) + \frac{1}{2}\rho u_i^2(x) \tag{8.34}$$

Taking the derivative in the x-direction yields

$$\frac{\partial p}{\partial x} = \frac{dp_i(x)}{dx} = -\rho u_i(x)\frac{du_i(x)}{dx} \tag{8.35}$$

Hence the Prandtl boundary layer equations reduce to

$$\begin{cases} \frac{\partial u}{\partial x} + \frac{\partial v}{\partial y} = 0 \\ \rho u\frac{\partial u}{\partial x} + \rho v\frac{\partial u}{\partial y} = \rho u_i(x)\frac{du_i(x)}{dx} + \mu\frac{\partial^2 u}{\partial y^2} \end{cases} \tag{8.36}$$

This is a nonlinear system of two equations in two unknowns, (u, v) with a source term function of $u_i(x)$ provided by the inviscid flow. It is of parabolic type and can be marched in the flow direction as long as $u(x, y) \geq 0$.

The initial/boundary conditions are the following:

- at $x = 0$ $u(0, y) = U$, $v(0, y) = 0$, a uniform, undisturbed velocity profile is given,
- at $y = 0$, $u(x, 0) = v(x, 0) = 0$, along the solid wall,
- at $y = \delta(x)$, $u(x, \delta(x)) = u_i(x)$, at the edge of the boundary layer.

8.1.6 Boundary Layer over a Flat Plate

In this special case, the inviscid flow solution is simply $u_i(x) = U$ and $p_i(x) = p_\infty$. The pressure gradient vanishes and there is no source term in the boundary layer equations. From the order of magnitude analysis one concludes that

$$\frac{\delta(x)}{x} = O\left(\frac{1}{\sqrt{Re_x}}\right) \quad \Rightarrow \quad \delta(x) = C\frac{x}{\sqrt{\frac{\rho U x}{\mu}}} = \frac{C}{\sqrt{\frac{\rho U}{\mu}}}\sqrt{x} \tag{8.37}$$

The boundary layer has a parabolic growth. The constant $C \simeq 5.0$. A self-similar solution has been found and calculated accurately by Blasius in 1908, see Ref. [3]. Other approximate solutions have been obtained with von Karman integral equation, see below.

8.1.7 Numerical Method for the Solution of Boundary Layer Equations

A Cartesian mesh system is defined, with constant steps Δx and Δy and indices i and j in the x- and y-directions, respectively. This is an initial/boundary value problem. Before the obstacle, a uniform flow is specified $u_{1,j} = U,\ v_{1,j} = 0,\ \ j = 1, jx$. The pressure gradient or source term is specified from the inviscid flow solution. The continuity equation is used to solve for v and the momentum equation for u. Consider the x-momentum equation:

$$\rho\bar{u}\frac{\partial u}{\partial x} + \rho\bar{v}\frac{\partial u}{\partial y} = \rho u_i \frac{du_i}{dx} + \mu\frac{\partial^2 u}{\partial y^2} \tag{8.38}$$

where $\bar{u}$ and $\bar{v}$ are evaluated from previous iteration (for initial guess, extrapolate from previous line). Two boundary conditions are specified for u, i.e.

$$y_1 = 0,\ u_{i,1} = 0,\quad y_{jx} = h,\ u_{i,jx} = U,\quad i = 2, \ldots \tag{8.39}$$

h is the ordinate of the upper boundary, $h > \delta(l)$, large enough to include the boundary layer.

Central finite differences are used for the first and second partial derivatives of u in y

$$\left(\frac{\partial u}{\partial y}\right)_{i,j} \simeq \frac{u_{i,j+1} - u_{i,j-1}}{2\Delta y}, \quad \textit{and}$$

$$\left(\frac{\partial^2 u}{\partial y^2}\right)_{i,j} \simeq \frac{u_{i,j+1} - 2u_{i,j} + u_{i,j-1}}{\Delta y^2}, \quad i = 2, \ldots,\ j = 2, \ldots, jx - 1 \tag{8.40}$$

Backward differences are used for the first derivative of u in x

$$\left(\frac{\partial u}{\partial x}\right)_{i,j} \simeq \frac{u_{i,j} - u_{i-1,j}}{\Delta x}, \quad i = 2, \ldots,\ j = 2, \ldots, jx - 1 \tag{8.41}$$

A tridiagonal solver is used to obtain $u_{i,j}$ for line $i = 2, \ldots$

Note that, as long as $u_{i,j} > 0$, the tridiagonal matrix is diagonally dominant, which insures stability with respect to round-off errors.

Consider next the continuity equation to solve for v.

One boundary condition is specified

$$y_1 = 0, \; v_{i,1} = 0, \quad i = 2, \ldots \tag{8.42}$$

The following scheme is used

$$\left(\frac{\partial v}{\partial y}\right)_{i,j} \simeq \frac{v_{i,j+1} - v_{i,j}}{\Delta y} = -\frac{1}{2}\left(\frac{u_{i,j+1} - u_{i-1,j+1} + u_{i,j} - u_{i-1,j}}{\Delta x}\right),$$

$$i = 2, \ldots, \; j = 1, \ldots, jx - 1 \tag{8.43}$$

This equation is marched upward along the vertical line, solving for $v_{i,j+1}$.

The process is repeated, i.e. one iterates on u and v on line i until convergence (3 or 4 iterations). Then, one marches to the next line $i + 1$. The domain is swept once in the flow direction.

8.1.8 Boundary Layer Thicknesses

Let $\delta^*(x)$ denote the boundary layer displacement thickness. It is defined as the distance the wall should be displaced to compensate for the deficit of mass flow rate of the boundary layer compared to the local uniform inviscid flow $u_i(x)$, see Fig. 8.6.

Mathematically this means

$$\int_0^{\delta(x)} \rho(x, y)u(x, y)dy = \rho_i(x)u_i(x)\left(\delta(x) - \delta^*(x)\right) \quad \Rightarrow$$

$$\frac{\delta^*(x)}{\delta(x)} = \int_0^1 \left(1 - \frac{\rho(x, y)u(x, y)}{\rho_i(x)u_i(x)}\right) d\left(\frac{y}{\delta(x)}\right) \tag{8.44}$$

Note: the definition of displacement thickness is valid for compressible as well as incompressible flows. δ^* will be used in viscous/inviscid interaction.

Let $\theta(x)$ represent the boundary layer momentum thickness. It is related to the drag due to friction as $D'_f = \rho_i(x)u_i^2(x)\theta(x)$.

As will be shown later, by application of control volume (CV) theorems, the drag due to friction can be evaluated in terms of the velocity profile exiting the CV.

$$-\int_0^{\delta(x)} \rho(x, y)u^2(x, y)dy + \int_0^{\delta(x)} \rho(x, y)u(x, y)u_i(x)dy = \rho_i(x)u_i^2(x)\theta(x) \quad \Rightarrow$$

$$\theta(x) = \int_0^{\delta(x)} \frac{\rho(x, y)u(x, y)}{\rho_i(x)u_i(x)}\left(1 - \frac{u(x, y)}{u_i(x)}\right) dy \tag{8.45}$$

or,

$$\frac{\theta(x)}{\delta(x)} = \int_0^1 \frac{\rho(x, y)u(x, y)}{\rho_i(x)u_i(x)} \left(1 - \frac{u(x, y)}{u_i(x)}\right) d\left(\frac{y}{\delta(x)}\right) \tag{8.46}$$

Note: for incompressible flow: $\frac{\rho}{\rho_i} = 1$. Energy and enthalpy thicknesses can similarly be defined.

The skin friction coefficient is defined as

$$C_f = \frac{\tau}{\frac{1}{2}\rho_\infty U_\infty^2} = \frac{\mu \frac{\partial u}{\partial y}}{\frac{1}{2}\rho_\infty U_\infty^2} = \frac{2}{Re_l} \frac{\partial\left(\frac{u}{U_\infty}\right)}{\partial\left(\frac{y}{l}\right)} \tag{8.47}$$

8.2 Compressible Viscous Fluid Flow

8.2.1 Viscous Stresses and Constitutive Relations

For compressible flows, there is a second viscosity coefficient λ. Stresses are related to strain rates as follows:

$$\sigma_{1,1} = 2\mu\frac{\partial u}{\partial x} + \lambda\left(\frac{\partial u}{\partial x} + \frac{\partial v}{\partial y}\right) \tag{8.48}$$

$$\sigma_{2,2} = 2\mu\frac{\partial v}{\partial y} + \lambda\left(\frac{\partial u}{\partial x} + \frac{\partial v}{\partial y}\right) \tag{8.49}$$

$$\sigma_{1,2} = \sigma_{2,1} = \mu\left(\frac{\partial u}{\partial x} + \frac{\partial v}{\partial y}\right) \tag{8.50}$$

According to Stokes hypothesis: $\lambda = -\frac{2}{3}\mu$.

8.2.2 Navier-Stokes Equations for 2-D Compressible Flows

Assuming λ and μ are constant

Conservation of mass:

$$\frac{\partial \rho u}{\partial x} + \frac{\partial \rho v}{\partial y} = 0 \tag{8.51}$$

Conservation of x-momentum:

$$\rho u \frac{\partial u}{\partial x} + \rho v \frac{\partial u}{\partial y} = -\frac{\partial p}{\partial x} + \mu \left(\frac{\partial^2 u}{\partial x^2} + \frac{\partial^2 u}{\partial y^2} \right) + \frac{\mu}{3} \left(\frac{\partial^2 u}{\partial x^2} + \frac{\partial^2 v}{\partial x \partial y} \right) \tag{8.52}$$

Conservation of y-momentum:

$$\rho u \frac{\partial v}{\partial x} + \rho v \frac{\partial v}{\partial y} = -\frac{\partial p}{\partial y} + \mu \left(\frac{\partial^2 v}{\partial x^2} + \frac{\partial^2 v}{\partial y^2} \right) + \frac{\mu}{3} \left(\frac{\partial^2 u}{\partial x \partial y} + \frac{\partial^2 v}{\partial y^2} \right) \tag{8.53}$$

The system is completed with the equation of energy for temperature T and the equation of state (for a perfect gas) $p = \rho R T$.

8.2.3 Energy Equation for Compressible Viscous Flows

There are six types of energy important in our analysis (we exclude, chemical, electromagnetic and nuclear energies):

Mechanical energy		Thermodynamic energy
Kinetic energy		Internal energy
Potential energy		Heat
	← Work done by pressure →	
	← Work done by friction →	

For inviscid flow, the Euler equations have the following energy balance:

$$\frac{\partial \rho u H}{\partial x} + \frac{\partial \rho v H}{\partial y} = 0 \tag{8.54}$$

where $H = h + \frac{1}{2}\left(u^2 + v^2\right) + potential\ energy$, and $h = e + \frac{p}{\rho}$. Here we have ignored heat and viscous dissipation.

For the Navier-Stokes equations, the energy equation is

$$\frac{\partial \rho u H}{\partial x} + \frac{\partial \rho v H}{\partial y} = \frac{\partial}{\partial x}\left(k \frac{\partial T}{\partial x} \right) + \frac{\partial}{\partial y}\left(k \frac{\partial T}{\partial y} \right) + (\nabla . \mathbf{V} . \boldsymbol{\tau}) \tag{8.55}$$

where k is the heat conductivity coefficient, and the viscous dissipation term is given by:

$$(\nabla . \mathbf{V} . \boldsymbol{\tau}) = \frac{\partial}{\partial x}\left(u \left(\sigma_{1,1} + \sigma_{1,2} \right) \right) + \frac{\partial}{\partial y}\left(v \left(\sigma_{1,2} + \sigma_{2,2} \right) \right) \tag{8.56}$$

corresponding to the rate of work done by viscous forces.

8.2.4 Boundary Layer Equations for Compressible Flows

We use order of magnitude analysis to simplify the energy equation. The results are unchanged for the other equations:

$$\frac{\partial \rho u}{\partial x} + \frac{\partial \rho v}{\partial y} = 0 \tag{8.57}$$

$$\rho u \frac{\partial u}{\partial x} + \rho v \frac{\partial u}{\partial y} = -\frac{dp_i}{dx} + \frac{\partial}{\partial y}\left(\mu \frac{\partial u}{\partial y}\right) \tag{8.58}$$

$$\frac{\partial p}{\partial y} = 0 \tag{8.59}$$

$$\rho u \frac{\partial H}{\partial x} + \rho v \frac{\partial H}{\partial y} = \frac{\partial}{\partial y}\left(k \frac{\partial T}{\partial y}\right) + \frac{\partial}{\partial y}\left(\mu u \frac{\partial u}{\partial y}\right) \tag{8.60}$$

The nondimensional form of the compressible boundary layer equations read

$$\frac{\partial \bar{\rho}\bar{u}}{\partial \bar{x}} + \frac{\partial \bar{\rho}\bar{v}}{\partial \bar{y}} = 0 \tag{8.61}$$

$$\bar{\rho}\bar{u} \frac{\partial \bar{u}}{\partial \bar{x}} + \bar{\rho}\bar{v} \frac{\partial \bar{u}}{\partial \bar{y}} = -\frac{d\bar{p}_i}{d\bar{x}} + \frac{1}{Re_l}\frac{\partial}{\partial \bar{y}}\left(\bar{\mu} \frac{\partial \bar{u}}{\partial \bar{y}}\right) \tag{8.62}$$

$$\bar{\rho}\bar{u} \frac{\partial \bar{H}}{\partial \bar{x}} + \bar{\rho}\bar{v} \frac{\partial \bar{H}}{\partial \bar{y}} = \frac{1}{P_r Re_l}\frac{\partial}{\partial \bar{y}}\left(\bar{k} \frac{\partial \bar{T}}{\partial \bar{y}}\right) + \frac{1}{Re_l}\frac{\partial}{\partial \bar{y}}\left(\bar{\mu}\bar{u} \frac{\partial \bar{u}}{\partial \bar{y}}\right) \tag{8.63}$$

where, $\bar{x} = x/l$, $\bar{y} = y/l$, $\bar{u} = u/U$, $\bar{v} = v/U$, $\bar{p}_i = p_i/\rho_\infty U^2$, $\bar{\mu} = \mu/\mu_\infty$, and $\bar{k} = k/k_\infty$. Furthermore $H = C_p T + \frac{1}{2}u^2$, $\bar{H} = H/U^2$, $\bar{T} = C_p T/U^2$, $Re_l = \rho_\infty U l/\mu_\infty$, and $P_r = \mu_\infty C_p/k_\infty$.

The boundary conditions are the same as for incompressible flow with the addition of boundary conditions for the temperature, i.e. at the wall either T_w or $\frac{\partial T}{\partial y}|_w$ is given, and at the edge of the boundary layer $T_i(x)$ is given.

The numerical approach is the extension of the incompressible flow case.

- Solve for u from the x-momentum equation, assuming the coefficients are known.
- Solve for v from the continuity equation, assuming ρ and u are known.
- Solve for T from the energy equation, using central differences for $\frac{\partial T}{\partial y}$ and $\frac{\partial^2 T}{\partial y^2}$ and backward differences for $\frac{\partial T}{\partial x}$, and assuming ρ, u, v are known. Use $p_i = \rho RT$ to calculate ρ.
- Iterate on the line until convergence.
- March with the flow ($u > 0$).

8.2.5 Further Simplifications for the Case $P_r = 1$

If $P_r = 1$ (i.e. $\mu C_p = k$), the energy equation in the boundary layer approximation becomes

$$\rho u \frac{\partial H}{\partial x} + \rho v \frac{\partial H}{\partial y} = \frac{\partial}{\partial y}\left(\mu C_p \frac{\partial T}{\partial y}\right) + \frac{\partial}{\partial y}\left(\mu \frac{\partial}{\partial y}\left(\frac{u^2}{2}\right)\right) = \frac{\partial}{\partial y}\left(\mu \frac{\partial H}{\partial y}\right) \quad (8.64)$$

This simplified equation admits a solution $H = H_\infty$, provided the boundary conditions on the temperature is consistent with this solution:

$$C_p T + \frac{1}{2} u^2 = H_\infty \quad (8.65)$$

At the wall $u = 0$, hence $T_w = H_\infty / C_p$; also,

$$C_p \frac{\partial T}{\partial y} + u \frac{\partial u}{\partial y} = 0, \quad at\ the\ wall \quad \frac{\partial T}{\partial y}|_w = 0 \quad (8.66)$$

The energy equation is replaced by Bernoulli's law inside the boundary layer:

$$C_p T + \frac{1}{2} u^2 = H_\infty \quad or \quad \frac{\gamma}{\gamma - 1} \frac{p}{\rho} + \frac{1}{2} u^2 = H_\infty \quad (8.67)$$

where $p = p_i(x)$.

8.2.6 Determination of Drag

Assume that the undisturbed flow is aligned with the x-axis.

The drag per unit span, D' can be obtained by direct integration of the stresses around the profile as

$$D' = \oint_C p dy + \tau_w dx \quad (8.68)$$

where (dx, dy) is a small element of the profile contour C.

From momentum balance applied to a large rectangle surrounding the profile, see Fig. 8.7. Contributions on control surfaces 3 and 4 vanish when the control volume is expanded far away and contributions on 1 and 2 can be combined, resulting in the drag being obtained as an integral across the far wake:

$$D' = \int_{-\infty}^{\infty} \rho u(x, y)\,(U - u(x, y))\, dy \quad (8.69)$$

The drag can also be evaluated in terms of entropy (Oswatitsch formula [1]).

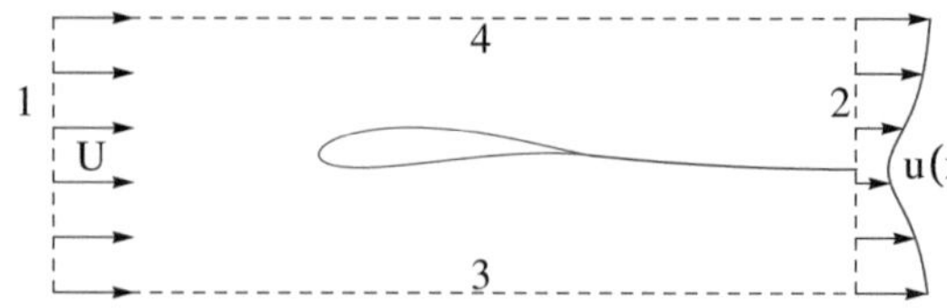

Fig. 8.7 Control volume for the evaluation of drag D'

Assume $P_r = 1$ and $H = H_\infty$. Using Crocco's relation for viscous flows at the far wake:

$$-U\frac{\partial u}{\partial y} = T_\infty\frac{\partial s}{\partial y} + viscous\ term \tag{8.70}$$

Ignoring the last term for high Reynolds number flows in the wake, the above equation gives

$$-Uu = T_\infty s + const. \tag{8.71}$$

Far from the wake, the flow returns to uniform conditions and the entropy to s_∞. Hence

$$U(U-u) = T_\infty(s - s_\infty) \tag{8.72}$$

The drag reads

$$D' = \frac{T_\infty}{U}\int_{-\infty}^{\infty}\rho u(s - s_\infty)dy \tag{8.73}$$

The work done by the drag $D'U$ is related to the losses in terms of entropy.

In the case of a flat plate, the drag is simply the friction drag.

In the case of an airfoil, the drag is the sum of friction drag and pressure drag (due to separated flow).

8.2.7 Unsteady Boundary Layer

The unsteady boundary layer equations are given by

$$\frac{\partial\rho}{\partial t} + \frac{\partial\rho u}{\partial x} + \frac{\partial\rho v}{\partial y} = 0 \tag{8.74}$$

$$\frac{\partial\rho u}{\partial t} + \rho u\frac{\partial u}{\partial x} + \rho v\frac{\partial u}{\partial y} = \frac{\partial\rho u_i}{\partial t} + \rho u_i\frac{\partial u_i}{\partial x} + \frac{\partial}{\partial y}\left(\mu\frac{\partial u}{\partial y}\right) \tag{8.75}$$

$$\frac{\partial \rho H}{\partial t} + \rho u \frac{\partial H}{\partial x} + \rho v \frac{\partial H}{\partial y} = \frac{\partial p_i}{\partial t} + \frac{\partial}{\partial y}\left(k \frac{\partial T}{\partial y}\right) + \frac{\partial}{\partial y}\left(\mu u \frac{\partial u}{\partial y}\right) \tag{8.76}$$

where $p = \rho RT$ and $\mu = \mu(T)$.

The boundary conditions at the solid wall ($y = 0$) are

$$u(x, 0) = u_w(t), \quad v(x, 0) = 0, \quad T = T_w(x, t) \tag{8.77}$$

As a simple example of unsteady boundary layer, Liepmann and Roshko [2] discuss the Rayleigh's problem representing the diffusion of vorticity, see also Schlichting [3] and Howarth [4].

8.3 Historical and Classical Works

The history of boundary layer is very rich and interesting. Starting from Prandtl's original work (1904) and his student Blasius and von Karman and others (in the field of turbulent flows as well), the theory has developed over the years. Many researchers, all over the world, contributed to this subject. In Göttingen, Germany, where Prandtl worked, there was a conference in 2004, celebrating 100 years anniversary of boundary layer theory (the proceedings is published by Springer Verlag [5]).

In these notes, some topics will be discussed with historical remarks. Transformations of subsonic boundary layer equations to the incompressible ones show that compressibility effects do not change the basic phenomena. Singularities at separation (and attachment) points as well as trailing edge show the limitation of boundary layer equations and the importance of coupling viscous and inviscid flows (as suggested by Prandtl himself through introducing the concept of displacement thickness). Treatment of separated flows is important for practical applications. Axisymmetric flow transformation to 2-D like equations (Mangler) and three dimensional flows over swept wings are also discussed.

Some of these works are of limited values today, thanks to the power of our computers. Nevertheless, it is still useful to study the ideas behind them.

8.3.1 Blasius Solution for the Flat Plate (No Pressure Gradient)

In incompressible flow, the self-similar solution due to Blasius is based on the idea that the u-velocity profile is of the form

$$\frac{u(x, y)}{U} = f\left(\frac{y}{\delta(x)}\right) = f(\eta) \tag{8.78}$$

where, as seen before, $\delta(x) \sim \sqrt{\frac{\mu x}{\rho U}}$, and $\eta = y\sqrt{\frac{\rho U}{\mu x}}$. If one introduces the stream function $\psi(x, y) = U\delta(x)F(\eta)$, the governing equation for $F(\eta)$ is:

$$FF'' + 2F''' = 0 \tag{8.79}$$

The boundary conditions are as follows

- at $\eta = 0$: $F = F' = 0$,
- as $\eta \to \infty$: $F' \to 1$.

The numerical solution of Blasius yields the results

$$\frac{\delta(x)}{x} = \frac{5.0}{\sqrt{Re_x}} \tag{8.80}$$

$$\frac{\delta^*(x)}{x} = \frac{1.7203}{\sqrt{Re_x}} \tag{8.81}$$

$$\frac{\theta(x)}{x} = \frac{0.664}{\sqrt{Re_x}} \tag{8.82}$$

$$C_f(x) = \frac{0.664}{\sqrt{Re_x}} \tag{8.83}$$

$$C_{Df} = \frac{1.328}{\sqrt{Re_l}} \tag{8.84}$$

Note that $C_{Df} = 2C_f(l) = 2\theta(l)/l$.

8.3.2 Flow Past a Wedge (Falkner/Skan)

The inviscid solution is of the form $U(x) = Cx^m$. The governing equation for the self-similar solution for the stream function reads

$$F''' + FF'' + \beta(1 - F'^2) = 0 \tag{8.85}$$

where $\beta = \frac{2m}{m+1}$ is the wedge angle, and with boundary conditions

- at $\eta = 0$: $F = F' = 0$,
- as $\eta \to \infty$: $F' \to 1$.

For more on this, we refer to Hartee numerical analysis and Stewartson analysis.

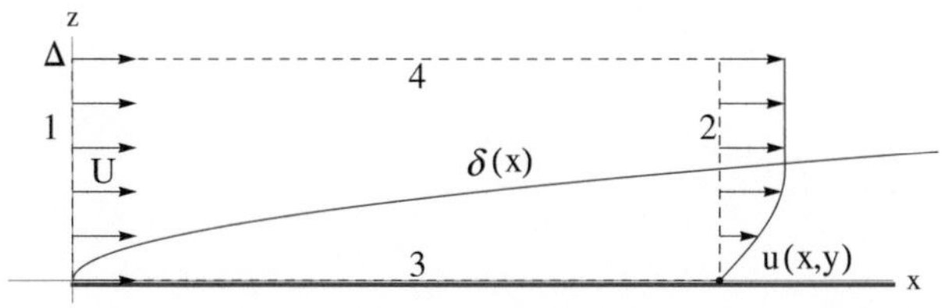

Fig. 8.8 Control volume for application of momentum theorem

Special cases:

- Flat plate: $m = 0,\ \beta = 0$.
- Stagnation point: $m = 1,\ \beta = 1$.

When $\beta < 0$ ($m < 0$), separation occurs.

8.3.3 von Karman Integral Momentum Equation

Consider first the flat plate and incompressible flow (no pressure gradient)

The conservation theorems applied to the control volume defined in Fig. 8.8 give the following results.

Conservation of mass:

$$-\rho U \Delta + \int_0^{\Delta} \rho u(l, y) dy + \int_0^{l} \rho v(l, y) dx = 0 \tag{8.86}$$

Conservation of x-momentum:

$$-\rho U^2 \Delta + \int_0^{\Delta} \rho u^2(l, y) dy + \int_0^{l} \rho U v(l, y) dx = -\int_0^{l} \tau_w(x) dx = -D'_f \tag{8.87}$$

Multiply the mass equation by U and subtract the x-momentum to obtain

$$D'_f = \int_0^{\Delta} \rho \left(U u(l, y) - u^2(l, y) \right) dy = \rho U^2 \int_0^{\Delta} \frac{u(l, y)}{U} \left(1 - \frac{u(l, y)}{U} \right) dy$$

$$= \rho U^2 \int_0^{\delta(l)} \frac{u(l, y)}{U} \left(1 - \frac{u(l, y)}{U} \right) dy = \rho U^2 \theta(l) \tag{8.88}$$

Note that

$$D'_f = \int_0^{l} \tau_w(x) dx \tag{8.89}$$

Replacing l by x

$$\int_0^x \tau_w(\xi)d\xi = \rho U^2\theta(x) \quad \Rightarrow \quad \tau_w(x) = \rho U^2 \frac{d\theta(x)}{dx}$$

where $\tau_w(x) = \mu \frac{\partial u}{\partial y}|_w$.

Example of application of von Karman's integral equation.

Assume that the velocity profile is a parabola:

$$\frac{u(x,y)}{U} = a + b\frac{y}{\delta(x)} + c\left(\frac{y}{\delta(x)}\right)^2 \tag{8.90}$$

- At $\frac{y}{\delta(x)} = 0, \quad \frac{u(x,0)}{U} = 0 \quad \Rightarrow \quad a = 0$
- At $\frac{y}{\delta(x)} = 1, \quad \frac{u(x,0)}{U} = 1 \quad \Rightarrow \quad b + c = 1$

and $\frac{\partial u(x,\delta(x))}{\partial y} = 0 \quad \Rightarrow \quad b + 2c = 0$. Hence $b = 2$ and $c = -1$, giving

$$\frac{u(x,y)}{U} = 2\frac{y}{\delta(x)} - \left(\frac{y}{\delta(x)}\right)^2 \tag{8.91}$$

Next, one can evaluate the boundary layer thicknesses and the wall shear stress from their definitions:

$$\frac{\delta^*(x)}{\delta(x)} = \int_0^1 \left(1 - \frac{u(x,y)}{U}\right) d\left(\frac{y}{\delta(x)}\right) = \int_0^1 (1 - 2\eta + \eta^2)d\eta = \frac{1}{3} \tag{8.92}$$

$$\frac{\theta(x)}{\delta(x)} = \int_0^1 \frac{u(x,y)}{U}\left(1 - \frac{u(x,y)}{U}\right) d\left(\frac{y}{\delta(x)}\right) = \int_0^1 (2\eta - \eta^2)(1 - 2\eta + \eta^2)d\eta = \frac{2}{15} \tag{8.93}$$

$$\tau_w(x) = \mu\frac{\partial u(x,0)}{\partial y} = 2\frac{\mu U}{\delta(x)} \tag{8.94}$$

Substitution in von Karman equation gives

$$\tau_w(x)dx = \rho U^2 d\theta \quad \Rightarrow \quad 2\frac{\mu U}{\delta(x)}dx = \rho U^2 \frac{2}{15} d\delta(x) \tag{8.95}$$

Integration gives

$$\frac{\delta^2(x)}{2} = 15\frac{\mu x}{\rho U} + const. \tag{8.96}$$

The constant is determined by the condition $\delta(0) = 0$ to give

$$\delta(x) = 5.5\sqrt{\frac{\mu x}{\rho U}} \tag{8.97}$$

The following result

$$\frac{\delta(x)}{x} = \frac{5.5}{\sqrt{Re_x}} \tag{8.98}$$

$$\frac{\delta^*(x)}{x} = \frac{1.83}{\sqrt{Re_x}} \tag{8.99}$$

$$\frac{\theta(x)}{x} = \frac{0.73}{\sqrt{Re_x}} \tag{8.100}$$

$$C_f(x) = \frac{0.73}{\sqrt{Re_x}} = \frac{\theta(x)}{x} \tag{8.101}$$

$$C_{Df} = 2\frac{\theta(l)}{l} = 2C_f(l) = \frac{1.46}{\sqrt{Re_l}} \tag{8.102}$$

These results can be compared with Blasius solution.

8.3.4 Derivation of von Karman Integral Equation for Compressible Flows with Pressure Gradient

Consider the equation of conservation of mass

$$\frac{\partial \rho u}{\partial x} + \frac{\partial \rho v}{\partial y} = 0 \tag{8.103}$$

Integrating across the boundary layer yields

$$\int_0^{\delta(x)} \left\{ \frac{\partial \rho u}{\partial x} + \frac{\partial \rho v}{\partial y} \right\} dy = \int_0^{\delta(x)} \frac{\partial \rho u}{\partial x} dy + \rho_i v_i = 0 \tag{8.104}$$

Exchanging integration and differentiation on the first term gives

$$\frac{d}{dx}\int_0^{\delta(x)} \rho u dy - \rho_i u_i \frac{d\delta}{dx} + \rho_i v_i = 0 \tag{8.105}$$

Now, introducing the displacement thickness, one gets

$$\frac{d}{dx}\left\{\rho_i u_i \delta - \rho_i u_i \int_0^{\delta(x)} \left(1 - \frac{\rho u}{\rho_i u_i}\right) dy\right\} - \rho_i u_i \frac{d\delta}{dx} + \rho_i v_i = 0 \tag{8.106}$$

Upon simplification one obtains

$$\frac{d}{dx}(\rho_i u_i)\,\delta - \frac{d}{dx}\left(\rho_i u_i \delta^*\right) + \rho_i v_i = 0 \tag{8.107}$$

Proceeding similarly with the x-momentum

$$\frac{\partial \rho u^2}{\partial x} + \frac{\partial \rho u v}{\partial y} = -\frac{dp_i}{dx} + \frac{\partial \tau}{\partial y} \tag{8.108}$$

$$\int_0^{\delta(x)} \frac{\partial \rho u^2}{\partial x} dy + \rho_i u_i v_i = -\frac{dp_i}{dx}\delta - \tau_w \tag{8.109}$$

$$\frac{d}{dx}\int_0^{\delta(x)} \rho u^2 dy - \rho_i u_i^2 \frac{d\delta}{dx} + \rho_i u_i v_i = -\frac{dp_i}{dx}\delta - \tau_w \tag{8.110}$$

Introducing the momentum thickness

$$\frac{d}{dx}\left\{u_i \int_0^{\delta(x)} \rho u dy - \rho_i u_i^2 \int_0^{\delta(x)} \frac{\rho u}{\rho_i u_i}\left(1 - \frac{u}{u_i}\right) dy\right\} - \rho_i u_i^2 \frac{d\delta}{dx} + \rho_i u_i v_i$$
$$= -\frac{dp_i}{dx}\delta - \tau_w \tag{8.111}$$

$$\frac{d}{dx}\left\{u_i \int_0^{\delta(x)} \rho u dy\right\} - \frac{d}{dx}\left(\rho_i u_i^2 \theta\right) - \rho_i u_i^2 \frac{d\delta}{dx} + \rho_i u_i v_i = -\frac{dp_i}{dx}\delta - \tau_w \tag{8.112}$$

With displacement thickness

$$\frac{d}{dx}\left\{u_i\left(\rho_i u_i \delta - \rho_i u_i \delta^*\right)\right\} - \frac{d}{dx}\left(\rho_i u_i^2 \theta\right) - \rho_i u_i^2 \frac{d\delta}{dx} + \rho_i u_i v_i = -\frac{dp_i}{dx}\delta - \tau_w \tag{8.113}$$

After some simplification one finds

$$\frac{d}{dx}\left(\rho_i u_i^2\right)\delta - \frac{d}{dx}\left(\rho_i u_i^2 \delta^*\right) - \frac{d}{dx}\left(\rho_i u_i^2 \theta\right) + \rho_i u_i v_i = -\frac{dp_i}{dx}\delta - \tau_w \tag{8.114}$$

Furthermore, the pressure gradient term can be replaced by the acceleration term to give

$$\frac{d}{dx}\left(\rho_i u_i^2\right)\delta - \frac{d}{dx}\left(\rho_i u_i^2 \delta^*\right) - \frac{d}{dx}\left(\rho_i u_i^2 \theta\right) + \rho_i u_i v_i = \rho_i u_i \frac{du_i}{dx}\delta - \tau_w \quad (8.115)$$

This equation can be subtracted from the mass conservation multiplied by u_i which yields the final result

$$\frac{d}{dx}\left(\rho_i u_i^2 \theta\right) + \rho_i u_i \frac{du_i}{dx}\delta^* = \tau_w \quad (8.116)$$

8.3.5 Transformations of Boundary Layer Equations

von Mises Transformation

The von Mises transformation is a change of independent variables, from (x, y) to (X, ψ), where $X = x$ and ψ is the stream function such that

$$\frac{\partial \psi}{\partial x} = -\rho v, \quad \frac{\partial \psi}{\partial y} = \rho u \quad (8.117)$$

Using the chain rule one gets

$$\frac{\partial u}{\partial x} = \frac{\partial u}{\partial X}\frac{\partial X}{\partial x} + \frac{\partial u}{\partial \psi}\frac{\partial \psi}{\partial x} = \frac{\partial u}{\partial X} - \rho v \frac{\partial u}{\partial \psi} \quad (8.118)$$

$$\frac{\partial u}{\partial y} = \frac{\partial u}{\partial X}\frac{\partial X}{\partial y} + \frac{\partial u}{\partial \psi}\frac{\partial \psi}{\partial xy} = \rho u \frac{\partial u}{\partial \psi} \quad (8.119)$$

The mass conservation equation is automatically satisfied by the introduction of the stream function. Substitution in the x-momentum yields

$$\rho u \frac{\partial u}{\partial x} + \rho v \frac{\partial u}{\partial y} = \frac{\partial}{\partial y}\left(\mu \frac{\partial u}{\partial y}\right) \quad (8.120)$$

$$\rho u \frac{\partial u}{\partial X} - \rho u \rho v \frac{\partial u}{\partial \psi} + \rho u \rho v \frac{\partial u}{\partial \psi} = \rho u \frac{\partial}{\partial \psi}\left(\mu \rho u \frac{\partial u}{\partial \psi}\right) \quad (8.121)$$

After simplification, there remains

$$\frac{\partial u}{\partial X} = \frac{\partial}{\partial \psi}\left(\mu \rho u \frac{\partial u}{\partial \psi}\right) \quad (8.122)$$

This is completed with the energy equation

$$\rho u \frac{\partial H}{\partial x} + \rho u v \frac{\partial H}{\partial y} = \frac{\partial}{\partial y}\left(k \frac{\partial T}{\partial y}\right) + \frac{\partial}{\partial y}\left(\mu u \frac{\partial u}{\partial y}\right) \tag{8.123}$$

yielding

$$\frac{\partial H}{\partial X} = \frac{\partial}{\partial \psi}\left(k \rho u \frac{\partial T}{\partial \psi}\right) + \frac{\partial}{\partial \psi}\left(\mu \rho u^2 \frac{\partial u}{\partial \psi}\right) \tag{8.124}$$

Notice that the solution is not regular at $y = 0$ where $\psi = 0$. To see the behavior at the wall, let $u = a(x)y + \cdots$ and $\psi = \frac{1}{2}\rho_w a(x) y^2 + \cdots$, $a(x) > 0$. The leading term reads

$$u = \sqrt{\frac{2a(x)\psi}{\rho_w}}, \quad \Rightarrow \quad \frac{\partial u}{\partial \psi} = \sqrt{\frac{a(x)}{2\rho_w \psi}} \tag{8.125}$$

The shear stress at the wall is finite since

$$\tau_w = \mu \frac{\partial u}{\partial y}|_{y=0} = \lim_{y \to 0} \mu \rho u \frac{\partial u}{\partial \psi} = \mu a(x) \tag{8.126}$$

Use of Natural Coordinates

Changing coordinates from (x, y) to (s, n), the following relations hold

$$q \frac{\partial u}{\partial s} = u \frac{\partial u}{\partial x} + v \frac{\partial u}{\partial y} \tag{8.127}$$

$$q \frac{\partial u}{\partial n} = u \frac{\partial u}{\partial y} - v \frac{\partial u}{\partial x} \tag{8.128}$$

where $q = \sqrt{u^2 + v^2}$. In boundary layer flow, $q \simeq u$ and $\frac{\partial u}{\partial y} \simeq \frac{\partial u}{\partial n}$, hence the equations reduce to

$$u \frac{\partial u}{\partial s} = \frac{\partial}{\partial n}\left(\mu \frac{\partial u}{\partial n}\right) \tag{8.129}$$

$$u \frac{\partial H}{\partial s} = \frac{\partial u}{\partial s} = \frac{\partial}{\partial n}\left(k \frac{\partial T}{\partial n}\right) + \frac{\partial}{\partial n}\left(\mu u \frac{\partial u}{\partial n}\right) \tag{8.130}$$

Howarth/Dorodnitcyn Transformation

Let $d\bar{y} = \rho dy$ and $\bar{x} = x$, then it follows that

$$u = \frac{1}{\rho} \frac{\partial \psi}{\partial y} = \frac{\partial \psi}{\partial \bar{y}}, \quad v = -\frac{1}{\rho}\left(\frac{\partial \psi}{\partial \bar{x}} + u \frac{\partial \bar{y}}{\partial x}\right) \tag{8.131}$$

$$\frac{\partial u}{\partial x} = \frac{\partial u}{\partial \bar{x}} + \frac{\partial u}{\partial \bar{y}}\frac{\partial \bar{y}}{\partial x}, \quad \frac{\partial u}{\partial y} = \rho\frac{\partial u}{\partial \bar{y}} \tag{8.132}$$

Therefore the x-momentum reads

$$\rho u\left(\frac{\partial u}{\partial \bar{x}} + \frac{\partial u}{\partial \bar{y}}\frac{\partial \bar{y}}{\partial x}\right) + \rho v\rho\frac{\partial u}{\partial \bar{y}} = \rho\frac{\partial}{\partial \bar{y}}\left(\mu\rho\frac{\partial u}{\partial \bar{y}}\right) \tag{8.133}$$

or, in terms of the stream function

$$\frac{\partial \psi}{\partial \bar{y}}\frac{\partial^2 \psi}{\partial x \partial \bar{y}} - \frac{\partial \psi}{\partial x}\frac{\partial^2 \psi}{\partial \bar{y}^2} = \frac{\partial}{\partial \bar{y}}\left(\mu\rho\frac{\partial u}{\partial \bar{y}}\right) \tag{8.134}$$

Special Case

If $\mu\rho = \mu_\infty\rho_\infty = const.$, then the compressible and incompressible flow equations are the same, using the von Mises or Howarth/Dorodnitcyn transformations in the case of zero pressure gradient. Note that in boundary layer flow, $\frac{p}{p_i} = 1$, hence $\frac{\rho T}{\rho_i T_i} = 1$. So we require

$$\frac{\mu}{\mu_\infty} = \frac{T}{T_\infty} \tag{8.135}$$

to have the similarity between compressible and incompressible boundary layer.

For this case

$$C_f = \frac{\tau_w}{\frac{1}{2}\rho_\infty U_\infty^2} = \frac{\mu\frac{\partial u}{\partial y}|_w}{\frac{1}{2}\rho_\infty U_\infty^2} = \frac{\mu\rho\frac{\partial u}{\partial \bar{y}}|_w}{\frac{1}{2}\rho_\infty U_\infty^2} = \frac{\mu_\infty\rho_\infty\frac{\partial u}{\partial \bar{y}}|_w}{\frac{1}{2}\rho_\infty U_\infty^2} \tag{8.136}$$

Since $u(\bar{y})$ is the same for compressible and incompressible flows, C_f is independent of Mach number for this special case.

Effect of Mach Number on Displacement Thickness

Consider a flat plate and compressible boundary layer with the assumptions

- no pressure gradient
- the Prandtl number $Pr = 1$
- $\frac{\mu}{\mu_\infty} = \frac{T}{T_\infty}$

From the definitions of δ_c^*, θ and δ^*:

$$\delta_c^* = \int_0^{\delta_c}\left(1 - \frac{\rho u}{\rho_i u_i}\right)dy \tag{8.137}$$

$$\theta = \int_0^{\delta}\frac{u}{u_i}\left(1 - \frac{u}{u_i}\right)dy \tag{8.138}$$

$$\delta^* = \int_0^\delta \left(1 - \frac{u}{u_i}\right) dy \tag{8.139}$$

with $H = H_\infty$ one can write

$$C_p T + \frac{1}{2}u^2 = H_\infty = C_p T_\infty + \frac{1}{2}U_\infty^2 \tag{8.140}$$

which reduces to

$$\frac{T}{T_\infty} + \frac{\gamma - 1}{2} M_\infty^2 \left(\frac{u}{U_\infty}\right)^2 = 1 + \frac{\gamma - 1}{2} M_\infty^2 \tag{8.141}$$

Now, using either the von Mises or Howarth/Dorodnitcyn transformation, we find (after some manipulations)

$$\delta_c^* = \delta^* + \frac{\gamma - 1}{2} M_\infty^2 \left(\delta^* + \theta\right) \tag{8.142}$$

Dividing by x and rearranging

$$\frac{\delta_c^* - \delta^*}{x} = \frac{\gamma - 1}{2} M_\infty^2 \left(\frac{\delta^*}{x} + \frac{\theta}{x}\right) \tag{8.143}$$

Substitution of known results yields

$$\frac{\delta_c^* - \delta^*}{x} = \frac{1.192(\gamma - 1) M_\infty^2}{\sqrt{Re_x}} \tag{8.144}$$

8.3.6 Flow Separation

Near a separation point, the flow streamlines bifurcate as shown in Fig. 8.9. From the x-momentum equation, applied at the wall where $u = v = 0$

$$\rho u \frac{\partial u}{\partial x} + \rho v \frac{\partial u}{\partial y} = -\frac{dp}{dx} + \frac{\partial}{\partial y}\left(\mu \frac{\partial u}{\partial y}\right) \Rightarrow \frac{dp}{dx} = \frac{\partial}{\partial y}\left(\mu \frac{\partial u}{\partial y}\right)|_{y=0} \tag{8.145}$$

In the presence of adverse pressure gradient (i.e. $\frac{dp}{dx} > 0$), the curvature of the velocity profile is positive $\frac{\partial^2 u}{\partial y^2}|_{y=0} > 0$. Integrating the x-momentum equation near the wall yields (freezing the coefficients at the wall values)

$$\frac{dp}{dx} y = \mu \frac{\partial u}{\partial y} - \tau_w \tag{8.146}$$

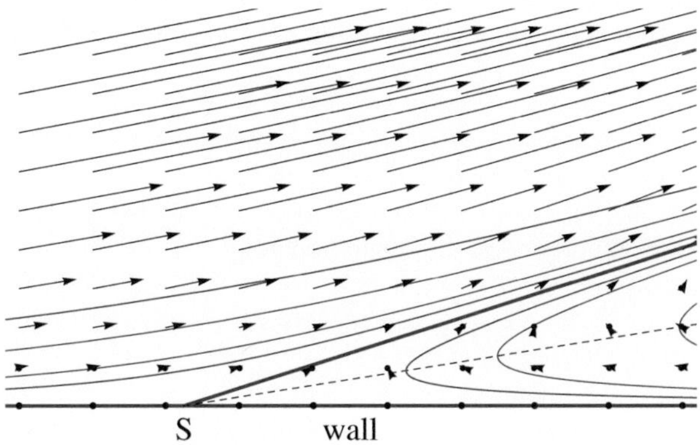

Fig. 8.9 Flow field near a separation point

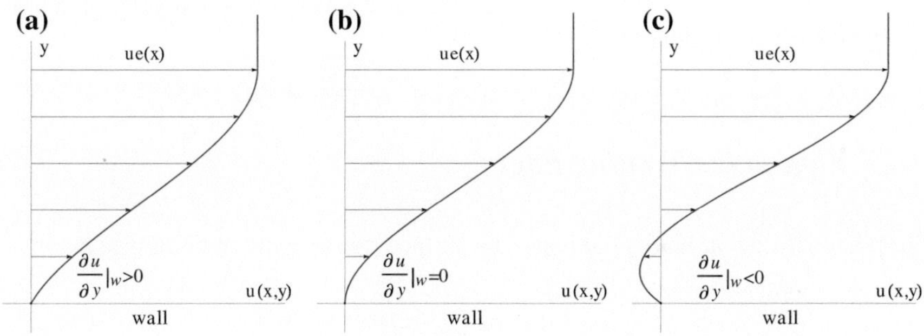

Fig. 8.10 u-velocity profiles near a separation point, **a** before separation, **b** at separation, **c** after separation

Integrating again gives

$$\frac{dp}{dx}\frac{y^2}{2} = \mu u - \tau_w y \tag{8.147}$$

At the separation point ($\tau_w(x_s) = 0$), the velocity profile behaves as

$$u(x_s, y) = \frac{dp}{dx}\frac{y^2}{2\mu} \tag{8.148}$$

See Fig. 8.10.

On the other hand, differentiating the momentum equation twice and using conservation of mass yields

$$\frac{1}{2}\frac{\partial \tau^2}{\partial x} = \nu \frac{\partial^4 u}{\partial y^4} \tag{8.149}$$

If $\frac{\partial^4 u}{\partial y^4}$ is not zero at the point of separation, the integration of the above equation implies

$$\frac{\partial u}{\partial y}|_{y=0} \sim |x - x_s|^{\frac{1}{2}} \tag{8.150}$$

Thus, in the neighborhood of the separation point

$$x - x_s = u^2 f(y) + O(u^3) \tag{8.151}$$

i.e. u has a square root singularity.

This singularity is removable, if the displacement thickness rather than the inviscid velocity is specified at the edge of the boundary layer. In viscous/inviscid interaction procedures, when they converge, the velocity in the viscous layer is bounded. The local accuracy of the calculations will suffer since v is not much smaller than u as in boundary layer assumption.

8.3.7 Flow at the Trailing Edge

For potential flow around a flat plate, the surface pressure distribution is given by

$$C_{p\,u,l} = \mp 2 \sin \alpha \sqrt{\frac{c - x}{x}} \tag{8.152}$$

hence, we expect the boundary layer response to this adverse pressure gradient near the trailing edge to result possibly in separation, since the pressure gradient is a forcing function in the momentum equation.

Even for a flat plate at zero angle of attack and ignoring the pressure gradient of the outer inviscid flow, at the trailing edge the skin friction is discontinuous as shown in Fig. 8.11.

The displacement thickness will have discontinuity in the derivative at the trailing edge. The inviscid pressure over the augmented body will not be regular in x.

More sophisticated treatment is required to account for separation, trailing edge singularity and curvature of the wake.

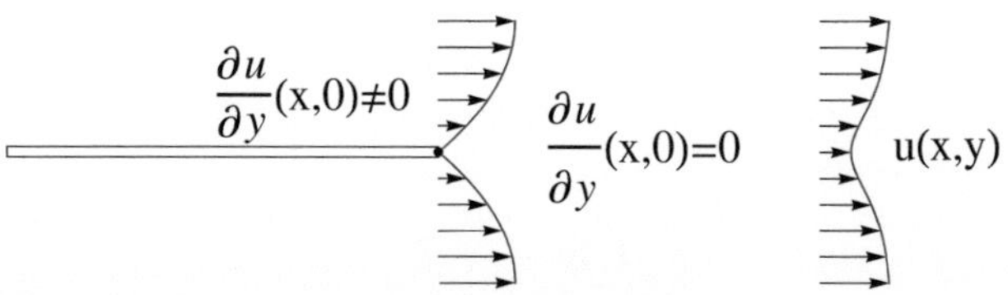

Fig. 8.11 Boundary layer in the vicinity of the trailing edge

8.3.8 Three Dimensional Boundary Layers

Consider a flat wing, away from the tips, at zero angle of attack in incompressible flow. The boundary layer equations are

$$\frac{\partial u}{\partial x} + \frac{\partial v}{\partial y} + \frac{\partial w}{\partial z} = 0 \tag{8.153}$$

$$\rho u \frac{\partial u}{\partial x} + \rho v \frac{\partial u}{\partial y} + \rho w \frac{\partial u}{\partial z} = -\frac{dp}{dx} + \frac{\partial}{\partial y}\left(\mu \frac{\partial u}{\partial y}\right) \tag{8.154}$$

$$\rho u \frac{\partial w}{\partial x} + \rho v \frac{\partial w}{\partial y} + \rho w \frac{\partial w}{\partial z} = -\frac{dp}{dz} + \frac{\partial}{\partial y}\left(\mu \frac{\partial w}{\partial y}\right) \tag{8.155}$$

The same result can be found that $\frac{\partial p}{\partial y} \simeq 0$, hence the pressure is given by the inviscid flow solution.

The boundary conditions at the solid surface are $u = v = w = 0$.

At the outer edge of the boundary layer, $u = u_i$ and $w = w_i$. The v-component at the edge carries the influence of the boundary layer on the inviscid flow. It is not known a priori and is part of the solution.

The initial condition, upstream of the wing is simply uniform flow $u = U_\infty$.

For infinite swept wings with the z-axis in the direction of span (Fig. 8.12)

$$\frac{\partial p}{\partial z} = 0, \quad \frac{\partial u}{\partial z} = 0, \quad \frac{\partial w}{\partial z} = 0 \tag{8.156}$$

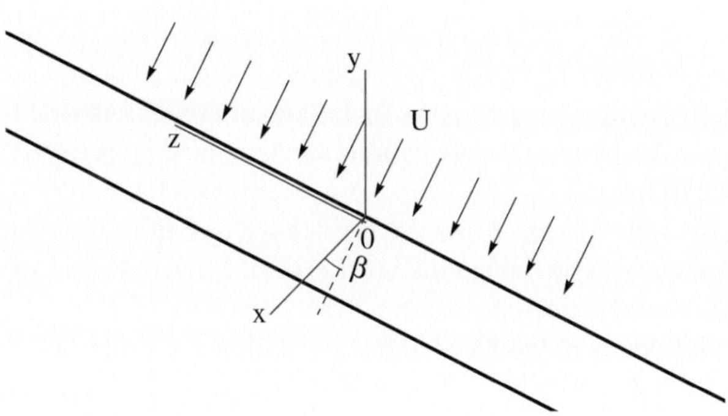

Fig. 8.12 Infinite wing at sweep angle β

hence, the boundary layer equations reduce to ($\mu = const.$)

$$\frac{\partial u}{\partial x} + \frac{\partial v}{\partial y} = 0 \tag{8.157}$$

$$\rho u \frac{\partial u}{\partial x} + \rho v \frac{\partial v}{\partial y} = -\frac{dp}{dx} + \mu \frac{\partial^2 u}{\partial y^2} \tag{8.158}$$

$$\rho u \frac{\partial w}{\partial x} + \rho v \frac{\partial w}{\partial y} = \mu \frac{\partial^2 w}{\partial y^2} \tag{8.159}$$

Note that u and v are independent of w, the so-called *independence principle*. This is true only for incompressible flow where $\rho = const.$

The general three-dimensional boundary layers are covered in Howarth [4] and Stewartson [6]

Finally, the sources of the fundamental materials in this section are reviewed. The analysis of the relative motion near a point is discussed by Batchelor [7].

The theory of boundary layer is covered in Schlichting [3], Rosenhead [8] and White [9].

Applications to low speed aerodynamics are discussed in Moran [10]. Compressible viscous flows are studied in Liepmann and Roshko [2] and in Stewartson [6].

Stability of viscous flows as well as transitional and turbulent flows are not covered here and the reader is referred to literature for these topics.

8.4 Summary of Chapter 8

This chapter starts with a discussion of vorticity versus strain rate, the derivation of viscous stresses and of the Navier-Stokes equations for 2-D incompressible flows, followed by Prandtl boundary layer theory including numerical solution for the boundary layer over a flat plate and calculations of boundary layers displacement and momentum thicknesses.

Next, compressible viscous flows are considered and the corresponding Navier-Stokes and boundary layer equations are derived. Determination of drag is also discussed.

Historical and classical works of Blasius solution for a flat plate and Falkner-Skan solution for flow past a wedge, as well as the von Karman Integral Momentum equation (with streamwise pressure gradient) and transformation of boundary layer equations (due to von Mises and also Howarth and Dorodnitcyn) are considered together with comments on flow separation, flow at the trailing edges and three-dimensional boundary layers.

Turbulent boundary layers are not covered in this book.

8.5 Problems

8.5.1

Consider the stream function in the upper half plane

$$\psi(x, y) = \frac{1}{6}y^3 - \frac{1}{2}mxy^2, \quad y \geq 0$$

where $m > 0$ is a fixed parameter.

Find the shape and make a sketch of the streamlines of this flow. Some remarkable streamlines correspond to $\psi = 0$. Show that $y = 0$ could be a solid wall.

Calculate the velocity components (u, v) and show, with arrows, the direction of the flow. Indicate with dots, the location of flow reversal line, $u = 0$.

Calculate the vorticity $\omega = \partial v/\partial x - \partial u/\partial y$ and indicate with dotted line the location $\omega = 0$.

Calculate the wall shear stress $\tau_w(x)$.

Sketch some velocity profiles $u(x, y)$ at three fixed $x-$locations, $x = -0.5, x = 0$ and $x = 0.5$.

Why is this not a real viscous flow solution (Hint: look for a pressure field).

References

1. Oswatitsch, K.: Gas Dynamics. Academic Press, New York (1956)
2. Liepmann, H., Roshko, A.: Elements of Gas Dynamics. Wiley, New York (1957)
3. Schlichting, H.: Boundary Layer Theory. McGraw Hill, New York (1979)
4. Howarth, L.: Laminar boundary layers. Fluid Dynamics I. Encyclopedia of Physics, vol. VIII/I. Springer, Berlin (1959)
5. Meier, G.E., et al. (eds.): IUTAM Symposium on One Hundred Years of Boundary Layer Research. Springer, New York (2004)
6. Stewartson, K.: The Theory of Laminar Boundary Layers in Compressible Fluids. Oxford University Press, Oxford (1964)
7. Batchelor, G.K.: An Introduction to Fluid Dynamics. Cambridge University Press, New York (1967)
8. Rosenhead, L. (ed.): Laminar Boundary Layers. Oxford University Press, Oxford (1963)
9. White, F.M.: Viscous Fluid Flow. McGraw Hill, New York (1974)
10. Moran, J.: Theoretical and Computational Aerodynamics. Wiley, New York (1984)

Chapter 9
Viscous/Inviscid Interaction Procedures

In the previous chapters, small disturbance potential flow theories for 2-D and 3-D, incompressible and compressible, steady and unsteady flows over aerodynamic configurations are studied. The main results are the surface pressure distributions and the associated lift and "inviscid" drag coefficients. Also, boundary layer theory for the corresponding flows is discussed and the main result is the prediction of the skin friction drag, assuming the surface pressure distributions are known from the inviscid calculations (Fig. 9.1).

The feedback effects of boundary layers on the inviscid flows can be represented through the displacement thickness. This concept, for both laminar and turbulent flows, allows uncoupled calculations at least for the cases with no separation. The mutual strong interaction between the thin viscous layer over the body and in the wake, for high Reynolds numbers, and the main inviscid flow is the subject of this chapter. Beside the displacement thickness methods, other approaches based on domain decomposition techniques will be discussed as well. A challenging problem pertinent to this chapter concerns the shock wave/boundary layer interaction, an example of which is shown in Fig. 9.2 for an 18 % thick biconvex circular arc airfoil, from [1].

9.1 Viscous/Inviscid Interaction Procedures Based on Displacement Thickness Concept

Consider steady 2-D flows. For high Reynolds numbers, an inviscid model can be used everywhere except in a thin viscous layer over the body and in its wake. The latter is needed since the inviscid flow theory allows a slip velocity at the solid surface and the physical "no slip" boundary condition can not be imposed, in general.

If the thin viscous layer is not introduced, the inviscid flow profile, extended to the wall, will have more mass flow rate than the corresponding viscous flow profile as shown in the figure, see Fig. 9.3.

J.J. Chattot and M.M. Hafez, *Theoretical and Applied Aerodynamics*,
DOI 10.1007/978-94-017-9825-9_9

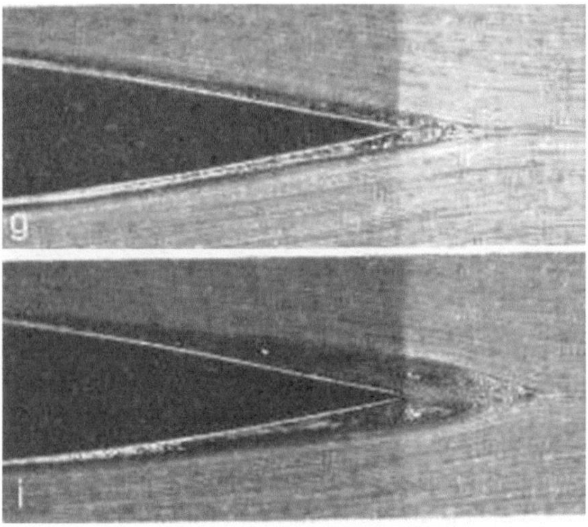

Fig. 9.1 Hydrodynamic visualization, courtesy H. Werlé ONERA—the French Aerospace Lab, ONERA RA 1987-3

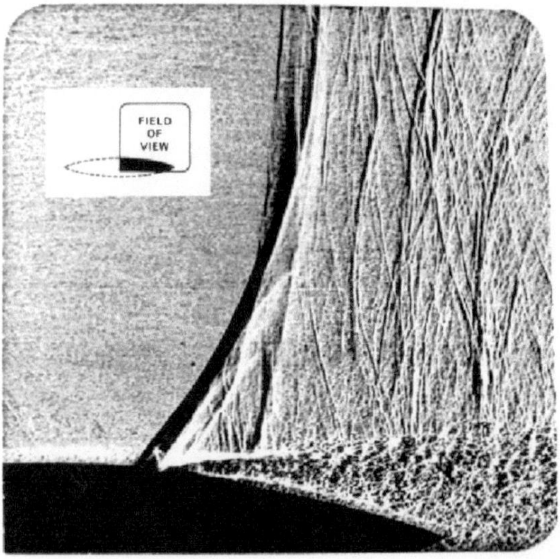

Fig. 9.2 Shock-induced separation, $M = 0.785$, $Re = 10^7$, from McDevitt Ref. [1]

It is assumed that the viscous stress contributions are negligible outside the thin layer of thickness δ, compared to the pressure and inertia forces.

As discussed before in the previous chapter on boundary layers, Prandtl introduced the concept of displacement thickness, where the body is displaced such that the inviscid flow profile over the displaced body will have the same mass flow rate as the viscous flow profile over the original body. The equation for such a displacement thickness δ^*, is given by

$$\delta^* = \int_0^\delta \left(1 - \frac{\rho u}{\rho_i u_i}\right) dy \tag{9.1}$$

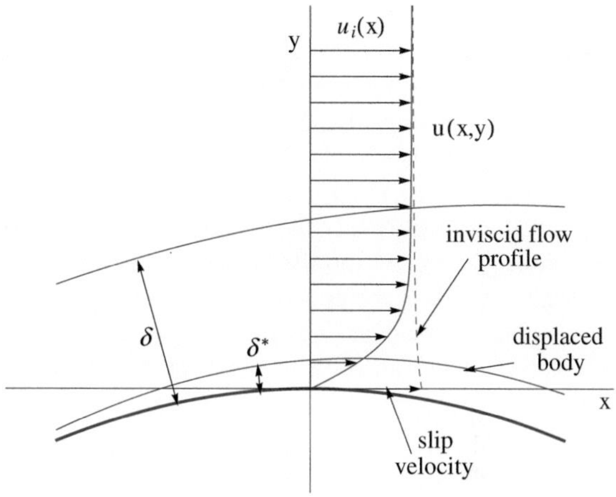

Fig. 9.3 Displacement thickness

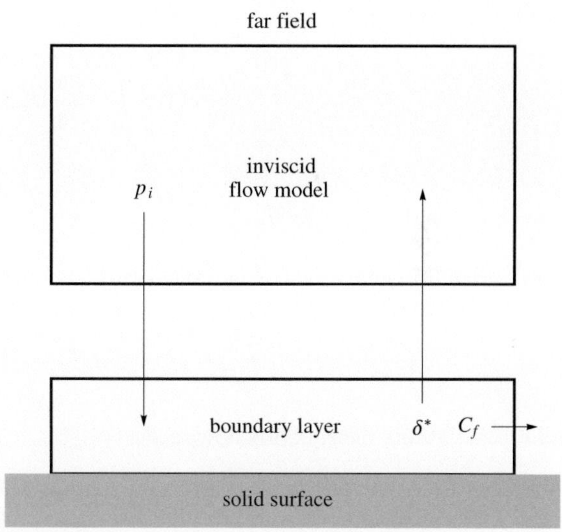

Fig. 9.4 Viscous/inviscid interaction

For incompressible flows, $\rho = \rho_i = const$. Notice also, both δ and δ^* are changing from one station to another, in general. However, for a self-similar solution of boundary layer over a flat plate without pressure gradient, u/u_i is a function of $\eta = y/\delta$ only, hence the ratio δ^*/δ becomes a constant. In the non similar case, u_i changes in the streamwise direction as well.

Now the viscous/inviscid interaction can be described by the following diagram. See Fig. 9.4.

The boundary layer and inviscid flow calculations are coupled. Boundary layer codes require the inviscid surface pressure distributions from the inviscid flow codes, while the inviscid flow calculations are performed over the displaced body (and wake). In general, the two calculations must be performed simultaneously.

For some cases, mainly attached flows, the interaction is weak, and the two calculations are decoupled and the final results are obtained via iterations. To start the procedure, $\delta^*(x)$ is assumed known (for example through ignoring the pressure gradient) and the cycle is started until convergence. This is not always the case. A limit cycle may occur and hence the results are not satisfactory. The convergence problems are critical for the viscous effects to be properly accounted for.

The feedback of the boundary layer to the inviscid flow can be imposed through a transpiration boundary condition to avoid changing the grid in the inviscid flow calculations from iteration to iteration. Using small disturbance formulation, the same grid is used for all iterations. Instead of imposing the no penetration boundary condition on the displaced body (i.e. the inviscid flow is tangent to the displaced body), an equivalent condition on the original body is derived using the Taylor series expansion as follows.

Integrating across the boundary layer and using the Leibnitz rule, one gets in terms of local coordinates (x, y):

$$\int_0^\delta \frac{\partial \rho u}{\partial x} dy + \int_0^\delta \frac{\partial \rho v}{\partial y} dy = 0 \tag{9.2}$$

$$\frac{d}{dx} \int_0^\delta \rho u \, dy - \rho_i u_i \frac{\partial \delta}{\partial x} + \rho_i v_i = 0 \tag{9.3}$$

The definition of displacement thickness can be written

$$\int_0^\delta \rho u \, dy = \rho_i u_i \left(\delta - \delta^*\right) \tag{9.4}$$

after substitution, the equation now reads

$$\begin{aligned} \rho_i u_i \frac{d}{dx} \left(\delta - \delta^*\right) + \left(\delta - \delta^*\right) \frac{d \rho_i u_i}{dx} - \rho_i u_i \frac{\partial \delta}{\partial x} + \rho_i v_i \\ = \delta \frac{d \rho_i u_i}{dx} - \frac{d}{dx} \left(\rho_i u_i \delta^*\right) + \rho_i v_i = 0 \end{aligned} \tag{9.5}$$

A Taylor expansion of (ρv) provides the transpiration mass flux to first order as

$$(\rho v)_0 = (\rho v)_i - \delta \frac{d\,(\rho v)_i}{dy} + O(\delta^2) = \rho_i v_i + \delta \frac{d \rho_i u_i}{dx} + O(\delta^2) \simeq \frac{d}{dx} \left(\rho_i u_i \delta^*\right) \tag{9.6}$$

Hence, if a flux $(\rho v)_0$ is blown at the original surface given by the above equation, the inviscid flow will be tangent to the displaced body.

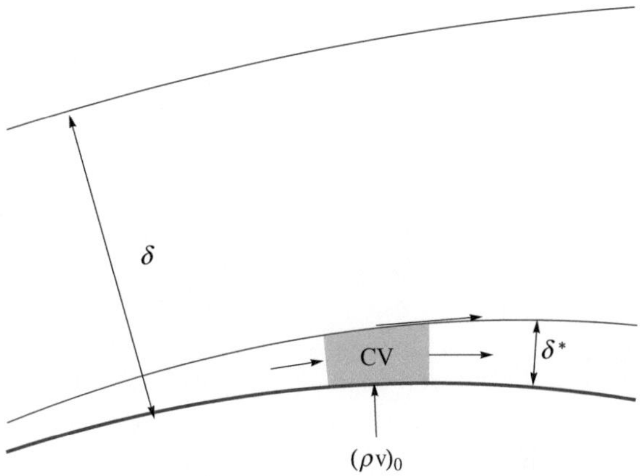

Fig. 9.5 Control volume for application of transpiration boundary condition

The above statement is a consequence of conservation of mass for the inviscid flow applied to the control volume shown in Fig. 9.5. In the above argument, the variation of $\rho_i u_i$ across the boundary layer is ignored.

In the following, four formulations are considered:

(1) Coupling integral formulations of both inviscid and viscous flow regions.
(2) Coupling the numerical solution of the partial differential equations of the inviscid flow with integral equations of boundary layers.
(3) Coupling the numerical solution of the partial differential equations of the viscous flow with the integral equations of the outer inviscid flows.
(4) Coupling numerical solutions of partial differential equations of both the inviscid and viscous flow regions.

After these four formulations are covered, other approaches based on domain decomposition will be addressed in latter part of this chapter.

9.1.1 *Coupling Integral Formulations of both Inviscid and Viscous Flow Regions*

9.1.1.1 Analysis of Internal Flows in Variable Area Ducts

The governing equations for the quasi-one dimensional flows in the inviscid core are given by the conservation laws, assuming perfect gas:
continuity

$$\frac{\partial \rho A}{\partial t} + \frac{\partial \rho u A}{\partial x} = 0 \tag{9.7}$$

momentum

$$\frac{\partial \rho u A}{\partial t} + \frac{\partial (\rho u^2 + p)A}{\partial x} = p\frac{dA}{dx} \tag{9.8}$$

energy

$$\frac{\partial \rho E A}{\partial t} + \frac{\partial \rho u H A}{\partial x} = 0 \tag{9.9}$$

where $H = E + p/\rho = \gamma p/(\gamma - 1)\rho + u^2/2$.

$H = C_p T$ is the total enthalpy, $E = C_v T$ the total internal energy, T the total temperature and $A(x)$ denotes the effective cross-sectional area. The average inviscid velocity in the slowly varying area duct, u, denotes also the velocity u_i at the edge of the boundary layer.

The above equations can be solved with artificial viscosity or using upwind schemes. For the latter, see Chattot [2].

For the steady case, the time dependent terms vanish and the continuity and energy equations reduce to the integrals of motion in terms of the algebraic relations:

$$\rho u A = \dot{m} = const. \tag{9.10}$$

$$\frac{\gamma}{\gamma - 1}\frac{p}{\rho} + \frac{u^2}{2} = H = const. \tag{9.11}$$

The momentum equation can be rewritten in one variable, say u (or p) by eliminating the other two variables using the two algebraic relations. Again, artificial viscosity or upwind scheme can be used for this nonlinear scalar ordinary differential equation, see Hafez et al. [3].

On the other hand, the von Karman integral momentum equation for the steady compressible boundary layer is given by:

$$\frac{1}{R^n}\frac{d}{dx}\left(\rho u^2 R^n \theta\right) + \rho u \delta^* \frac{du}{dx} = \rho u^2 \frac{C_f}{2} \tag{9.12}$$

where

$$\begin{cases} n = 0 & for\ planar\ flow \\ n = 1 & for\ axisymmetric\ flow \end{cases}$$

In the above equation, δ^* is the displacement thickness, and θ is the momentum thickness, while C_f is the skin friction coefficient.

A special case is the incompressible flow through a channel of width b, where ρ is constant and the effective area is simply $A(x) = A_{geo}(x) - 2b\delta^*(x)$.

where A_{geo} is the geometric area. Moreover, $uA = \dot{m}/\rho = C = const.$, hence

$$u(x) = u_i(x) = \frac{C}{A_{geo}(x) - 2b\delta^*(x)} \tag{9.13}$$

The problem thus reduces to solving the integral momentum equation. There are however three variables in this equation: δ^*, θ and C_f. Following Pohlhausen's approximate method, a form of the velocity profile in the boundary layer is assumed, where

$$\frac{u}{u_i} = f(\eta) \quad and \quad \eta = \frac{y}{\delta(x)} \tag{9.14}$$

where $\delta(x)$ is the total thickness of the boundary layer and y is the distance from the duct wall. The function f should satisfy some boundary conditions on u:

$$at\ y = 0, \quad u = 0, \quad \frac{\partial^2 u}{\partial y^2} = -\frac{u_i}{\nu}\frac{du_i}{dx}, \quad \frac{d^3 u_i}{dx^3} = 0, \ldots \tag{9.15}$$

$$as\ y \to \infty, \quad u \to u_i, \quad \frac{\partial u}{\partial y} \to 0, \quad \frac{\partial^2 u}{\partial y^2} \to 0 \tag{9.16}$$

The conditions at infinity are only approached asymptotically. These conditions will be applied at $y = \delta$ instead.

The conditions on $f(\eta)$ become:

$$f(0) = 0, \quad f''(0) = -\Lambda, \quad f'''(0) = 0, \ldots \tag{9.17}$$

$$f(1) = 1, \quad f'(1) = f''(1) = f'''(1) = 0 \tag{9.18}$$

where

$$\Lambda = \frac{\delta^2}{\nu}\frac{du_i}{dx} \tag{9.19}$$

From the assumed velocity profile, δ^*, θ and C_f are related to δ as follows:

$$\delta^* = \delta \int_0^1 (1 - f)d\eta \tag{9.20}$$

$$\theta = \delta \int_0^1 (f - f^2)d\eta \tag{9.21}$$

$$C_f = \frac{2\nu}{u_i \delta} f'(0) \tag{9.22}$$

Pohlhausen used a fourth order polynomial for $f(\eta)$, satisfying five conditions, two at $y = 0$ and three at $y = \delta$. Hence

$$\frac{u}{u_i} = f(\eta) = 2\eta - 2\eta^3 + \eta^4 + \frac{1}{6}\Lambda\eta(1-\eta)^3 \tag{9.23}$$

With this form, the following relations can be obtained:

$$\frac{\delta^*}{\delta} = \frac{3}{10} - \frac{\Lambda}{120}, \quad \frac{\theta}{\delta} = \frac{1}{63}\left(\frac{37}{5} - \frac{\Lambda}{15} - \frac{\Lambda^3}{144}\right), \quad \frac{C_f u_i \delta}{2\nu} = 4 + \frac{\Lambda}{3} \tag{9.24}$$

Thus, a nonlinear ordinary differential equation in Λ must be solved numerically. A step by step technique can be used in a standard way.

Obviously, other forms of the velocity profile can be chosen. Variations of Pohlhausen method are available in the literature. Notably is Thwaites' method which is based on an empirical relation to simplify the algebra.

Methods based on two equations, integral momentum equation and integral mechanical energy equation, are known to produce more accurate results. In particular, a family of profiles with two parameters are introduced in the work of Wieghardt [4], see also Walz [5], Truckenbrodt [6] and Tani [7].

In their work, the mechanical energy equation is derived by multiplying the momentum equation by u and upon integration across the boundary layer, an integral relation is obtained in terms of a mechanical energy thickness.

The final results for incompressible flow is

$$\frac{d}{dx}\left(u_i^3\vartheta\right) = 2\nu\int_0^\infty \left(\frac{\partial u}{\partial y}\right)^2 dy \tag{9.25}$$

where

$$\vartheta = \int_0^\infty \frac{u}{u_i}\left(1 - \frac{u^2}{u_i^2}\right)^2 dy \tag{9.26}$$

Alternatively, Listsianski and Dorodnitsyn used their method of integral relations, where the boundary layer is divided into strips and approximates are obtained via weighted averaging over each strip. For more details see Holt [8].

In this regard, one should mention that over the last century, boundary layer equations have been solved by the methods of weighted residuals, including the method of collocation, the method of moments and the Galerkin method. See for example Schlichting [9], Rosenhead [10].

For compressible flows, an integral based on the total energy equation is used, see Ref. [11].

Following Stewartson [11], a total temperature is defined as:

$$T = t + \frac{u^2}{2C_p} \tag{9.27}$$

where t represents the static temperature. In terms of the total temperature, the energy equation reads:

$$\rho u \frac{\partial T}{\partial x} + \rho v \frac{\partial T}{\partial y} = \frac{\partial}{\partial y}\left[\frac{\mu}{P_r}\frac{\partial}{\partial y}\left(T + \frac{P_r - 1}{2C_p}u^2\right)\right] \tag{9.28}$$

Upon integration w.r.t. y, one obtains

$$\frac{d}{dx}\int_0^\infty \rho u\,(T - T_i)\,dy = -\frac{\mu}{P_r}\frac{\partial T}{\partial y}|_{y=0} \tag{9.29}$$

where

$$T_i = t_i + \frac{u_i^2}{2C_p} \tag{9.30}$$

P_r is the Prandtl number. If the wall is insulated (i.e. $\partial T/\partial y = 0$ at $y = 0$), the equation is integrated to give (see [12])

$$\int_0^\infty \rho u\,(T - T_i)\,dy = const. \tag{9.31}$$

For the special case, $P_r = 1$, Crocco and Busemann noticed independently, that $T = const$. The total temperature is thus constant everywhere across the boundary layer and the wall temperature is given by:

$$t_w = t_i + \frac{u_i^2}{2C_p} = t_i\left(1 + \frac{\gamma - 1}{2}M_i^2\right) = T_i \tag{9.32}$$

In general, the total energy thickness is defined as:

$$\varepsilon = \int_0^\infty \frac{\rho u}{\rho_i u_i}\left(1 - \frac{T}{T_i}\right)dy \tag{9.33}$$

The integral mechanical energy equation becomes:

$$\frac{d\varepsilon}{dx} = -\frac{q_w}{\rho_i u_i T_i} \tag{9.34}$$

where q_w is the heat flux through the wall

$$q_w = k \frac{\partial t}{\partial y}|_{y=0} \tag{9.35}$$

Gruschwitz [13] introduced a method similar to that of Pohlhausen to calculate compressible flows with arbitrary Prandtl numbers, using both the momentum and total energy integral equations and in this case a temperature profile is assumed satisfying the temperature boundary conditions.

On the other hand, Oswatitsch [14] developed a simpler method assuming $P_r = 1$. Since the pressure is constant across the boundary layer, one can show that, see [15]

$$\frac{\rho_i}{\rho} = \frac{t}{t_i} = 1 + \frac{\gamma - 1}{2} M_i^2 \left(1 - \frac{u^2}{u_i^2}\right) \tag{9.36}$$

Hence, only the momentum integral equation for compressible flow with a quartic polynomial for the velocity in terms of Λ, may be used, where the density and the temperature distributions are related to the velocity profile.

The above development is for steady laminar flows. For steady and unsteady compressible turbulent flows, the reader is referred to [16], where several empirical formulae are employed.

9.1.1.2 Analysis of Incompressible Flows over Thin Airfoils Using Integral Momentum Equation for Boundary Layer

Consider thin symmetric airfoil (at zero angle of attack). The perturbation velocity of the inviscid flow can be approximated by:

$$u(x, 0) = \frac{U}{\pi} \int_0^c \frac{y'(\xi)}{x - \xi} d\xi \tag{9.37}$$

where $y(x)$ is the equation of the upper surface of the profile.

To account for the viscous effects, the shape of the airfoil is augmented by the displacement thickness as $y(x) + \delta^*(x)$.

The integral momentum equation can be solved coupled with the above integral where $u_i = U + u$.

To account for the wake, an integral from the trailing edge to the far field is added to the above formula, namely

$$u_w(x, 0) = \frac{U}{\pi} \int_0^c \frac{\delta_w^{*'}(\xi)}{x - \xi} d\xi \tag{9.38}$$

The velocity profile in the wake must satisfy a different boundary condition at $y = 0,\ x > c$, namely $\partial u / \partial y = 0$, rather than $u = 0$.

The extension to linearized subsonic flows is straightforward. The Prandtl/Glauert transformation can be used to modify the inviscid velocity distribution together with Oswatitsch's modification of Pohlhausen method for compressible boundary layer as discussed before.

The finite plate at zero angle of attack is a special case, where $y(x) = 0$.

For thin airfoils at angle of attack, the viscous layers on the top and the bottom are different. Moreover, the inviscid flow analysis requires solution of an integral equation, not just an evaluation of an integral as in the symmetric case.

9.1.1.3 Analysis of Supersonic Flows over Thin Airfoils Using Integral Momentum Equation for Boundary Layer

Again, the body is augmented with the displacement thickness and the Oswatitsch method may be used to solve the integral momentum equation for the compressible boundary layer. The inviscid flow solution, using d'Alembert formula (according to Ackeret thin airfoil theory), is given by:

$$u(x, 0^+) = -\frac{U}{\beta}(f'^+(x) - \alpha), \quad u(x, 0^-) = \frac{U}{\beta}(f'^-(x) - \alpha) \tag{9.39}$$

where $\beta = \sqrt{M_0^2 - 1}$ and α is the incidence.

For attached supersonic flows, the solution is obtained with a marching scheme in the x-direction, in one sweep. Iteration are still needed at each step. The case of angle of attack can be easily treated by splitting the top and bottom regions, since there is no interaction between the two.

9.1.2 Coupling the Numerical Solution of the Partial Differential Equations of the Inviscid Flow with Integral Equations of Boundary Layers

9.1.2.1 Inviscid Potential Flow Field Calculations and Integral Momentum Equations of Boundary Layer

For small disturbance potential flows, the nonlinear governing equation is given by the conservation form

$$\frac{\partial F}{\partial x} + \frac{\partial^2 \phi}{\partial y^2} = 0 \tag{9.40}$$

where

$$F = (1 - M_0^2)\frac{\partial \phi}{\partial x} - \frac{\gamma + 1}{2} M_0^2 \frac{\partial^2 \phi}{\partial x^2} \tag{9.41}$$

This equation is valid for transonic flows with shocks as well as pure sub- and super-sonic flows, including the incompressible case ($M_0 = 0$).

The body is augmented with the displacement thickness and the boundary condition becomes, for the upper surface

$$\frac{\partial \phi}{\partial y}(x, 0^+) = \frac{d}{dx}\left(y_u + \delta^*\right) \tag{9.42}$$

Together with the integral momentum equation, the discrete equations can be solved using standard relaxation methods. To help the convergence, artificial time and viscosity terms are added in terms of the displacement thickness

$$\alpha \frac{\partial \rho_e u_e \delta^*}{\partial t} + \frac{\partial \rho_e u_e \theta}{\partial x} + \rho_e u_e \delta^* \frac{du_e}{dx} = \rho_e u_e^2 \frac{C_f}{2} + \frac{d}{dx}\left(\epsilon \frac{d\delta^*}{dx}\right) \tag{9.43}$$

The parameters α and ϵ may be chosen guided by numerical analysis. Horizontal line relaxation with block pentadiagonal solver for the potential and δ^* at the lines above and below the airfoil (at $y = 0$), will allow tighter coupling for the case of boundary layer separation.

Moses et al. [17], solved the integral momentum equation and the integral mechanical energy equation simultaneously with the inviscid potential flow equation to simulate separated flow in a diffuser.

In their work, the velocity profile is assumed to be a function of two parameters

$$\frac{u}{u_e} = f\left(\frac{y}{\delta}, \beta\right) \tag{9.44}$$

where δ is the thickness of the boundary layer and β is a profile shape parameter.

The integral equations can be written as a system of two ordinary differential equations

$$a_{11}\frac{d\delta}{dx} + a_{12}\frac{d\beta}{dx} = b_{11}\frac{du_e}{dx} + b_{12} \tag{9.45}$$

$$a_{21}\frac{d\delta}{dx} + a_{22}\frac{d\beta}{dx} = b_{12}\frac{du_e}{dx} + b_{22} \tag{9.46}$$

For attached flows, with a given velocity u_e, these two equations can be solved numerically with a stepwise procedure in the downstream direction.

For separated flows, the equations are singular at the separation point; the determinant of the coefficient matrix, $A = \left(a_{i,j}\right)$, is zero. Solutions exist only for a specific pressure gradient and in this case, the solution is not unique. Moreover, the calculation is unstable for reverse flows. To avoid the difficulties associated with separation, the edge velocity, u_e, is treated as unknown and the discrete inviscid equations are solved, augmented with the two equations for δ and β, using vertical line relaxation, sweeping in the main flow direction.

Moses et al. [17] represented the inviscid flow in terms of stream function, hence the governing equation is

$$\nabla^2 \psi = 0 \tag{9.47}$$

and the boundary condition is

$$\psi = const. \tag{9.48}$$

on the displaced wall, say on the upper surface, $y = y_u + \delta^*$. The coupling between the inviscid flow and the boundary layer is imposed through the relation at the displaced wall

To start the calculations, quasi-one dimensional problem is solved to provide a good initial guess. The development of boundary layer through separation and attachment is successfully simulated and the edge velocity, u_e, the displacement thickness, δ^*, and the skin friction coefficient, C_f, are accurately calculated.

9.1.2.2 Coupling Euler Calculations with Boundary Layer Integral Momentum Equations

As discussed before, the displacement thickness is determined such that the mass flow rate over the displaced body, at a given station, based on an equivalent inviscid flow will be equal to the mass flow rate of the viscous flow profile over the original body. Let $y = y_u$ represent the ordinate of the upper surface and in short $\delta^* = \delta_u^*$ represent the displacement thickness along the upper surface of the airfoil.

If the Euler equations are used to represent the inviscid flow, instead of the potential flow equation, then additional considerations are necessary, since beside the continuity equation, momentum and energy equations are involved.

Le Balleur [18], has considered the general problem of matching and in the following his defect formulation will be discussed. The reader is also referred to Johnston and Soskol [19], Murman and Bussing [20] and Whitfield et al. [21] for more applications of one and two dimensional problems. Three dimensional flows over wings are considered in Ref. [22].

In the Le Balleur defect formulation, the equivalent inviscid flow problem and the viscous flow problem have an overlap region, starting from the wall and the wake center line and extends to cover the whole viscous layer. Outside the viscous layer, the viscous stresses are assumed negligible compared to the inertia terms.

For steady 2-D flows, at a given station $x = const.$, the inviscid flow equations are

$$\frac{\partial \mathbf{f}}{\partial x} + \frac{\partial \mathbf{g}}{\partial y} = \mathbf{0} \tag{9.49}$$

where

$$\mathbf{f} = \begin{bmatrix} \rho u \\ \rho u^2 + p \\ \rho uv \\ \rho uH \end{bmatrix}, \quad and \quad \mathbf{g} = \begin{bmatrix} \rho v \\ \rho uv \\ \rho v^2 + p \\ \rho vH \end{bmatrix} \tag{9.50}$$

and

$$H = \frac{\gamma}{\gamma - 1}\frac{p}{\rho} + \frac{u^2 + v^2}{2} \tag{9.51}$$

while the viscous flow equations are given by

$$\frac{\partial \mathbf{F}}{\partial x} + \frac{\partial \mathbf{G}}{\partial y} = \mathbf{S} \tag{9.52}$$

where $\mathbf{S}$ denotes the viscous and heat transfer terms in Navier-Stokes or thin layer approximation equations.

The asymptotic matching condition is

$$\lim_{y \to 0} (\mathbf{f} - \mathbf{F}) = \mathbf{0} \tag{9.53}$$

The defect formulation takes into consideration the variation of inviscid flow variables, including the pressure, across the boundary layer. The defect equation in vector form reads

$$\frac{\partial (\mathbf{f} - \mathbf{F})}{\partial x} + \frac{\partial (\mathbf{g} - \mathbf{G})}{\partial y} = -\mathbf{S} \tag{9.54}$$

Integration across the layer gives for example

$$\rho v(x, 0) = \frac{d}{dx}\int_0^\infty (\rho u - \bar{\rho}\bar{u})\, dy \tag{9.55}$$

where the bar denotes the viscous flow variables.

To first order approximation, the normal momentum equation reduces to $\bar{p} = p$ across the viscous layer and the integral equations for the continuity, streamwise momentum equations are

$$\rho v(x,0) = \frac{d}{dx}\left(\rho q \delta^*\right) \tag{9.56}$$

$$\rho u v(x,0) = \frac{d}{dx}\left(\rho q^2 (\theta + \frac{u}{q}\delta^*)\right) - \rho q^2 \frac{Cf}{2} \tag{9.57}$$

where $q^2 = u^2 + v^2$.

$$\delta \rho q(x,0) = \int_0^\infty (\rho u - \bar{\rho}\bar{u})\,dy \simeq \int_0^h (\rho u - \bar{\rho}\bar{u})\,dy \tag{9.58}$$

$$\left(\theta + \frac{u}{q}\delta^*\right)\rho q^2(x,0) = \int_0^\infty \left(\rho u^2 - \bar{\rho}\bar{u}^2\right)dy \simeq \int_0^h \left(\rho u^2 - \bar{\rho}\bar{u}^2\right)dy \tag{9.59}$$

and $h > \delta$.

Le Balleur considered also the effect of curvature of the wake which is not discussed here.

Following Whitfield, the energy equation, using no porosity and no slip at adiabatic surface, one obtains

$$\rho v H(x,0) \simeq \frac{d}{dx}\int_0^h \left(\rho u H - \bar{\rho}\bar{u}\bar{H}\right)dy \tag{9.60}$$

To avoid these extra calculations, an approximation consistent with boundary layers with $P_r = 1$ is used, where $H = \bar{H} = H_0$, hence

$$\rho v H(x,0) \simeq H_0 \frac{d}{dx}\int_0^h \left(\rho u H - \bar{\rho}\bar{u}\bar{H}\right)dy \tag{9.61}$$

which is the same as the equation obtained from the integration of the continuity equation.

Murman and Bussing [20] applied Le Balleur's defect formulation for compressible flow in a convergent-divergent nozzle and used boundary layer approximations for the viscous flows. The matching conditions used are

$$u = \bar{u},\ H = \bar{H},\ p = \bar{p},\ \ and\ \ v = \bar{v},\ \ at\ \ h > \delta \tag{9.62}$$

The first three conditions are used in the boundary layer calculations. Therefore, one additional condition must be specified to enforce the matching. The boundary conditions for Euler equations will be derived using the fourth condition in three cases: the equivalent inviscid flow at the boundary layer thickness $y = y_u + \delta$; or at the original body $y = y_u$; or at the displaced body $y = y_u + \delta^*$.

Integrating the continuity equation from the body surface to δ yields

$$\frac{d}{dx}\int_0^{\delta} \bar{\rho}\bar{u} dy - (\bar{\rho}\bar{u})_\delta \frac{d\delta}{dx} + (\bar{\rho}\bar{v})_\delta = 0 \tag{9.63}$$

Using the standard definition of δ^*

$$\delta^* = \int_0^{\delta} \left(1 - \frac{\bar{\rho}\bar{u}}{(\rho u)_\delta}\right) dy \tag{9.64}$$

and enforcing the fourth matching condition

$$(\bar{\rho}\bar{v})_\delta = (\rho v)_\delta \tag{9.65}$$

and since $(\bar{\rho}\bar{u})_\delta = (\rho u)_\delta$, the above equation reduces to

$$(\rho v)_\delta = (\rho u)_\delta \frac{d\delta}{dx} - \frac{d}{dx}\left((\rho u)_\delta (\delta - \delta^*)\right) \tag{9.66}$$

This is the entrainment equation, where the second term represents the entrained mass flow rate. The second term is usually negative and it is the only boundary condition at $y = \delta$ required for the Euler equations calculations (subsonic boundary condition).

Instead, if the continuity equation is integrated from δ^* to δ and enforcing the matching conditions, one obtains

$$(\rho v)_{\delta^*} = (\rho u)_{\delta^*} \frac{d\delta^*}{dx} + \frac{d}{dx}\left((\rho u)_\delta \int_{\delta^*}^{\delta} \left(1 - \frac{\rho u}{(\rho u)_\delta}\right) dy\right) \tag{9.67}$$

If the gradient of the equivalent inviscid flow is negligible in the region $\delta^* \leq y \leq \delta$, then the last term drops and the flow is tangent to the δ^* surface. In this case, the no penetration surface boundary condition is imposed for the Euler calculations.

If the inner boundary condition of the inviscid flow is located at the original surface, the standard definition of δ^* is used, and the transpiration boundary condition reads

$$(\rho v)_0 = \frac{d}{dx}\left((\rho u)_\delta \delta^*\right) \tag{9.68}$$

In most regions, the body will be an inflow boundary, which means that three characteristics are pointing into the domain and one points out. The above matching condition provides only one. The inviscid flow is not uniquely determined! (For potential flow, the total pressure and total enthalpy are constant and the transpiration boundary condition uniquely determines the flow).

To determine the inviscid flow governed by the Euler equations, Johnson and Sockol integrated the momentum and energy equations from the body surface to δ to get

$$(\rho u v)_0 = u_\delta \frac{d}{dx}\left((\rho u)_\delta\, \delta^*\right) + \frac{d}{dx}\int_0^\delta \left(\left(\rho u^2 + p\right) - \left(\rho u^2 + p\right)_\delta\right) dy + \delta\left(\rho u \frac{\partial u}{\partial x} + \frac{\partial p}{\partial x}\right)_\delta \quad (9.69)$$

$$(\rho v H)_0 = H_\delta \frac{d}{dx}\left((\rho u)_\delta\, \delta^*\right) + \frac{d}{dx}\int_0^\delta \left((\rho u H) - (\rho u H)_\delta\right) dy + \delta\left(\rho u \frac{\partial H}{\partial x}\right)_\delta \quad (9.70)$$

Only for weak interaction and isoenergetic external flows, the second and third terms in the above equations can be ignored. In this case, one can use $u_0 = u_\delta$ and $H_0 = H_\delta$, i.e. assigning the edge values of u and H to the transpired mass (this strategy is however valid for strong interaction regions as discussed by Le Balleur [18]).

Whitfield and Thomas [21] used a reflection boundary condition for their Euler calculations, namely

$$\frac{\partial u}{\partial y} = 0, \quad and \quad \frac{\partial H}{\partial y} = 0 \quad (9.71)$$

at the body surface.

Again, these conditions are valid only for weak interactions.

Murman and Bussing [20], simulated compressible flow through a nozzle using the following quasi-one dimensional equations

$$\frac{d\,(\rho u A)}{dx} = 2\frac{d\,(\rho u \delta^*)}{dx} \quad (9.72)$$

$$\frac{d\left((\rho u^2 + p)A\right)}{dx} = 2\frac{d}{dx}\left(\rho u^2(\delta^* + \theta)\right) - C_f \quad (9.73)$$

$$\frac{d\,(\rho u H A)}{dx} = 2\frac{d}{dx}\left(\rho u H(\delta^* + \theta_H)\right) + St \quad (9.74)$$

where

$$\theta_H = \int_0^\delta \frac{\bar{\rho}\bar{u}}{\rho u}\left(1 - \frac{\bar{H}}{H}\right) dy \quad (9.75)$$

and St is the Stanton number.

The right hand sides represent sources of mass, momentum and energy for the equivalent inviscid core flow. Murman and Bussing calculated the inviscid flow based on the above equations and compared the results using inviscid flow with equivalent area, $A^* = A - 2\pi R\delta^*$, and both calculations are in good agreement, confirming the conclusions regarding the weak interaction case. The strong interaction case is still a problem and other approaches will be considered later following Cousteix and Mauss [23]. (Le Balleur used his defect formulation only for potential flows; Euler equations were not considered).

Le Balleur [18] studied the stability of the coupling procedures using harmonic analysis for both subsonic and supersonic flows, with attached and separated boundary layers and introduced a new semi-inverse iterative method for steady and unsteady transonic airfoils with shocks. To put his work in perspective, comparisons among the available methods and their limitations are useful. The standard direct method, discussed before, does not converge for separated flows in general. Following Catherall and Mangler [24], the inverse methods have been used by many researchers. However, such methods may not converge for attached flows. Figure 9.6 depicts the inverse and direct coupling procedures.

In a fully inverse method, the boundary layer equations are solved with a given displacement thickness and the outer edge velocity is calculated and fed to the inviscid flow calculations. To avoid the difficulties of both methods, Carter [25] introduced a semi-inverse method. In this method, starting with a displacement thickness both a direct inviscid flow calculation and an inverse viscous flow calculation are performed separately, then a correction formula, based on the difference on the inviscid surface pressure and the viscous pressure at the edge of the boundary layer, is used to update δ^*, see Fig. 9.7.

At convergence, p_i and p_v are the same.

Le Balleur pointed out the limitations of Carter's method and proposed an important modification based on his harmonic analysis, namely the correction formula is changed to

$$\delta^*_{new} = \delta^*_{old} + f_1 \left(\frac{dp_i}{dx} - \frac{dp_v}{dx} \right), \quad for\ M < 1 \tag{9.76}$$

$$\delta^*_{new} = \delta^*_{old} + f_2 \left(\frac{d^2 p_i}{dx^2} - \frac{d^2 p_v}{dx^2} \right), \quad for\ M > 1 \tag{9.77}$$

Underrelaxation of δ^* is used to guarantee convergence for all cases. Before we discuss the Fourier analysis and stability of the coupling methods, two other methods are outlined. Brune et al. [26] introduced the fully simultaneous method which is shown in Fig. 9.8.

The potential flow calculations should provide the relation

$$u_P = u_P(\delta^*_P) \tag{9.78}$$

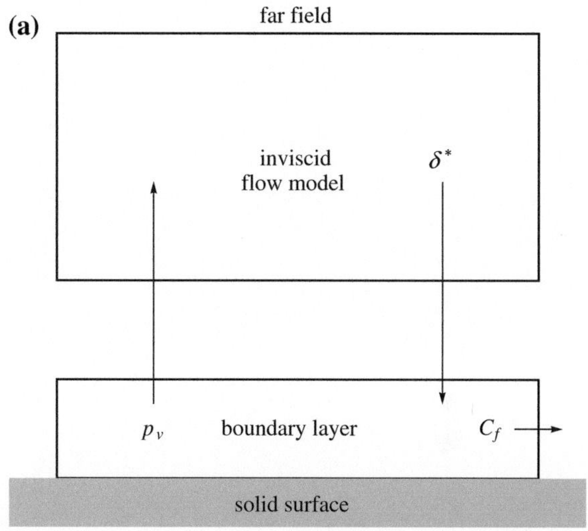

Fig. 9.6 Coupling procedures: **a** inverse method; **b** direct method

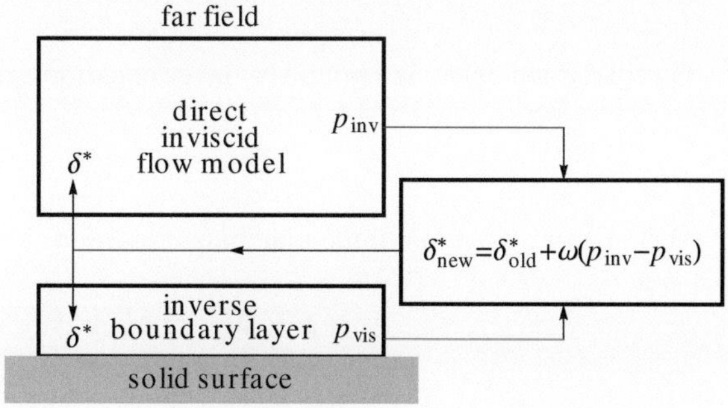

Fig. 9.7 Semi-inverse method

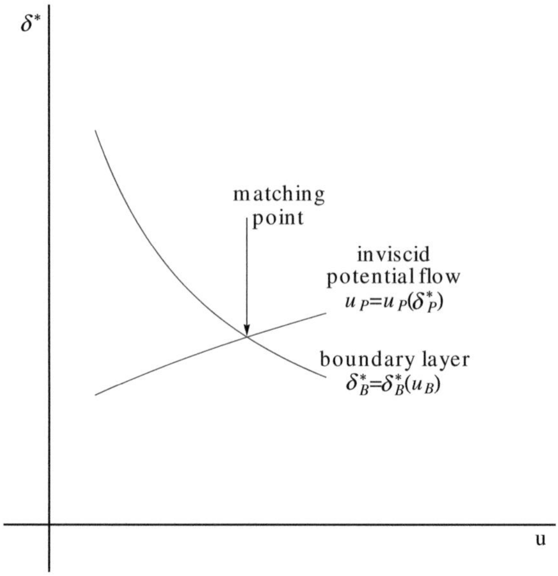

Fig. 9.8 Brune et al. [26] coupling procedure

while the boundary layer calculation gives

$$\delta_B^* = \delta_B^*(u_B) \tag{9.79}$$

The matching conditions are

$$\delta_P^* = \delta_B^* \quad and \quad u_P = u_B \tag{9.80}$$

In general, the potential equations are non linear. Upon linearization around an initial guess, the change of u_P, Δu_P, can be obtained in terms of change of δ_P^*, $\Delta\delta_P^*$, through an influence matrix, A_P

$$\Delta u_P = A_P \Delta\delta_P^* \tag{9.81}$$

Similarly, the perturbed boundary layer equations gives

$$\Delta u_B = A_B \Delta\delta_B^* \tag{9.82}$$

A first approximation to the matching conditions is

$$\delta_P^* + \Delta\delta_P^* = \delta_B^* + \Delta\delta_B^* \tag{9.83}$$

$$u_P + \Delta u_P = u_B + \Delta u_B \tag{9.84}$$

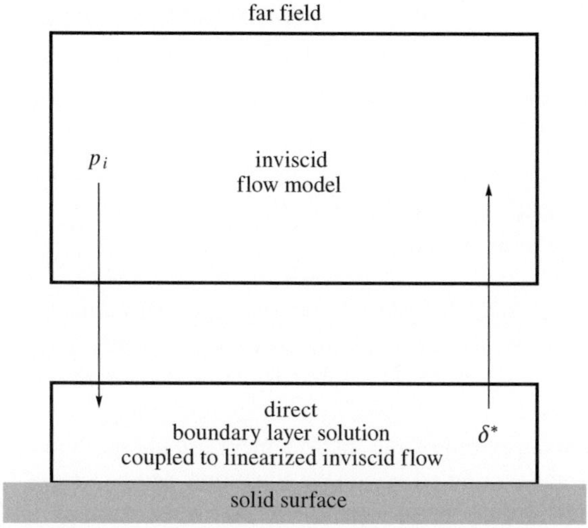

Fig. 9.9 Quasi-simultaneous coupling method of Veldman

The above four relations are solved simultaneously to obtain the corrections to u_P and δ_B^* and the process is continued until convergence.

To simplify the above procedure, Veldman [27] introduced the Quasi-simultaneous method as in the following Fig. 9.9.

Veldman demonstrated the advantage of his procedure in a series of papers. Cebeci [28] also used similar procedures. Careful treatment is necessary for the case of transonic flows, since the linearized inviscid flow equations are of mixed type. Ignoring this feature, as in some published works, may lead into problems. A comprehensive review is published in Ref. [30].

The following quasi-simultaneous coupling method has been successfully used by Wigton and Holt [29] for transonic flows Fig. 9.10.

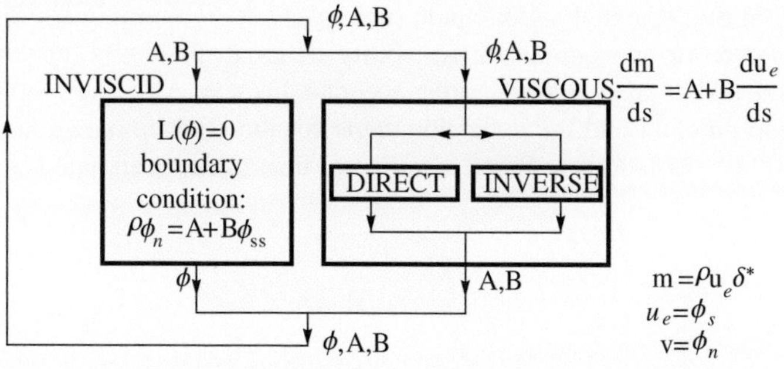

Fig. 9.10 Quasi-simultaneous coupling method of Wigton for transonic flows

9.1.3 Coupling the Numerical Solution of the Partial Differential Equations of the Viscous Flow with the Integral Equations of the Outer Inviscid Flows

9.1.3.1 Coupling the Numerical Solution of Boundary Layer Equations with Integral Representation of the Inviscid Flows: Incompressible Flows

For incompressible flows, the outer flow region is approximated by integrals over the body and the wake in terms of the shape of the body and the displacement thickness, as discussed before. The boundary layer equations are discretized using finite difference methods. The velocity u at the outer edge of the boundary layer is related to the distribution of the displacement thickness. Iterative methods must be used to solve these coupled nonlinear equations. Many sweeps of the boundary layer are required and upstream effects are recovered through the coupling with the inviscid flow integral representation. Separated flows can be simulated using upwinding scheme for the convective terms of the streamwise momentum equation. This procedure is used by many authors, see for example Lock and Williams [30]. A similar procedure is used also in triple deck theory, see Stewartson's review [31].

9.1.3.2 Coupling the Numerical Solution of Boundary Layer Equations with Integral Representation of the Inviscid Flows: Supersonic Flows

For supersonic flows, the inviscid outer region can be represented by the Ackeret formula. The finite difference approximations of the boundary layer equations can be solved by marching in the direction of the flow in case of no separation. local iterations at each step are still needed. This procedure was used in the sixties by Reynher and Flugge-Lotz [32] to study shock wave/ boundary layer interaction. For separated flows, they switch off the uu_x term in the momentum equation since u is negative but very small in the separated region. Such strategy (FLARE approximation) has been adopted by others to avoid numerical instability of the calculations of separated flows, although the results are at best first order accurate this way. Also, the simple relation between the pressure and the deflection angle for the outer flow region limits the validity of the calculations to Small disturbance linearized supersonic aerodynamic problems.

9.1.4 *Coupling Numerical Solutions of Partial Differential Equations of both the Inviscid and Viscous Flow Regions*

9.1.4.1 Coupling the Numerical Solution of Boundary Layer Equations with the Numerical Solution of the Inviscid Flow Equations

In this case, some coupling procedures discussed before as inverse and semi-inverse or simultaneous iterations are used. For incompressible flows, Moses et al. [17] introduced a simple effective method where the inviscid flow and viscous flow equations are solved coupled along vertical lines extending from the boundary layer to the outer inviscid flow region, governed by elliptic equation for the streamfunction (or potential function). hence, the line relaxation method of the inviscid flow equation is augmented at each streamwise location with the discrete boundary layer equations, coupled through the displacement thickness and many sweeps are required for convergence. The parabolic nature of the boundary layer equations in the case of attached flows, is not utilized in this method. However, the method can be used for the simulation of separated flows.

Also, the method can be extended to transonic flows where the inviscid flow equations are nonlinear and of mixed-type for both potential and Euler formulations. In this regard, one should mention the successful method introduced by Drela and Giles [33].

Higher order boundary layer equations, including curvature effects and pressure variation normal to the main flow direction, have been used and have led to the triple deck theory (see Stewartson [31], Neiland [34] and Sychev [35]). The latter has been proved to be successful for the trailing edge and separated laminar flows. Attempts to extend the theory to turbulent flows are faced with the problem of turbulence modeling. Extensions to unsteady flows are in progress (see Ryzhov [36]).

9.2 Viscous/Inviscid Interaction Procedures Based on Domain Decomposition Techniques

In these zonal procedures, the flow field is divided into two regions. To patch the two regions, an overlap layer, where the equations on both sides are valid, must exist.

Quarteroni and Valli [37] studied direct coupling of the Euler and Navier-Stokes equations. The motivation is clear. Away from the body, the grid is usually coarse and the viscous terms are not resolved and are not needed anyway, hence they are switched off. It turns out that errors can be reflected at the interface. Some fixes are possible however, but the general case is complicated.

To demonstrate some of these ideas, we will use stream function/vorticity formulation for incompressible two-dimensional steady laminar flow. The Navier-Stokes equations in Cartesian coordinates are given by

$$\frac{\partial^2 \psi}{\partial x^2} + \frac{\partial^2 \psi}{\partial y^2} = -\omega \tag{9.85}$$

$$u\frac{\partial \omega}{\partial x} + v\frac{\partial \omega}{\partial y} = \frac{1}{Re}\left(\frac{\partial^2 \omega}{\partial x^2} + \frac{\partial^2 \omega}{\partial y^2}\right) \tag{9.86}$$

where

$$u = \frac{\partial \psi}{\partial y}, \quad v = -\frac{\partial \psi}{\partial x} \tag{9.87}$$

The boundary conditions on a solid surface are given in terms of ψ (ω is unknown there), namely $\psi = const$ and $\partial\psi/\partial n = 0$, where n is the normal direction to the surface. For multi-connected domain, special treatment is required to assign the level of ψ on each closed body.

For high Reynolds number flows, upwind differences are used for the convective terms of the vorticity equation. Moreover, thin layer approximations (or partially parabolized Navier-Stokes equations) may be used, where the second derivatives in the tangential direction of the main flow in the vorticity equation, ω_{xx}, is neglected.

Notice, neglecting ψ_{xx} compared to ψ_{yy} as well, will lead to the boundary layer approximations and the equations become parabolic, where upstream effect is not possible. hence, the name of partially parabolized equations, where only ω_{xx} is neglected.

Obviously, the flow in the whole domain can be simulated using a stretched grid and iterative methods of the discrete equations coupled, for example successive line iteration can be used with 2×2 block tridiagonal solver, marching in the flow direction. Several sweeps are required for convergence.

Now, one can split the domain in two regions. in the outer one, vorticity vanishes and the vorticity equation is not needed. The question is where should the interface be? One can calculate the viscous terms, from the previous iteration, and if they are relatively small, they are switched off. Here, for convenience, the two regions are fixed a priori using rough boundary layer estimates.

To see the effect of switching off the vorticity in the outer region, the same numerical method described before for the whole domain is used. Comparison of accuracy of the results (i.e. friction drag) and the convergence of iterations for different sizes of the viscous region may assess the merit of this zonal approach (see Halim and Hafez [38], Tang and Hafez [39]), Hafez and Guo [40].

To take advantage of the different scales involved, one may solve the viscous and inviscid flow problems separately, hence a coupling procedure is needed. Two possibilities will be discussed.

In the first method, the viscous flow calculations provide ψ_v as boundary condition for the inviscid flow calculations, while the inviscid flow calculations provide in return ψ_i and $(\psi_n)_i$ for the viscous flow calculations, as shown in Fig. 9.11.

Again, the size of the overlap region must be determined based on the accuracy of the results and the convergence of iterations.

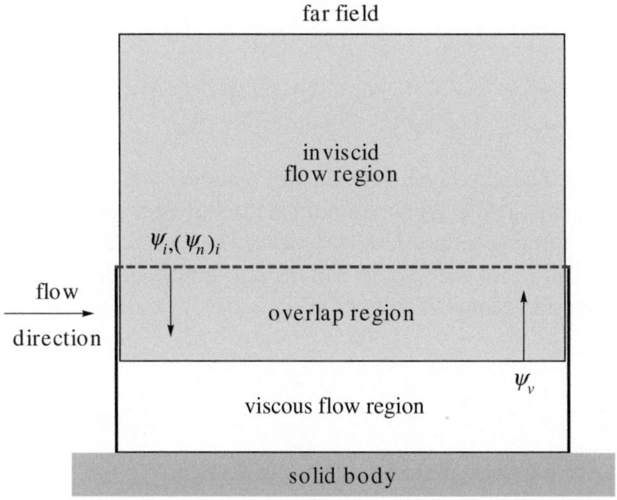

Fig. 9.11 Sketch for method 1

In the second method, a coupling formula is introduced along the interface. The only requirement for the coupling formula is to be consistent (i.e. it does not upset the converged solution), see for example Carter's formula [25] or semi-inverse incompressible flow example used by Cebeci [28] for transonic flow calculations. It is argued, however, a "better" coupling formula can be introduced if the flow physics is taken into consideration.

An application to our simple problem is shown in Fig. 9.12.

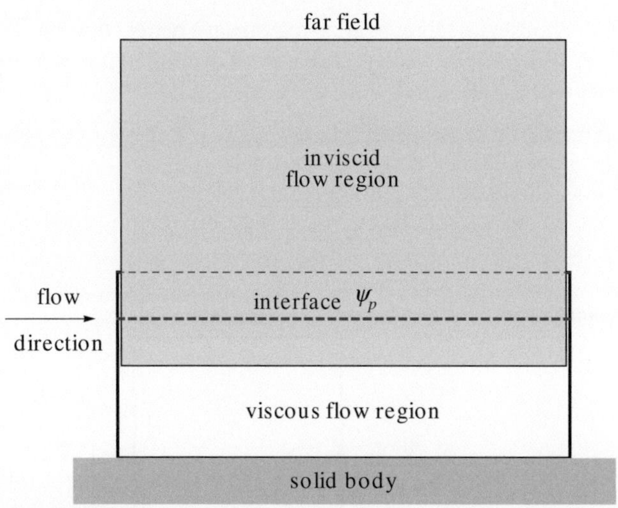

Fig. 9.12 Sketch for method 2

According to Carter, a coupling formula for a semi-inverse implicit procedure could be

$$\psi_{i,p}^{n+1} = \psi_{i,p}^{n} + \alpha(u_i - u_v) \tag{9.88}$$

where u_i is the velocity component calculated from the inviscid flow region and u_v is the velocity component calculated from the viscous flow region.

In the above formula, α is a free parameter. However, since the interface must lie in a region where both the viscous and inviscid flow equations are valid, one can choose the coupling formula as follows

$$\psi_{i,p}^{n+1} - \psi_{i,p}^{n} = \frac{\psi_{i+1,p}^{n+1} - 2\psi_{i,p}^{n+1} + \psi_{i-1,p}^{n+1}}{\Delta x^2} + \frac{\psi_{i,p+1}^{n} - 2\psi_{i,p}^{n} + \psi_{i,p-1}^{n}}{\Delta y^2} \tag{9.89}$$

The first term on the right-hand-side is evaluated at level $n+1$, leading to an implicit formula for $\psi_{i,p}$ and requires a tridiagonal solver.

In general, the interface equation may be the governing equation for ψ, hence the coupling formula may be given for low speed flow, by

$$(1 - M^2)\frac{\partial^2 \delta\psi}{\partial x^2} + \beta\frac{\partial \delta\psi}{\partial x} + \alpha\delta\psi = -\left[\frac{\partial}{\partial x}\left(\frac{1}{\bar{\rho}}\frac{\partial \psi}{\partial x}\right) + \frac{\partial}{\partial y}\left(\frac{1}{\bar{\rho}}\frac{\partial \psi}{\partial y}\right)\right] \tag{9.90}$$

where $\delta\psi$ is the correction, M is the local Mach number and $\bar{\rho}$ is the artificial density. Notice α should be zero for supersonic flow regions, according to the Fourier analysis!

In the above method, the overlap region consists of only three horizontal lines. The updated values of ψ_p is used for both the inviscid and viscous flow regions. The latter requires also $\psi_{i,p+1}$ to calculate $u_i = (\psi_{i,p+1} - \psi_{i,p})/\Delta y$.

In general, methods with larger overlap region have better convergence. In the following, methods using all the viscous region as an overlap region will be considered and their merits will be discussed.

Consider the same simple problem with the following splitting, method 3, Fig. 9.13:

In this method, the streamfunction equation is solved in the whole domain, assuming ω is known (in the viscous layer) and using only the no penetration boundary

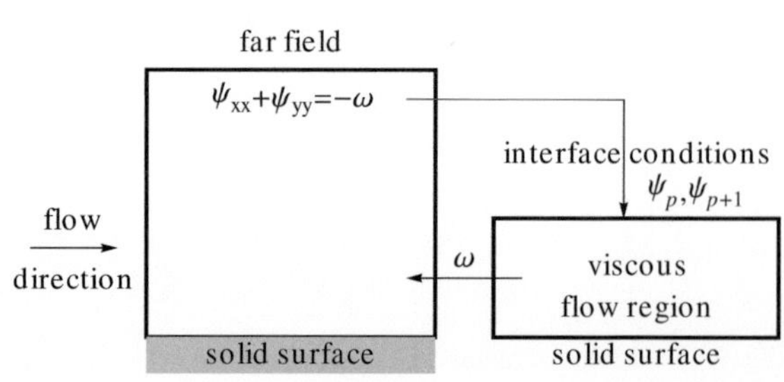

Fig. 9.13 Sketch for method 3

condition on the solid surface ($\psi = const$). The no slip condition, $\psi_n = 0$, is not imposed on this step of the calculations. The viscous flow is solved as usual with the two interface conditions and the no slip and no penetration conditions at the solid surface. The vorticity calculated, in the viscous flow region, is used as a forcing function distributed over part of the domain for the first calculations. The overlap region, now, is the whole viscous flow region. Moreover, the forcing function, ω, is distributed over the whole viscous flow region. An example of this method is given in [39] for the trailing edge flow and good comparison with the full Navier-Stokes equations and triple deck theory calculations is shown.

The same example is solved in [41], using potential formulation augmented with velocity correction in the viscous layer. The governing equations are

$$u = \frac{\partial \phi}{\partial x} + \bar{u} + U \tag{9.91}$$

$$v = \frac{\partial \phi}{\partial y} \tag{9.92}$$

The continuity equation becomes

$$\frac{\partial^2 \phi}{\partial x^2} + \frac{\partial^2 \phi}{\partial y^2} = -\frac{\partial \bar{u}}{\partial x} \tag{9.93}$$

In the momentum equation

$$u\frac{\partial u}{\partial x} + v\frac{\partial u}{\partial y} = -\frac{\partial p}{\partial x} + \frac{1}{Re}\frac{\partial^2 u}{\partial y^2} \tag{9.94}$$

the pressure term is approximated via linearized Bernoulli's law, $p = p_0 - \rho U u$. Notice, $u = \phi_x$ varies with y inside the viscous flow region. This formulation is similar to Cousteix' recent analysis [23].

The augmented potential equation is solved everywhere with the forcing function $-\bar{u}_x$, distributed in the viscous layer. Only the no penetration $\phi_n = 0$ boundary condition at the solid surface is imposed. The momentum equation is solved for u with the pressure gradient given in terms of ϕ_x as a forcing function and the no slip boundary condition at the solid surface and $u = \phi_x + U$ at the interface (i.e. $\bar{u} = 0$ there).

A more general formulation is adopted in [41] for the simulation of viscous flow over a rotating cylinder, where

$$u = \frac{\partial \phi}{\partial x} + \bar{u} + U \tag{9.95}$$

$$v = \frac{\partial \phi}{\partial y} + \bar{v} \tag{9.96}$$

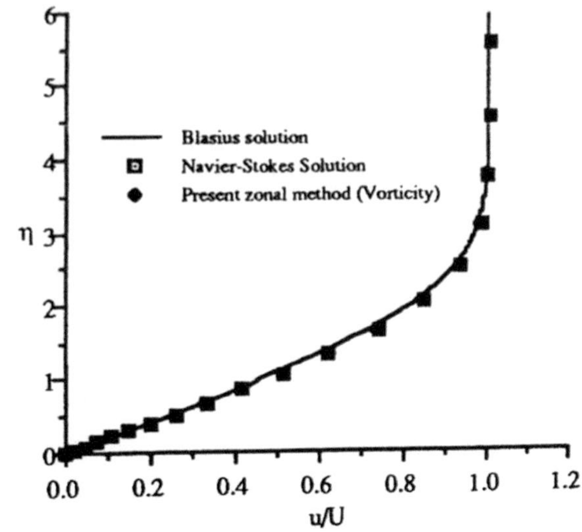

Fig. 9.14 Flat plate ($Re = 100{,}000$): streamwise velocity at mid-plate ($x/L = 0.5$)

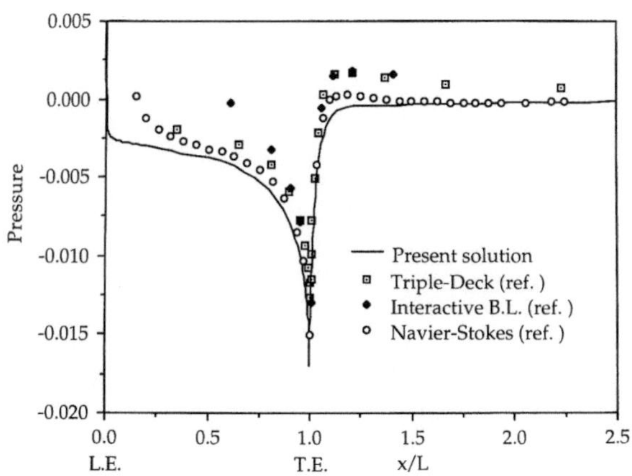

Fig. 9.15 Flat plate ($Re = 100{,}000$): pressure distribution along plate and wake

Both the $x-$ and the $y-$momentum equations are used to update $\bar{u}$ and $\bar{v}$. The pressure terms are still evaluated based on Bernoulli's law and in terms of ϕ_x and ϕ_y.

The results for finite flat plate and rotating cylinders are shown in Figs. 9.14, 9.15 and 9.16.

Finally, the approximation of the pressure in terms of ϕ_x and ϕ_y, using Bernoulli's law can be removed. The pressure can be evaluated by integrating the normal momentum equation inwards, from the interface to the solid body and thus the formulation, in the viscous flow region, becomes completely equivalent to the Navier-Stokes equations. This variable fidelity formulation has been used to simulate shock wave/laminar boundary layer interaction and the comparison with results of Navier-Stokes equa-

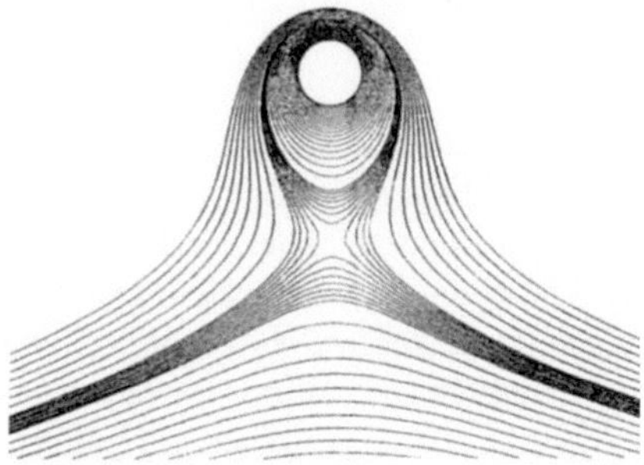

Fig. 9.16 Streamlines for the rotating cylinder ($Re = 200$)

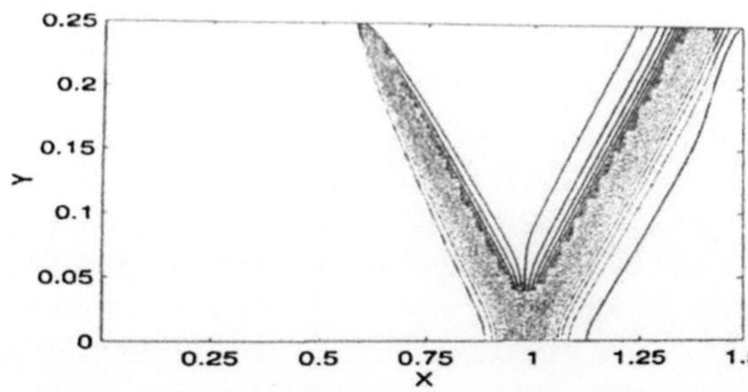

Fig. 9.17 Pressure contours (potential flow)

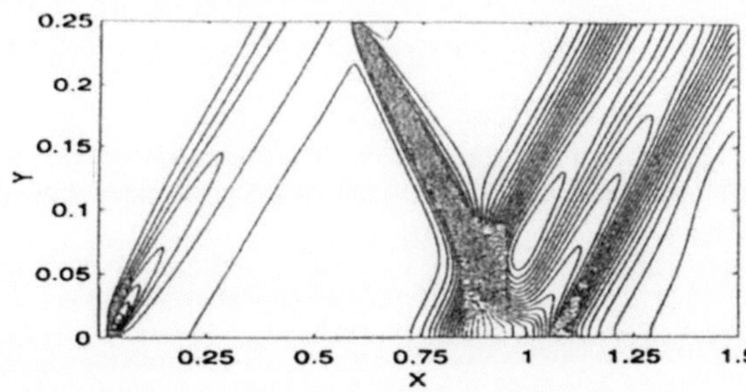

Fig. 9.18 Pressure contours (viscous flow—$Re = 2.96 \times 10^5$)

tions is given in Ref. [42]. Figures 9.17 and 9.18 from Ref. [42] show the wall pressure distribution and the Mach contours respectively.

The same method has been used to simulate transonic flows with shocks over airfoils and wings, see Hafez and Wahba [43–45]. It is claimed that the domain decomposition technique allows to use multigrid and upwind schemes more efficiently.

Viscous/inviscid interaction procedures for unsteady flows are discussed in Ref. [46, 47].

9.3 Summary of Chapter 9

Two main approaches are discussed in this chapter. The first one is based on the displacement thickness concept and the inviscid flow and boundary layer calculations are coupled through the displacement thickness and the streamwise pressure distribution. The inviscid flow and boundary layer calculations can be based on integral formulations or finite discrete approximation techniques and the four possibilities are covered. In the case where both calculations are based on integral formulations, quasi-one dimensional flow in a nozzle is studied via Pollhausen method, extended to compressible flow by Gruschwitz and by Oswatisch, as well as incompressible and supersonic flows over thin airfoils. Also, the potential flow calculation based on the streamfunction formulation, for two-dimensional flows, is coupled with the integral momentum equation. Coupling Euler calculations with boundary layer calculations are discussed in details, using several ideas.

Finally, coupling the numerical solution of the field equations of inviscid and boundary layer equations is briefly reviewed.

In the second approach, heterogeneous domain decomposition is used to couple potential and Euler calculations with solutions of boundary layer or thin layer approximation of the Navier-Stokes equations. Streamfunction/vorticity formulation is described to demonstrate the concept, and potential formulation augmented with vorticity correction terms is used for compressible flow calculations.

References

1. McDevitt, J.B.: Supercritical flow about a thick circular-arc airfoil. NASA TM 78549 (1979)
2. Chattot, J.-J.: Computational Aerodynamics and Fluid Dynamics: An Introduction. Springer, New York (2002)
3. Hafez, M., Palaniswamy, S. Mariani, P.: Calculations of transonic flows with shocks using Newton's method and direct solver. Part II. AIAA paper 88-0226 (1988)
4. Wieghardt, K.: Uber einen Energiesatz zur Brechnung Laminarer Grenzschichten. Ingen-Arch. **16**, 23–243 (1948)
5. Walz, A.: Application of Wieghardt's Energy Theorem to Velocity Profiles of One-Parameter in Laminar Boundary Layers. Rep. Aero. Res. Coun, London, No 10133. (1946)
6. Truckenbrodt, E.: An approximate method for the calculation of the laminar and turbulent boundary layer by simple quadrature for two dimensional and axially symmetric flow. JAS **19**, 428–429 (1952)
7. Tani, I.: On the approximate solution of the laminar boundary layer equations. JAS **21**, 487–495 (1954)
8. Holt, M.: Numerical Methods in Fluid Dynamics. Springer, New York (1984)
9. Schlichting, H.: Boundary Layer Theory. McGraw Hill, New York (1979)
10. Rosenhead, L. (ed.): Laminar Boundary Layers. Oxford University Press, Oxford (1963)
11. Swafford, T., Huddleston, D., Busby, J., Chesser, B.L.: Computation of Steady and Unsteady Quasi-One-Dimensional Viscous/Inviscid Interacting Internal Flows at Subsonic. Transonic and Supersonic Mach Numbers. Final Report NASA Lewis Research Center (1992)
12. Stewartson, K.: The Theory of Laminar Boundary Layers in Compressible Fluids. Oxford University Press, Oxford (1964)

13. Gruschwitz, E.: Calcul Approché de la Couche Limite Laminaire en Ecoulement Compressible sur une Paroi Non-Conductrice de la Chaleur, ONERA, Publication No. 47 (1950)
14. Oswatitsch, K.: Gas Dynamics. Academic Press, New York (1956)
15. Liepmann, H., Roshko, A.: Elements of Gas Dynamics. Wiley, New York (1957)
16. Whitfield, D., Swafford, T., Jackocks, J.: Calculations of turbulent boundary layers with separation and viscous-inviscid interaction. AIAA J. **19**(10), 1315 (1981)
17. Moses, H., Jones, R., O'Brien, W.: Simultaneous solution of the boundary layer and freestream with separated flow. AIAA J. **16**(1), 419–426 (1978)
18. Le Balleur, J.-Cl.: Numerical flow calculation and viscous-inviscid interaction techniques, (1978)
19. Johnston, W., Sockol, P.: Matching procedure for viscous-inviscid interactive computations. AIAA J. **17**(6), 661–665 (1979)
20. Murman, E., Bussing, T.: On the coupling of boundary-layer and Euler equations. Numerical and Physical Aspects of Aerodynamic Flows II., Springer, New York (1984)
21. Whitfield, D., Thomas, J.L.: Viscous-inviscid interaction using euler and inverse boundary-layer equations. Computational Methods in Viscous Flows. Pineridge Press, Swansea, (1984)
22. Wigton, L., Yoshihara, H.: Viscous-inviscid interaction with a three-dimensional inverse boundary layer code. Numerical and Physical Aspects of Aerodynamics. Springer (1984)
23. Cousteix, J., Mauss, J.: Asymptotic Analysis and Boundary Layer. Springer, New York (2007)
24. Catherall, D., Mangler, K.W.: The integration of the two-dimensional laminar boundary-layer equations past a point of vanishing skin friction. J. Fluid Mech. **26**, 163–182 (1966)
25. Carter, J.: A new boundary-layer inviscid iteration technique for separated flow. AIAA paper 79-1450 (1979)
26. Brune, G. W., Rubbert, P., Nark, T.: A new approach to inviscid flow/boundary layer matching. AIAA paper 74-601 (1974)
27. Veldman, A.: New, quasi-simultaneous method to calculate interactive boundary layer. AIAA J. **19**(1), 79–85 (1981)
28. Cebeci, T.: An Engineering Approach in Calculations of Aerodynamic Flows. Springer, New York (1999)
29. Wigton, L. and Holt, M.: Viscous-inviscid interaction in transonic flow. AIAA paper 81-1003 (1981)
30. Lock, R.C., Williams, B.R.: Viscous-inviscid interactions in external aerodynamics. Prog. Aerosp. Sci. **24**(2), 51–171 (1987)
31. Stewartson, K.: Multistructured boundary layers on flat plates and related bodies. Adv. Appl. Mech. **14**, 145–239 (1974)
32. Reynher, T., Flugge-Lotz, I.: The interaction of a shock wave with a laminar boundary layer. Int. J. Nonlinear Mech. **3**(2), 173–199 (1968)
33. Drela, M., Giles, M.D.: Viscous-inviscid analysis of transonic and low Reynolds number airfoils. AIAA J. **25**(10), 1347–1355 (1987)
34. Neiland, V.: Flow behind the boundary layer separation point in a supersonic stream. Fluid Dyn. **6**, 378–384 (1971)
35. Sychev, V.: On laminar separation. Fluid Dyn. **7**, 407–417 (1972)
36. Ryzhov, O.: Nonlinear Stability and Transition, to be Published by World Scientific. (2014)
37. Quarteroni, A., Valli, A.: Domain Decomposition Methods for Partial Differential Equations. Oxford University Press, Oxford (1999)
38. Halim, A., Hafez, M.: Calculation of Separation Bubbles using Boundary-layer Type Equations. Computational Methods in Viscous Flows. Pineridge Press, Swansea (1984)
39. Tang, C., Hafez, M.: Numerical simulation of steady compressible flows using zonal formulation. Part I Part II, Comput. Fluids **7–8**, 989–1001 (2001)
40. Hafez, M. and Guo, W.: Simulation of steady compressible flows based on Cauchy/Riemann equations and Crocco's relations. Part I and Part II, Int. J. Numer. Methods Fluids, **25**, 1–12 and, **26**, 325–344 (1998)
41. Hafez, M., Wahaba, E.: Incompressible viscous steady flow over finite plate and rotating cylinder with suction. Comput. Fluid Dyn. J. **13**(3), 571–584 (2004)

42. Hafez, M., Wahaba, E.: Hierarchical formulations for transonic flow simulations. Comput. Fluid Dyn. J. **11**(4), 377–382 (2003)
43. Hafez, M., Wahaba, E.: Numerical simulations of transonic aerodynamic flows based on hierarchical formulation. Int. J. Numer. Methods Fluids, **47**(6–7), 491–516 (2005)
44. Hafez, M., Wahaba, E.: Simulations of viscous transonic flows over lifting airfoils and wings. Comput. Fluids **36**(1), 32–52 (2006)
45. Hafez, M., Wahaba, E.: Viscous/inviscid interaction procedures for compressible aerodynamic flow simulations. Comput. Fluids **35**(7), 755–761 (2006)
46. Shatalov, A., Nakajima, M., Hafez, M.: Simulation of unsteady incompressible flows using Helmoltz-velocity decomposition. Comput. Fluid Dyn. J. **14**(3), 246–254 (2005)
47. Shatalov, A., Nakajima, M., Hafez, M.: Numerical simulations of unsteady laminar incompressible flows using viscous/inviscid interaction procedures. Comput. Fluid Dyn. J. **15**(3), 246–254 (2005)

Part II
Special Topics

Chapter 10
Wind Turbine and Propeller Aerodynamics—Analysis and Design

Wind turbines and propellers are very similar from the aerodynamics point of view, the former extracting energy from the wind, the latter putting energy into the fluid to create a thrust. The main part of this chapter will be devoted to wind turbine analysis and design, as this is currently a major area of research. But much of the theory and numerics is applicable to propellers. Section 10.9 will discuss some results pertinent to propellers. By convention, the power absorbed by a wind turbine rotor will be negative, whereas, that provided by the power plant of a propeller driven system will be positive.

10.1 Introduction—the Different Types of Wind Turbines

Wind-driven machines can be classified according to the orientation of their axis relative to the wind direction. Cross wind-axis machines have their axis in a plane perpendicular to the incoming wind velocity vector; wind-axis machines have their axis parallel to the incoming air flow. This fundamental difference impacts the study of these two types of machines: under the simplest of assumptions of constant wind speed, constant rotation speed and isolated rotor (neglecting support interference), the flow past a wind-axis machine, considered in a frame rotating with the blades is steady, whereas, for a cross wind-axis machine it is always unsteady. Unsteady flows are more complex and costly to analyze analytically and numerically. For this reason we will focus our attention on wind-axis machines. Sources of unsteadiness for wind-axis machines are yaw, when the wind direction is not aligned with the rotor axis, tower interference, earth boundary layer, wind gusts and, ultimately, blade deflection. This will be considered in Sect. 10.6 on unsteady flow. Prior to that, Sect. 10.2 will discuss the general 1-D conservation theorems, commonly called actuator disk theory. Section 10.3 will introduce the vortex model based on Goldstein "airscrew theory" with the treatment of the vortex sheets and the derivation of the torque and thrust in the spirit of the Prandtl lifting line theory. A discretization of the

J.J. Chattot and M.M. Hafez, *Theoretical and Applied Aerodynamics*,
DOI 10.1007/978-94-017-9825-9_10

vortex sheet as a lattice is proposed for the application of the Biot-Savart law. The next section, Sect. 10.4, deals with the design of the optimum rotor, discusses the minimum energy condition of Betz and gives a more general result for the optimum condition. Section 10.5 is devoted to the analysis problem, that is of finding the solution of the flow past a given rotor. A technique for handling high incidences and stalled flow on the blades is detailed and illustrated with an example. In Sect. 10.6 the extension of the method to unsteady flow is presented. The key issues are discussed. The effects of yaw and tower interference are assessed and the limits of the method shown by comparisons with experiments. Section 10.7 discusses the hybrid method of coupling a Navier-Stokes code with the vortex model as a means to achieve high fidelity modeling of viscous effects. This is followed, in Sect. 10.8, by perspectives and further development currently under way.

10.1.1 Aerodynamics Forces—Lift and Drag

The aerodynamic forces acting on an obstacle are decomposed into lift and drag components, $\vec{F} = \vec{D} + \vec{L}$, the drag being a force aligned with the incoming velocity as seen by the body and the lift being a force perpendicular to it. This decomposition is made in a frame of reference attached to the blade element, as pertinent to a moving machine. This is depicted in Fig. 10.2. The wind velocity $\vec{V}$, and the velocity of the element $\vec{V}_e$ in the absolute frame give the velocity of the fluid relative to the obstacle as

$$\vec{V}_r = \vec{V} - \vec{V}_e \tag{10.1}$$

The drag is aligned and in the direction of $\vec{V}_r$. The lift is rotated 90° opposite the circulation Γ developed inside the profile (Fig. 10.1).

Some wind machines are drag-driven, such as the Savonius wind turbine, whereas some are lift-driven as the Darrieus and the wind-axis turbines. The drag-driven machines are less efficient as the velocity of the device is always less than the wind velocity. Such systems are more useful as propulsive than power producing systems. The 15th century galleon such as Columbus "Santa Maria" belongs to the category of drag-driven systems. The modern racing sail boats use their sails as aircraft wings and are examples of lift-driven systems. A drag-driven device rotating about the y-axis (into the paper in Fig. 10.2) will correspond to $\vec{V}_e \simeq V_e \vec{i}$ where $V_e < V$ and the lift force is either zero or does not contribute to the power $P = \vec{F}.\vec{V} \simeq DV$. A lift-driven machine rotating about the x-axis (wind-axis turbine) will correspond to $\vec{V}_e \simeq -V_e \vec{k}$ and with $(L/D)_{max} \simeq 10$, the lift force will be the component contributing primarily to power generation.

Fig. 10.1 Tower Mill at Wijk bij Duurstede by J. van Ruysdael, (from https://en.wikipedia.org/wiki/File:The_Windmill_at_Wijk_bij_Duurstede_1670_Ruisdael.jpg Rijksmuseum Amsterdam)

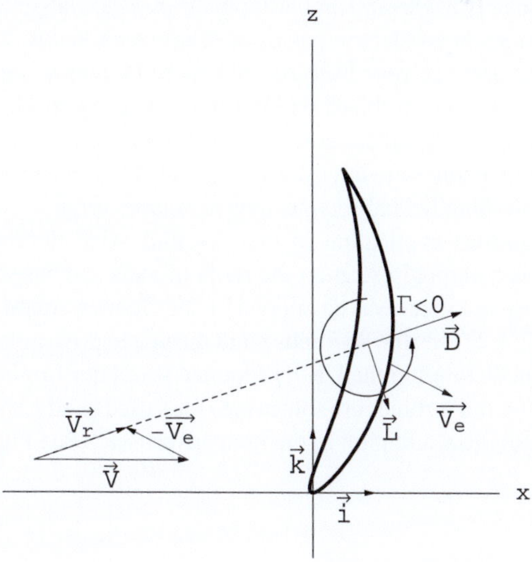

Fig. 10.2 Lift and drag forces acting on a moving blade element

10.1.2 Savonius Rotor

According to Wikipedia (https://en.wikipedia.org/w/index.php?title=Savonius_wind_turbine&oldid=555948336):

"Savonius wind turbines are a type of vertical-axis wind turbine (VAWT), used for converting the power of the wind into torque on a rotating shaft. They were invented by the Finnish engineer Sigurd J. Savonius in 1922. Johann Ernst Elias Bessler (born 1680) was the first to attempt to build a horizontal windmill of the Savonius type in the town of Furstenburg in Germany in 1745. He fell to his death whilst construction was under way. It was never completed but the building still exists.

Savonius turbines are one of the simplest turbines. Aerodynamically, they are drag-type devices, consisting of two or three scoops. Looking down on the rotor from above, a two-scoop machine would look like an *S* shape in cross section. Because of the curvature, the scoops experience less drag when moving against the wind than when moving with the wind. The differential drag causes the Savonius turbine to spin. Because they are drag-type devices, Savonius turbines extract much less of the wind's power than other similarly-sized lift-type turbines. Much of the swept area of a Savonius rotor is near the ground, making the overall energy extraction less effective due to lower wind speed at lower heights.

Savonius turbines are used whenever cost or reliability is much more important than efficiency. For example, most anemometers are Savonius turbines, because efficiency is completely irrelevant for that application. Much larger Savonius turbines have been used to generate electric power on deep-water buoys, which need small amounts of power and get very little maintenance. Design is simplified because, unlike horizontal axis wind turbines (HAWTs), no pointing mechanism is required to allow for shifting wind direction and the turbine is self-starting. Savonius and other vertical-axis machines are not usually connected to electric power grids. They can sometimes have long helical scoops, to give smooth torque.

The most ubiquitous application of the Savonius wind turbine is the Flettner Ventilator which is commonly seen on the roofs of vans and buses and is used as a cooling device. The ventilator was developed by the German aircraft engineer Anton Flettner in the 1920s. It uses the Savonius wind turbine to drive an extractor fan. The vents are still manufactured in the UK by Flettner Ventilator Limited.

Small Savonius wind turbines are sometimes seen used as advertising signs where the rotation helps to draw attention to the item advertised". See Fig. 10.3.

10.1.3 Darrieus Rotor

According to Wikipedia (https://en.wikipedia.org/w/index.php?title=Darrieus_wind_turbine&oldid=581321514):

"The Darrieus wind turbine is a type of vertical axis wind turbine (VAWT) used to generate electricity from the energy carried in the wind. The turbine consists of a

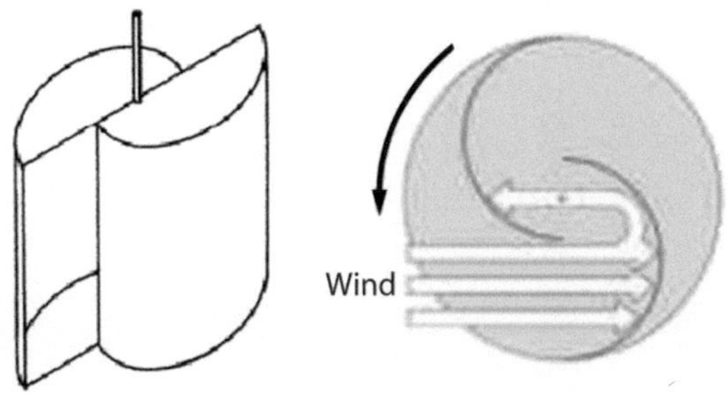

Fig. 10.3 Schematic drawings and operation of a two-scoop Savonius turbine, (from http://commons.wikimedia.org/wiki/File:Savonius_turbine.svg Author: Jaganath, http://commons.wikimedia.org/wiki/File:Savonius_Querschnitt.png Author Schargel)

number of aerofoils vertically mounted on a rotating shaft or framework. This design of wind turbine was patented by Georges Jean Marie Darrieus, a French aeronautical engineer in 1931.

The Darrieus type is theoretically just as efficient as the propeller type if wind speed is constant, but in practice this efficiency is rarely realized due to the physical stresses and limitations imposed by a practical design and wind speed variation. There are also major difficulties in protecting the Darrieus turbine from extreme wind conditions and in making it self-starting.

In the original versions of the Darrieus design, the aerofoils are arranged so that they are symmetrical and have zero rigging angle, that is, the angle that the aerofoils are set relative to the structure on which they are mounted. This arrangement is equally effective no matter which direction the wind is blowing—in contrast to the conventional type, which must be rotated to face into the wind.

When the Darrieus rotor is spinning, the aerofoils are moving forward through the air in a circular path. Relative to the blade, this oncoming airflow is added vectorially to the wind, so that the resultant airflow creates a varying small positive angle of attack (AoA) to the blade. This generates a net force pointing obliquely forwards along a certain 'line-of-action'. This force can be projected inwards past the turbine axis at a certain distance, giving a positive torque to the shaft, thus helping it to rotate in the direction it is already travelling in. The aerodynamic principles which rotates the rotor are equivalent to that in autogiros, and normal helicopters in autorotation.

As the aerofoil moves around the back of the apparatus, the angle of attack changes to the opposite sign, but the generated force is still obliquely in the direction of rotation, because the wings are symmetrical and the rigging angle is zero. The rotor spins at a rate unrelated to the windspeed, and usually many times faster. The energy arising from the torque and speed may be extracted and converted into useful power by using an electrical generator.

The aeronautical terms lift and drag are, strictly speaking, forces across and along the approaching net relative airflow respectively, so they are not useful here. We really want to know the tangential force pulling the blade around, and the radial force acting against the bearings.

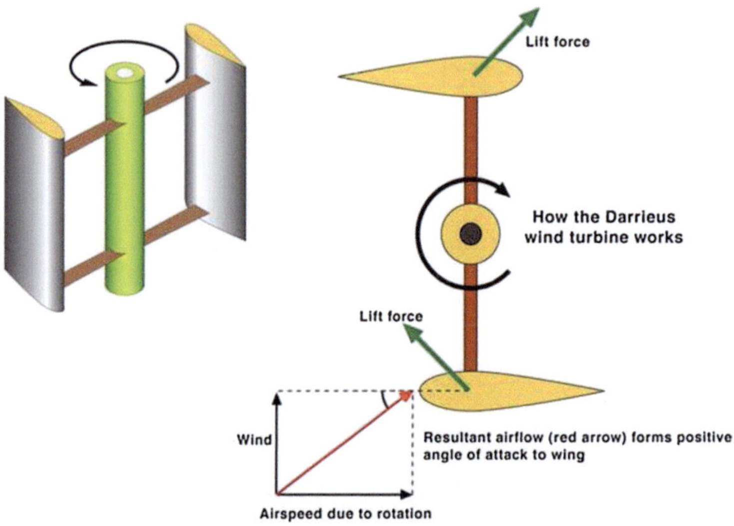

Fig. 10.4 Schematic of a Darrieus turbine: perspective (*left*); *top view* (*right*) (from https://en.wikipedia.org/wiki/File:Darrieus.jpg Author: Grahamuk)

When the rotor is stationary, no net rotational force arises, even if the wind speed rises quite high—the rotor must already be spinning to generate torque. Thus the design is not normally self starting. It should be noted though, that under extremely rare conditions, Darrieus rotors can self-start, so some form of brake is required to hold it when stopped. See. Fig. 10.4.

One problem with the design is that the angle of attack changes as the turbine spins, so each blade generates its maximum torque at two points on its cycle (front and back of the turbine). This leads to a sinusoidal (pulsing) power cycle that complicates design. In particular, almost all Darrieus turbines have resonant modes where, at a particular rotational speed, the pulsing is at a natural frequency of the blades that can cause them to (eventually) break. For this reason, most Darrieus turbines have mechanical brakes or other speed control devices to keep the turbine from spinning at these speeds for any lengthy period of time…

…In overall comparison, while there are some advantages in Darrieus design there are many more disadvantages, especially with bigger machines in MW class. The Darrieus design uses much more expensive material in blades while most of the blade is too near of ground to give any real power. Traditional designs assume that wing tip is at least 40 m from ground at lowest point to maximize energy production and life time. So far there is no known material (not even carbon fiber) which can meet cyclic load requirements".

Fig. 10.5 NREL turbine in NASA Ames 80′ × 120′ wind tunnel, from Ref [1]

10.1.4 Horizontal Axis Wind Turbine

In the rest of the chapter, the attention will be devoted to horizontal axis wind turbines (HAWTs) which represent, today, the main type of wind power machines that are developed and installed in all parts of the world. Much efforts have been done in understanding better the flow past HAWT both experimentally and analytically. A major wind tunnel campaign has been carried out by the National Renewable Energy Laboratory (NREL) at the NASA Ames Research Center large 80′ × 120′ wind tunnel facility [1]. They used a two-bladed rotor of radius $R = 5\,\text{m}$ with blades equipped with the S809 profile. A picture of the wind turbine in the wind tunnel is shown in Fig. 10.5. Note in particular the well defined evolution of the tip vortex visualized with smoke emitted from the tip of one blade. One can count seven or eight turns with regular spacing, despite the dramatic distortion due to the encounter with the tower when the blade tip vortex passes in front of it.

10.2 General 1-D Conservation Theorems—Actuator Disk Theory

As a first-order approximation, the flow past a horizontal axis wind turbine (HAWT) is treated as quasi-one dimensional. The stream tube captured by the rotor is described by its cross-section $A(x)$ and the velocity $\overrightarrow{V} = (U + u, v, w)$ representing the

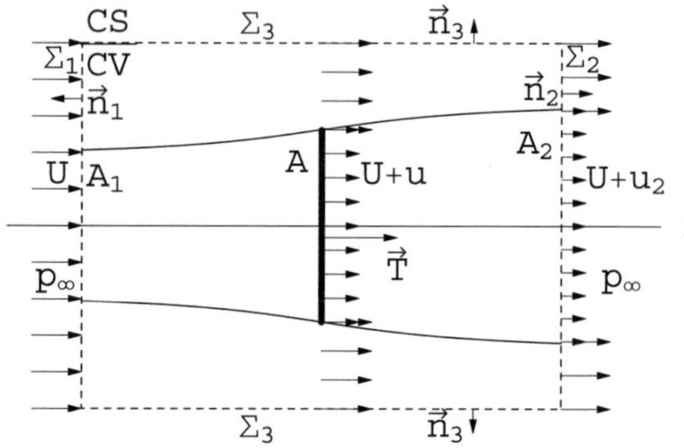

Fig. 10.6 Control volume/control surface for conservation theorems

uniform flow plus a perturbation (u, v, w) in cylindrical coordinates. At the rotor disk, the velocity is continuous, but the pressure is allowed to jump, as if the rotor had infinitely many blades and was equivalent to a porous disk. The stream tube boundary downstream of the disk allows for a discontinuity in axial velocity between the flow inside and outside the stream tube, called slip stream. As a further simplification, the flow rotation behind the rotor is neglected, hence the azimuthal velocity component is zero $(w = 0)$.

Consider a cylindrical control volume/control surface Σ_3 with large diameter, containing the captured stream tube and bounded by two disks far upstream Σ_1 and far downstream Σ_2 located in what is called the Trefftz plane, see Fig. 10.6. At the upstream boundary, the incoming flow is uniform and undisturbed with velocity U aligned with the turbine axis, and pressure p_∞. Downstream, at the Trefftz plane, the pressure returns to p_∞, but the velocity inside the stream tube is less than the undisturbed value U outside of it. The flow is considered steady, incompressible $(\rho = const.)$ and inviscid. Application of the conservation laws to the control volume/control surface yields the results known as the Rankine-Froude theory [2, 3] or actuator disk theory.

The conservation of mass (volume flow) reads:

$$\int_{CS} (\vec{V}.\vec{n})dA = 0 \tag{10.2}$$

which can be broken into the various control surface contributions as

$$(-U)\Sigma_1 + (U)(\Sigma_2 - A_2) + (U + u_2)A_2 + \int_{\Sigma_3} (\vec{V}.\vec{n})dA = 0 \tag{10.3}$$

The cross section of the control volume is constant, hence $\Sigma_1 = \Sigma_2$. After simplification, this reduces to

$$u_2 A_2 + \int_{\Sigma_3} (\vec{V}.\vec{n})dA = 0 \tag{10.4}$$

which means that the deficit or excess of volume flow through the exit section is compensated by the flow across the lateral surface of the cylinder. The conservation of x-momentum reads

$$\int_{CS} \rho(U+u)(\vec{V}.\vec{n})dA = -\int_{CS} pn_x dA - T \tag{10.5}$$

where T represents the thrust exerted by the flow on the rotor/actuator disk. This is expanded as

$$\rho U(-U)\Sigma_1 + \rho(U)^2(\Sigma_2 - A_2) + \rho(U+u_2)^2 A_2 + \int_{\Sigma_3} \rho U(\vec{V}.\vec{n})dA = -T \tag{10.6}$$

Note that, since pressure is constant and equal to p_∞ on all control surfaces, it does not contribute to the momentum balance. The last term on the left-hand-side can be evaluated using the result of mass conservation. Solving for T:

$$T = -\rho(U+u_2)u_2 A_2 = -\dot{m}u_2 \tag{10.7}$$

where $\dot{m} = \rho(U+u)A = const.$ represents the mass flow rate in the stream tube and $A = \pi R^2$, R denotes the radius of the rotor. Note that, since $T > 0$, then $u_2 < 0$ for a turbine.

The steady energy equation applies to the stream tube with one-dimensional entrance and exit sections:

$$\int_{A_1} \left(\frac{p}{\rho} + \frac{V^2}{2} + gz\right) d\dot{m} = \int_{A_2} \left(\frac{p}{\rho} + \frac{V^2}{2} + gz\right) d\dot{m} - \dot{W}_s \tag{10.8}$$

$\dot{W}_s = P$ represents the shaft work or power extracted ($P < 0$) from the flow by the rotor. Solving for P

$$P = \left(\frac{p_\infty}{\rho} + \frac{(U+u_2)^2}{2}\right)\dot{m} - \left(\frac{p_\infty}{\rho} + \frac{U^2}{2}\right)\dot{m} \tag{10.9}$$

After simplifying the above expression one finds

$$P = \left(U + \frac{u_2}{2}\right)\dot{m}u_2 \tag{10.10}$$

The power is also given by $P = -T(U+u) = (U+u)\dot{m}u_2$. From this we conclude that $u_2 = 2u$.

Results of this analysis can be summarized in terms of the rotor/actuator disk parameters

$$T = -2\rho A(U+u)u \tag{10.11}$$

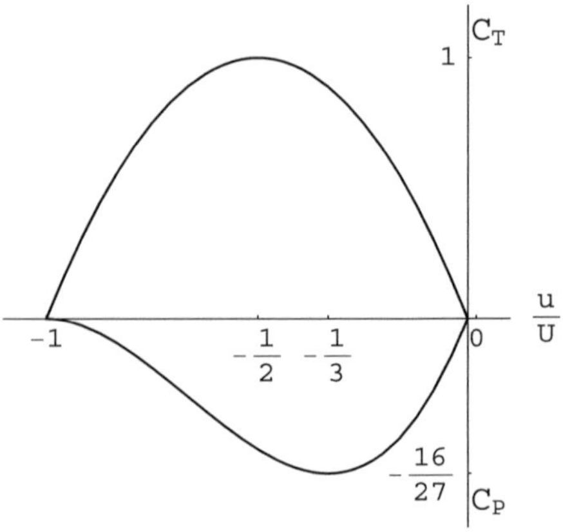

Fig. 10.7 Thrust and power coefficients versus axial induced velocity at the wind turbine rotor plane

$$P = 2\rho A(U + u)^2 u \tag{10.12}$$

If one introduces the thrust and power coefficients

$$C_T = \frac{2T}{\rho U^2 A} = -4\left(1 + \frac{u}{U}\right)\frac{u}{U}, \quad C_P = -\eta = \frac{2P}{\rho U^3 A} = 4\left(1 + \frac{u}{U}\right)^2 \frac{u}{U} \tag{10.13}$$

With change in sign, the power coefficient is also interpreted as the efficiency, η. These are shown in Fig. 10.7 in terms of $\frac{u}{U}$.

The maximum value of the power coefficient $|C_{Pmax}| = 16/27 = 0.59$ obtained for $\frac{u}{U} = -\frac{1}{3}$ is known as the Betz limit [4]. The maximum thrust on the tower corresponds to $\frac{u}{U} = -\frac{1}{2}$, for which the final wake velocity is zero.

More elaborate models have been developed from this simple one-dimensional model with the addition of rotation in the wake by Joukowski [5] and others, but this requires making some empirical assumptions in order to be able to solve the equations. Instead of expanding more along these lines, we prefer devoting our time to a more realistic model, called the vortex model.

10.3 Vortex Model and Strip Theory—The Goldstein Model

10.3.1 General Comments

The next level of modelization is the extension of the Prandtl lifting line theory, developed by Prandtl [6] for studying the inviscid flow past finite wings, to the flow past a rotor. The fundamental difference is in the shape of the vortex sheet. Behind a wing, the vortex sheet, often called "wake" even though the model is inviscid, can be assumed to be a surface generated by semi-infinite lines or vortices, originating at the

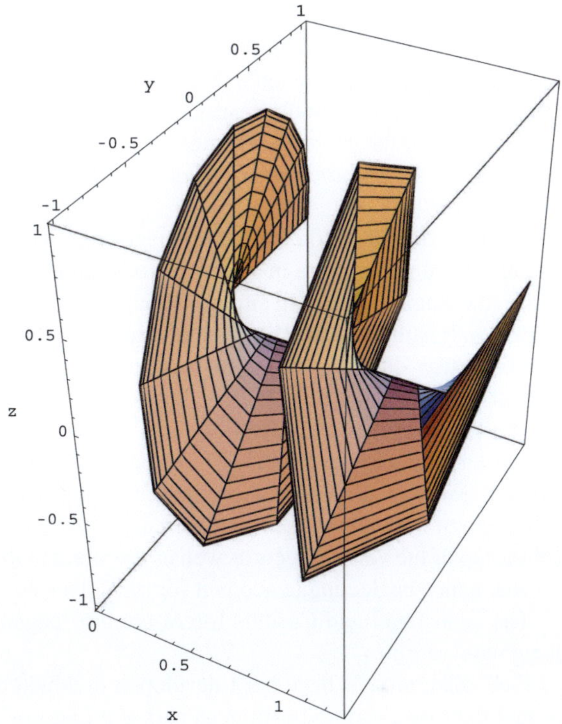

Fig. 10.8 Vortex sheet from blade 1 and discretization as a vortex lattice

sharp trailing edge of the wing and parallel to the incoming velocity vector. These vortices, according to the Biot-Savart formula, induce at the wing, represented as a single line (the lifting line), a velocity component w, perpendicular to the incoming flow, called "downwash". In contrast, behind a rotor, the vortex sheets are modeled as perfect helices, obtained by rotating and translating a segment of straight line perpendicular and in the same plane as the axis. Each blade is represented by such a segment and generates an individual vortex sheet. The rotation corresponds to the rotor speed of rotation Ω. The translation velocity varies from $U + u$ at the rotor plane, to $U + 2u$ far downstream, as shown in the one-dimensional flow theory. These helicoidal surfaces are made of helicoidal vortices that induce, according to the Biot-Savart formula, velocity components that affect the local incidences of the blade elements through the axial and azimuthal contributions. This is the Goldstein model [7]. See Fig. 10.8.

The simplifications associated with this model are important to understand. The fundamental choice is to prescribe the shape of the vortex sheets as perfect helices. This is often described as prescribed wake model or rigid wake model. The same is true of the wake model in the Prandtl lifting line theory, which is aligned with the undisturbed flow, yet yields very reasonable results. From a physical point of view, vortex sheets are surfaces of discontinuity of the velocity vector, wetted on both sides by the fluid and find an equilibrium position such that the pressure is continuous across them. From a practical point of view, if the prescribed vortex sheets are close to the actual ones, the evaluation of the induced velocities will be accurate to first-order, as is the case of Prandtl's theory and in the transfer of the tangency condition from the actual profile to the nearby axis in small disturbance theory of thin airfoils. In this method, the prescribed vortex structure satisfies an "equilibrium condition" by matching the absorbed rotor power with the power deficit in the far field, as will be seen later.

Another simplification consists in neglecting the rolling-up of the sheets edges, a well-known phenomena in aircraft aerodynamics associated with tip vortices that can create a hazard for small aircrafts flying in the tip vortices wake of large transport airplanes. Results indicate that keeping the sheets flat does not affect the accuracy of the simulation since the rolling-up does not change the vorticity content of the wake, only displaces it slightly.

The average axial velocity decreases behind the rotor. As a consequence, the stream tube diameter increases. The helicoidal vortex sheets would be expected to expand, however, this is not accounted for and possibly does not need to, since the rolling up of the sheets edges may partially counter this effect. It is believed that the rolling-up of the vortex sheets as well as the stream tube expansion are second-order effects, hence neglecting to account for them does not jeopardize the results.

The centrifugal and Coriolis forces are also neglected, which is acceptable for large wind turbines.

Free wake models have been developed in which the vortex sheets are allowed to find their equilibrium position as part of a Lagrangian iterative process in which vorticity is shed by the blades and convected downstream. Experience has shown that these methods are unstable. The vortex filaments tend to become disorganized and chaotic downstream of the rotor. Artificial dissipation is needed to prevent the calculation from diverging, which defeats the purpose of keeping the wake dissipation free, as expected with an inviscid model. Another aspect is that the number of vorticity laden particles increases with each time step and the run time per iteration increases quickly to excessive values as N^3, where N is the number of particles. In the end, the free wake models are neither reliable nor efficient.

10.3.2 Adimensionalization—Discretization of the Vortex Sheets

Consider a rotor rotating at speed Ω, placed in an incoming uniform flow with velocity $\mathbf{V} = (U, 0, 0)$ aligned with the turbine axis. We neglect the tower interference on the rotor flow. In a frame of reference attached to the rotor, the flow is steady. The tip speed ratio is defined as $TSR = \Omega R/U = 1/adv$, which is also the inverse of the advance ratio, *adv* used as main parameter for propellers. Other dimensionless quantities are defined and used below:

$$\overline{x} = Rx,\ \overline{y} = Ry,\ \overline{z} = Rz \tag{10.14}$$

$$\overline{c} = Rc \tag{10.15}$$

$$\overline{u} = Uu,\ \overline{v} = Uv,\ \overline{w} = Uw \tag{10.16}$$

$$\overline{\Gamma} = UR\Gamma \tag{10.17}$$

$$T = \frac{\pi}{2}\rho U^2 R^2 C_T,\ \tau = \frac{\pi}{2}\rho U^2 R^3 C_\tau,\ P = \frac{\pi}{2}\rho \Omega^3 R^5 p_\tau,\ L' = \frac{1}{2}\rho U^2 R c C_l \tag{10.18}$$

The bar "-" values correspond to dimensional variables, when the same letter is used. T is the thrust on the tower (N), C_T the thrust coefficient, τ is the torque $(N.m)$, C_τ the torque coefficient, P the power (W), p_τ a dimensionless power coefficient, L' is the local lift force per unit span, and C_l is the local lift coefficient.

Each vortex sheet such as that in Fig. 10.8 is discretized as a lattice, shown unrolled in Fig. 10.9, to better explain the numerical scheme.

The circulation $\Gamma_{i,j}$ is located at the grid points of the lattice (indicated by the circles). The blade is represented by a lifting line along the y-axis for blade 1 (thick solid line) and $\Gamma_{1,j}$ represents the circulation inside the blade. The small elements of vorticity $\overrightarrow{dl}$ are located between the grid points and correspond either to trailed vorticity (in the x-direction in the presentation of Fig. 10.10) or shed vorticity (in the y-direction) that is present only when the flow is unsteady. When the flow is steady, as assumed here, the circulation is constant along a vortex filament, i.e. $\Gamma_{i,j} = \Gamma_{1,j},\ \forall i = 2, \ldots, ix$. The vorticity is given in terms of the differences in

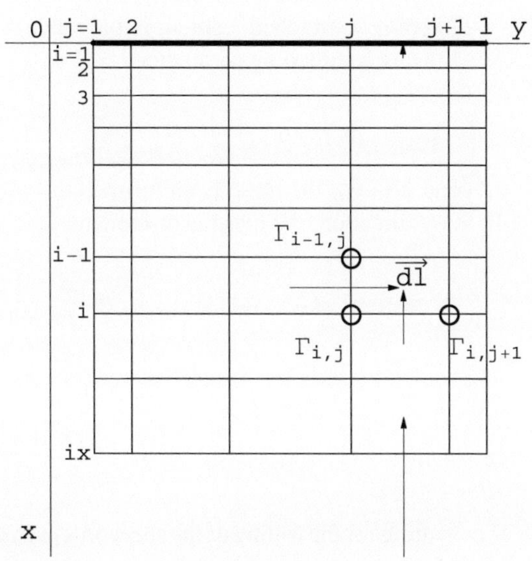

Fig. 10.9 Schematic representation of a vortex sheet

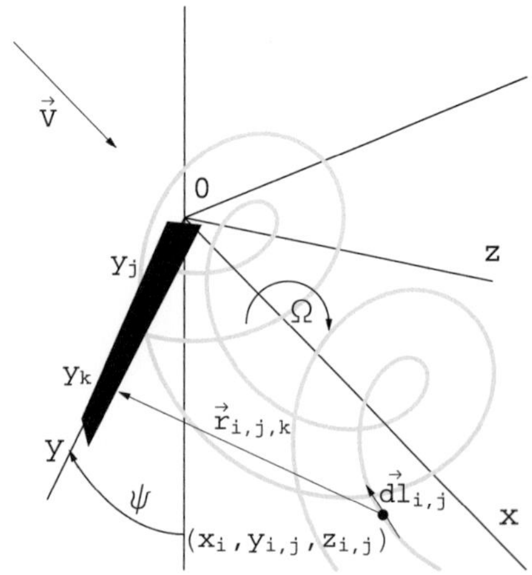

Fig. 10.10 Blade 1 and two trailing vortices—Schematic of Biot-Savart calculation

circulation as $\Delta\Gamma_{i,j} = \Gamma_{i,j+1} - \Gamma_{i,j}$ for the trailed vorticity and $\Delta\Gamma_{i,j} = \Gamma_{i,j} - \Gamma_{i-1,j}$ (=0 here) for shed vorticity. The vortex element orientation is indicated by the arrow which represents positive $\Delta\Gamma_{i,j}$ but the vector components account for the rolling of the sheet on a perfect helix. Let x_i, y_i, z_i represent the tip vortex of blade 1

$$y_i = \cos\left(\frac{x_i}{adv_i}\right), \quad z_i = \sin\left(\frac{x_i}{adv_i}\right) \tag{10.19}$$

The advance ratio is made to vary linearly from $adv_1 = (V + u)/\Omega R$ at the rotor plane $i = 1$ to $adv_{ix} = (V + 2u)/\Omega R$ after a few turns of the helix, then remains constant thereafter to the Trefftz plane $i = ix$. The mesh distribution in the x-direction is stretched from an initial value dx_1 at the blade, to some location $x_{str} \simeq 1$, using a stretching parameter $s > 1$ as

$$x_i = x_{i-1} + dx_{i-1}, \quad dx_i = s dx_{i-1}, \quad i = 2, \ldots, ix \tag{10.20}$$

Beyond $x = x_{str}$ the mesh is uniform all the way to the Trefftz plane $x_{ix} \simeq 20$. In the y-direction, the mesh is discretized with a cosine distribution

$$y_j = y_0 + \frac{1}{2}(1 - y_0)\cos\theta_j, \quad \theta_j = \frac{j-1}{jx-1}\pi, \quad j = 1, \ldots, jx \tag{10.21}$$

The trailed vortices are located between the mesh lines according to

$$\eta_j = y_0 + \frac{1}{2}(1 - y_0)\cos\theta_j, \quad \theta_j = \frac{2j-1}{2(jx-1)}\pi, \quad j = 1, \ldots, jx - 1 \tag{10.22}$$

Accounting for the rolling of the sheet on a perfect helix, a small element of trailed vorticity has components

$$\vec{dl} = \{dx_i, dy_{i,j}, dz_{i,j}\}$$

$$= \left\{-\frac{1}{2}(x_{i+1} - x_{i-1}), -\frac{1}{2}(y_{i+1} - y_{i-1})\eta_j, -\frac{1}{2}(z_{i+1} - z_{i-1})\eta_j\right\} \quad (10.23)$$

whereas, a small element of shed vorticity has components

$$\vec{dl} = \{dx_i, dy_{i,j}, dz_{i,j}\} = \left\{0, \frac{1}{2}(y_{i-1} + y_i)(\eta_j - \eta_{j-1}), \frac{1}{2}(z_{i-1} + z_i)(\eta_j - \eta_{j-1})\right\} \quad (10.24)$$

Application of the Biot-Savart formula provides the influence coefficients for the small elements, which are accumulated and stored in multidimensional arrays.

A particular treatment is made for the trailed elements closest to the lifting line and at the Trefftz plane. The former extend only half a cell with origin at $\frac{1}{2}(x_1 + x_2)$ (see Fig. 10.9), and the latter extend from ∞ to $\frac{1}{2}(x_{ix-1} + x_{ix})$, where a remainder accounts for the contribution downstream of the Trefftz plane. Note that in the case of the three-bladed rotor, the lifting lines themselves induce velocities on the other two (but not on themselves as straight blades). The same discretization applies for the calculation of the lifting line contributions for a three-bladed rotor, but now the small vortex element is located along the lifting line, having components

$$\vec{dl} = \{0, \eta_j - \eta_{j-1}, 0\}, \; j = 2, \ldots, jx - 1 \quad (10.25)$$

in the case of blade 1, and $\Delta\Gamma_{i,j} = \Gamma_{1,j}$ represents the circulation inside the blade.

10.3.3 Biot-Savart Law—Induced Velocities

The Biot-Savart formula reads

$$d\vec{V}_{i,j,k} = \frac{\Delta\Gamma_{i,j}}{4\pi} \frac{\vec{dl} \wedge \vec{r}_{i,j,k}}{|\vec{r}_{i,j,k}|^3} \quad (10.26)$$

where $d\vec{V}_{i,j,k}$ is the induced velocity of the small element $\vec{dl}$ at a point k along a blade. $\vec{r}_{i,j,k}$ is the distance vector having its origin at the vortex element and its extremity at point k. See Fig. 10.10.

For a three-bladed rotor, the u-component for a blade is given by

$$u_j = \sum_{k=1}^{jx-1} (\Gamma_{k+1} - \Gamma_k) a_{k,j} + \sum_{k=1}^{jx-1} \Gamma_k \tilde{a}_{k,j} \quad (10.27)$$

where the first term corresponds to influence of the vortex filaments k of each blade on the control point j on the lifting line and the tilde "~" term corresponds to the influence of the two other lifting line elements at y_k on the control point j. Similarly, the w-components are obtained from the circulation with the influence coefficients $c_{k,j}$ and, with straight blades and zero coning angle, $\widetilde{c}_{k,j} = 0$

$$w_j = \sum_{k=1}^{jx-1} (\Gamma_{k+1} - \Gamma_k) c_{k,j} \tag{10.28}$$

Given the induced velocity components, the flow at the blade element can be analyzed and the local lift found. This is an approximation that amounts to neglecting the effect of the neighboring elements and considers each element in isolation, as if it were part of an infinite blade. Mathematically, this corresponds to neglecting the derivatives in the span direction compared to the other derivatives: $\frac{\partial}{\partial y} \ll \frac{\partial}{\partial x}, \frac{\partial}{\partial z}$. This is appropriate, as has been found, for large aspect ratio wings and blades, that is those for which the chord is small compared to the span. This approach is called "strip theory".

10.3.4 Forces and Moment

The flow is assumed to be inviscid, $C_d = 0$. At the blade element j the flow configuration is as depicted in Fig. 10.11.

The local incidence is

$$\alpha_j = \phi_j - t_j = \arctan\left(\frac{1 + u_j}{\frac{y_j}{adv} + w_j}\right) - t_j \tag{10.29}$$

where ϕ_j is called the flow angle and t_j is the angle of twist at the element. The local lift coefficient is obtained from the result of thin airfoil theory or from an experimental or numerical profile lift curve $C_l(\alpha)$. In thin airfoil theory, the lift is given by

$$C_l(\alpha) = 2\pi(\alpha + 2\frac{d}{c}) \tag{10.30}$$

$\frac{d}{c}$ is the profile mean camber defined as

$$\frac{d}{c} = \frac{1}{2\pi} \int_0^1 d'(x) \frac{1 - 2\frac{x}{c}}{\sqrt{\frac{x}{c}(1 - \frac{x}{c})}} \frac{dx}{c} \tag{10.31}$$

where $d(x)$ is the equation of the camber line. According to the Kutta-Joukowski lift theorem, the lift force per unit span is perpendicular to the incoming flow, its

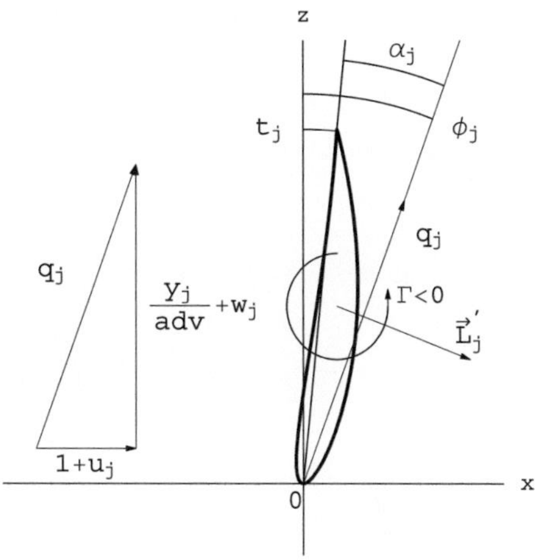

Fig. 10.11 Blade element flow configuration

magnitude is $L' = \rho \overline{q} \overline{\Gamma}$ and its orientation is 90° from the incoming flow direction, rotating opposite to the circulation. $q = \frac{\overline{q}}{U} = \sqrt{(1+u)^2 + \left(\frac{y}{adv} + w\right)^2}$ is the normalized magnitude of the incoming flow velocity vector. This can be expressed mathematically as

$$\vec{L}' = \rho U \vec{q} \wedge U R \Gamma \vec{j} \tag{10.32}$$

where dimensionless quantities have been introduced. $\vec{j}$ is the unit vector along the y-axis and $\Gamma < 0$ in Fig. 10.11. Upon a dimensionalization of the lift, the circulation is obtained as

$$\Gamma_j = -\frac{1}{2} q_j c_j C_l(\alpha_j) \tag{10.33}$$

The contribution of the blade element to thrust and torque can now be derived. Projection in the x-direction gives

$$dT_j = L'_j \cos\phi_j R dy_j = -\rho U q_j \cos\phi_j U R \Gamma_j R dy_j = -\rho U^2 R^2 \Gamma_j \left(\frac{y_j}{adv} + w_j\right) dy_j \tag{10.34}$$

or, upon integration

$$T = -\rho U^2 R^2 \int_{y_0}^{1} \Gamma(y) \left(\frac{y}{adv} + w(y)\right) dy \tag{10.35}$$

In dimensionless form

$$C_T = \frac{T}{\frac{1}{2}\rho U^2 \pi R^2} = -\frac{2}{\pi} \int_{y_0}^{1} \Gamma(y) \left(\frac{y}{adv} + w(y)\right) dy \tag{10.36}$$

Projection of the lift force in the z-direction and accounting for the distance from the axis, the torque contribution reads

$$d\tau_j = L'_j \sin\phi_j Ry_j Rdy_j = \rho U q_j \sin\phi_j UR\Gamma_j Ry_j Rdy_j = \rho U^2 R^3 \Gamma_j (1 + u_j) y_j dy_j \tag{10.37}$$

which integrates to give

$$\tau = \rho U^2 R^3 \int_{y_0}^{1} \Gamma(y)\,(1 + u(y))\, y dy \tag{10.38}$$

or, in dimensionless form

$$C_\tau = \frac{\tau}{\frac{1}{2}\rho U^2 \pi R^3} = \frac{2}{\pi} \int_{y_0}^{1} \Gamma(y)\,(1 + u(y))\, y dy \tag{10.39}$$

These results are for one blade. Total thrust and torque must account for all the blades.

10.3.5 Wake Equilibrium Condition

The power absorbed by the rotor is also estimated by using the torque evaluation and the speed of rotation as

$$P = \tau \Omega \tag{10.40}$$

The vortex sheets are considered in "equilibrium" when the power absorbed by the rotor matches the power deficit $\dot{W}_s$ in the far wake obtained from actuator disk theory, i.e. with the current notation

$$P = \tau \Omega = 2\pi \rho U^3 R^2 (1 + u)^2 u \tag{10.41}$$

This is a cubic equation for the average axial induced velocity u at the rotor

$$4(1 + u)^2 u = TSR\, C_\tau = \frac{C_\tau}{adv} = C_P \tag{10.42}$$

The solution procedure consists in recalculating the solution with an updated vortex structure until it satisfies the equilibrium condition to acceptable accuracy ($|\Delta C_\tau / C_\tau| \leq 10^{-3}$).

10.4 Aerodynamic Design of a Rotor Blade—Betz Minimum Energy Condition

10.4.1 Formulation

The minimum energy condition of Betz [8] refers to the optimum conditions which lead to the minimum loss of energy in the slipstream. It applies equally to propellers and wind turbines. For the latter, the optimum distribution of circulation $\Gamma(y)$ minimizes the torque (negative) for a given thrust on the tower. The simple argument of Betz is that, in this case, an elementary force $\delta C_T = -\frac{2}{\pi}\Gamma(y)\left(\frac{y}{adv} + w(y)\right)dy$ should produce a constant elementary torque $\delta C_\tau = \frac{2}{\pi}\Gamma(y)\,(1 + u(y))\,ydy$, independent of y. Hence the minimum energy condition of Betz reads

$$\frac{(1 + u(y))\,y}{\frac{y}{adv} + w(y)} = y\,\tan\phi(y) = const. \tag{10.43}$$

This is true only for vanishing circulation and induced velocities, as will be shown later. By analogy, lets consider the problem of the optimum wing loading in the Prandtl lifting line theory. Given that the lift and induced drag are

$$C_L = 2\int_{-1}^{1}\Gamma(y)dy, \quad C_{Di} = -\int_{-1}^{1}\Gamma(y)w(y)dy \tag{10.44}$$

where $w(y)$ is the downwash (negative),

$$w(y) = -\frac{1}{4\pi}\int_{-1}^{1}\frac{\Gamma'(\eta)d\eta}{y-\eta} \tag{10.45}$$

Munk [9] showed that, for the optimum loading, the downwash w has to be constant, else, moving a small element of lift (elementary horseshoe vortex) $\delta C_L = 2\delta\Gamma dy$ from its current position y_1 to a position y_2 such that $\delta w = w(y_2) - w(y_1) > 0$, would not change the total lift, yet would change the induced drag by $\delta C_{Di} = -2\delta\Gamma\delta w dy < 0$, a decrease in total drag, indicating that the distribution is not optimal. In other words, for an elementary $\delta C_L = 2\Gamma dy$, the induced drag $\delta C_{Di} = -2\Gamma(y)w(y)dy$ should be independent of y, which implies that $w(y) = const$. A more mathematical proof consists in considering the objective function

$$F(\Gamma) = C_{Di}(\Gamma) + \lambda C_L(\Gamma) \tag{10.46}$$

consisting of the induced drag to be minimized for a given lift, where λ is the Lagrange multiplier. Taking the Frechet derivative yields the minimization equation that must hold for any change $\delta\Gamma$ of circulation. The Frechet derivative can be defined as the limiting process

$$\frac{\partial F}{\partial \Gamma}(\delta\Gamma) = \lim_{\theta \to 0} \frac{d}{d\theta} F(\Gamma + \theta\delta\Gamma) \tag{10.47}$$

The result is linear in $\delta\Gamma$. It represents the derivative of F with respect to Γ, in the direction of $\delta\Gamma$. Here we obtain:

$$\frac{\partial F}{\partial \Gamma}(\delta\Gamma) = -\int_{-1}^{1} \left(\delta\Gamma(y)w(y) + \Gamma(y)\frac{\partial w}{\partial \Gamma}(\delta\Gamma)\right) dy + 2\lambda \int_{-1}^{1} \delta\Gamma(y) dy \tag{10.48}$$

But the kernel of $w(\Gamma)$ is antisymmetric. Hence, integration by part twice using the fact that $\Gamma(\pm 1) = \delta\Gamma(\pm 1) = 0$, yields the identity

$$\Gamma(y)\frac{\partial w}{\partial \Gamma}(\delta\Gamma) = \delta\Gamma w(\Gamma) \tag{10.49}$$

Indeed, consider the second term in the integral above. By definition of the Frechet derivative of $w(\Gamma)$ in the direction $\delta\Gamma$

$$I = -\int_{-1}^{1} \Gamma(y)\frac{\partial w}{\partial \Gamma}(\delta\Gamma) dy = -\int_{-1}^{1} \Gamma(y) \left(-\frac{1}{4\pi} \int_{-1}^{1} \frac{\delta\Gamma'(\eta) d\eta}{y - \eta}\right) dy \tag{10.50}$$

$$I = \frac{1}{4\pi} \int_{-1}^{1} \int_{-1}^{1} \frac{\Gamma(y)\delta\Gamma'(\eta) d\eta dy}{y - \eta} \tag{10.51}$$

A first integration by parts in η yields

$$I = \frac{1}{4\pi} \int_{-1}^{1} \left\{ \left[\frac{\Gamma(y)\delta\Gamma(\eta)}{y - \eta}\right]_{-1}^{1} - \int_{-1}^{1} \frac{\Gamma(y)\delta\Gamma(\eta) d\eta}{(y - \eta)^2} \right\} dy \tag{10.52}$$

The integrated term vanishes to leave after a second integration by parts in y

$$I = -\frac{1}{4\pi} \int_{-1}^{1} \left\{ \left[-\frac{\delta\Gamma(\eta)\Gamma(y)}{y - \eta}\right]_{-1}^{1} + \int_{-1}^{1} \frac{\delta\Gamma(\eta)\Gamma'(y) dy}{y - \eta} \right\} d\eta \tag{10.53}$$

Again the integrated term vanishes and the second term, thanks to the antisymmetry of the kernel $\frac{1}{y-\eta}$, can be interpreted as

$$I = -\int_{-1}^{1} \delta\Gamma(\eta) \left\{-\frac{1}{4\pi} \int_{-1}^{1} \frac{\Gamma'(y) dy}{\eta - y}\right\} d\eta = -\int_{-1}^{1} \delta\Gamma(\eta) w(\eta) d\eta \tag{10.54}$$

Finally, the minimization equation reduces to

$$\frac{\partial F}{\partial \Gamma}(\delta\Gamma) = -2 \int_{-1}^{1} \delta\Gamma(y) \left(w(y) - \lambda\right) dy, \ \forall \delta\Gamma \tag{10.55}$$

The solution is simply $w(y) = \lambda = const.$, as obtained by Munk with his simpler argument. The antisymmetry of the kernel simply indicates that two elementary horseshoe vortices of same intensity placed at y_1 and y_2 induce downwash that are opposite at the other horseshoe location. This is not the case with a wind turbine wake, because two vortices along the helix will have different shapes depending on their initial y-location at the blade, hence induce velocities u or w are not related. Let y_v, z_v be the equation of the vortex sheet of blade 1

$$\begin{cases} y_v = \eta \cos\left(\frac{x}{adv(x)}\right) \\ z_v = \eta \sin\left(\frac{x}{adv(x)}\right) \end{cases} \tag{10.56}$$

A vortex filament corresponds to a fixed value of η. Two such vortices are shown in Fig. 10.10.

The kernel for u and w does not have the antisymmetry property for the helicoidal vortex sheet as can be seen from the formulae

$$u(y) = \int_{y_0}^{1} \int_{0}^{\infty} \frac{\Gamma'(\eta)}{4\pi} \frac{(y - y_v)\frac{dz_v}{dx} + z_v\frac{dy_v}{dx}}{\left[x^2 + (y - y_v)^2 + z_v^2\right]^{\frac{3}{2}}} dx d\eta \tag{10.57}$$

$$w(y) = -\int_{y_0}^{1} \int_{0}^{\infty} \frac{\Gamma'(\eta)}{4\pi} \frac{(y - y_v) + x\frac{dy_v}{dx}}{\left[x^2 + (y - y_v)^2 + z_v^2\right]^{\frac{3}{2}}} dx d\eta \tag{10.58}$$

The objective function for the optimum turbine blade reads

$$F(\Gamma) = C_\tau(\Gamma) + \lambda C_T(\Gamma) \tag{10.59}$$

stating that the search is for minimum torque (negative) for a given thrust on the tower at a given tip speed ratio and blade root location y_0. The minimization equation is derived for one blade as

$$\frac{\partial F}{\partial \Gamma}(\delta\Gamma) = \frac{2}{\pi} \int_{y_0}^{1} \left\{ \delta\Gamma\,(1 + u(y))\,y + \Gamma(y)\frac{\partial u}{\partial \Gamma}(\delta\Gamma) \right\} dy$$

$$-\frac{2\lambda}{\pi} \int_{y_0}^{1} \left\{ \delta\Gamma \left(\frac{y}{adv} + w(y) \right) + \Gamma(y)\frac{\partial w}{\partial \Gamma}(\delta\Gamma) \right\} dy = 0, \ \forall \delta\Gamma \tag{10.60}$$

The terms such as $\Gamma(y)\frac{\partial u}{\partial \Gamma}(\delta\Gamma)$ cannot be transformed in this case, however, choosing $\delta\Gamma = \Gamma$ yields the following identity

$$\int_{y_0}^{1} \Gamma \left\{ (1 + 2u(y))\,y - \lambda\left(\frac{y}{adv} + 2w(y) \right) \right\} dy = 0 \tag{10.61}$$

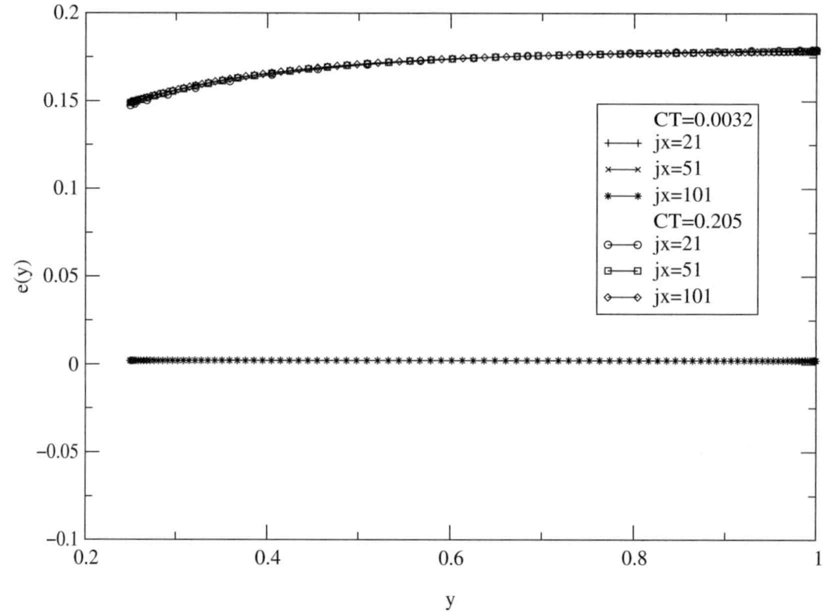

Fig. 10.12 Verification of Betz condition for two design examples at $TSR = 2.9$

the interpretation of which is that the optimum solution satisfies the minimum energy condition of Betz, in the Trefftz plane, "in the average", with Γ as weighting factor. For a tip speed ratio $TSR = 2.9$, the design of a two bladed-rotor at $(C_l)_{opt} = 0.8979$ with very light loading demonstrates that the minimum energy condition of Betz is verified. The thrust coefficient is $C_T = 0.0032$, the corresponding value of the Lagrange multiplier is $\lambda = 0.3433$ and the power coefficient $C_P = 0.0031$. For a higher loading, corresponding to the two-bladed NREL rotor at the same TSR and $C_T = 0.205$, one finds $\lambda = 0.254$ and $C_P = 0.1768$. The deviation from the Betz condition, $e(y) = (y \tan\phi(y) - \lambda)/\lambda$, is shown in Fig. 10.12. As can be seen, for lightly loaded rotor, the Betz condition is satisfied, but it is no longer the case when loading is more realistic.

10.4.2 Discretization

The discrete formulation is now described. The influence coefficients are calculated as

$$a_{j,k} \text{ or } \widetilde{a}_{j,k} = \frac{1}{4\pi} \sum_{i=2}^{ix} \frac{(y_k - y_{i,j})dz_{i,j} + z_{i,j}dy_{i,j}}{\left[x_i^2 + (y_k - y_{i,j})^2 + z_{i,j}^2\right]^{\frac{3}{2}}} \tag{10.62}$$

$$c_{j,k} = -\frac{1}{4\pi}\sum_{i=2}^{ix}\frac{(y_k - y_{i,j})dx_i + x_i dy_{i,j}}{\left[x_i^2 + (y_k - y_{i,j})^2 + z_{i,j}^2\right]^{\frac{3}{2}}} \tag{10.63}$$

with the notation introduced earlier and in Fig. 10.10. The contributions to $a_{j,k}$ and $c_{j,k}$ of the vortex filament part that is beyond the Trefftz plane are approximated by the following remainders

$$\delta a_{j,k} = \frac{1}{4\pi}\int_{x_{ix}}^{\infty}\frac{(y - y_v)\frac{dz_v}{dx} + z_v\frac{dy_v}{dx}}{\left[x^2 + (y - y_v)^2 + z_v^2\right]^{\frac{3}{2}}}dx \simeq -\frac{1}{4\pi}\frac{y_k - 2\eta_j\cos\left(\frac{x_{ix}}{adv_{ix}}\right)}{2x_{ix}^2} + O\left(\frac{1}{x_{ix}^3}\right) \tag{10.64}$$

$$\delta c_{j,k} = -\frac{1}{4\pi}\int_{x_{ix}}^{\infty}\frac{(y - y_v) + x\frac{dy_v}{dx}}{\left[x^2 + (y - y_v)^2 + z_v^2\right]^{\frac{3}{2}}}dx \simeq \frac{1}{4\pi}\frac{-2\eta_j\sin\left(\frac{x_{ix}}{adv_{ix}}\right)}{2x_{ix}^2} + O\left(\frac{1}{x_{ix}^3}\right) \tag{10.65}$$

The torque and thrust coefficients are discretized as

$$C_\tau = \frac{2}{\pi}\sum_{k=2}^{jx-1}\Gamma_k\,(1 + u_k)\,y_k(\eta_k - \eta_{k-1}) \tag{10.66}$$

$$C_T = -\frac{2}{\pi}\sum_{k=2}^{jx-1}\Gamma_k\left(\frac{y_k}{adv} + w_k\right)(\eta_k - \eta_{k-1}) \tag{10.67}$$

Each contributes to the minimization equation

$$\frac{\partial C_\tau}{\partial \Gamma_j} = \frac{2}{\pi}\left(1 + u_j\right)y_j(\eta_j - \eta_{j-1})$$

$$+\frac{2}{\pi}\sum_{k=2}^{jx-1}\Gamma_k\left(a_{j-1,k} - a_{j,k} + \tilde{a}_{j,k}\right)y_k(\eta_k - \eta_{k-1}),\quad j = 2,\ldots,jx-1 \tag{10.68}$$

$$\frac{\partial C_T}{\partial \Gamma_j} = -\frac{2}{\pi}\left(\frac{y_j}{adv} + w_j\right)(\eta_j - \eta_{j-1})$$

$$-\frac{2}{\pi}\sum_{k=2}^{jx-1}\Gamma_k\left(c_{j-1,k} - c_{j,k}\right)(\eta_k - \eta_{k-1}),\quad j = 2,\ldots,jx-1 \tag{10.69}$$

Boundary conditions complete the formulation with $\Gamma_1 = \Gamma_{jx} = 0$
The minimization equation

$$\frac{\partial F(\Gamma)}{\partial \Gamma_j} = \frac{\partial C_\tau}{\partial \Gamma_j} + \lambda \frac{\partial C_T}{\partial \Gamma_j} = 0, \quad j = 2, \ldots, jx - 1 \tag{10.70}$$

is a linear, non-homogeneous system for the Γ_j's that can be solved by relaxation. Let $\Delta\Gamma_j = \Gamma_j^{n+1} - \Gamma_j^n$ be the change of circulation between iterations n and iteration $n+1$, and ω the relaxation factor. The iterative process reads

$$2\left[\lambda\left(c_{j-1,k} - c_{j,k}\right) - \left(a_{j-1,j} - a_{j,j} + \widetilde{a}_{j,j}\right) y_j\right](\eta_j - \eta_{j-1})\frac{\Delta\Gamma_j}{\omega}$$
$$= \left[\left(1 + u_j\right) y_j - \lambda\left(\frac{y_j}{adv} + w_j\right)\right](\eta_j - \eta_{j-1})$$
$$+ \sum_{k=2}^{jx-1} \Gamma_k \left[\left(a_{j-1,k} - a_{j,k} + \widetilde{a}_{j,k}\right) y_k - \lambda\left(c_{j-1,k} - c_{j,k}\right)\right](\eta_k - \eta_{k-1}) \tag{10.71}$$

With $jx = 101$ and $\omega = 1.8$ the solution converges in a few hundreds iterations. Note that if $\lambda = adv$ the system is homogeneous and the solution is $\Gamma_j = 0, \ \forall j$, which corresponds to zero loading of the rotor that rotates freely and does not disturb the flow. This is called "freewheeling". For a given λ the solution corresponds to a certain thrust. In order to find the value of the Lagrange multiplier that will correspond to the desired value of the thrust coefficient, say $C_{T\,target}$, C_τ and C_T are decomposed into linear and bilinear forms in terms of the Γ_j's

$$C_\tau = C_{\tau 1} + C_{\tau 2} = \frac{2}{\pi}\sum_{j=2}^{jx-1} \Gamma_j y_j (\eta_j - \eta_{j-1}) + \frac{2}{\pi}\sum_{j=2}^{jx-1} \Gamma_j u_j y_j (\eta_j - \eta_{j-1}) \tag{10.72}$$

$$C_T = C_{T1} + C_{T2} = -\frac{2}{\pi}\sum_{j=2}^{jx-1} \Gamma_j \frac{y_j}{adv}(\eta_j - \eta_{j-1}) - \frac{2}{\pi}\sum_{j=2}^{jx-1} \Gamma_j w_j (\eta_j - \eta_{j-1}) \tag{10.73}$$

$C_{\tau 1}$ and C_{T1} are homogeneous of degree one. $C_{\tau 2}$ and C_{T2} are homogeneous of degree two. Using the properties of homogeneous forms, the summation of the minimization equations multiplied each by the corresponding Γ_j results in the identity

$$\sum_{j=2}^{jx-1} \Gamma_j \left(\frac{\partial C_\tau}{\partial \Gamma_j} + \lambda \frac{\partial C_T}{\partial \Gamma_j}\right) = C_{\tau 1} + 2C_{\tau 2} + \lambda\left(C_{T1} + 2C_{T2}\right)$$

$$= \sum_{j=2}^{jx-1} \Gamma_j \left\{ (1+2u_j)y_j - \lambda \left(\frac{y_j}{adv} + 2w_j \right) \right\} (\eta_j - \eta_{j-1}) = 0 \tag{10.74}$$

which is the discrete analog of the optimum condition derived earlier.

If one assumes that the optimum distributions of circulation for different $C_{T\,target}$ vary approximately by a multiplication factor, then knowing the solution Γ_j for a particular value of λ (say $\lambda = 0$) allows to find a new value of the Lagrange multiplier as follows. Change the Γ_j's to new values $\kappa \Gamma_j$. In order to satisfy the constraint one must have

$$C_{T\,target} = \kappa C_{T1} + \kappa^2 C_{T2} \tag{10.75}$$

Solving for κ yields

$$\kappa = \frac{\sqrt{C_{T1}^2 + 4C_{T2}C_{T\,target}} - C_{T1}}{2C_{T2}} \tag{10.76}$$

The new estimate for λ is obtained from the above identity as

$$\lambda = -\frac{C_{\tau 1} + 2\kappa C_{\tau 2}}{C_{T1} + 2\kappa C_{T2}} \tag{10.77}$$

This procedure is repeated two or three times to produce the desired solution. Once the optimum circulation is obtained, the chord distribution is given by

$$c_j = -\frac{2\Gamma_j}{q_j\,(C_l)_{opt}} \tag{10.78}$$

where $(C_l(\alpha))_{opt}$ corresponds to the point α on the polar such that C_l/C_d is maximum. The twist distribution results from

$$t_j = \phi_j - \alpha \tag{10.79}$$

10.4.3 Viscous Correction

The viscous effects are considered a small deviation of the inviscid solution at the design point. In particular, the design lift distribution along the blade is such that $C_l(y) < C_{l\,max}$. The viscous torque and viscous thrust coefficients, $C_{\tau v}$ and C_{Tv} are added to their inviscid counterparts. For one blade

$$C_{\tau v} = \frac{1}{\pi} \int_{y_0}^{1} q(y) \left(\frac{y}{adv} + w(y) \right) C_{dv} c(y) y dy \tag{10.80}$$

$$C_{Tv} = \frac{1}{\pi} \int_{y_0}^{1} q(y)\,(1 + u(y))\, C_{dv} c(y) dy \tag{10.81}$$

In discrete form

$$C_{\tau v} = \frac{1}{\pi} \sum_{k=2}^{jx-1} q_k \left(\frac{y_k}{adv} + w_k \right) C_{dv\,k} c_k y_k (\eta_k - \eta_{k-1}) \tag{10.82}$$

$$C_{Tv} = \frac{1}{\pi} \sum_{k=2}^{jx-1} q_k\,(1 + u_k)\, C_{dv\,k} c_k (\eta_k - \eta_{k-1}) \tag{10.83}$$

where the 2-D viscous drag coefficient, $C_{dv\,k}$, is approximated locally by a parabola [10]. The viscous polar, C_l versus C_{dv} is obtained from experimental measurements or numerical simulation, as a data set $\{C_{dv\,m}, C_{l\,m}, \alpha_m\}$ where m is the index corresponding to the α-sweep. The viscous drag is locally given by

$$C_{dv}(C_l) = (C_{d0})_m + (C_{d1})_m C_l + (C_{d2})_m C_l^2 \tag{10.84}$$

The optimization proceeds along lines similar to the inviscid case. First an inviscid solution is obtained. Then the viscous correction is performed. κ is defined as before, with C_{T1} replaced by $C_{T1} + C_{Tv1}$, C_{T2} by $C_{T2} + C_{Tv2}$ and $C_{T\,target}$ replaced by $C_{T\,target} - C_{T0}$ and λ reads

$$\lambda = -\frac{C_{\tau 1} + C_{\tau v1} + 2\kappa(C_{\tau 2} + C_{\tau v2})}{C_{T1} + C_{Tv1} + 2\kappa(C_{T2} + C_{Tv2})} \tag{10.85}$$

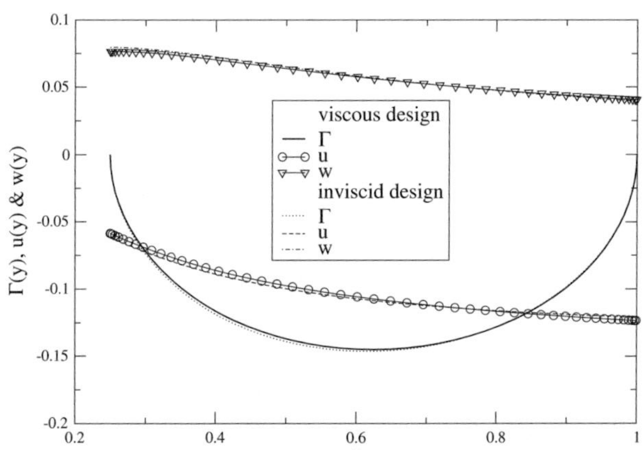

Fig. 10.13 Circulation and induced velocity for optimum blade at $TSR = 2.9$ and $C_T = 0.205$

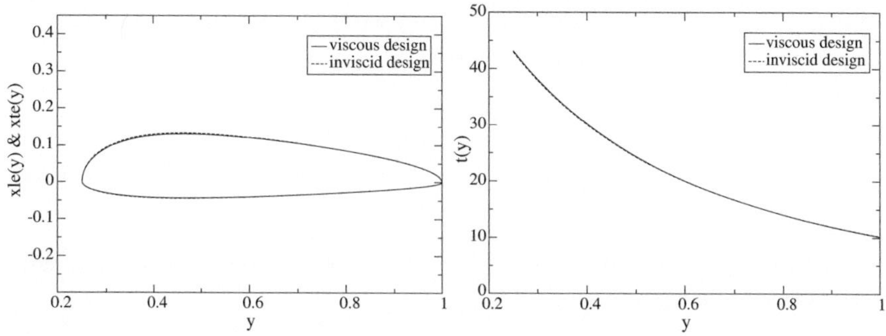

Fig. 10.14 Chord and twist distributions for optimum blade at $TSR = 2.9$ and $C_T = 0.205$

where the viscous contributions have been decomposed into 3 terms, independent, linear and quadratic functions of Γ as

$$C_{\tau v} = C_{\tau v0} + C_{\tau v1} + C_{\tau v2} \tag{10.86}$$

$$C_{Tv} = C_{Tv0} + C_{Tv1} + C_{Tv2} \tag{10.87}$$

The inviscid and viscous distributions are compared in Figs. 10.13 and 10.14, for the $TSR = 2.9$ and $C_{T\,target} = 0.205$. As can be seen, the effect of viscosity on the geometry is very small, however the efficiency drops 3.5 % from $\eta = 0.177$ to $\eta = 0.171$.

10.5 Analysis of the Flow Past a Given Rotor

In this section we consider the problem of calculating the flow past a given rotor. This is particularly useful to assess the performance of an optimum rotor at off-design conditions. Here, the blade sections will have variable working conditions in terms of $\alpha(y)$, with the possibility, on part of the blade, for the incidences to be larger than the incidence of maximum lift, i.e. $\alpha > (\alpha)_{Cl\,max}$. In other words, part of the blade may be stalled. This will typically happen at small *TSR*'s.

10.5.1 Formulation

With reference to Fig. 10.11, given a data set $\{C_{dv\,m}, C_{l\,m}, \alpha_m\}$ that characterizes the blade profile viscous polar, the governing equation simply reads

$$\Gamma(y) = -\frac{1}{2}q(y)c(y)C_l\,(\alpha(y)) \tag{10.88}$$

where $\alpha(y) = \phi(y) - t(y)$. The chord $c(y)$ and twist $t(y)$ distributions are given. The incoming velocity is $q(y) = \sqrt{(1+u(y))^2 + (\frac{y}{adv} + w(y))^2}$ and Γ is the unknown circulation ($\Gamma < 0$). In discrete form this reads

$$\Gamma_j = -\frac{1}{2} q_j c_j C_l(\alpha_j), \quad j = 2, \ldots, jx - 1 \tag{10.89}$$

When $\alpha_j < (\alpha)_{Cl\,max}$, then $\frac{dC_{lj}}{d\alpha} > 0$ and the simple, partial Newton linearization gives a converging algorithm

$$\Gamma_j^{n+1} = \Gamma_j^n + \Delta\Gamma_j = -\frac{1}{2} q_j c_j \left(C_l(\alpha_j) + \frac{dC_{lj}}{d\alpha} \Delta\alpha_j \right), \quad j = 2, \ldots, jx - 1 \tag{10.90}$$

where

$$\Delta\alpha_j = \frac{(\frac{y_j}{adv} + w_j)\Delta u_j - (1 + u_j)\Delta w_j}{q_j^2} \tag{10.91}$$

Δu_j and Δw_j are given by the induced velocity coefficients associated with $\Delta\Gamma_j = \Gamma_j^{n+1} - \Gamma_j^n$.

$$\left\{ 1 + \frac{c_j}{2q_j} \frac{dC_{lj}}{d\alpha} \left((1 + u_j)(c_{j,j} - c_{j-1,j}) - (\frac{y_j}{adv} + w_j)(a_{j,j} - a_{j-1,j} + \tilde{a}_{j,j}) \right) \right\} \frac{\Delta\Gamma_j}{\omega}$$
$$= -\frac{1}{2} q_j c_j C_l(\alpha_j) - \Gamma_j^n \tag{10.92}$$

ω is the relaxation factor and can be chosen up to $\omega = 1.8$. It is found that the coefficients $a_{j,j} - a_{j-1,j} + \tilde{a}_{j,j} < 0$ and $c_{j,j} - c_{j-1,j} > 0$ so that the linearization contributions reinforce the diagonal of the iterative matrix underlying the relaxation method [11]. With $jx = 101$ this typically converges in a few hundreds of iterations.

10.5.2 Algorithm for High Incidences—Regularization of the Solution

In relation to Fig. 10.11, with decrease in *TSR* or equivalently with increase in *adv*, the incidence of the blade element increases. When $\alpha_j \geq (\alpha)_{Cl\,max}$, the sign of the lift slope changes to $\frac{dC_{lj}}{d\alpha} \leq 0$, which destabilizes the algorithm. This simply reflects that multiple values of α exist for a given C_l, therefore it is necessary to regularize the solution to make it unique. This can be done by adding an artificial viscosity term to the right-hand-side:

$$\Gamma_j = -\frac{1}{2}q_j c_j C_l(\alpha_j) + \mu_j(\Gamma_{j+1} - 2\Gamma_j + \Gamma_{j-1}), \quad j = 2, \ldots, jx - 1 \quad (10.93)$$

where $\mu_j \geq 0$ is the artificial viscosity coefficient. A test case with exact solution, based on the lifting line theory of Prandtl and a 2-D lift coefficient given analytically by $C_l(\alpha) = \pi \sin 2\alpha$, for a wing with elliptic planform and equipped with a symmetric profile, has shown that this approach gives excellent results for the complete range of $\alpha's$, from zero to $\frac{\pi}{2}$ [12]. The artificial viscosity coefficient is given by

$$\mu_j \geq \sup\left\{-\frac{c_j}{4q_j}\frac{dC_{lj}}{d\alpha}\left((1+u_j)(c_{j,j} - c_{j-1,j}) - (\frac{y_j}{adv} + w_j)(a_{j,j} - a_{j-1,j} + \widetilde{a}_{j,j})\right); 0\right\} \quad (10.94)$$

and the equation now reads

$$\left\{1 + \frac{c_j}{2q_j}\frac{dC_{lj}}{d\alpha}\left((1+u_j)(c_{j,j} - c_{j-1,j}) - (\frac{y_j}{adv} + w_j)(a_{j,j} - a_{j-1,j} + \widetilde{a}_{j,j})\right) + 2\mu_j\right\}\frac{\Delta\Gamma_j}{\omega}$$

$$= -\frac{1}{2}q_j c_j C_l(\alpha_j) - \Gamma_j^n + \mu_j(\Gamma_{j+1}^n - 2\Gamma_j^n + \Gamma_{j-1}^{n+1}) \quad (10.95)$$

Now, the relaxation factor needs to be reduced to a value less than one, say $\omega = 0.3$, depending on the polar smoothness. The need for the smoothing term is highlighted with a calculation of the two-bladed NREL rotor [1] at $TSR = 3.8$. The S809 profile that equips the blade is represented by the viscous polar calculated with XFOIL [13], for which one finds $(\alpha)_{Cl\,max} = 17.5°$. At this low *TSR* the blade is stalled from $y = 0.35$ to almost $y = 0.6$. The blade elements working conditions for the converged solution obtained with the artificial viscosity term are shown in Fig. 10.15 with the profile polar.

The circulation and incidence distributions, although converged for $\mu = 0$, show large oscillations which are unphysical, Fig. 10.16.

10.5.3 Validation

A series of calculations has been performed with the design and analysis codes in order to assess their capabilities and performances.

First, the analysis code has been used to predict the power captured by the NREL rotor for a range of *TSR* values and compare the theoretical results with the experimental measurements performed in the NASA Ames 80′ × 120′ wind tunnel [1]. The range in wind velocities is from 5 to 20 m/s. The results are shown in Fig. 10.17, the large circles corresponding to the experiments and the small squares to the vortex model prediction. As can be seen, thc model predicts the power coefficient accurately

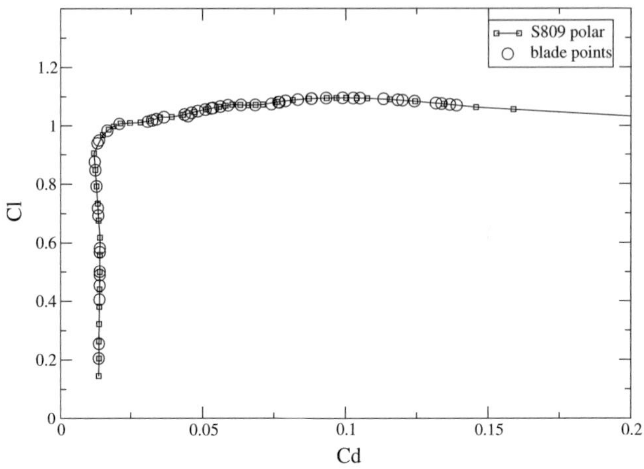

Fig. 10.15 Viscous polar at $Re = 500{,}000$ and blade working conditions at $TSR = 3.8$

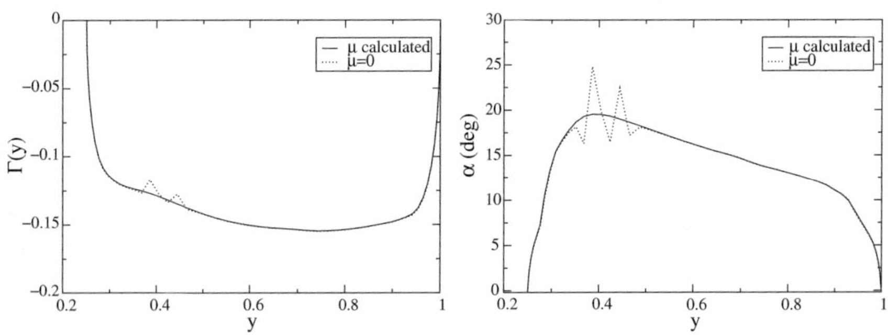

Fig. 10.16 Circulation and incidence distributions for the $S809$ blade at $TSR = 3.8$

for high *TSR* values, when the flow is attached, but for $TSR < 3.3$ much of the blade is stalled and the prediction becomes poor.

Secondly, for four values of the tip speed ratio, $TSR = 2.9, 3.3, 3.8, 4.6$, four rotors are designed with the optimization code each at the *TSR* and at the same C_T coefficient corresponding to the value of the NREL rotor thrust on the tower. These results are shown as large squares in Fig. 10.17. In all cases the optimum rotor improves the power capture of the rotor by a significant amount. Then, for each of these rotors, the geometry has been kept fixed and the tip speed ratio varies to investigate off-design performance. The dotted lines in Fig. 10.17 show the results of this study. It is clear that the optimum rotor does not perform well at high *TSR* compared to the NREL rotor. However it must be noted that the power captured in that area is small and the benefit is really for low *TSR*'s, because the power increases proportionally to V^3 and most of the energy resides in $TSR < 3.3$. It must be also

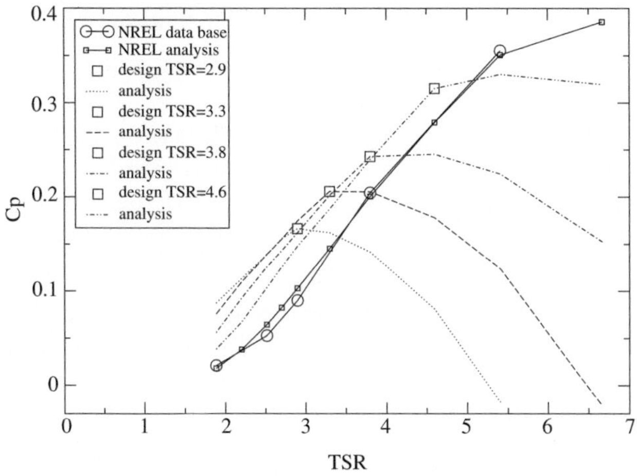

Fig. 10.17 Optimum rotors and analysis comparison with NREL rotor

noted that rotor blades have a variable pitch that can be used to compensate for the change in wind speed, a feature that has not been used in this study.

The analysis code performs well and is efficient and reliable when the flow is attached. This is consistent with the results obtained with the Prandtl lifting line theory applied to wings of large aspect ratios. When significant amount of separation occurs on a blade, the results are no longer accurate. The viscous effects tend to introduce strong blade-wise gradients that are not predictable with a 2-D viscous polar and strip theory. Attempts have been made at modifying the polar to account for these effects, using the experimental data as a guide, but the results remain limited to the particular rotor and wind velocity to which the viscous polar have been tuned. A more productive approach is the hybrid method, to be described later, which consists in using a Navier-Stokes solver to capture the viscous effects in the blade near field and the vortex model for the far field and boundary conditions of the Navier-Stokes code.

10.6 Unsteady Flow Simulation

The most commonly considered sources of unsteadiness in wind turbine flow are yaw and tower interference. Wind turbines can pivot on the tower to face the incoming wind. The control system is designed to correct for non zero yaw deviations, but yaw cannot be maintain to zero at all times. It is therefore important to understand the effect of yaw on the forces and moments in the blades in order to estimate fatigue life of the rotor structure subjected to this low frequency solicitations. For the same reason, it is necessary to estimate the rapid load changes that occur when a blade is passing in front of or behind the tower, as this is both a source of fatigue but also of noise.

For this analysis, we have to take into account the following new requirements:

- the blades have different loading and circulation distributions, $\Gamma_{i=1,j,n}$, $n = 1, 2$ *or* 3 for a two- or three-bladed rotor
- the circulation on the vortex sheets varies with both i and j along the vortex lattices
- the blades shed vorticity that is convected downstream with the flow
- the incoming flow contributes to the axial and azimuthal components in the rotating frame.

In other words, the circulation is no longer constant along a vortex filament. For this reason, each small element of the vortex lattice carries vorticity that varies in time and space and its individual contribution can no longer be accumulated with those along the same vortex filament. For a three-bladed rotor, the induced u-velocity component is calculated as

$$u_{j,1} = \sum_{k=1}^{jx-1} (\Gamma_{i,k+1,1} - \Gamma_{i,k,1}) a_{i,k,j,1} + \sum_{k=1}^{jx-1} (\Gamma_{i,k+1,2} - \Gamma_{i,k,2}) a_{i,k,j,2}$$
$$+ \sum_{k=1}^{jx-1} (\Gamma_{i,k+1,3} - \Gamma_{i,k,3}) a_{i,k,j,3} + \sum_{k=2}^{jx-1} (\Gamma_{i+1,k,1} - \Gamma_{i,k,1}) a_{i,k,j,4}$$
$$+ \sum_{k=2}^{jx-1} (\Gamma_{i+1,k,2} - \Gamma_{i,k,2}) a_{i,k,j,5} + \sum_{k=2}^{jx-1} (\Gamma_{i+1,k,3} - \Gamma_{i,k,3}) a_{i,k,j,6}$$
$$+ \sum_{k=2}^{jx-1} \Gamma_{1,k,2} \tilde{a}_{k,j,2} + \sum_{k=2}^{jx-1} \Gamma_{1,k,3} \tilde{a}_{k,j,3} \qquad (10.96)$$

where the first three terms correspond to the trailed vorticity influence of the vortex sheets of blades 1, 2 and 3 on blade 1, the next three terms are the contributions of the shed vorticity, and the last two tilde "~" terms correspond to the influence of the lifting lines 2 and 3 on blade 1. Similar formulae hold for blades 2 and 3 due to the symmetry in the geometry of the blades and vortex sheets. Similarly, the w-components influence coefficients are stored in arrays $c_{i,j,k,n}$, $n = 1, \ldots, 6$.

Note that in this approach, the vortex sheets are not displaced or distorted by yaw or tower interference, which limits the yaw angle to approximately 20° maximum. The vortex sheet is that which corresponds to zero yaw and no tower modelization and is called the "base helix". It is depicted in Fig. 10.18.

One of the nice features of this approach is that the flow becomes periodic shortly after the initial shed vorticity has crossed the Trefftz plane, that is for a dimensionless time $T \geq \frac{x_{ix}}{(1+2u)}$ or a number of iterations $n \geq \frac{x_{ix}}{(1+2u)\Delta t} = \frac{x_{ix}}{\Delta x}$.

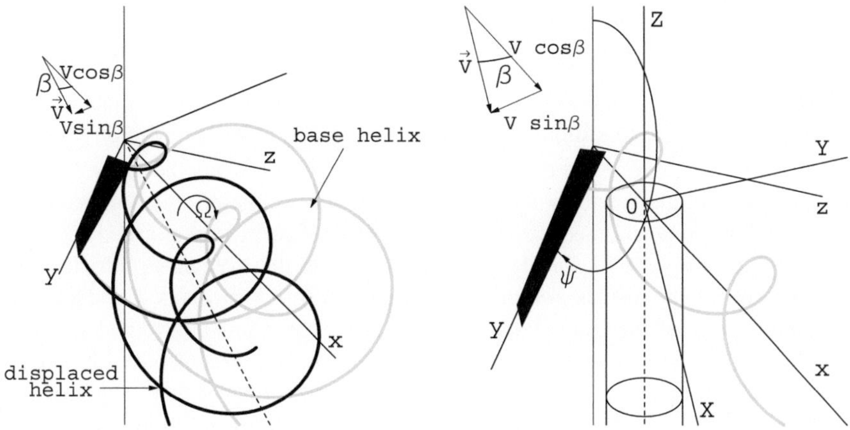

Fig. 10.18 Displaced helix, base helix and coordinate system for the treatment of unsteady flows

10.6.1 Effect of Yaw

The procedure to calculate the circulation at the blades is unchanged. Once the induced velocities are obtained, the local incidence of the blade is found and the steady viscous polar searched for lift and drag. The local blade element working conditions now depend on the yaw angle β and the azimuthal angle ψ measured to be zero when blade 1 is in the vertical position. The lift provides the circulation according to the Kutta-Joukowski lift theorem

$$\Gamma_{1,j,1} = \frac{1}{2} q_{j,1} c_j C_{l\,j,1} \tag{10.97}$$

where $q_{j,1}$ is the local velocity magnitude, c_j the chord and $C_{l\,j,1}$ the local lift coefficient for blade 1.

The new equation that governs the unsteady flow physics is the convection of the circulation along the wake

$$\frac{\partial \Gamma}{\partial t} + (1 + 2u)\frac{\partial \Gamma}{\partial x} = 0 \tag{10.98}$$

where $2u$ is the average axial velocity at the Trefftz plane as given by the actuator disk theory and is used from the rotor disk, but could be made to vary as is done for the pitch of the helix. This equation is solved with a two-point semi-implicit scheme:

$$\theta \frac{\Gamma_i^{n+\frac{\nu}{\nu+1}} - \Gamma_i^n}{\Delta t} + (1-\theta)\frac{\Gamma_{i-1}^{n+\frac{\nu}{\nu+1}} - \Gamma_{i-1}^n}{\Delta t}$$

$$+\theta(1+2u)\frac{\Gamma_i^{n+\frac{\nu}{\nu+1}}-\Gamma_{i-1}^{n+\frac{\nu}{\nu+1}}}{x_i-x_{i-1}}+(1-\theta)(1+2u)\frac{\Gamma_i^n-\Gamma_{i-1}^n}{x_i-x_{i-1}}=0 \quad (10.99)$$

This scheme is unconditionally stable for $\theta \geq \frac{1}{2}$ and reduces to Crank-Nicolson scheme for $\theta = \frac{1}{2}$. Here we use $\theta = \frac{1}{2}$. The index ν represents the inner iteration loop, needed to satisfy, within one time step, both the circulation equation inside the blades and the convection equation along the vortex sheets to a prescribed accuracy $\epsilon = 10^{-5}$. Let $\sigma = (1+2u)\Delta t/(x_i - x_{i-1})$ denote the Courant-Friedrich-Lewy (CFL) number. The distribution of points along the vortex sheets is stretched from the trailing edge of the blade, where $\Delta x_1 = 2.0 \times 10^{-3}$ to a constant step $\Delta x = (1+2u)\Delta t$ from approximately $x = 1$ to the Trefftz plane, such that the CFL number is one along that uniform mesh region. This feature, together with $\theta = \frac{1}{2}$ provides the nice property that the circulation is convected without dissipation or distortion all the way to the Trefftz plane (perfect shift property) [14].

10.6.2 Tower Interference Model

Two configurations are considered in this section, one in which the rotor is upstream of the tower and the other in which it is downwind of the tower. These situations are quite different from a physical point of view. In the former, the inviscid model is appropriate and the tower interference consists in a blockage effect due to the flow slowing down and stagnating along the tower. In the latter, the blade crosses the viscous wake of the tower, a region of rotational flow with low stagnation pressure, the extent of which as well as steadiness characteristics depend on the Reynolds number.

In the upwind configuration, the tower interference is modeled as a semi-infinite line of doublets aligned with the incoming flow. One can show that the velocity potential for such a flow is given by

$$\Phi(X,Y,Z) = \frac{Vr^2}{2}\frac{X}{X^2+Y^2}\left(1-\frac{Z}{\sqrt{X^2+Y^2+Z^2}}\right) \quad (10.100)$$

where $r = \overline{r}/R$ is the dimensionless tower radius. The coordinate system is aligned with the wind, the X-axis being aligned and in the direction of the incoming flow, the Z-axis being vertical oriented upward and the Y-axis completing the direct orthonormal reference system. See Fig. 10.18.

In the downwind configuration, an empirical model due to Coton et al. [15] is used. The above potential model of the tower is applied almost everywhere, but when a blade enters the narrow region $|Y| \leq 2.5\,r$ corresponding to the tower shadow, the tower induced axial velocity is set to $\Phi_X = -0.3V$ and the other two components are set to zero.

The velocity field induced by the tower is given by $\nabla\Phi = (\Phi_X, \Phi_Y, \Phi_Z)$. In the rotating frame, a point on blade 1 has coordinates $(0, y_j, 0)$. Let $d = \overline{d}/R$ be the dimensionless distance of the rotor from the tower axis. A negative value indicates upwind configuration, while a positive value indicates downwind case. In the fixed frame of reference, such a point has coordinates given by

$$\begin{pmatrix} X \\ Y \\ Z \end{pmatrix} = \begin{pmatrix} \cos\beta & -\sin\beta & 0 \\ \sin\beta & \cos\beta & 0 \\ 0 & 0 & 1 \end{pmatrix} \begin{pmatrix} d \\ y_j \sin\psi \\ y_j \cos\psi \end{pmatrix} \tag{10.101}$$

Using the same transformation, the tower-induced velocities at the blade can be calculated with the help of the components of the unit vectors in the rotating frame:

$$\begin{pmatrix} \vec{i} \\ \vec{j} \\ \vec{k} \end{pmatrix} = \begin{pmatrix} \cos\beta & \sin\beta & 0 \\ -\sin\beta\sin\psi & \cos\beta\sin\psi & \cos\psi \\ \sin\beta\cos\psi & -\cos\beta\cos\psi & \sin\psi \end{pmatrix} \begin{pmatrix} \vec{I} \\ \vec{J} \\ \vec{K} \end{pmatrix} \tag{10.102}$$

where $(\vec{i}, \vec{j}, \vec{k})$ and $(\vec{I}, \vec{J}, \vec{K})$ are the unit vectors in the rotating frame and the wind frame respectively. For blade 1 the induced velocities are $u_{j,1} = (\nabla\Phi.\vec{i})$ and $w_{j,1} = (\nabla\Phi.\vec{k})$. Application of this formula to the incoming flow $\vec{V} = (1, 0, 0)$ provides the axial and azimuthal components of the unperturbed flow in the rotating frame, again for blade 1

$$u = (\vec{V}.\vec{i}) = \cos\beta \tag{10.103}$$

$$w = (\vec{V}.\vec{k}) = \sin\beta\cos\psi \tag{10.104}$$

10.6.3 Validation

The helicoidal vortex model has been applied to a large number of cases, steady and unsteady, from the two-bladed rotor NREL S Data Sequence, in order to explore the domain of validity and assess the accuracy of this approach [16], see Table 10.1. In the experiments, the rotation speed of the turbine was held constant at 72 rpm, thus velocity was the main control parameter to vary the *TSR*. As velocity increases and *TSR* decreases, stall affects a larger and larger part of the blades. In steady flow conditions ($\beta = 0$), incipient separation occurs around 8 m/s and then spreads to the whole blade above 15 m/s. This is shown in Fig. 10.19 where the left graph indicates with dots the points along the blade where stall occurs. The right plot shows the steady power output comparison with the experimental data.

As can be seen, the agreement with the power output is excellent until $V = 8$ m/s, that is until there is separated flow on the blades.

Table 10.1 Flow velocities and yaw angles analyzed as part of code validation

Velocity (m/s)\Yaw angle (deg)	0	5	10	20	30	45
5	X	X	X	X	X	X
6	X		X		X	
7	X	X	X	X	X	
8	X		X		X	
9	X		X		X	
10	X	X	X	X	X	
11	X		X			
12	X					
13	X					
14	X					
15	X	X	X		X	
16	X					
17	X					
18	X					
19	X					
20	X					
21	X					
22	X					
23	X					
24	X					
25	X					

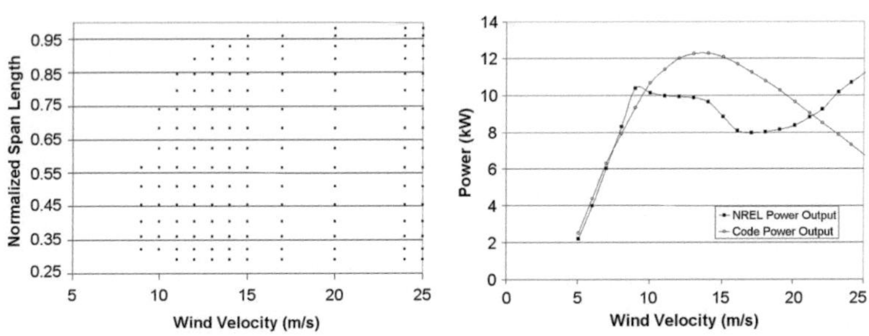

Fig. 10.19 Spanwise stall and steady power output versus wind speed

The effect of yaw on power is analyzed. The flow at yaw is periodic and the average power is calculated and compared with the average power given in the NREL data base. The results are shown in Fig. 10.20. Good results are obtained with the vortex model for yaw angles of up to $\beta = 20°$. This is consistent with the approximate treatment of the vortex sheets.

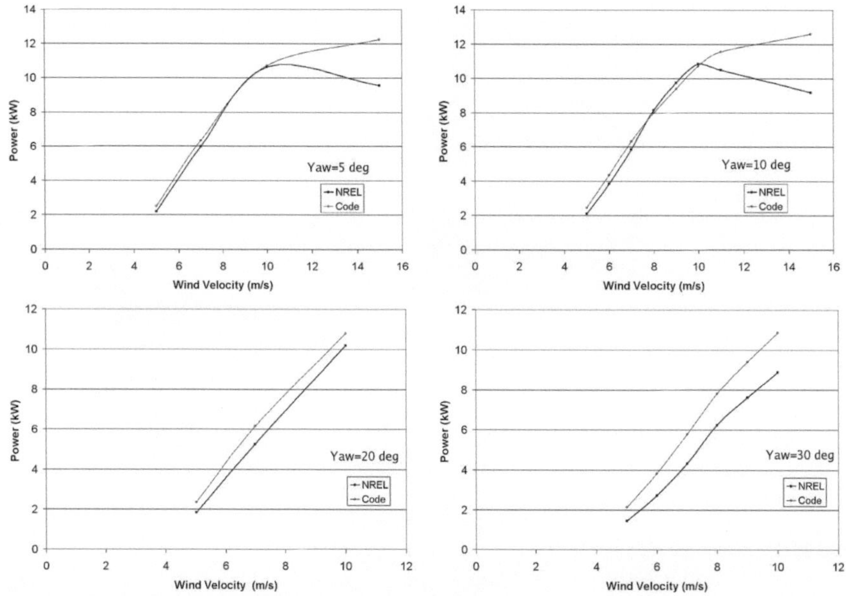

Fig. 10.20 Average power output versus wind speed at $\beta = 5, 10, 20$ and $30°$

10.7 Hybrid Method

Euler and Navier-Stokes solvers employ dissipative schemes to ensure the stability of the computation, either built into the numerical scheme or added artificially as a supplemental viscosity term. The dissipative property of those schemes works well with discontinuities such as shock waves, but has an adverse effect on discontinuities such as contact discontinuities and vortex sheets. It has been known for a long time that these schemes dissipate contact discontinuities associated with the shock-tube problem and vortex sheets trailing a finite wing, helicopter rotor, propeller and wind turbine. In the case of turbine flow, the influence of the vortex sheets is rapidly lost, which has an effect on the blade working conditions by not accounting for induced velocities contributed by the lost part of the wake.

On the other hand, the vortex model described earlier can maintain the vortex sheets as far downstream as needed, but cannot handle large span-wise gradients resulting in particular from viscous effects and separation.

In the rest of this section, we will present some results of the hybrid method which consists in coupling a Navier-Stokes solver with the vortex model, the former being limited to a small region surrounding the blades where the more complex physics can be well represented, and the latter being responsible to carry the circulation to the far field and impress on the Navier-Stokes' outer boundary the induced velocities

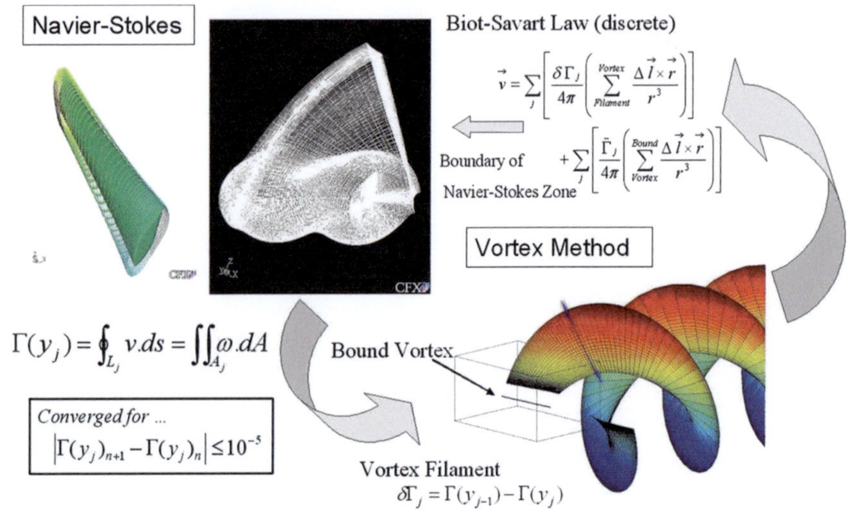

Fig. 10.21 Coupling methodology

accurately calculated by the Biot-Savart law. This new approach has been found to be both more accurate and more efficient than full domain Navier-Stokes calculations, by combining the best capabilities of the two models.

The coupling procedure is described in Fig. 10.21. It consists in calculating the Navier-Stokes solution with boundary conditions that only account for incoming flow and rotating frame relative velocities at the outer boundary of the domain. The solution obtained provides, by path integration, the circulation at each vortex location along the trailing edge of the blade and the Biot-Savart law allows to update the velocity components at the outer boundary of the Navier-Stokes region by adding the wake-induced contributions. This closes the iteration loop with a new Navier-Stokes solution. It takes approximately five cycles for the circulation to be converged with $|\Delta\Gamma| < 10^{-5}$.

As an ideal test case for the hybrid methodology, an optimized turbine blade of a two-bladed rotor was designed with the optimization vortex code and analyzed with the coupled solver at the same design conditions for which the flow is attached and viscous effects are minimum. The results of inviscid and viscous calculations are compared with the vortex code on the global thrust, bending moment, torque and power as well as spanwise distributions in Table 10.2. The global results are in excellent agreement with less than 1 % difference in power in the inviscid simulation and 7 % in the viscous one. The comparison is also excellent for torque and bending moment distributions as shown in Fig. 10.22 for inviscid and viscous cases.

The true benefit of the hybrid code resides in the simulation of the flow past a rotor when blades experience significant viscous effects that trigger large spanwise gradients. The strip theory assumption no longer holds and the flow becomes truly three-dimensional on the blade and cannot be accurately described by 2-D viscous

Table 10.2 Comparison of global results on test case

	Inviscid	Inviscid	Viscous	Viscous
	Vortex model	Hybrid solver	Vortex model	Hybrid solver
Thrust (N)	509.6	508.3	472.4	458.6
Tangential force (N)	183.6	179.9	163.3	150.8
Bending moment (N.m)	1803.1	1814.8	1670.2	1636.4
Torque (N.m)	588.8	583.8	519.6	485.5
Power (kW)	8.879	8.804	7.835	7.321

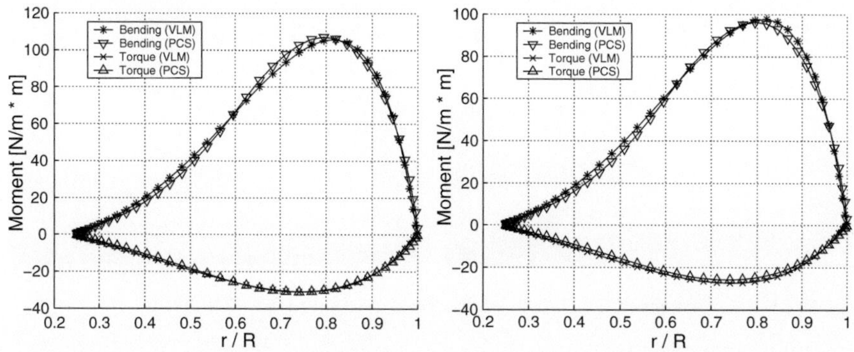

Fig. 10.22 Vortex model (VLM) and hybrid code (PCS) comparison: inviscid (*left*) and viscous (*right*) torque and bending moment distributions

polars. The NREL test results have been a remarkable resource for evaluation of code prediction capabilities, ranging from simpler, steady, attached flow cases at low speed, to more challenging separated flows at higher speeds, as well as unsteady flow in presence of yaw and tower interaction. The hybrid method has been very successful at capturing the formation of a separation bubble on the suction side of the blade at about 9 m/s, located at the 30 % blade span, that evolves into a "tornado vortex" rotating counter to the root vortex, see Fig. 10.23. The view is of the upper surface of the blade (suction side), with the blade root on the left and the blade tip on the right. At 7 m/s the flow is fully attached. The cross-section by a plane perpendicular to the wake, a short distance of the trailing edge, displays the root vortex rotating clockwise and tip vortex rotating counter-clockwise, visualized with the spanwise velocity component v. As the incoming velocity increases to 9 m/s, the wake of the blade, not visible before, now can be seen spanning the whole blade and a more defined structure appears at the 30 % location. The latter evolves into a strong vortex at 10 m/s whose counter-clockwise sense of rotation is opposite to that of the root vortex. At the higher speed of 11 m/s, the whole blade is stalled, yet the tornado vortex, as it has been called, has grown and has possibly produced another vortex at its lower right, rotating clockwise [17].

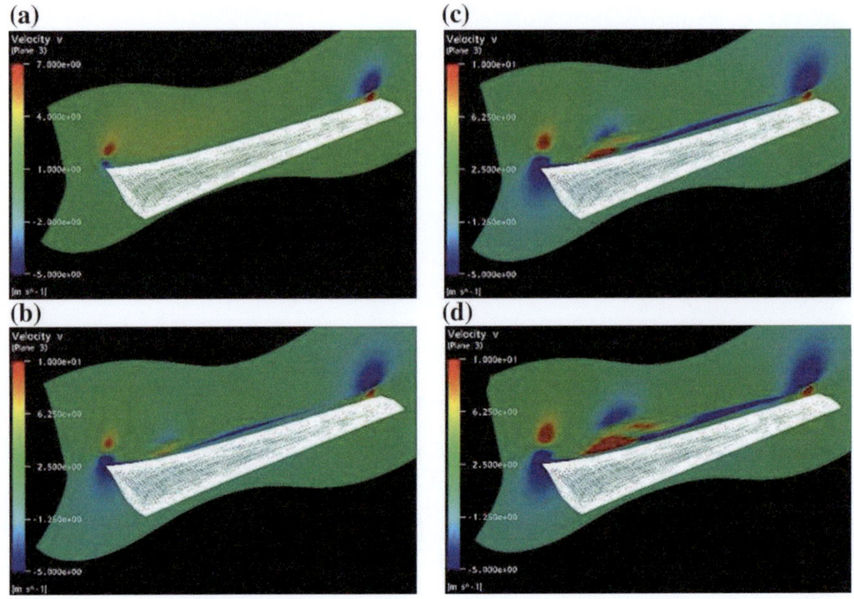

Fig. 10.23 Development of a well-defined viscous feature above the NREL blade

10.8 Perspectives

The design of blades is still a major task for manufacturers. Aerodynamics and structures are intimately associated in this process, to account for fabrication cost, transportation/installation, maintenance and fatigue life. Steady flow modelization remains the primary preoccupation of the designers, while unsteady effects contribute to the overall system performance assessment.

The importance of unsteadiness and its modelization in wind turbine aerodynamics is greater than ever before, in large part due to the larger size of the blades and their greater "softness" or flexibility. Advanced designs are contemplated that incorporate blade sweep and winglets with a view to improving the performance and power capture [18]. Composite materials allow for these increases in size and complexity, but also decreased stiffness. The question of fatigue has therefore become more relevant as the blades are subject to more vibrations, bending and twisting.

The vortex model has the potential of providing useful capabilities for the simulation of aeroelastic phenomena. Some work has already been carried out in this direction [19]. More work needs to be done to obtain a realistic simulation tool that is cheap and reliable, and operates directly in the time domain, in contrast to the many approaches based on eigen-frequencies and eigen-modes in the frequency domain with highly simplified aerodynamics. The hybrid method would be a natural candidate for the extension to fluid/structure interaction when viscous effects are important.

10.9 Propellers

The actuator disk theory, applied to a propeller, yields the same results for the thrust and the power coefficients, except that the fluid accelerates behind the propeller so that the inductance is now positive, $u_2 = 2u \geq 0$ and the thrust coefficient is negative, while the power coefficient is positive

$$C_T = \frac{2T}{\rho U^2 A} = -4\left(1 + \frac{u}{U}\right)\frac{u}{U}, \quad C_P = \eta = \frac{2P}{\rho U^3 A} = 4\left(1 + \frac{u}{U}\right)^2 \frac{u}{U} \tag{10.105}$$

This is shown in Fig. 10.24. Note that the thrust and power vary monotonically with u, and do not have extremum values. The vortex model for the analysis of a propeller is identical to that of the wind turbine. The difference resides in the fact that the engine or motor power is given: typically, the engine characteristic is given, i.e. power versus rpm, and can be quite different for a gas engine and an electric motor. The characteristic curve for an electric motor is linear, whereas, for a gas engine it looks like the semi-cubic in Fig. 10.25. For a given geometry, at a given advance ratio, the power coefficient p_τ is sought by iteration, as with wind turbine analysis, until the values match at the rotor and the far wake. As a result, the power P and the

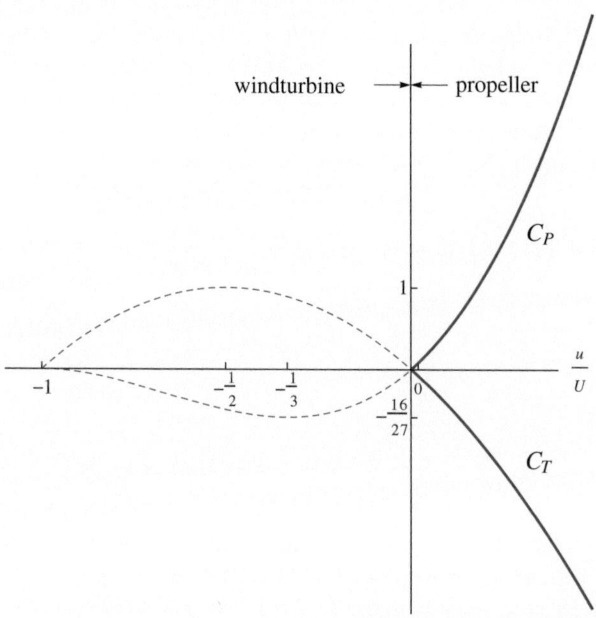

Fig. 10.24 Thrust and power coefficients versus axial induced velocity at the propeller rotor plane

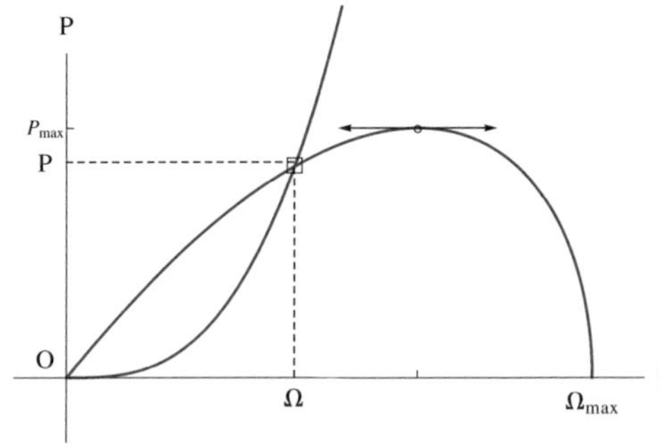

Fig. 10.25 Characteristic power versus rpm curve for a gas engine and matching with analysis result

rotational speed Ω are obtained, Fig. 10.25. The incoming flow velocity U and the induced axial velocity u, are then found.

The situation is slightly different for the design of a propeller, [20]. Given a rotor diameter, velocity and power (or torque) at a specified rpm, it is possible to design the rotor that will produce the maximum thrust (negative) in absolute value by minimization of the objective function

$$F(\Gamma) = C_T(\Gamma) + \lambda C_\tau(\Gamma) \tag{10.106}$$

p_τ is known and there is no need to iterate the vortex structure which is determined once for all.

Goldstein [7], found an analytical solution to the optimum propeller that satisfies the Betz condition for lightly loaded rotors. He showed that the solution to the partial differential equation and boundary conditions, in the ultimate wake, can be mapped onto a two-dimensional problem, using the geometrical invariance of the flow along helices of same pitch as the vortex sheets. In dimensionless form the problem reads

$$(1+\mu^2)\frac{\partial^2\phi}{\partial\xi^2} + \left(\mu\frac{\partial}{\partial\mu}\right)^2\phi = 0, \quad 0 \le \xi = \theta - \frac{x}{adv} \le \pi, \quad 0 \le \mu = \frac{r}{adv} < \infty \tag{10.107}$$

$$\frac{\partial\phi}{\partial\xi}(0\ or\ \pi, \mu) = -\frac{\mu^2}{1+\mu^2}, \quad 0 \le \mu \le \frac{1}{adv} \tag{10.108}$$

$$\mu\frac{\partial\phi}{\partial\mu} \to 0\ as\ \mu \to 0, \quad \phi \to 0\ as\ \mu \to \infty \tag{10.109}$$

where θ is the polar angle measured from the y-axis, in cylindrical coordinates about the x-axis. ϕ is the perturbation potential, made dimensionless with Uu_2/Ω, where u_2 is the induced axial velocity in the Trefftz plane. The solution in the (ξ, μ)-plane can be expanded in terms of modified Bessel functions.

The finite difference solution on a Cartesian mesh system, $\mu_j = (j-1)\Delta\mu$, $\xi_i = (i-1)\Delta\xi$, is readily obtained by relaxation with the five-point scheme for

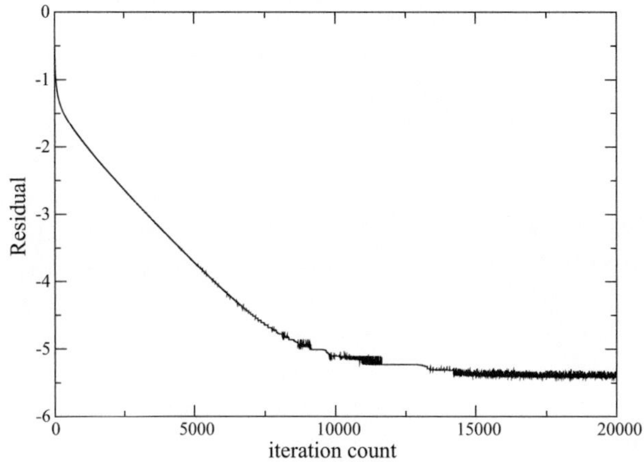

Fig. 10.26 Convergence history of the numerical solution of the Goldstein problem

interior points

$$(1+\mu_j^2)\frac{\phi_{i+1,j}-2\phi_{i,j}+\phi_{i-1,j}}{\Delta\xi^2}$$

$$+\frac{\mu_j}{\Delta\mu}\left(\mu_{j+\frac{1}{2}}\frac{\phi_{i,j+1}-\phi_{i,j}}{\Delta\mu}-\mu_{j-\frac{1}{2}}\frac{\phi_{i,j}-\phi_{i,j-1}}{\Delta\mu}\right)=0,\ i=2,\ldots,ix-1,\ j=2,\ldots,jx-1 \tag{10.110}$$

and boundary conditions

$$\frac{\phi_{2,j}-\phi_{1,j}}{\Delta\xi}=-\frac{\mu_j^2}{1+\mu_j^2},\quad \frac{\phi_{ix,j}-\phi_{ix-1,j}}{\Delta\xi}=-\frac{\mu_j^2}{1+\mu_j^2},\quad j=1,\ldots,j_1 \tag{10.111}$$

$$-3\phi_{i,1}+4\phi_{i,2}-\phi_{i,3}=0,\quad \phi_{i,jx}=0,\quad i=2,\ldots,ix-1 \tag{10.112}$$

$j_1, 1<j_1<jx$, is the index that denotes the tip of the blade located at $\mu=1/adv$. The convergence rate depends on the mesh aspect ratio $\Delta\mu/\Delta\xi$. With a mesh of $ix=51$ and $jx=201$, extending to $\mu_{max}=2/adv$, the convergence history for a relaxation factor $\omega=0.5$ is shown in Fig. 10.26. The circulation is obtained as the jump in ϕ across the sheet

$$\Gamma_j=\phi_{1,j}-\phi_{ix,j},\quad 1\le j\le j_1 \tag{10.113}$$

The normalized circulation $\bar{\Gamma}\Omega/(\pi U u_2)$, for a two-bladed rotor and for a sequence of advance ratios, is compared with the analytic solution of Goldstein in Fig. 10.27.

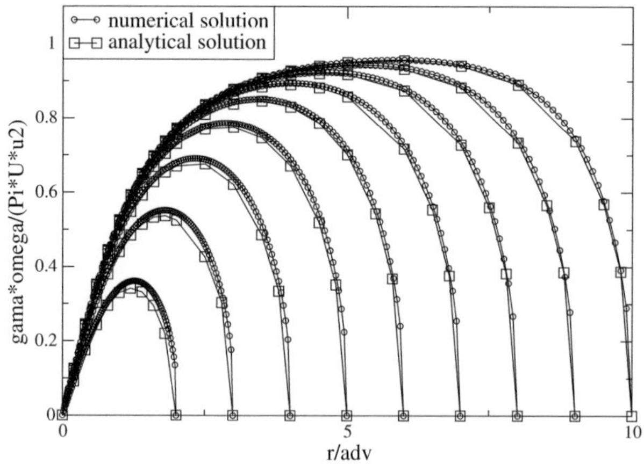

Fig. 10.27 Circulation distribution—comparison with Goldstein analytic solution

10.10 Problems

10.10.1

Using the result of the actuator disk theorems that the average induced velocity at the Trefftz plane is twice that at the rotor, $u_2 = 2u$, show that the application of Bernoulli's equation upstream and downstream of the disk provides the thrust force on the rotor as $T = -(p^+ - p^-)A$ where $p^+ - p^- =< p >$ represents the pressure jump across the actuator disk. Conclude.

10.10.2

As shown in Fig. 10.6, make sketches of the 1-D stream tubes and sketch the blade element operating conditions in the rotating frame as in Fig. 10.11, in the following cases (assume symmetric airfoil profile for simplicity):

(i) $\frac{u}{U} = 0$
(ii) $\frac{u}{U} = -\frac{1}{2}$
(iii) $\frac{u}{U} > 0$

and indicate for each case if the flow situation corresponds to propeller, freewheeling or turbine.

10.10.3

By application of the Biot-Savart formula, show that the axial velocity along the x-axis, induced by the vortex sheets in the Trefftz plane, is twice that induced in the plane of the rotor. Hint: consider a single vortex filament $\eta = const.$ and assume a perfect helix with constant pitch such as given by

$$\begin{cases} y_v = \eta \cos\left(\frac{x}{adv_{ix}} + \varphi\right) \\ z_v = \eta \sin\left(\frac{x}{adv_{ix}} + \varphi\right) \end{cases}$$

10.10.4

Derive the remainders $\delta a_{j,k}$ and $\delta c_{j,k}$ given in Sect. 10.4.2.

10.10.5

Consider the scheme for the convection equation:

$$\frac{\partial \Gamma}{\partial t} + (1 + 2u)\frac{\partial \Gamma}{\partial x} = 0$$

given in paragraph 6.1, with $\theta = \frac{1}{2}$.

(i) Show that the scheme is unconditionally stable
(ii) Show that, on a uniform mesh $\Delta x = (1 + 2u)\Delta t$ the scheme has the "perfect shift" property.

Acknowledgments One of the authors (JJC), acknowledges that part of the material in this chapter was originally published in the International Journal of Aerodynamics, Ref. [21].

References

1. Hand, M.M., Simms, D.A., Fingersh, L.J., Jager, D.W., Cotrell, J.R., Schreck, S., Larwood, S.M.: Unsteady Aerodynamics Experiment Phase VI: Wind Tunnel Test Configurations and Available Data Campaigns, NREL/TP-500-29955 (2001)
2. Rankine, W.J.: Trans. Inst. Nav. Archit. **6**, 13 (1865)
3. Froude, R.E.: Trans. Inst. Nav. Archit. **30**, 390 (1889)
4. Betz, A.: Wind Energie und Ihre Ausnutzung durch Windmühlen. Göttingen, Vandenhoeck (1926)
5. Joukowski, N.E.: Travaux du Bureau des Calculs et Essais Aéronautiques de l'Ecole Supérieure Technique de Moscou (1918)
6. Prandtl, L., Betz, A.: Vier Abhandlungen zur Hydro- und Aero-dynamik, Selbstverlag des Kaiser Wilhelminstituts fur Strömungsforshung. Göttingen Nachr. Göttingen, Germany (1927)
7. Goldstein, S.: On the vortex theory of screw propellers. Proc. R. Soc. Lond. Ser. A **123**, 440–465 (1929)
8. Betz, A.: Schraubenpropeller mit geringstem Energieverlust, Nach der Kgl. Gesellschaft der Wiss. zu Göttingen, Math.-Phys. Klasse, pp. 193–217; reprinted in Vier Abhandlungen zur Hydro- und Aero-dynamik, by L. Prandtl and A. Betz, Göttingen, 1927 (reprint Ann Arbor: Edwards Bros. 1943), pp. 68–92 (1919)
9. Munk, M.M.: The Minimum Induced Drag of Aerofoils, NACA report. 121 (1921)
10. Chattot, J.-J.: Optimization of wind turbines using helicoidal vortex model. J. Sol. Energy Eng. Spec. Issue: Wind Energy **125**(4), 418–424 (2003)
11. Chattot, J.-J.: Computational Aerodynamics and Fluid Dynamics: An Introduction. Scientific Computation. Springer, Berlin (2004). ISBN 3-540-43494-1, Second Printing

12. Chattot, J.-J.: Analysis and design of wings and wing/winglet combinations at low speeds. Comput. Fluid Dyn. J., Spec. Issue, 13(3) (2004)
13. Drela, M.: XFOIL: an analysis and design system for low reynolds number airfoils. In: Mueller, T.J. (ed.) Low Reynolds Number Aerodynamics. Lecture Notes in Engineering, vol. 54, pp. 1–12. Springer, Berlin (1989)
14. Chattot, J.-J.: Helicoidal vortex model for steady and unsteady flows. Comput. Fluids **35**, 733–741 (2006)
15. Coton, F.N., Wang, T., Galbraith, R.A. McD: An examination of key aerodynamics modeling issues raised by the NREL blind comparison. AIAA paper no. 0038 (2002)
16. Hallissy, J.M., Chattot, J.-J.: Validation of a helicoidal vortex model with the NREL unsteady aerodynamic experiment. Comput. Fluid Dyn. J. Spec. Issue **14**(3), 236–245 (2005)
17. Schmitz, S., Chattot, J.-J.: Method for aerodynamic analysis of wind turbines at peak power. J. Propuls. Power **23**(1), 243–246 (2007)
18. Chattot, J.-J.: Effects of blade tip modifications on wind turbine performance using vortex model. Comput. Fluids **38**(7), 1405–1410 (2008)
19. Chattot, J.-J.: Helicoidal vortex model for wind turbine aeroelastic simulation. Comput. Struct. **85**, 1072–1079 (2007)
20. Chattot, J.-J.: Optimization of propellers using helicoidal vortex model. Comput. Fluid Dyn. J. **10**(4), 429–438 (2002)
21. Chattot, J.-J.: Wind turbine aerodynamics: analysis and design. Int. J. Aerodyn. **1**(3/4), 404–444 (2011). http://www.inderscience.com/jhome.php?jcode=ijad

Chapter 11
Glider and Airplane Design

11.1 Introduction

This chapter deals with simple aspects of airplane design and stability analysis which have been motivated by one of the authors (JJC) participation over the last 15 years, as faculty advisor of a team of undergraduate students, in the Society of Automotive Engineering (SAE) Aero-Design West competition. This is a heavy lifter competition. Each year SAE publishes new rules for the competition, enticing students to design a different airplane. In the "regular class", the power plant and engine fuel are imposed and the airplane must take-off within a 200 ft runway. Some other constraints change from year to year, such as maximum wing span, or maximum lifting surfaces area, etc. In the "advanced class", the engine displacement is specified, but the engine make and fuel are the team's choice. A take-off distance is either imposed or targeted with a flight raw score depending heavily on how close to 200 ft take-off occurs. A round counts when the airplane lands within a 400 ft marked area of the runway without having lost any part. Then, the payload is removed from the payload bay and is weighted to attribute points to the flight. During these years, basic prediction tools were developed to help students design and predict their airplane performance. They consist of three simple computer models, a rapid prototyping code, a ground acceleration code and a flight equilibrium code. Furthermore, each year, the SAE competition has served as an on-going project in the applied aerodynamics class, culminating in a design project, as last assignment for the quarter, that adheres to the SAE rules of that year. This was a source of motivation for the students as many of the team members of the Advanced Modeling Aeronautics Team (AMAT) are taking the class. Since 2006, a double element wing was used, which attracted a lot of interest and proved to produce a very high lift. This chapter describes the various models implemented, with a view to helping students with design and estimation of airplane performance. First, the aerodynamic model of a classical configuration is developed, with the main wing in the front and the tail behind it. This is an easier model to study than the "canard" configuration in which small lifting surfaces are placed near the front of the fuselage and the main wing is moved further back, due to the more complex interaction of the canard on the main wing.

J.J. Chattot and M.M. Hafez, *Theoretical and Applied Aerodynamics*,
DOI 10.1007/978-94-017-9825-9_11

11.2 Model of a Classical Wing/Tail Configuration

As starting point, one shall assume incompressible, steady, attached flow on the fuselage and the tail. The main wing, of medium to large aspect ratio, either uses the simple linear model presented below, or a more accurate model based on the Prandtl Lifting Line theory. When the latter model is not available, the linear model for the wing is used.

11.2.1 Linear Model for the Main Wing

Consider a rectangular, untwisted wing of span b_m and chord c_m, equipped with an airfoil of constant relative camber, d_m/c_m. The wing aspect ratio is $AR_m = b_m/c_m$. Let α be the geometric incidence, i.e. the angle between the incoming flow vector **U** and the fuselage axis $\mathbf{Ox_1}$ which serves as reference for the angles. The coordinate system attached to the airplane is composed of the $(\mathbf{Ox_1}, \mathbf{Oy_1}, \mathbf{Oz_1})$ axes , with $\mathbf{Ox_1}$ oriented downstream, $\mathbf{Oy_1}$ aligned with the right wing and $\mathbf{Oz_1}$completing the direct coordinate frame. Accounting for the induced incidence, assumed constant along the span (elliptic loading), and building up on the results from thin airfoil theory, the lift coefficient is given by

$$C_{Lm} = \frac{2\pi}{1+\frac{2}{AR_m}}\left(\alpha + t_m + 2\frac{d_m}{c_m}\right) \tag{11.1}$$

t_m is the setting angle of the wing on the fuselage. The linear model comprises the lift slope and the $\alpha = 0$ lift coefficient as

$$\frac{dC_{Lm}}{d\alpha} = \frac{2\pi}{1+\frac{2}{AR_m}}, \quad C_{Lm0} = \frac{2\pi}{1+\frac{2}{AR_m}}\left(t_m + 2\frac{d_m}{c_m}\right) \tag{11.2}$$

The reference area for the wing lift is the wing area $A_m = b_m c_m$. At small incidences, the friction drag is estimated with a flat plate formula

$$C_{Dm0} = \begin{cases} 2\frac{1.328}{\sqrt{Re_{cm}}}, & Re_{cm} < 5.0 \times 10^5 \\ 2\frac{0.031}{Re_{cm}^{1/7}}, & Re_{cm} > 5.0 \times 10^5 \end{cases} \tag{11.3}$$

where Re_{cm} is the Reynolds number based on the wing chord. The factor 2 accounts for the two sides of the wing. The reference area for the wing friction drag is A_m (Fig. 11.1).

The induced drag is given by the classical formula

$$C_{Dim} = \frac{C_{Lm}^2}{\pi e_m AR_m} \tag{11.4}$$

Fig. 11.1 Wright Flyer III, (from http://wright.nasa.gov/wilbur.htm)

e_m is the wing efficiency or efficiency factor. For a medium aspect ratio wing ($AR \simeq 10$) a value $e_m = 0.9$ is acceptable, but values that are higher or even larger than one are possible if winglets are added. The total drag of the main wing is thus $C_{Dm} = C_{Dm0} + C_{Dim}$.

The moment coefficient of the main wing about the origin of the coordinate system, located at the nose of the airplane, results also from thin airfoil theory as

$$C_{M,om} = C_{M,acm} - \frac{x_{acm}}{c_{am}} C_{Lm} = -\pi \frac{d_m}{c_m} - \frac{x_{acm}}{c_{am}} \left(\frac{dC_{Lm}}{d\alpha} \alpha + C_{Lm0} \right) \quad (11.5)$$

where $C_{M,acm} = -\pi \frac{d_m}{c_m}$ is the moment about the aerodynamic center of the wing (function of the mean camber), x_{acm} is the location of the quarter-chord of the wing and c_{am} is the mean aerodynamic chord (here $c_{am} = c_m$). The linear decomposition of the moment reads

$$\frac{dC_{M,om}}{d\alpha} = -\frac{x_{acm}}{c_{am}} \frac{dC_{Lm}}{d\alpha}, \quad C_{M,om0} = -\pi \frac{d_m}{c_m} - \frac{x_{acm}}{c_{am}} C_{Lm0} \quad (11.6)$$

The reference "volume" for the wing moment is $A_m c_{am}$.

11.2.2 Linear Model for the Tail

With the classical configuration, the tail is influenced by the downwash of the main wing. The downwash varies from w_w at the wing ($x = 0$) to $w_T = 2w_w$ in the Trefftz plane ($x = +\infty$). The coefficient of downwash, $k_i = w_t/|w_w|$ at the tail location is between -1 and -2. A study of the downwash induced by the vortex sheet of an elliptically loaded wing, provides a quantification of this effect, as shown in Fig. 11.2.

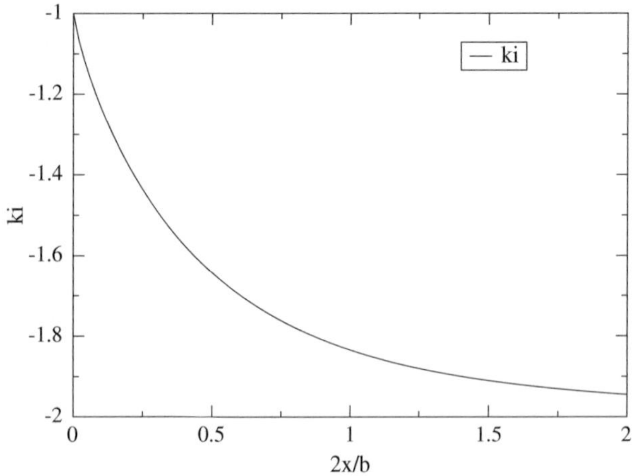

Fig. 11.2 Coefficient of downwash as a function of reduced distance from the main wing

One also assumes a rectangular tail with an elliptic loading. This is a valid approximation since the tail has a low aspect ratio $AR_t = b_t/c_t$ where b_t and c_t are the tail span and chord, respectively. The tail mean camber is d_t. The tail lift coefficient is given by

$$C_{Lt} = 2\pi \left(\alpha + \alpha_{im} + \alpha_{it} + t_t + 2\frac{d_t}{c_t} \right) \tag{11.7}$$

where α_{im} represents the induced incidence due to the wing downwash at the tail lifting line and α_{it} that due to the tail downwash on itself. Using the results of Prandtl Lifting Line

$$\alpha_{im} = k_i \frac{|w_w|}{U} = k_i \frac{C_{Lm}}{\pi AR_m}, \quad \alpha_{it} = \frac{w_t}{U} = -\frac{C_{Lt}}{\pi AR_t} \tag{11.8}$$

Hence, solving for C_{Lt} gives

$$C_{Lt} = \frac{2\pi}{1 + \frac{2}{AR_t}} \left(\alpha + \frac{k_i}{\pi AR_m} \left(\frac{dC_{Lm}}{d\alpha} \alpha + C_{Lm0} \right) + t_t + 2\frac{d_t}{c_t} \right) \tag{11.9}$$

The linear decomposition for the tail lift consists of

$$\frac{dC_{Lt}}{d\alpha} = \frac{2\pi}{1 + \frac{2}{AR_t}} \left(1 + \frac{k_i}{\pi AR_m} \frac{dC_{Lm}}{d\alpha} \right)$$

$$C_{Lt0}(t_t) = \frac{2\pi}{1 + \frac{2}{AR_t}} \left(t_t + 2\frac{d_t}{c_t} + \frac{k_i}{\pi AR_m} C_{Lm0} \right) \tag{11.10}$$

Note that the tail setting angle t_t, which controls the airplane flight operation, does not appear in the tail lift slope. The reference area for the tail lift is the tail area $A_t = b_t c_t$.

The tail drag is the sum of the friction drag C_{Dt0} and the induced drag C_{Dit} calculated with formulae similar to those used for the wing. The area of reference for the tail drag is A_t.

The moment coefficient for the tail can be decomposed as

$$\frac{dC_{M,ot}}{d\alpha} = -\frac{x_{act}}{c_{at}}\frac{dC_{Lt}}{d\alpha}, \quad C_{M,ot0}(t_t) = -\pi\frac{d_t}{c_t} - \frac{x_{act}}{c_{at}}C_{Lt0}(t_t) \tag{11.11}$$

Again, we note that the moment slope for the tail is independent of the tail setting angle.

11.2.3 Aerodynamic Coefficients for the Fuselage

According to slender body theory, the fuselage does not contribute to lift.

The drag coefficient, C_{Df0} is calculated with the flat plate formula at Reynolds number $Re_f = \rho U l_{ref}/\mu$, where l_{ref} is the fuselage length. The plate width is estimated to be h_f such that the fuselage wetted area is approximately $h_f l_{ref}$.

The fuselage moment coefficient is now obtained. In slender body theory, the lift of a truncated body of revolution, with base area $A = \pi R^2$ as reference area, is given by

$$C_L = 2\alpha \tag{11.12}$$

A truncated fuselage that extends from the nose to x will have a lift force given by

$$L = \frac{1}{2}\rho U^2 A(x)\, 2\alpha \tag{11.13}$$

A small section of the fuselage from x to $x + dx$ will have a lift force dL (see Fig. 11.3)

$$dL = \rho U^2 \alpha\, A'(x) dx \tag{11.14}$$

that will contribute to the pitching moment as

$$dM_{,of} = -x\, dL = -\rho U^2 \alpha A'(x)\, x dx \tag{11.15}$$

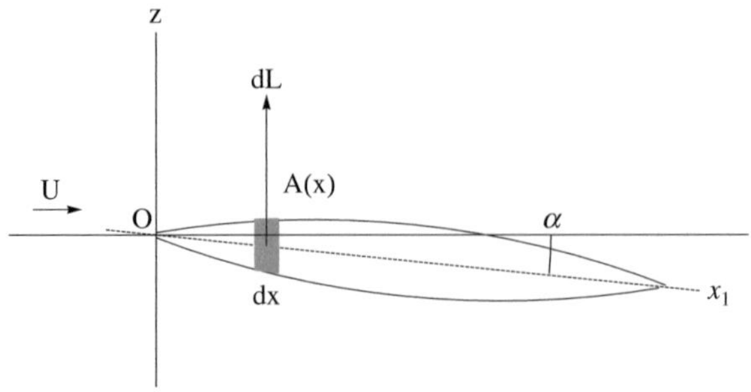

Fig. 11.3 Moment of a body of revolution at incidence

The minus sign is due to the convention of a positive nose up moment. The integration yields

$$M_{,of} = -\rho U^2 \alpha \int_0^{l_{ref}} A'(x)\, x dx = -\rho U^2 \alpha \left\{ [A(x)\, x]_0^{l_{ref}} - \int_0^{l_{ref}} A(x) dx \right\} \tag{11.16}$$

In the integration by parts, the first term cancels out since the fuselage has zero area at its extremities, and the integral term represents the fuselage volume ϑ_f. Hence the fuselage moment reduces to

$$M_{,of} = \rho U^2 \vartheta_f \alpha \tag{11.17}$$

In dimensionless form, using the maximum cross section area A_f as reference area, one gets

$$C_{M,of} = \frac{2\vartheta_f}{A_f l_{ref}} \alpha \tag{11.18}$$

Note that the fuselage moment slope $dC_{M,of}/d\alpha$ is positive, which is a destabilizing moment. In general, the fuselage contribution to the moment is small.

11.2.4 Global Aerodynamic Coefficients for the Glider

The aerodynamic coefficients for the complete configuration cannot be obtained by summing the individual coefficients of each component of the glider, since the reference areas and reference lengths are different. But forces and moments can be summed, and not accounting for the dynamic pressure, one can write

$$A_{ref} C_L = A_m C_{Lm} + A_t C_{Lt} \tag{11.19}$$

$$A_{ref} C_D = A_m C_{Dm} + A_t C_{Dt} + A_f C_{Df} \tag{11.20}$$

$$A_{ref} l_{ref} C_{M,o} = A_m c_{am} C_{M,om} + A_t c_{at} C_{M,ot} + A_f l_{ref} C_{M,of} \tag{11.21}$$

The global aerodynamic model for the lift and moment coefficients, is linear. If one chooses as area of reference the sum of the wing and tail areas, the global lift coefficient becomes

$$C_L = \frac{A_m C_{Lm} + A_t C_{Lt}}{A_m + A_t} \tag{11.22}$$

and the global moment coefficient

$$C_{M,o} = \frac{A_m c_{am} C_{M,om} + A_t c_{at} C_{M,ot} + A_f L_{ref} C_{M,of}}{(A_m + A_t) l_{ref}} \tag{11.23}$$

It is easy to see that the lift and moment slopes are thus combinations of the wing, tail and fuselage slopes and do not depend on the tail setting angle t_t. But the $\alpha = 0$ coefficients of lift and moment depend on t_t.

11.3 Rapid Prototyping

Rapid prototyping is a process by which the airplane main lifting component, the wing, can be sized and its weight estimated. For a composite wing, given the span and the chord, the weight depends on its volume, and the density of the foam used to build the wing core, and its surface area, and the density of the carbon fiber/glass fiber materials to make the wing skin. The weight of the other components, fuselage, tail and landing gear are also estimated, and by subtraction from the gross weight at take-off, the payload prediction is obtained and a flight raw score calculated. This code is a "cookie-cutter" for the wing as it estimates the best chord size for a given span and wing profile lift characteristics. The following assumptions are made to simplify as much as possible the analysis:

(i) rectangular wing with span b and chord c
(ii) elliptic circulation (the lower the aspect ratio AR, the more appropriate it is)
(iii) drag coefficient defined by

$$C_D = C_{D0} + \frac{C_L^2}{\pi e AR} \tag{11.24}$$

(iv) engine thrust $T(V) = T_0 - KV$
(v) rolling friction is neglected (although it had been implemented when taking-off from grass field)

The aspect ratio is defined as $AR = b^2/S$ where S is the wing area and reduces to $AR = b/c$ for a rectangular wing. The zero-lift drag coefficient C_{D0} is estimated from the airplane wetted area and parasitic drags (landing gear primarily). The efficiency factor, e, is close to unity ($e \cong 0.9$) for a wing of medium aspect ratio, $AR = 10$ or so, but could be closer or even larger than one with winglets. The lift coefficient C_L depends on the setting angle of the wing on the fuselage which determines the wing incidence during the rolling phase. The propeller is selected to match the engine and to produce a high static thrust T_0 in (N) at high nominal rpm, measured with a fish scale. As the airplane accelerates, the thrust decreases. For simplicity, we use the actuator disk theory to estimate $T(V)$ at low speeds. A more accurate approach would be to measure the thrust of the propeller placed in a wind tunnel or to use a propeller code (if the engine power were known as a function of rpm). Let U_b be the average axial induced velocity at the rotor, then power and thrust are given by

$$P = 2\pi\rho R^2(V + U_b)^2 U_b \tag{11.25}$$

$$T = 2\pi\rho R^2(V + U_b)U_b, \quad \Rightarrow T_0 = 2\pi\rho R^2 U_b^2 \tag{11.26}$$

Assuming that the power remains constant at low speeds, the condition $dP = 0$ reduces to

$$\frac{dU_b}{dV} = -\frac{2U_b}{V + 3U_b} \tag{11.27}$$

On the other hand, the change in thrust at low speeds reads

$$\frac{dT}{dV} = 2\pi\rho R^2\left(U_b + (V + 2U_b)\frac{dU_b}{dV}\right) = -2\pi\rho R^2\frac{V + U_b}{V + 3U_b}U_b \simeq -\frac{2\pi}{3}\rho R^2 U_b \tag{11.28}$$

Upon elimination of U_b one finds

$$K = \left|\frac{dT}{dV}\right| = \frac{\sqrt{2\pi}}{3}R\sqrt{\rho T_0} \tag{11.29}$$

with T_0 and K, the model can be further developed. Newton' second law reads

$$M\frac{d^2x}{dt^2} = M\frac{dV}{dt} = M\frac{dV}{dx}\frac{dx}{dt} = MV\frac{dV}{dx} = T_0 - KV - \frac{1}{2}\rho V^2 A_{ref} C_D \tag{11.30}$$

The last term corresponds to the total aerodynamic drag, i.e. viscous drag and induced drag, $D = \frac{1}{2}\rho V^2 A_{ref} C_D$. One can separate the variables and write

$$\frac{MVdV}{T_0 - KV - \frac{1}{2}\rho A_{ref} C_D V^2} = dx \tag{11.31}$$

Let V_1 and V_2 be the roots of the quadratic equation for V

$$V_{1,2} = \frac{-K \pm \sqrt{K^2 + 2\rho A_{ref} C_D T_0}}{\rho A_{ref} C_D} \tag{11.32}$$

The positive root, V_2, corresponds to the maximum velocity obtained as asymptotic limit over an infinite rolling distance (V_1 has no physical significance). The equation can now be written as

$$\frac{V dV}{(V - V_1)(V - V_2)} = \frac{1}{V_1 - V_2}\left(\frac{V_1}{V - V_1} - \frac{V_2}{V - V_2}\right) dV = -\frac{1}{2}\rho A_{ref} C_D \frac{dx}{M} \tag{11.33}$$

Upon integration from $V = 0$ for a finite rolling distance (say, $L = 55\,m = 180\,\text{ft}$ to allow for rotation before take-off) one finds

$$\frac{1}{V_1 - V_2}\left(V_2 \ln\left(\frac{V_2 - V}{V_2}\right) - V_1 \ln\left(\frac{V - V_1}{-V_1}\right)\right) = \frac{1}{2}\rho V^2 A_{ref} C_D \frac{L}{M} \tag{11.34}$$

This is an implicit equation for the take-off velocity V. It can be solved iteratively as

$$V^{(n+1)} = V_2\left[1 - \left(\frac{-V_1}{V^{(n)} - V_1}\right)^{\left(\frac{-V_1}{V_2}\right)} \exp\left\{-\frac{L}{MV_2}\sqrt{K^2 + 2\rho A_{ref} C_D T_0}\right\}\right] \tag{11.35}$$

where $n = 0, 1, \ldots$ is the iteration index. An initial guess is $V^{(0)} = V_2$.

Given a maximum lift coefficient C_{Lmax} for the wing at take-off, the drag D at take-off is known and the climb angle β can be calculated as

$$\beta = \frac{T(V) - D}{Mg} \tag{11.36}$$

The wing span being given (and changed if the rules allow), the code performs a loop on the wing chord by steps of 1 cm. For each chord size, the wing weight is calculated and the gross weight of the airplane estimated. The take-off velocity is then calculated, using the above iterative formula. Finally the climb angle is obtained and the loop on the chord is terminated when the climb angle is $\beta \geq 3°$. This value is chosen to allow the airplane to clear the runway, but also to account for uncertainties. The flight score is predicted from the payload.

Fine tuning of the configuration is carried out with the next level of modelisation.

11.4 Acceleration Phase

The acceleration phase is very critical as it determines the take-off velocity $V_{t.o.}$ at the point where the pilot rotates the airplane to give it the incidence needed to begin climbing. The velocity V depends primarily on the power plant and the mass of the airplane. To a lesser extent it depends on the parasitic drags, aerodynamic and rolling friction drags. This is the reason why a high wing lift coefficient C_{Lmax} is needed in order to lift the highest possible weight. These can be modeled in more details with the acceleration code as

$$M\frac{d^2x}{dt^2} = T[V] - \frac{1}{2}\rho\left(\frac{dx}{dt}\right)^2 A_{ref}(C_{D0} + C_{Di}) - a_k W_{app} \tag{11.37}$$

In this formulation, the thrust $T\,[V]$ can be a more elaborate function of the velocity, say from wind tunnel measurements, although the previous linear model can be used when more detailed data is not available. The zero-lift drag C_{D0} will vary with velocity via the Reynolds number $R_e = \rho V l_{ref}/\mu$ based on the air density and dynamic viscosity, the velocity and the fuselage length l_{ref}. In general we will assume turbulent flow on most of the wetted area of the airplane. Therefore the viscous drag will be estimated as

$$C_{D0}(R_e) = C_{Dfref}\left(\frac{V_{ref}}{V}\right)^{\frac{1}{5}} \tag{11.38}$$

The reference drag coefficient, C_{Dfref} is estimated using the friction drag of each streamlined component and tables for blunt elements such as the landing gear of the airplane. The reference velocity is chosen typically as $V_{ref} = 20\,\text{m/s}$. Accounting for the main wing, the tail, the fuselage and the landing gear, adding the different contributions with their corresponding reference areas, gives

$$A_{ref}C_{Dfref} = A_m C_{Dm} + A_t C_{Dft} + A_f C_{Dff} + A_g C_{Dfg} \tag{11.39}$$

The induced drag is primarily due to the main wing because of the high lift coefficient, even during the rolling phase. The induced drag of the tail is neglected. The induced drag reads as previously

$$C_{Di} = \frac{C_L^2}{\pi e AR} \tag{11.40}$$

The rolling friction is modeled with a coefficient of friction a_k as

$$F_r = a_k W_{app} \tag{11.41}$$

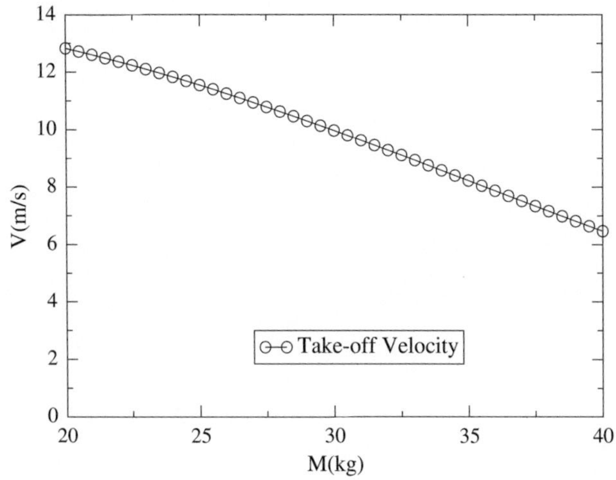

Fig. 11.4 Take-off velocity versus total airplane mass

a_k depends on the materials in contact and for rubber on asphalt a value $a_k = 0.03$ is used. The normal force acting on the tires is the apparent weight of the airplane, where

$$W_{app} = Mg - \frac{1}{2}\rho V^2 A_{ref} C_L \tag{11.42}$$

This model is transformed into a system of 2 first-order ordinary differential equations in 2 unknowns x and V as

$$\begin{cases} \frac{dx}{dt} = V \\ \frac{dV}{dt} = \left(T[V] - \frac{1}{2}\rho V^2 A_{ref}(C_{D0}[V] + C_{Di}) - a_k W_{app}[V]\right)/M \end{cases} \tag{11.43}$$

Integration is carried out with a 4th-order Runge-Kutta scheme and a time step $\Delta t = 0.01$ s. The total mass M is varied and a relation $V_{t.o.}(M)$ is obtained for the take-off velocity achievable for different airplane masses. This result is important as it will determine the feasibility of taking-off and flying a round when running the equilibrium model. For an open class entry, the results are presented in Fig. 11.4.

11.5 Longitudinal Equilibrium

In this model, we consider the airplane flying at uniform velocity along a straight trajectory contained in a vertical plane. This corresponds to a steady situation, with no acceleration. The roll and yaw are both zero. Two coordinate systems are introduced: the aerodynamic coordinate system in which the x-axis is aligned with the airplane

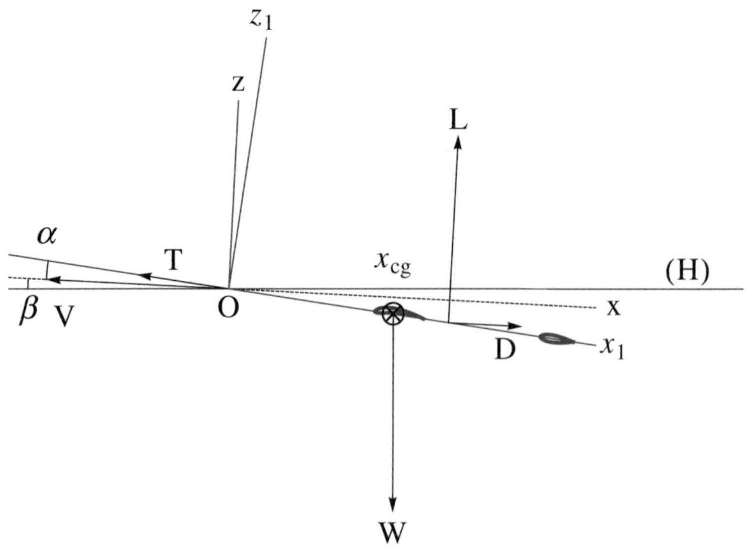

Fig. 11.5 Coordinate systems for the study of longitudinal equilibrium

velocity vector but oriented in the opposite direction and the z-axis upward; the other coordinate system is attached to the airplane, with the x_1-axis along the fuselage axis and the z_1-axis upward. The origin of the coordinate systems is placed at the nose of the airplane, see Fig. 11.5.

The incidence angle is the angle $\alpha = (Ox, Ox_1)$, the slope angle is $\beta = (H, Ox)$, both positive in the figure. The dimensionless coefficients, thrust, weight, lift, drag and moment coefficients are defined as

$$C_T = \frac{T(V)}{\frac{1}{2}\rho V^2 A_{ref}}, \quad C_W = \frac{W}{\frac{1}{2}\rho V^2 A_{ref}}, \quad C_L = \frac{L}{\frac{1}{2}\rho V^2 A_{ref}}$$

$$C_D = \frac{D}{\frac{1}{2}\rho V^2 A_{ref}}, \quad C_{M,o} = \frac{M,o}{\frac{1}{2}\rho V^2 A_{ref} l_{ref}} \tag{11.44}$$

The lift force includes contributions from the main wing and the tail. The latter could be positive or negative, depending on the flight conditions. Note that the fuselage does not contribute to lift, according to slender body theory. The drag is evaluated with the best estimations of zero-lift drag, using flat plate formula for the wetted areas of streamlines elements, except for the main wing, and drag tables for the landing gear. The wing viscous drag is included in the wing viscous polar: The 2-D viscous polar has been obtained for a range of incidences corresponding to attached and separated flows with the well-known program XFOIL (or MSES for multi-element airfoils) of Dr. Mark Drela; the 3-D polar is derived from it, using Prandtl Lifting

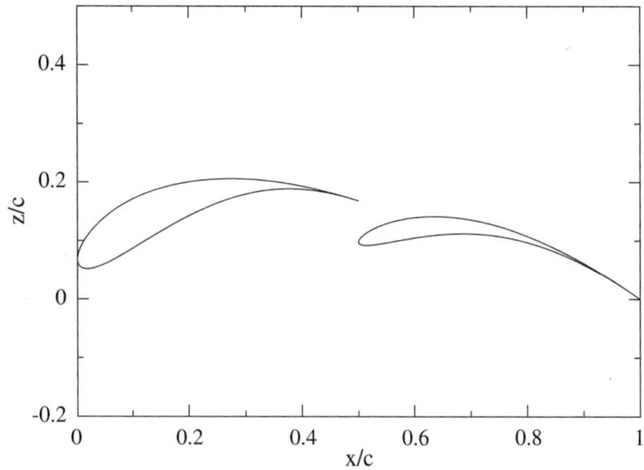

Fig. 11.6 Double element geometry

Line theory and the 2-D viscous polar at each span station. The 3-D polar is extended beyond stall. The induced drags of the main wing and the tail are included as well. The moment coefficient refers to the nose of the airplane. The geometry of the main wing double element profile is shown in Fig. 11.6, and the corresponding main wing 2-D and 3-D polars for the double element airfoil and wing (AMAT 2006) are shown in Fig. 11.7 and compared with the 2-D polar of the Selig1223 airfoil at Reynolds number 200,000. It is interesting to note the high maximum lift coefficient of the Selig 1223 with $C_{l,max} = 2.1$ at this low Reynolds number and the even larger value achieved by the double element airfoil with $C_{l,max} = 3.1$. Note that the 2-D viscous

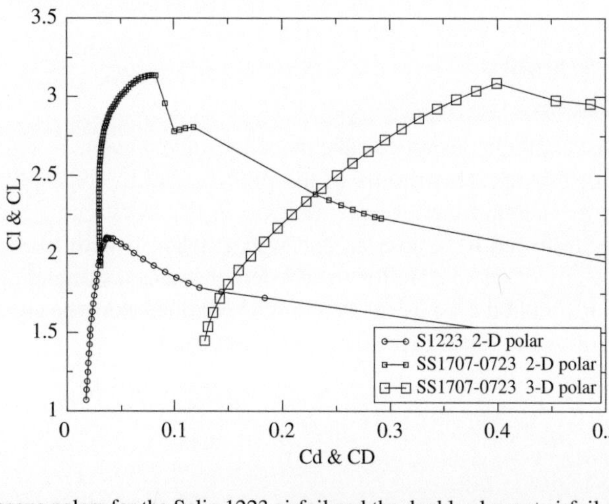

Fig. 11.7 Viscous polars for the Selig 1223 airfoil and the double element airfoil and wing

drag of the double element is comparable to that of the S1223 drag near maximum lift up to $C_l = 2.8$. Another noteworthy point to make is the large induced drag of the AMAT wing which is a direct result of the medium wing aspect ratio $AR = 8.75$, indicating that at $C_L = 2.8$, 88 % of the wing drag is induced drag.

The tail setting angle t_t is the parameter that controls the equilibrium solution (trimmed equilibrium).

The longitudinal equilibrium equations consist of two equations for the forces and one for the moment. They read

$$\begin{cases} C_L + C_T \sin(\alpha + \tau) - C_W \cos\beta = 0 \\ C_D - C_T \cos(\alpha + \tau) + C_W \sin\beta = 0 \\ C_{M,o} + \frac{x_{cg}}{l_{ref}} C_W \cos(\alpha + \beta) = 0 \end{cases} \tag{11.45}$$

Here, the propeller thrust makes an angle τ with the airplane axis. There are three unknowns, the equilibrium incidence, α_{eq}, velocity, V_{eq} and slope, β_{eq}. This is a highly nonlinear system, especially near the stall angle (maximum lift) since multiple values of α exist for a given C_L. A combination of fixed-point iterations until close enough to the solution, followed by Newton's iterations are needed to converge, when the solution exists, depending on the trim angle t_t.

The equilibrium code is used concurrently with the acceleration result $V_{t.o.}(M)$ to verify the feasibility of taking-off with a given total mass M. By "pulling on the stick", or equivalently, by rotating the tail flap in the upward direction, the equilibrium incidence increases and the equilibrium velocity decreases. If it is possible to find a tail setting angle such that $V_{eq} \leq V_{t.o.}$ take-off can be achieved provided $\beta \geq 3°$ and the incidence is not too close to the incidence of maximum lift. If all these conditions are fulfilled, it is possible to increase the mass of the airplane by adding payload weight thus increasing the possible score.

11.6 Static Stability

One of the most common design mistake made by student teams is a misplacement of the center of gravity. This results in the pilot having to constantly act on the tail deflection to maintain the airplane aloft or, in the worst case, ground rolling is immediately followed by a nose up and stall at take-off with likely loss of the airplane. The study of static stability clearly demonstrates a sufficient requirement for stability. For simplicity, we use a linear model obtained from the above model by assuming the following linear behavior of the coefficients

$$C_L(\alpha) = \frac{dC_L}{d\alpha}\alpha + C_{L0}(t_t), \quad C_{M,o}(\alpha) = \frac{dC_{M,o}}{d\alpha}\alpha + C_{M,o0}(t_t) \tag{11.46}$$

Here the lift and moment slopes are assumed constant as well as the $\alpha = 0$ lift and moment coefficients which depend only on the tail flap setting angle t_t. We further assume small angles α and β and neglect the thrust in the first equation, Eq. (11.45). The system reduces to

$$\begin{cases} C_L - C_W = 0 \\ C_D - C_T + \beta C_W = 0 \\ C_{M,o} + \frac{x_{cg}}{l_{ref}} C_W = 0 \end{cases} \tag{11.47}$$

Substitution of C_W from the first equation into the third one yields an equation for α_{eq}

$$C_{M,o}(\alpha_{eq}) + \frac{x_{cg}}{l_{ref}} C_L(\alpha_{eq}) = C_{M,cg}(\alpha_{eq}) = 0 \tag{11.48}$$

representing the moment about the center of gravity. This is easily solved with the linear model as

$$\alpha_{eq}(t_t) = -\frac{C_{M,o0}(t_t) + \frac{x_{cg}}{l_{ref}} C_{L0}(t_t)}{\frac{dC_{M,o}}{d\alpha} + \frac{x_{cg}}{l_{ref}} \frac{dC_L}{d\alpha}} \tag{11.49}$$

The equilibrium incidence is controlled by the tail setting angle.

The equilibrium velocity is obtained from the definition of C_W and the first equation, now that the lift $C_{L,eq} = C_L(\alpha_{eq})$ is known

$$V_{eq}(t_t) = \sqrt{\frac{W}{\frac{1}{2}\rho A_{ref} C_{L,eq}}} \tag{11.50}$$

Finally, knowing the velocity and incidence, the coefficients of drag $C_{D,eq}$ and thrust $C_{T,eq}$ can be calculated and the slope of the trajectory evaluated as

$$\beta_{eq}(t_t) = \frac{C_{T,eq} - C_{D,eq}}{C_{L,eq}} \tag{11.51}$$

Note that if $C_{T,eq} < C_{D,eq}$ the slope is negative and the trajectory is down. This is the case for a glider or when the engine is turned off since $C_T = 0$. This simple system also shows that for a glider, if one neglects the change in Reynolds number hence in $C_{D0}(Re)$, adding or subtracting mass at the center of gravity does not affect the equilibrium incidence α_{eq} nor the slope angle β_{eq}, and only affects the velocity on the trajectory which will change proportionally to $\sqrt{W}$.

Considering the moment of the aerodynamic forces at the center of gravity eliminates the action of the weight. It is easy to show that, for static stability, one needs to satisfy the inequality

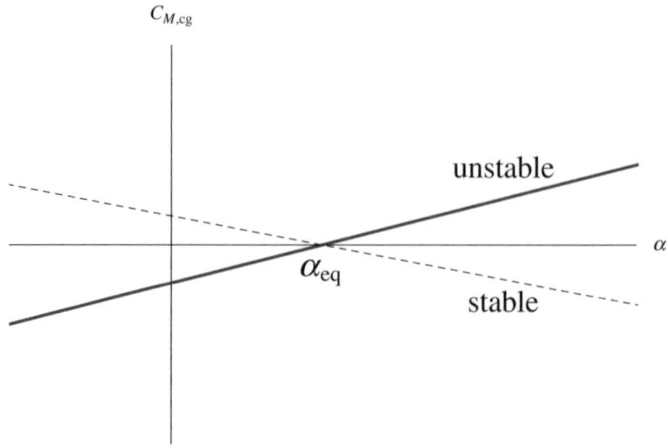

Fig. 11.8 Stable and unstable pitching moment curves

$$\frac{dC_{M,cg}(\alpha)}{d\alpha} < 0 \tag{11.52}$$

Indeed, if a perturbation, say a gust of wind, deviates the incidence from the equilibrium incidence by a $\Delta\alpha > 0$ or nose up, a negative pitching or nose down moment $\Delta C_{M,cg} < 0$ is needed to restore the equilibrium incidence, and vice-versa, Fig. 11.8.

Another important point for the aerodynamic static stability of an airplane is the aerodynamic center (or neutral point) of the configuration. It is located between the aerodynamic center of the main wing and the aerodynamic center of the tail, proportionally to their areas and lift slopes as given by the linear model (neglecting the small fuselage contribution)

$$\frac{x_{ac}}{l_{ref}} = \frac{A_m \frac{dC_{Lm}}{d\alpha}\frac{x_{acm}}{l_{ref}} + A_t \frac{dC_{Lt}}{d\alpha}\frac{x_{act}}{l_{ref}}}{A_m \frac{dC_{Lm}}{d\alpha} + A_t \frac{dC_{Lt}}{d\alpha}} \tag{11.53}$$

Note that its location is independent of t_t. It is defined as the point about which the moment of the aerodynamic forces is independent of incidence, i.e. it satisfies

$$\frac{dC_{M,o}}{d\alpha} + \frac{x_{ac}}{l_{ref}}\frac{dC_L}{d\alpha} = 0 \tag{11.54}$$

Taking the derivative of the moment equation about the center of gravity, $C_{M,cg}(\alpha)$ and satisfying the static stability inequality reads

$$\frac{dC_{M,o}}{d\alpha} + \frac{x_{cg}}{l_{ref}}\frac{dC_L}{d\alpha} < 0 \tag{11.55}$$

Fig. 11.9 UC Davis entry at the 2006 aero design west competition

Elimination of $dC_{M,o}/d\alpha$ between the two yields the following inequality for stability

$$\left(\frac{x_{cg}}{l_{ref}} - \frac{x_{ac}}{l_{ref}}\right)\frac{dC_L}{d\alpha} < 0 \tag{11.56}$$

Since the lift slope is positive, the condition reduces to

$$\frac{x_{cg}}{l_{ref}} - \frac{x_{ac}}{l_{ref}} < 0 \tag{11.57}$$

In other words, the center of gravity must be in front of the aerodynamic center. This is one of the most important design requirements, that was fulfilled as always in the model airplane of 2006, Fig. 11.9. The static margin defines the distance $x_{a.c.} - x_{c.g.}$ as a percentage of the reference length as

$$SM = 100\left(\frac{x_{ac} - x_{cg}}{l_{ref}}\right) \tag{11.58}$$

11.7 Winglet Design

Winglets are small aerodynamic surfaces, placed at the tip of wings, to improve their aerodynamic efficiency. Many different wingtip device geometries have been proposed, including raked tip, blended winglet, canted winglet, up/down winglet, spiroid, tip feathers, tip fence, etc., see Bertin and Cummings Ref. [1]. In a seminal 1921 paper "The Minimum Induced Drag of Aerofoil" [2], M.M. Munk, using inviscid flow and lifting line theory, showed that the most efficient winglets are placed at 90° from the main wing. Such winglets, in the upward position (placed on the suction side of the wing), have been studied by JJC in a recent paper [3], using inviscid flow including a viscous correction. The reader is referred to the paper for the theoretical details. An optimization/analysis code has been developed to help students design winglets. For a given dimensionless winglet height a_t (in reference to half span $b/2$) and target lift coefficient $C_{L,target}$, the optimization code calculates the optimum distribution of circulation along the wing span and the winglet, as a generalization of the lifting line theory to non-planar wings. The downwash is generalized as the normal wash, i.e. the component of the induced velocity normal to the dihedral shape of the wing. The 2-D polars for the main wing and the winglet can be different, corresponding to different profiles, with high camber for the wing and lower or no camber for the winglet (the choice made here). Once the circulation is known, a design can be selected since, as in inviscid flow, there is an infinite number of possible geometries that will produce the given circulation. One of the simplest strategies is to achieve a constant local lift coefficient C_l by designing the chord distribution according to the local circulation and the root chord or the local desired lift coefficient $C_l(\alpha_{opt})$ (corresponding to maximum C_l/C_d), as

$$c(s) = c(0)\frac{\Gamma(s)}{\Gamma(0)} \tag{11.59}$$

or

$$c(s) = \frac{2\Gamma(s)}{UC_l(\alpha_{opt})} \tag{11.60}$$

s is the curvilinear abscissa. For the main wing, α_{opt} represents the effective incidence, i.e. $\alpha_{opt} = \alpha + \alpha_i$, where α is the geometric incidence and α_i the induced incidence. On the winglets, which are equipped with a symmetric profile, $\alpha_{opt} = \beta = C_{l,winglet}/2\pi$, represents the toe-in angle, since the induced normal wash is zero. In Fig. 11.10, a comparison of the optimum distributions of circulation and normal wash is presented for a simple wing and a wing with 20 % winglets, for a given lift coefficient $C_L = 2.0$. The following comments can be made: the total lift is the same for both wings. The simple wing has an elliptic loading, with a high maximum root value, decreasing to zero at the tips, whereas the wing with winglets has a flatter distribution with lower root circulation and with higher loading at the winglet root. The downwash is constant for both wings, with a lower absolute value for the wing with winglets, which explains the lower induced drag given by

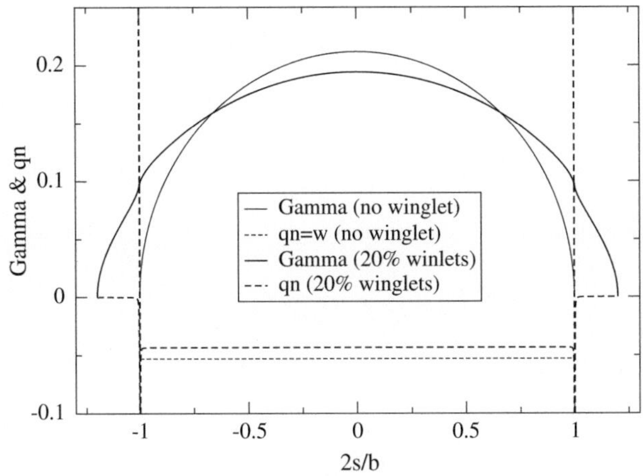

Fig. 11.10 Circulation and normal wash distributions for wings with and without winglets

$$A_{ref}C_{Di} = -\frac{1}{U^2}\int_{-\frac{S}{2}}^{\frac{S}{2}} \Gamma(s)q_n(s)ds \tag{11.61}$$

where $S = (1 + a_t)b/2$ represents the curvilinear length of the non planar wing and $q_n(s)$ is the normal wash such that $q_n(s) = w(s)$ along the main wing, $q_n(s) = v(s)$ on the left winglet and $q_n(s) = -v(s)$ on the right winglet. The winglets carry a load which does not contribute to lift, since lift is given by

$$A_{ref}C_L = \frac{2}{U}\int_{-\frac{S}{2}}^{\frac{S}{2}} \Gamma(s)\frac{dy}{ds}ds \tag{11.62}$$

where $y(s)$ represents the dihedral shape of the non planar wing, and $dy/ds = 0$ on the winglets. The winglets do not contribute to induced drag either since $q_n(s) = 0$ on the winglets. One should also note the flow singularity at the wing/winglet junction which shows as very large values of q_n and requires a fine mesh system to capture the solution (200 points are used in this calculation). The induced drag of the wing with winglet, $C_{Di} = 0.0868$, is compared to that of the simple wing, $C_{Di} = 0.106$, which provides an 18 % decrease in induced drag and corresponds to an efficiency factor $e = 1.22$. The winglet geometry is shown in Fig. 11.11, where the chord distribution is normalized with the winglet root chord c_{rw} and the winglet height with $a_t b/2$. The winglet corresponds to the area located below the curve, the root being the bottom edge of the box and the trailing edge the right edge of the box.

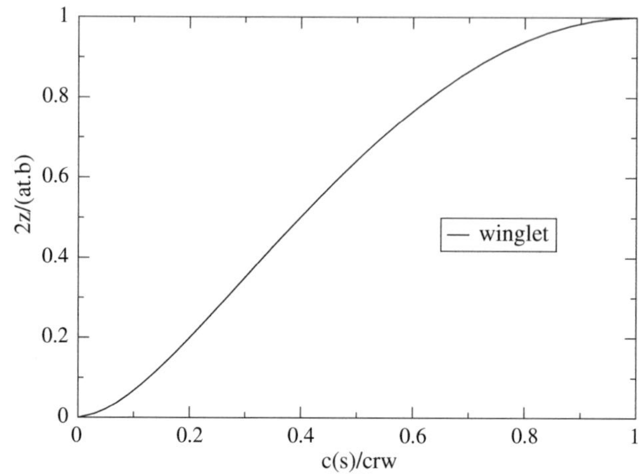

Fig. 11.11 Winglet geometry

11.8 Trimming the Glider for Maximum Distance

One of the glider design goal consists in achieving the largest distance on the ground, given an initial release point and altitude, assuming a perfectly quiet atmosphere. The situation is depicted in Fig. 11.12a. The glider is released from a height H and will land at point at a distance L. The slope of the trajectory is β, hence the objective is to minimize $|\beta| = D/L$. With the simple equilibrium model, and the drag being given by

$$C_D = C_{D0} + \frac{C_L^2}{\pi e AR} \tag{11.63}$$

it is equivalent to minimize D/L or C_D/C_L as

$$f(C_L) = \frac{C_D}{C_L} = \frac{C_{D0}}{C_L} + \frac{C_L}{\pi e AR},$$

$$\Rightarrow f' = -\frac{C_{D0}}{C_L^2} + \frac{1}{\pi e AR} = 0, \quad \Rightarrow C_L = \sqrt{\pi e AR\, C_{D0}} \tag{11.64}$$

so that the corresponding drag is $C_D = 2C_{D0}$. Geometrically, this corresponds to the point of contact of the tangent through the origin with the parabola, see Fig. 11.12b.

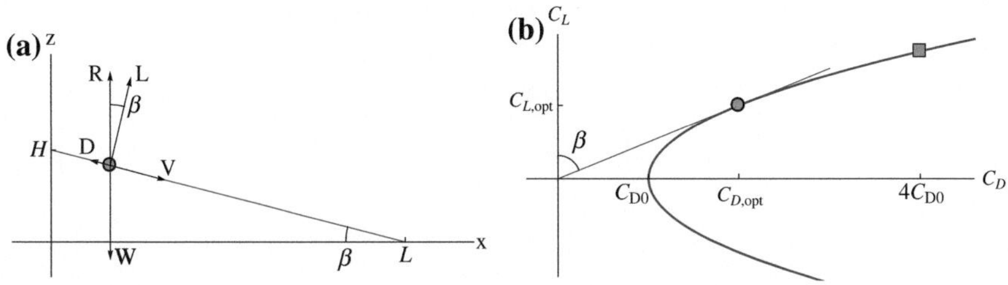

Fig. 11.12 **a** Sketch of glider trajectory; **b** optimum points on polar (not to scale)

11.9 Trimming the Glider for Maximum Duration

Another strategy for a glider is to stay airborne the longest time, or maximum duration. This is achieved by minimizing the speed of descend, i.e. for small angles, $|w| \cong V\,|\beta|$. As pointed out earlier, the equilibrium speed V is proportional to the square root of the weight $\sqrt{W}$, but also to the inverse square root of the lift coefficient $1/\sqrt{C_L}$. Hence $|w| \simeq C_D/C_L^{3/2}$. One obtains

$$f(C_L) = \frac{C_{D0}}{C_L^{3/2}} + \frac{C_L^{1/2}}{\pi eAR},$$

$$\Rightarrow f' = -\frac{3}{2}\frac{C_{D0}}{C_L^{5/2}} + \frac{1}{2}\frac{C_L^{-1/2}}{\pi eAR} = 0, \quad \Rightarrow C_L = \sqrt{3\pi eAR\, C_{D0}} \qquad (11.65)$$

and the corresponding value of drag is $C_D = 4C_{D0}$. This is represented by the square in Fig. 11.12b. Note that to change from maximum distance to maximum duration, the incidence is increased and the velocity decreased. Analysis of the results show that the ratio of the maximum duration time to the maximum distance time is $3^{3/4}/2 = 1.14$, a 14 % increase in time. The distance achieved is however reduced, in comparison to the maximum distance calculated above, in the ratio $\sqrt{3}/2 = 0.87$, a 13 % reduction.

11.10 Classical Versus Canard Configurations

A classical configuration consists of the main wing in the front and the tail in the back. In a "canard" configuration, the two lifting elements are reversed (see for example the Wright Brothers' flyer). The common wisdom is that in a classical configuration, the tail has a negative lift, whereas with a canard design, the forward control surfaces have positive lift. This makes the canard a desirable configuration for a heavy lift airplane. Many teams have used this design configuration, but only

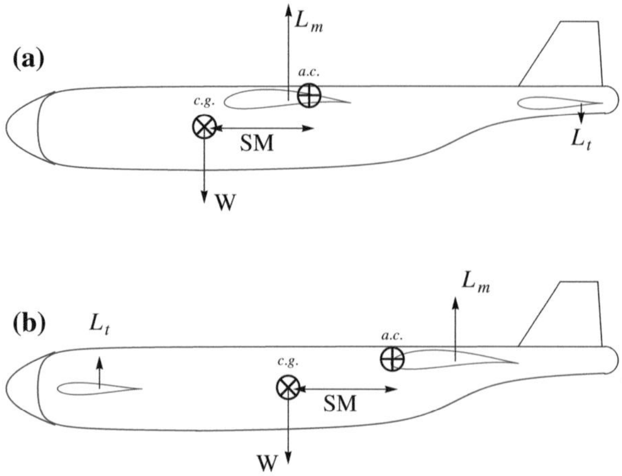

Fig. 11.13 **a** Classical versus **b** canard configurations

few have succeeded in having a stable airplane. See Fig. 11.13 for illustration of the difference in cruise. In Fig. 11.13a the moment of the main wing about the center of gravity is a nose down moment that needs to be balanced by a nose up moment from the tail. With the canard, Fig. 11.13b, the same situation requires a nose up moment from the forward lifting elements, which provide a positive lift.

However, it is possible to design a classical configuration in such a way that the tail will be lifting at take-off. The key point is to oversize the tail design, which is not a significant empty weight penalty, since it can be made out of light materials and it will lift more than its own weight. In this option, one requires that the center of gravity of the airplane be located aft of the main wing quarter-chord (main wing aerodynamic center). To do so, and still insure that the necessary static margin (SM) is preserved, the tail needs to be quite large, in order to move the aerodynamic center of the complete configuration sufficiently downstream. At take-off, when the airplane is fully loaded, the incidence is quite large, say 15° or so, and as we know from thin airfoil theory, the main wing lift force will move close to the main wing quarter-chord. As the airplane rotates to take-off, the lift force moves towards the quarter-chord and passes in front of the center of gravity which creates a nose up moment that must be balanced by the positive lift force of the tail, see Fig. 11.14a. But in cruise, the main wing lift force moves back past the center of gravity as the airplane speed has increased and the incidence decreased, requiring a negative lift force on the tail, Fig. 11.14b.

The equilibrium code is used to size the tail in order to achieve the above result. From the pilot point of view, flying such a configuration did not make any difference in handling qualities.

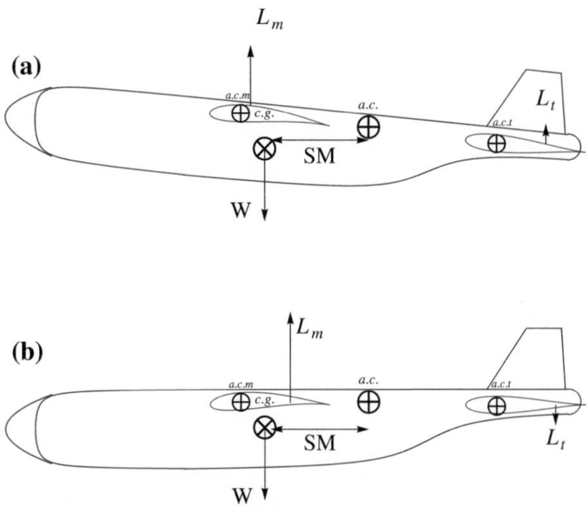

Fig. 11.14 Classical configuration with lifting tail take-off (**a**), and in cruise (**b**)

11.11 Conclusion

A simple approach to design a small remote control glider or airplane has been presented. It is based on a hierarchy of computer models, that help with the sizing of the wing, using a rapid prototyping code, estimating the take-off velocity with input and best estimate of rolling conditions, and finally, an equilibrium code that analyzes the equilibrium in various phases of the flight, from take-off to cruise and descent and provides the flight envelop in which the airplane is controllable with the tail as a function of the static margin. The design of winglets that contribute to improved efficiency by decreasing the induced drag, is also discussed along with trimming the glider for maximum distance or maximum duration. Lastly, it is shown that with a classic configuration, the wing and tail sizing and the center of gravity location can be found to make the tail a lifting tail at take-off, which is desirable for a heavy lifter airplane.

11.12 Problems

Aerodynamic Center of the Glider

Derive the formula

$$\frac{x_{ac}}{l_{ref}} = \frac{A_m \frac{dC_{Lm}}{d\alpha} \frac{x_{acm}}{l_{ref}} + A_t \frac{dC_{Lt}}{d\alpha} \frac{x_{act}}{l_{ref}}}{A_m \frac{dC_{Lm}}{d\alpha} + A_t \frac{dC_{Lt}}{d\alpha}}$$

from the linear aerodynamic model of Sect. 11.2.

Equilibrium Equation for the Moment

Show that the equilibrium equations can be combined to eliminate C_W from the moment equation to give

$$C_{M,o} + \frac{x_{cg}}{l_{ref}} (C_L \cos\beta - C_D \sin\beta + C_T \sin(\alpha + \beta + \tau)) \cos(\alpha + \beta) = 0$$

Global Aerodynamic Coefficients of a Glider

A glider has the following lift and moment characteristics in terms of the geometric angle of attack α (rd) and the tail setting angle t_t (rd):

$$C_L(\alpha, t_t) = 5.0\alpha + 0.5t_t + 0.5, \quad C_{M,o}(\alpha, t_t) = -1.3\alpha - 0.4t_t - 0.1$$

Find the effective aspect ratio AR of the glider (wing+tail) assuming ideal loading and

$$\frac{dC_L}{d\alpha} = \frac{2\pi}{1 + 2/AR}$$

Give the definition of the aerodynamic center.

Find the location $x_{a.c.}/l_{ref}$ of the aerodynamic center.

Find the location $x_{c.g.}/l_{ref}$ of the center of gravity, given a 6 % static margin of stability $SM = 6.0$.

Derive the expression for the moment about the aerodynamic center as a function of tail setting angle t_t, i.e. $C_{M,a.c.}(t_t)$.

Find an expression for the moment about the center of gravity $C_{M,c.g.}(\alpha, t_t)$.

Which condition holds for the moment about the center of gravity at equilibrium?

Find the equilibrium condition $\alpha_{eq}(t_t)$ and $C_{L,eq}(t_t)$ for the glider.

If the glider take-off lift coefficient is $C_L = 1.5$, find the equilibrium angle of attack α_{eq} and corresponding tail setting angle t_t.

Explain and sketch under which condition the tail can create positive lift at take-off.

Acknowledgments One of the authors (JJC), acknowledges that part of the material in this chapter was originally published in the International Journal of Aerodynamics, Ref. [4].

References

1. Bertin, J.J., Cummings, R.M.: Aerodynamics for Engineers, 5th edn., pp. 653–656 (2009)
2. Munk, M.M.: The Minimum Induced Drag of Aerofoils. NACA, Report 121 (1921)
3. Chattot, J.-J.: Low speed design and analysis of wing/winglet combinations including viscous effects. J. Aircr. **43**(2), 386–389 (2006)
4. Chattot, J.-J.: Glider and airplane design for students. Int. J. Aerodyn. **1**(2), 220–240 (2010). http://www.inderscience.com/jhome.php?jcode=ijad

Chapter 12
Introduction to Hypersonic Flows

In the previous chapters, compressibility effects are covered, particularly in the transonic regime where the Mach number is close to one. For very high Mach number flows, the situation is different, see Fig. 12.1. The details required for the analysis and design of hypersonic vehicles used in a space program, are both useful and interesting.

The study, in this chapter, is however very limited in its scope. it covers small disturbance theories, hypersonic similitude and similarity parameters as well as blunt body problems (with detached shocks) and theoretical and computational methods for both inviscid and viscous flows at very high velocities.

There are many references available in this field, old and new. For books, see [1–15], for review articles and proceedings, see [16–33] and for numerical methods and solutions, see [34–40]. The purpose here is to give an overview for the reader with emphasis on the main relevant concepts.

12.1 Regimes of Compressible Flow

For steady, inviscid and adiabatic flows, Bernoulli's law is

$$H = h + \frac{V^2}{2} = C \tag{12.1}$$

where C is constant along a streamline and $h = C_p T$ for a perfect gas.

The above relation can be rewritten in terms of the speed of sound ($a^2 = \gamma RT$) as

$$V^2 + \frac{2}{\gamma - 1} a^2 = const = V_{max}^2 \tag{12.2}$$

where V_{max} is the maximum possible velocity in the fluid (where the absolute temperature is zero) corresponding to the escape velocity when the fluid is expanded to vacuum (Fig. 12.1).

J.J. Chattot and M.M. Hafez, *Theoretical and Applied Aerodynamics*,
DOI 10.1007/978-94-017-9825-9_12

Fig. 12.1 Reentry capsule flow (from history.nasa.gov)

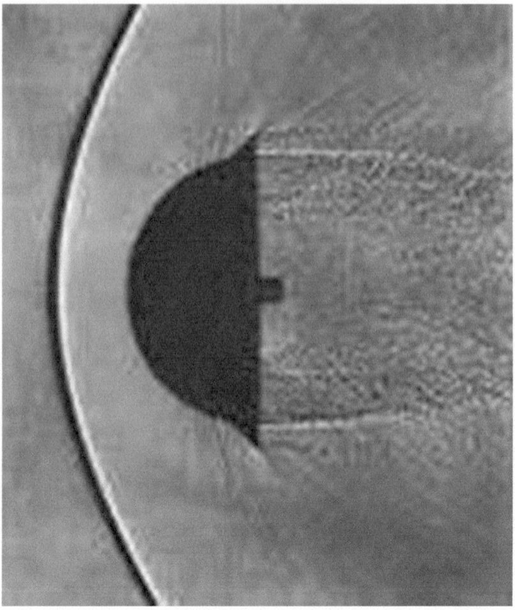

In the following sketch, the different regimes of compressible flow are identified and hence their physical characteristics are discussed, see Fig. 12.2.

(i) For incompressible flow $M \ll 1,\ V \ll a$ and $|\Delta a| \ll |\Delta V|$

Notice that

$$V\Delta V + \frac{2}{\gamma - 1} a\Delta a = 0 \tag{12.3}$$

Fig. 12.2 Regimes of compressible flow

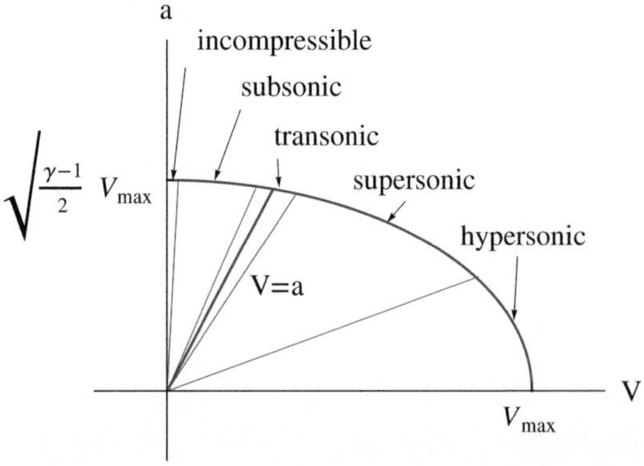

hence,

$$M = \frac{V}{a} = -\frac{2}{\gamma - 1}\frac{\Delta a}{\Delta V} \tag{12.4}$$

(ii) For subsonic flow $M < 1,\ V, a$ and $|\Delta a| < |\Delta V|$
(iii) For supersonic flow $M > 1,\ V > a$ and $|\Delta a| > |\Delta V|$
(iv) For transonic flow $M \simeq 1,\ V \simeq a$ and $|\Delta a| \simeq |\Delta V|$
(v) For hypersonic flow $M \gg 1,\ V \gg a$ and $|\Delta a| \gg |\Delta V|$

In hypersonic flow, changes in velocity are very small and the variations in Mach number are mainly due to changes in a.

12.2 Inviscid Theories of Hypersonic Flows

12.2.1 Hypersonic Potential Flow

In 1946, Tsien [41] derived similarity laws of hypersonic flows, following von Karman analysis of transonic similitude. Assuming isentropic, irrotational flow, Tsien used the full nonlinear potential equation with the tangency boundary condition at the solid surface (and not the linearized boundary condition). In the case of two-dimensions, let

$$u = U + \frac{\partial \varphi}{\partial x}, \quad v = \frac{\partial \varphi}{\partial y} \tag{12.5}$$

The governing equation for the perturbation potential is

$$\left(1 - \frac{u^2}{a^2}\right)\frac{\partial^2 \varphi}{\partial x^2} + 2\frac{uv}{a^2}\frac{\partial^2 \varphi}{\partial x \partial y} + \left(1 - \frac{v^2}{a^2}\right)\frac{\partial^2 \varphi}{\partial y^2} = 0 \tag{12.6}$$

where

$$a^2 = a_0^2 - \frac{\gamma - 1}{2}\left((u^2 + v^2) - U^2\right) \tag{12.7}$$

a_0 is the undisturbed speed of sound at flow velocity U.

For hypersonic flow, a_0 and $\partial\varphi/\partial x, \varphi/\partial y$ are small compared to U, hence retaining terms up to second order, the equation for φ becomes

$$\left(1 - (\gamma + 1)M_0\frac{1}{a_0}\frac{\partial \varphi}{\partial x} - \frac{\gamma - 1}{2}\frac{1}{a_0^2}\left(\frac{\partial \varphi}{\partial y}\right)^2 - M_0^2\right)\frac{\partial^2 \varphi}{\partial x^2} - 2M_0\frac{1}{a_0}\frac{\partial \varphi}{\partial y}\frac{\partial^2 \varphi}{\partial x \partial y}$$

$$+\left(1 - (\gamma - 1)M_0\frac{1}{a_0}\frac{\partial \varphi}{\partial x} - \frac{\gamma + 1}{2}\frac{1}{a_0^2}\left(\frac{\partial \varphi}{\partial y}\right)^2\right)\frac{\partial^2 \varphi}{\partial y^2} = 0 \tag{12.8}$$

Tsien [41] introduced the coordinate transformation

$$\xi = \frac{x}{c}, \quad \eta = \frac{y}{e} \tag{12.9}$$

where c is the chord and e is the thickness of the airfoil. He justified this choice since for hypersonic flow over a slender body, the variation of fluid velocity due to the presence of the body is limited to a narrow region close to the body.

The corresponding non-dimensional form for φ is

$$\varphi = \frac{a_0 c}{U} f(\xi, \eta) \tag{12.10}$$

The equation for f and boundary conditions are given by

$$\left(1 - (\gamma - 1)\frac{\partial f}{\partial \xi} - \frac{\gamma + 1}{2}\frac{1}{(M_0\tau)^2}\left(\frac{\partial f}{\partial \eta}\right)^2\right)\frac{\partial^2 f}{\partial \eta^2} = (M_0\tau)^2\frac{\partial^2 f}{\partial \xi^2} + 2\frac{\partial f}{\partial \eta}\frac{\partial^2 f}{\partial \xi \partial \eta} \tag{12.11}$$

$$as\ \xi \to -\infty \quad \frac{\partial f}{\partial \xi} = \frac{\partial f}{\partial \eta} = 0 \tag{12.12}$$

$$at\ the\ profile\ surface \quad \frac{\partial f}{\partial \eta} = (M_0\tau)^2\frac{d\bar{y}}{d\xi} \tag{12.13}$$

where $\tau = e/c$ and $\bar{y}$ is the dimensionless thickness distribution of a thin body at zero angle of attack.

In the above derivation, it is assumed that $\tau \ll 1$ and $M_0\tau$ is at least of $O(1)$.

It follows that the function $f(\xi, \eta)$ depends only on γ, K and $h(\xi)$, where $K = M_0\tau$ is the hypersonic similarity parameter.

From the exact isentropic relations, one obtains

$$\frac{p}{p_0} = \frac{\left(1 + \frac{\gamma-1}{2}M_0^2\right)^{\frac{\gamma}{\gamma-1}}}{\left(1 + \frac{\gamma-1}{2}M^2\right)^{\frac{\gamma}{\gamma-1}}} \tag{12.14}$$

hence, C_p in terms of f is given by

$$C_p = \frac{p - p_0}{\frac{1}{2}\rho_0 U^2} = \frac{2}{\gamma M_0^2}\left(\frac{p}{p_0} - 1\right) \tag{12.15}$$

$$\simeq \frac{2}{\gamma M_0^2}\left[\left(1 - (\gamma - 1)\frac{\partial f}{\partial \xi} - \frac{\gamma + 1}{2}\frac{1}{(M_0\tau)^2}\left(\frac{\partial f}{\partial \eta}\right)^2\right)^{\frac{\gamma}{\gamma-1}} - 1\right] \tag{12.16}$$

or

$$C_P = \tau^2 P(\gamma, K, \xi, \eta) = \frac{K^2}{M_0^2} P(\gamma, K, \xi, \eta) \tag{12.17}$$

Therefore

$$C_l = \frac{1}{c} \oint C_p dx = \frac{K^2}{M_0^2} L(\gamma, K) \tag{12.18}$$

$$C_d = \frac{1}{c} \oint C_p \frac{dy}{dx} dx = \frac{K^3}{M_0^3} D(\gamma, K) \tag{12.19}$$

Comparison with Ackeret's linearized supersonic theory, where

$$C_l \simeq \frac{\tau}{\sqrt{M_0^2 - 1}}, \quad C_d \simeq \frac{\tau^2}{\sqrt{M_0^2 - 1}} \tag{12.20}$$

shows that the linear theory fits in the hypersonic laws for $M_0 \gg 1$.

Hayes and Probstein [2] introduced the combined supersonic-hypersonic similarity law by replacing M_0 by $\sqrt{M_0^2 - 1}$, thus

$$K = \sqrt{M_0^2 - 1}\,\tau, \quad C_p = \frac{K^2}{M_0^2 - 1} P(\gamma, K, \xi, \eta) \tag{12.21}$$

It can be shown that the same $K = M_0 \tau$ is the similarity parameter for axisymmetric flows as well (see Tsien [41]).

12.2.2 Prandtl/Meyer Expansion for Hypersonic Flows

Recalling the incremental relation between turning angle and velocity changes in supersonic flow

$$\frac{d\theta}{\sqrt{M^2 - 1}} = -\frac{dV}{V} \tag{12.22}$$

and from Bernoulli's law

$$\frac{da}{a} = -\frac{\gamma - 1}{2} \frac{M dM}{1 + \frac{\gamma - 1}{2} M^2} \tag{12.23}$$

and using the definition of $M = V/a$, or

$$\frac{dV}{V} = \frac{dM}{M} + \frac{da}{a} \tag{12.24}$$

one obtains the celebrated relation

$$d\theta = -\frac{\sqrt{M^2-1}}{1+\frac{\gamma-1}{2}M^2}\frac{dM}{M} \tag{12.25}$$

For hypersonic flows, $M^2 \gg 1$, hence

$$d\theta = -\frac{2}{\gamma-1}\frac{dM}{M^2} \tag{12.26}$$

12.2.3 Hypersonic Flow over a Flat Plate at Angle of Attack

Linnell [42] obtained the lift coefficient using the Prandtl-Meyer expansion and oblique shock relations for high Mach number, where $K = M_0\alpha$.

For normal and oblique shock wave relations for high Mach numbers, and their limits as $M_0 \to \infty$, see Cox and Crabtree [3].

$$\frac{C_L}{\alpha^2} = \frac{C_D}{\alpha^3} = \frac{\gamma+1}{2} + \sqrt{\left(\frac{\gamma+1}{2}\right)^2 + \left(\frac{2}{K}\right)^2} + \frac{2}{\gamma K^2}\left[1 - \left(1 - \frac{\gamma-1}{2}K\right)^{2\gamma/(\gamma-1)}\right] \tag{12.27}$$

If $K \geq \frac{2}{\gamma-1}$, the last term is reduced to $\frac{2}{\gamma K^2}$.

The lift curves are parabolas and as $K \to \infty$, $C_L \to (\gamma+1)\alpha^2$.

For $M_0 < 3$, Ackeret theory gives reasonable results where $C_L = 4\alpha/\sqrt{M_0^2-1}$, and the lift curve is a straight line.

For profiles with thickness, Linnell reported the results for diamond airfoils at angle of attack.

For hypersonic flow, the pressure change across a shock of given turning angle is much greater the across an expansion wave of equal turning angle. To avoid a leading edge shock, it is desirable to operate with large angles of attack. Since most of the lift is furnished by the lower surface, it follows that expansion on the lower surface is undesirable and the best profiles have flat lower surfaces.

12.2.4 Power Series Expansion of Pressure Coefficients

Dorrance [43] obtained analytical formulas for two-dimensional profiles via expanding the expression of the pressure coefficient as power series in K where the change of flow angle $\Delta\omega$ is taken positive for compression and negative for expansion, hence for oblique shocks

$$\frac{C_p}{(\Delta\omega)^2} = \frac{\gamma+1}{2} + \sqrt{\left(\frac{\gamma+1}{2}\right)^2 + \left(\frac{2}{K}\right)^2} \simeq \frac{2}{K} + \frac{\gamma+1}{2} + \left(\frac{K+1}{4}\right)^2 + O(K^3) \tag{12.28}$$

For expansion fans, the expansion becomes

$$\frac{C_p}{(\Delta\omega)^2} = \frac{2}{K} + \frac{\gamma+1}{2} + \frac{\gamma+1}{6} K + O(K^2) \tag{12.29}$$

The first two terms of the two expressions are identical, the third terms differ slightly ($\simeq$10 %) for $K = O(1)$.

Dorrance worked out in a manner like that of the linear Ackeret theory, simple forms for different profile shapes. For symmetric double wedge, the results are

$$\frac{C_L}{\delta^2} = 2\frac{\alpha}{\delta}\left[\frac{2}{M_0\delta} + \frac{\gamma+1}{6} M_0\delta\left(3 + \left(\frac{\alpha}{\delta}\right)^2\right)\right] \tag{12.30}$$

$$\frac{C_D}{\delta^3} = \frac{4}{M_0\delta}\left(1 + \left(\frac{\alpha}{\delta}\right)^2\right) + \frac{\gamma+1}{3} M_0\delta\left[\left(\frac{\alpha}{\delta}\right)^4 + 6\left(\frac{\alpha}{\delta}\right)^2 + 1\right] \tag{12.31}$$

12.2.5 Bodies of Revolution

The applicability of hypersonic similarity for cylindrical bodies with conical or ogival noses is demonstrated in Ref. [44]. For fitness ratio

$$F = \textit{length of the nose section/maximum diameter of the nose section} \tag{12.32}$$

the similarity parameter, $K = M_0/F$

and for $F \geq 2$ and $M_0 \geq 2$, the validity of the hypersonic similarity was established.

Hypersonic Flow over Cones

For large M_0, the shock angle is close to the semi-vertex angle of the cone, θ, hence

$$C_p = \frac{2}{\gamma M_0^2}\left(\frac{p}{p_0} - 1\right) = (\gamma + 1)\,\theta^2 \tag{12.33}$$

therefore

$$\frac{p - p_0}{p_0} \simeq \frac{\gamma(\gamma+1)}{2} K^2 \tag{12.34}$$

12.2.6 Hypersonic Small Disturbance Theory

In 1954, Van Dyke [45] derived first order small disturbance equations for thin bodies moving at hypersonic speeds, where the hypersonic similarity law and Hayes' unsteady analogy appeared in the course of the development. He also derived a unified supersonic-hypersonic similarity rule valid for all speeds above the transonic zone, with strong shock waves and rotational motion effects included.

Starting with the Euler equations, representing conservation of mass, momentum and energy for steady perfect gas, in Cartesian coordinates, the problem is formulated as follows

$$\frac{\partial \rho u}{\partial x} + \frac{\partial \rho v}{\partial y} + \frac{\partial \rho w}{\partial z} = 0 \tag{12.35}$$

$$\frac{\partial \rho u^2}{\partial x} + \frac{\partial \rho uv}{\partial y} + \frac{\partial \rho uw}{\partial z} = -\frac{\partial p}{\partial x} \tag{12.36}$$

$$\frac{\partial \rho uv}{\partial x} + \frac{\partial \rho v^2}{\partial y} + \frac{\partial \rho vw}{\partial z} = -\frac{\partial p}{\partial y} \tag{12.37}$$

$$\frac{\partial \rho uw}{\partial x} + \frac{\partial \rho vw}{\partial y} + \frac{\partial \rho w^2}{\partial z} = -\frac{\partial p}{\partial z} \tag{12.38}$$

$$\frac{\partial \rho uH}{\partial x} + \frac{\partial \rho vH}{\partial y} + \frac{\partial \rho wH}{\partial z} = 0 \tag{12.39}$$

where

$$H = \frac{\gamma}{\gamma - 1}\frac{p}{\rho} + \frac{1}{2}\left(u^2 + v^2 + w^2\right) \tag{12.40}$$

with boundary condition

$$\mathbf{V}.\nabla B = 0, \quad at\ B = 0 \tag{12.41}$$

where $B(x, y, z) = 0$ represents the body shape.

The far field is uniform supersonic flow with $M_0 \gg 1$. The shock wave is described by $S(x, y, z) = 0$ and it is not known.

The shock jump conditions can be obtained from the weak solution of the conservation laws. Conservation of tangential momentum requires that the velocity component tangent to the shock surface be continuous. The tangential velocity component is

$$\mathbf{V}_t = (\mathbf{n} \wedge \mathbf{V}) \wedge \mathbf{n} \tag{12.42}$$

where $\mathbf{n}$ is the unit vector normal to the surface S, and

$$\frac{< u >}{\frac{\partial S}{\partial x}} = \frac{< v >}{\frac{\partial S}{\partial y}} = \frac{< w >}{\frac{\partial S}{\partial z}}, \quad at\ S = 0 \tag{12.43}$$

The component normal to the shock is

$$V_n = \mathbf{V}.\mathbf{n} = \frac{u\frac{\partial S}{\partial x} + v\frac{\partial S}{\partial y} + w\frac{\partial S}{\partial z}}{\sqrt{\left(\frac{\partial S}{\partial x}\right)^2 + \left(\frac{\partial S}{\partial y}\right)^2 + \left(\frac{\partial S}{\partial z}\right)^2}} \tag{12.44}$$

and the other jump conditions at $S = 0$ are

$$< \rho u > \frac{\partial S}{\partial x} + < \rho v > \frac{\partial S}{\partial y} + < \rho w > \frac{\partial S}{\partial z} = 0 \tag{12.45}$$

$$< \rho \frac{\left(u\frac{\partial S}{\partial x} + v\frac{\partial S}{\partial y} + w\frac{\partial S}{\partial z}\right)^2}{\left(\frac{\partial S}{\partial x}\right)^2 + \left(\frac{\partial S}{\partial y}\right)^2 + \left(\frac{\partial S}{\partial z}\right)^2} > + < p >= 0 \tag{12.46}$$

$$< H >= 0 \tag{12.47}$$

The second law of thermodynamics requires that entropy does not decrease across shock waves, hence

$$< \frac{p}{\rho^\gamma} >\geq 0, \quad at\ S = 0 \tag{12.48}$$

The body is assumed to be thin where the streamwise slope of its surface is everywhere small compared to unity. A small parameter τ is introduced which may be the thickness ratio or the angle of attack for inclined bodies.

From the tangency condition, all cross wind velocities are, in general, of order τ. From the approximate solution for thin plane wedge, the pressure coefficient and the streamwise velocity perturbation are of order τ^2.

Such considerations suggest introducing new independent and dependent variables as follows

$$\bar{x} = x, \quad \bar{y} = \frac{y}{\tau}, \quad \bar{z} = \frac{z}{\tau}, \quad \bar{B} = B(\bar{x}, \bar{y}, \bar{z}), \quad \bar{S} = S(\bar{x}, \bar{y}, \bar{z}) \tag{12.49}$$

$$u = U\left(1 + \tau^2 \bar{u}\right), \quad v = U\tau\bar{v}, \quad w = U\tau\bar{w}, \quad \rho = \rho_0\bar{\rho}, \quad p = \gamma M_0^2 p_0 \tau^2 \bar{p} \tag{12.50}$$

The above new quantities are dimensionless and of order unity for a body of unit length. The reduced first order equations in τ are

$$\frac{\partial \bar{\rho}}{\partial \bar{x}} + \frac{\partial \bar{\rho}\bar{v}}{\partial \bar{y}} + \frac{\partial \bar{\rho}\bar{w}}{\partial \bar{z}} = 0 \tag{12.51}$$

$$\frac{\partial \bar{\rho}\bar{u}}{\partial \bar{x}} + \frac{\partial \bar{\rho}\bar{u}\bar{v}}{\partial \bar{y}} + \frac{\partial \bar{\rho}\bar{u}\bar{w}}{\partial \bar{z}} = -\frac{\partial \bar{p}}{\partial \bar{x}} \tag{12.52}$$

$$\frac{\partial \bar{\rho}\bar{v}}{\partial \bar{x}} + \frac{\partial \bar{\rho}\bar{v}^2}{\partial \bar{y}} + \frac{\partial \bar{\rho}\bar{v}\bar{w}}{\partial \bar{z}} = -\frac{\partial \bar{p}}{\partial \bar{y}} \tag{12.53}$$

$$\frac{\partial \bar{\rho}\bar{w}}{\partial \bar{x}} + \frac{\partial \bar{\rho}\bar{v}\bar{w}}{\partial \bar{y}} + \frac{\partial \bar{\rho}\bar{w}^2}{\partial \bar{z}} = -\frac{\partial \bar{p}}{\partial \bar{z}} \tag{12.54}$$

$$\frac{\partial \bar{\rho}\bar{H}}{\partial \bar{x}} + \frac{\partial \bar{\rho}\bar{v}\bar{H}}{\partial \bar{y}} + \frac{\partial \bar{\rho}\bar{w}\bar{H}}{\partial \bar{z}} = 0 \tag{12.55}$$

with boundary conditions

$$\frac{\partial \bar{B}}{\partial \bar{x}} + \bar{v}\frac{\partial \bar{B}}{\partial \bar{y}} + \bar{w}\frac{\partial \bar{B}}{\partial \bar{z}} = 0, \quad at\ \bar{B} = 0 \tag{12.56}$$

or

$$\frac{\partial \bar{\rho}\bar{B}}{\partial \bar{x}} + \frac{\partial \bar{\rho}\bar{v}\bar{B}}{\partial \bar{y}} + \frac{\partial \bar{\rho}\bar{w}\bar{B}}{\partial \bar{z}} = 0 \tag{12.57}$$

and

$$\left\{ \begin{array}{l} \bar{u}, \bar{v}, \bar{w} \ \to 0 \\ \bar{p} \ \to \frac{1}{\gamma M_0^2 \tau^2} \\ \quad \bar{\rho} \ \to 1 \end{array} \right\} \quad as\ \bar{x}\ \to -\infty \tag{12.58}$$

Notice,

$$H = U^2\left(\frac{1}{2} + \tau^2 \bar{H}\right), \quad and \quad \bar{H} = \frac{\gamma}{\gamma - 1}\frac{\bar{p}}{\bar{\rho}} + \frac{1}{2}\left(2\bar{u} + \bar{v}^2 + \bar{w}^2\right) \tag{12.59}$$

The jump conditions at the shock wave become

$$\frac{< \bar{u} >}{\frac{\partial \bar{S}}{\partial \bar{x}}} = \frac{< \bar{v} >}{\frac{\partial \bar{S}}{\partial \bar{y}}} = \frac{< \bar{w} >}{\frac{\partial \bar{S}}{\partial \bar{z}}}, \quad at\ \bar{S} = 0 \tag{12.60}$$

$$< \bar{\rho} > \frac{\partial \bar{S}}{\partial \bar{x}} + < \bar{\rho}\bar{v} > \frac{\partial \bar{S}}{\partial \bar{y}} + < \bar{\rho}\bar{w} > \frac{\partial \bar{S}}{\partial \bar{z}} = 0 \tag{12.61}$$

$$< \bar{\rho}\frac{\left(\frac{\partial \bar{S}}{\partial \bar{x}} + \bar{v}\frac{\partial \bar{S}}{\partial \bar{y}} + \bar{w}\frac{\partial \bar{S}}{\partial \bar{z}}\right)^2}{\left(\frac{\partial \bar{S}}{\partial \bar{y}}\right)^2 + \left(\frac{\partial \bar{S}}{\partial \bar{z}}\right)^2} > + < \bar{p} >= 0 \tag{12.62}$$

$$\frac{\gamma}{\gamma - 1} < \frac{\bar{p}}{\bar{\rho}} > + \frac{1}{2} < \frac{\left(\frac{\partial \bar{S}}{\partial \bar{x}} + \bar{v}\frac{\partial \bar{S}}{\partial \bar{y}} + \bar{w}\frac{\partial \bar{S}}{\partial \bar{z}}\right)^2}{\left(\frac{\partial \bar{S}}{\partial \bar{y}}\right)^2 + \left(\frac{\partial \bar{S}}{\partial \bar{z}}\right)^2} >= 0 \tag{12.63}$$

$$< \frac{\bar{p}}{\bar{\rho}^{\gamma}} > \geq 0 \tag{12.64}$$

Notice also, that the parameters M_0 and τ of the full problem appear only in the upstream condition on $\bar{p}$, and not anywhere else, in the combination $M_0\tau$.

Based on the above formulation, ignoring all τ^2 terms, the hypersonic similarity is justified.

It is interesting to notice that the error of first order small disturbance theories are of $O\left(\tau^{2/3}\right)$, $O\left(\tau\right)$ and $O\left(\tau^2\right)$ for transonic, linearized supersonic and hypersonic flows respectively. A second order theory for hypersonic speeds is needed the least!

12.2.7 Unsteady Analogy

Van Dyke considered the shock as an internal boundary (shock fitting) and wrote the equations in non-conservative form for the smooth flow downstream of the shock wave.

$$\frac{\partial \bar{\rho}}{\partial \bar{x}} + \bar{\rho}\frac{\partial \bar{v}}{\partial \bar{y}} + \bar{v}\frac{\partial \bar{\rho}}{\partial \bar{y}} + \bar{\rho}\frac{\partial \bar{w}}{\partial \bar{z}} + \bar{w}\frac{\partial \bar{\rho}}{\partial \bar{z}} = 0 \tag{12.65}$$

$$\bar{\rho}\frac{\partial \bar{u}}{\partial \bar{x}} + \bar{\rho}\bar{v}\frac{\partial \bar{u}}{\partial \bar{y}} + \bar{\rho}\bar{w}\frac{\partial \bar{u}}{\partial \bar{z}} = -\frac{\partial \bar{p}}{\partial \bar{x}} \tag{12.66}$$

$$\bar{\rho}\frac{\partial \bar{v}}{\partial \bar{x}} + \bar{\rho}\bar{v}\frac{\partial \bar{v}}{\partial \bar{y}} + \bar{\rho}\bar{w}\frac{\partial \bar{v}}{\partial \bar{z}} = -\frac{\partial \bar{p}}{\partial \bar{y}} \tag{12.67}$$

$$\bar{\rho}\frac{\partial \bar{w}}{\partial \bar{x}} + \bar{\rho}\bar{v}\frac{\partial \bar{w}}{\partial \bar{y}} + \bar{\rho}\bar{w}\frac{\partial \bar{w}}{\partial \bar{z}} = -\frac{\partial \bar{p}}{\partial \bar{z}} \tag{12.68}$$

He also replaced the energy equation by the condition that the entropy is constant along streamlines (but not across the shock)

$$\frac{\partial}{\partial \bar{x}}\left(\frac{\bar{p}}{\bar{\rho}^{\gamma}}\right) + \bar{v}\frac{\partial}{\partial \bar{y}}\left(\frac{\bar{p}}{\bar{\rho}^{\gamma}}\right) + \bar{w}\frac{\partial}{\partial \bar{z}}\left(\frac{\bar{p}}{\bar{\rho}^{\gamma}}\right) = 0 \tag{12.69}$$

It is clear now, that the x-momentum equation for $\bar{u}$ is uncoupled from the others. The continuity, the y- and z-momentum equations and the entropy equation can be solved for $\bar{\rho}$, $\bar{v}$, $\bar{w}$ and $\bar{p}$, independently of $\bar{u}$. If required, $\bar{u}$ can be determined from Bernoulli's law

$$\frac{\gamma}{\gamma - 1}\frac{\bar{p}}{\bar{\rho}} + \frac{1}{2}\left(2\bar{u} + \bar{v}^2 + \bar{w}^2\right) = \frac{1}{(\gamma - 1)\, M_0^2\tau^2} \tag{12.70}$$

Hence, Van Dyke confirmed Hayes' observation that the reduced problem is completely equivalent to a full problem for unsteady flow in one less space dimension.

The latter is the unsteady motion in the $(\bar{y}, \bar{z})$ plane due to a moving solid boundary described by $\bar{B} = 0$, where $\bar{x}$ is interpreted as the time. For example, the steady hypersonic flow over a slender pointed body of revolution is equivalent to the problem of unsteady planar motion due to a circular cylinder whose radius varies with time, growing from zero at time $\bar{x} = 0$ (see also Goldsworthy [46]).

In fact, for two dimensional steady flows, the equations of motion reduce to those of unsteady one dimensional flow if one of the velocity components remains constant everywhere (see Bird [47]).

12.2.8 Unsteady Hypersonic Flows

Extension to unsteady motions involving small time dependent oscillations of a thin body exposed to a steady uniform stream is straightforward. The full problem has a total derivative including local time derivative term, i.e.

$$\frac{\partial}{\partial t} + u\frac{\partial}{\partial x} + v\frac{\partial}{\partial y} + w\frac{\partial}{\partial z} \tag{12.71}$$

with a reduced time introduced where $\bar{t} = Ut/L$, the continuity equation for example becomes

$$\frac{\partial \bar{\rho}}{\partial \bar{t}} + \frac{\partial \bar{\rho}}{\partial \bar{x}} + \frac{\partial \bar{\rho}\bar{v}}{\partial \bar{y}} + \frac{\partial \bar{\rho}\bar{w}}{\partial \bar{z}} = 0 \tag{12.72}$$

The equation for $\bar{u}$ remains uncoupled from the other equations. However, $\bar{u}$ can no longer be found in terms of the other variables since there is no counterpart of the Bernoulli's equation. If required, $\bar{u}$ can be obtained form solving the x-momentum equation.

Notice that $\bar{x}$ and $\bar{t}$ derivatives appear only in the combination

$$\frac{\partial}{\partial \bar{t}} + \frac{\partial}{\partial \bar{x}} \tag{12.73}$$

hence, introducing the coordinates

$$\hat{x} = x - Ut, \quad or \quad \hat{x} = \bar{x} - U\bar{t} \tag{12.74}$$

reduces the unsteady small disturbance problem to exactly the form of the steady problem. This means that Hayes' analogy remains valid for unsteady motion if the variation of the contour with time is taken into account. For more details on this point, see Hamaker and Wong [48], and Lighthill [49].

12.2.9 Unified Supersonic-Hypersonic Small Disturbance Theory

It is clear, in the case of the nonlinear transonic small disturbance theory, that subsonic and supersonic theories are special cases.

For hypersonic flow, a connection with the adjoining supersonic range will be advantageous. The difficulty arises in the continuity equation. In the hypersonic theory, it becomes

$$U\frac{\partial \rho}{\partial x}+\frac{\partial \rho v}{\partial y}+\frac{\partial \rho w}{\partial z}=0 \tag{12.75}$$

while in the linearized supersonic theory it reduces instead to

$$U\frac{\partial \rho}{\partial x}+\rho_0\left(\frac{\partial u}{\partial x}+\frac{\partial v}{\partial y}+\frac{\partial w}{\partial z}\right)=0 \tag{12.76}$$

The term $\rho_0\frac{\partial u}{\partial x}$ must be retained in linearized supersonic theory, and it must be neglected in the hypersonic theory in order to achieve similitude.

Van Dyke proved, however, that the small disturbance hypersonic theory covers the linearized supersonic theory if it is interpreted in accordance with the similarity rule of the latter. The solutions of the hypersonic small disturbance theory remain valid at small values of the parameter $M_0\tau$ provided that the latter is replaced by $\beta\tau$, where $\beta=\sqrt{M_0^2-1}$ and the pressure and density are scaled as follows

$$p=p_0\left(\gamma M_0^2\tau^2\bar{p}-\frac{1}{\beta^2}\right), \quad \textit{and} \quad \rho=\rho_0\left(\frac{M_0^2}{\beta^2}\bar{\rho}-\frac{1}{\beta^2}\right) \tag{12.77}$$

The pressure coefficient becomes

$$C_p=\frac{p-p_0}{\frac{1}{2}\rho_0U^2}=2\tau^2\left(\bar{p}-\frac{1}{\gamma\beta^2\tau^2}\right) \tag{12.78}$$

The error in the unified theory is $O\left(\tau^2\right)$ or $O\left(\tau/\beta\right)$ whichever is the greater. See Van Dyke [45].

12.2.10 Formulation of the Full and Reduced Problems in Terms of Stream Functions

For steady two dimensional and axisymmetric compressible flows as well as for unsteady one dimensional flows, a stream function can be introduced to satisfy the continuity equation identically.

For example, in 2-D case, using Cartesian coordinates, one can use

$$\rho u = \frac{\partial \psi}{\partial y}, \quad \textit{and} \quad \rho v = -\frac{\partial \psi}{\partial x} \tag{12.79}$$

The equation for ψ is given by

$$\frac{\partial}{\partial x}\left(\frac{\frac{\partial \psi}{\partial x}}{\rho}\right) + \frac{\partial}{\partial y}\left(\frac{\frac{\partial \psi}{\partial y}}{\rho}\right) = -\omega \tag{12.80}$$

where ω is the vorticity.

The equation for ω, for steady isoenergetic flow is given by Crocco equation

$$\frac{D}{Dt}\left(\frac{\omega}{\rho T}\right) = 0 \tag{12.81}$$

or, simply ω/p is constant along a streamline. The vorticity is related to the entropy gradient

$$\omega = \frac{p}{\rho V}\frac{\partial}{\partial \psi}\left(\frac{S}{R}\right) = p\frac{\partial}{\partial n}\left(\frac{S}{R}\right) \tag{12.82}$$

where n is normal to the streamline. Notice S is constant along a streamline but it jumps across a shock.

Hence, the application of such a formulation requires the identification of the shock shape and position to find the jump of entropy.

There is another problem, even for irrotational flow, the density is not a single valued function of the mass flux. Indeed, for a given flux, there are two values, one for subsonic and the other for supersonic conditions. This is obvious considering the compressible flow in a stream tube (or a nozzle). Moreover, there is a square root singularity at the sonic point. The density is not defined (imaginary) for flux values higher than the sonic flux. The stream function equation is, in general, of mixed type as can be easily shown from the equivalent non-conservative form

$$\left(1 - \frac{u^2}{a^2}\right)\frac{\partial^2 \psi}{\partial x^2} - 2\frac{uv}{a^2}\frac{\partial^2 \psi}{\partial x \partial y} + \left(1 - \frac{v^2}{a^2}\right)\frac{\partial^2 \psi}{\partial y^2} = -\rho^2 a^2 \left(\frac{1}{\gamma - 1} + M_0^2\right)\frac{d}{d\psi}\left(\frac{S}{c_p}\right) \tag{12.83}$$

where a is the speed of sound given by Bernoulli's law

$$\frac{a^2}{\gamma - 1} + \frac{1}{2}\left(u^2 + v^2\right) = H \tag{12.84}$$

There is another form of the stream function equation introduced by Crocco, where the derivatives of ψ is determined by the velocity component u and v only.

Let

$$u = \left(1 - V^2\right)^{-\frac{1}{\gamma-1}} \frac{\partial \Psi}{\partial y}, \quad and \quad v = -\left(1 - V^2\right)^{-\frac{1}{\gamma-1}} \frac{\partial \Psi}{\partial x} \tag{12.85}$$

The equation for ψ becomes

$$(a^2 - u^2)\frac{\partial^2 \Psi}{\partial x^2} - 2uv\frac{\partial^2 \Psi}{\partial x \partial y} + (a^2 - v^2)\frac{\partial^2 \Psi}{\partial y^2} = \frac{V^2 - a^2}{2\gamma}\left(1 - V^2\right)^{\frac{\gamma+1}{\gamma-1}} \frac{d}{d\Psi}\left(\frac{S}{c_p}\right) \tag{12.86}$$

In passing, Sears (see for example Ref. [2]) derived a first order perturbation equation, where

$$\Psi = \Psi_0 + \psi \tag{12.87}$$

and Ψ_0 is the stream function of the irrotational flow. The linearized equation for ψ, accounting for the rotational effects is given by

$$\left(1 - M_0^2\right)\frac{\partial^2 \psi}{\partial x^2} + \frac{\partial^2 \psi}{\partial y^2} = -\rho_0^2 a_0^2 \left(\frac{1}{\gamma - 1} + M_0^2\right)\frac{d}{d\psi}\left(\frac{S}{c_p}\right) \tag{12.88}$$

Pai found analytical solution for attached shocks to a plane ogive. He also found analytical solution to the axially symmetric case. He concluded that the first order linearized theory failed at certain points in the weak shock region and he suggested to use the method of characteristics to solve the full equations in the neighborhood of these points.

On the other hand, Kogan [50] obtained solutions for Crocco's stream function equations for both two dimensional and axisymmetric flows with shocks attached to the leading edges. His results for biconvex parabolic airfoil are in good agreement with the method of characteristics near the airfoil leading edge. using successive approximation, taking the flow behind a plane shock wave as zero-order, the shock wave profile obtained in each step is used to determine the domain and the boundary conditions of the next step.

For axially symmetric flow, the flow past a cone is used a zero-order approximation for the ogive problem. Successive approximations with appropriate boundary conditions are solved analytically. The first order solution for Ψ contains a $\frac{3}{2}$-power branch point at $\theta = \theta_0$ (the cone surface) and no logarithmic term as reported by Shen and Lin [51]. Shen and Lin results seem to give infinite pressure gradient at the ogive surface, although solutions by the method of characteristics give no such indication.

The stream function formulation for the reduced problem was used by Van Dyke in his hypersonic small disturbance theory.

Let,

$$\frac{\partial \psi}{\partial \bar{r}} = \bar{r}^{\sigma} \bar{\rho}, \quad and \quad \frac{\partial \psi}{\partial \bar{x}} = -\bar{r}^{\sigma} \bar{\rho}\bar{v} \tag{12.89}$$

Hence

$$\bar{v} = -\frac{\partial \psi}{\partial \bar{x}} / \frac{\partial \psi}{\partial \bar{r}}, \quad \text{and} \quad \bar{\rho} = \frac{1}{\bar{r}^{\sigma}} \frac{\partial \psi}{\partial \bar{r}} \tag{12.90}$$

where $\sigma = 0$ for plane flow and $\sigma = 1$ for axially symmetric flow. Substituting into the reduced momentum equations yields

$$\left(\frac{\partial \psi}{\partial \bar{r}}\right)^2 \frac{\partial^2 \psi}{\partial \bar{x}^2} - 2\frac{\partial \psi}{\partial \bar{x}} \frac{\partial \psi}{\partial \bar{r}} \frac{\partial^2 \psi}{\partial \bar{x} \partial \bar{r}} + \left(\frac{\partial \psi}{\partial \bar{x}}\right)^2 \frac{\partial^2 \psi}{\partial \bar{r}^2}$$

$$= \frac{1}{\bar{r}^{\gamma-1}} \left(\frac{\partial \psi}{\partial \bar{r}}\right)^{\gamma+1} \left(\gamma S \frac{\partial^2 \psi}{\partial \bar{r}^2} + S' \left(\frac{\partial \psi}{\partial \bar{r}}\right)^2\right), \quad \text{for plane flow} \tag{12.91}$$

$$= \frac{1}{\bar{r}^{\gamma-1}} \left(\frac{\partial \psi}{\partial \bar{r}}\right)^{\gamma+1} \left(\gamma S \left[\frac{\partial^2 \psi}{\partial \bar{r}^2} - \frac{1}{\bar{r}} \frac{\partial \psi}{\partial \bar{r}}\right] + S' \left(\frac{\partial \psi}{\partial \bar{r}}\right)^2\right), \quad \text{for axially symmetric flow} \tag{12.92}$$

where

$$S = \bar{p}/\bar{\rho}^{\gamma}, \quad \text{and} \quad \bar{p} = S\bar{\rho}^{\gamma} = S\left(\frac{1}{\bar{r}^{\sigma}} \frac{\partial \psi}{\partial \bar{r}}\right)^{\gamma} \tag{12.93}$$

Van Dyke obtained analytical solution for hypersonic flow over a two dimensional profile, by assuming a given shock wave and Mach number and calculating the corresponding body shape. He identified the thickness parameter τ with the shock angle. For a plane wedge, the flow field is conical, so the stream function has the form

$$\psi(\bar{x}, \bar{r}) = \bar{x} f(\theta), \quad \theta = \frac{r}{\tau x} = \frac{\bar{r}}{\bar{x}} \tag{12.94}$$

The variable θ varies from $b = \delta/\tau$ (δ is the semi vertex angle of the wedge), at the wedge to unity at the shock wave. The equation for f is given by

$$f'' \left(f^2 - \gamma S_0 \left(f'\right)^{\gamma+1}\right) = 0 \tag{12.95}$$

where

$$S_0 = \frac{2\gamma K_0^2 - (\gamma-1)}{\gamma(\gamma+1)K_0^2} \left(\frac{2 + (\gamma-1)K_0^2}{(\gamma+1)K_0^2}\right)^{\gamma} \tag{12.96}$$

with $K_0 = M_0 \tau$.

The shock wave boundary conditions give (since $s' = 1$, s represents the shock shape and S is the entropy)

$$f(1) = 1, \quad \text{and} \quad f'(1) = \frac{(\gamma+1)K_0^2}{2 + (\gamma-1)K_0^2} \tag{12.97}$$

The solution which corresponds to $f'' = 0$, is

$$f(\theta) = \frac{(\gamma+1)K_0^2\theta - 2(K_0^2-1)}{2+(\gamma-1)K_0^2} \tag{12.98}$$

Requiring f to vanish at the wedge surface gives

$$\frac{\delta}{\tau} = 2\frac{K_0^2-1}{(\gamma+1)K_0^2} \tag{12.99}$$

In terms of M_0 instead of $M_0\tau$, the results are

$$\frac{\tau}{\delta} = \frac{1}{b} = \frac{\gamma+1}{4} + \sqrt{\left(\frac{\gamma+1}{4}\right)^2 + \frac{1}{M_0^2\delta^2}} \tag{12.100}$$

The pressure coefficient on the wedge (and anywhere between the wedge and the shock) is found to be

$$C_p = \delta^2\left(\frac{\gamma+1}{2} + \sqrt{\left(\frac{\gamma+1}{2}\right)^2 + \frac{4}{M_0^2\delta^2}}\right) \tag{12.101}$$

Van Dyke's result is identical with Linnell's. The unified supersonic-hypersonic result is obtained by replacing the parameter $M_0\delta$ by $\beta\delta$, hence

$$C_p = \delta^2\left(\frac{\gamma+1}{2} + \sqrt{\left(\frac{\gamma+1}{2}\right)^2 + \frac{4}{\beta^2\delta^2}}\right) \tag{12.102}$$

The above result is thus valid for ordinary supersonic range as well.

Van Dyke obtained also results for slender circular cone. The flow field is conical and the stream function has the form

$$\psi(\bar{x},\bar{r}) = \bar{x}^2 f(\theta), \quad \theta = \frac{r}{\tau x} = \frac{\bar{r}}{\bar{x}} \tag{12.103}$$

The equation for f becomes

$$4ff'' - 2ff'^2 = \gamma S_0 \frac{(f')^{\gamma+1}}{\theta^{\gamma-1}}\left(f'' - \frac{f'}{\theta}\right) \tag{12.104}$$

and boundary conditions

$$f(1) = \frac{1}{2}, \quad and \quad f'(1) = \frac{(\gamma + 1)K_0^2}{2 + (\gamma - 1)K_0^2} \tag{12.105}$$

The tangency condition at the surface of the cone is $f(b) = 0$.

Van Dyke solved the nonlinear ordinary differential equation for f numerically, step by step, starting from the shock surface inward until f vanishes to produce the solution at the solid surface. The pressure coefficient is given by

$$C_p = 2\delta^2 \left(\frac{S_0}{b^2} \left(\frac{f'}{\theta} \right)^{\gamma} - \frac{1}{\gamma M_0^2 \delta^2} \right) \tag{12.106}$$

The surface pressure agrees well with the results of the full equations (with error $O\left(\delta^2\right)$ as expected).

Van Dyke calculated also the flow over plane ogive and ogive of revolution. Again, his results agree with Kogan's.

12.2.11 Similarity Solutions for Power Law Bodies

There is a certain class of self-similar motions for which the partial differential equations reduce to ordinary differential equations, thus leading to a great simplification and in some cases to analytical solutions.

The concept of self-similarity refers here to flow in which the flow variables between the shock and the body are similar to each other at different stations along the body.

The use of unsteady analogy or equivalence principle, implies that a number of self-similar solutions obtained for unsteady flow problems become available for calculating steady hypersonic flows.

Besides the wedge and the cone, another useful solution is for cylindrical pistons having a motion given by $r \sim t^m$, and from which the flow past power law bodies may be obtained, see Velesko et al. [52]. Other self-similar solutions include a piston expanding according to an exponential law, $r = r_0 e^{t/t0}$, see Sedov [53] and Gusev [54].

The analogy with steady hypersonic flows was developed by Grodzovskii [55], Chernyi [56] and Stanyukovich [57].

Lees and Kubota [58] and Mirles [59] obtained directly similarity solutions of the hypersonic small disturbance equation for axisymmetric bodies.

The earliest work on similar solutions of this type of problems was carried by Bechert [60] and Guderley [61]. Guderley derived the appropriate equations and boundary conditions and then reduced the problem to a single first order differential equation and applied his theory to the problem of implosions. In his work, he gave a detailed study of the various singular points of the governing differential equation.

Following Kubota [62], we shall consider steady solutions of the hypersonic small disturbance equations for axisymmetric bodies.

Let the nondimensional variables f, g, h for radial velocity, pressure and density be functions only of $\eta = \bar{r}/\bar{r}_s$, where $\bar{r}_s$ is the reduced shock radius r_s/τ, hence

$$\bar{v}(\bar{x}, \bar{r}) = \frac{d\bar{r}_s}{dx} f(\eta) \tag{12.107}$$

$$\bar{p}(\bar{x}, \bar{r}) = \frac{1}{\gamma} \left(\frac{d\bar{r}_s}{dx}\right)^2 g(\eta) \tag{12.108}$$

$$\bar{\rho}(\bar{x}, \bar{r}) = h(\eta) \tag{12.109}$$

It is assumed that the shock is strong, i.e. $\frac{\gamma-1}{2} M_0^2 \beta^2 \gg 1$.
The boundary conditions at the shock ($\eta = 1$) become

$$f(1) = \frac{2}{\gamma + 1} \tag{12.110}$$

$$g(1) = \frac{2\gamma}{\gamma + 1} \tag{12.111}$$

$$h(1) = \frac{\gamma + 1}{\gamma - 1} \tag{12.112}$$

Substituting in the small disturbance equations yields

$$(\eta - f)\, h' - hf' - \frac{hf}{\eta} = 0 \tag{12.113}$$

$$(\eta - f)\, f' - \frac{g'}{\gamma h} = \frac{r_s r_s''}{r_s'^2} f \tag{12.114}$$

$$(\eta - f) \left(\frac{g'}{g} - \gamma \frac{h'}{h}\right) = 2 \frac{r_s r_s''}{r_s'^2} \tag{12.115}$$

For a self-similar solution, the equations must be independent of r_s, hence

$$\frac{r_s r_s''}{r_s'^2} = constant = \alpha \tag{12.116}$$

The solution of the above equation is

$$r_s' = A r^{\alpha}, \quad or \quad r_s = \left(\frac{A}{m}\right)^m x^m \tag{12.117}$$

where $m = 1/(1-\alpha)$.

The shock shape is then

$$r_s = C_1 x^m \tag{12.118}$$

Since at the body

$$\frac{v_b}{v_s} = \frac{f(\eta_b)}{f(1)} \tag{12.119}$$

and for flow similarity η_b is constant, along the body $r_b/r_s = constant$, hence the body is given by $r_b = C_2 x^m$.

The ratio of the body to the shock radius and the variation of the flow variables between the shock and the body can be obtained from the solution of the differential equations.

It is possible, however, to find the values of m for which similar flows may exist by considering the drag of the body. Since the surface pressure p_b is proportional to $\bar{r}_s'$, the drag is given by

$$D = 2^\sigma \pi^\sigma \int_0^l r_b^\sigma p_b \left(\frac{dr_b}{dx}\right) dx = const. \int_0^{r_s} r_s^{\sigma+2\alpha} dr_s \tag{12.120}$$

D is finite if $-2\alpha \leq 1+\sigma$, or $m \geq 2/(3+\sigma)$, where $\sigma = 0$ for plane and $\sigma = 1$ for axisymmetric bodies.

The breakdown of similarity for $m < 2/(3+\sigma)$ does not imply that no flow exists, but only that there is no similarity solution.

Kubota integrated the system of differential equations numerically, for $1 \geq m \geq 2/(3+\sigma)$, with the corresponding boundary conditions. The drag coefficient for bodies of revolution has a minimum value at $m = 0.92$, assuming $\gamma = 1.4$. The distance between the shock and the body decreases as the value of γ approaches unity. (For $\gamma = 1$, the shock and the body coincide).

Mirles [59] developed an approximate method based on the asymptotic form of the flow in the neighborhood of the body surface. Both first and second order approximations are in agreement with the numerical solution obtained by Kubota.

The above similarity solutions for power law bodies are limited to the case of strong shocks ($\frac{\gamma-1}{2} M_0^2 \beta^2 \gg 1$).

For a finite Mach number, the solution may be expected to break down towards the rear of the slender body as the shock becomes weaker and the shock angle approaches the Mach angle. Also, the region near the nose of a power law body is excluded since the solutions are based on the use of the small disturbance equations. The similarity solutions for $m < 1$ predict a density which falls to zero at the body surface. Since the pressure is finite, the entropy will become infinite. Indeed, there is an entropy layer next to the surface in which the small disturbance equations are no longer valid and must be excluded from solutions obtained using the similarity method. For more details see Cox and Crabtree [3].

12.2.12 Newtonian Flows

The impact theory Newton proposed, is simply to calculate the normal force on a plate at angle θ in terms of the change of normal momentum, i.e. (Fig. 12.3)

$$N = A\rho_0 U^2 \sin^2\theta \tag{12.121}$$

hence

$$C_p = \frac{p - p_0}{\frac{1}{2}\rho_0 U^2} = 2\sin^2\theta \tag{12.122}$$

For a curved surface, the modified Newtonian theory is

$$C_p = C_{p,max}\sin^2\theta \tag{12.123}$$

where $C_{p,max}$ is the stagnation pressure coefficient. Indeed, as $M_0 \to \infty$, the flow across a normal shock gives $C_{p,max} = 2$. However, for finite M_0, $C_{p,max} < 2$. See Fig. 12.4.

On the other hand, the oblique shock relations are given by

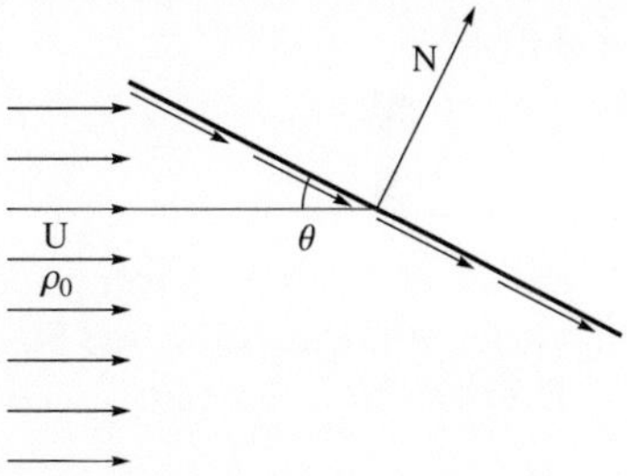

Fig. 12.3 Newton's impact theory

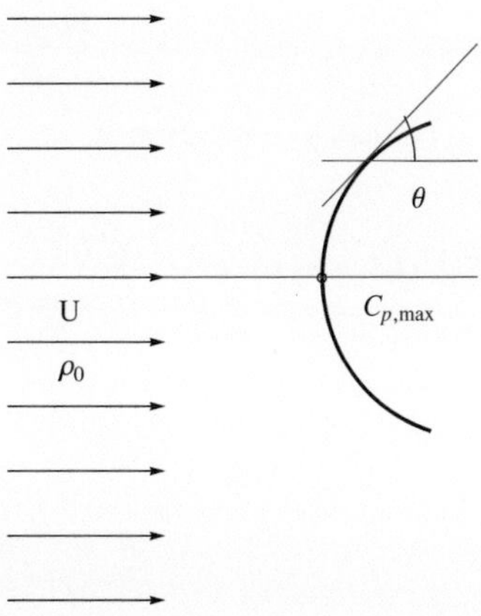

Fig. 12.4 Modified Newton's theory

$$\frac{\rho_1}{\rho_0} = \frac{(\gamma+1)M_0^2\sin^2\beta}{(\gamma-1)M_0^2\sin^2\beta+2}, \quad \Rightarrow \quad \frac{\rho_1}{\rho_0} = \frac{\gamma+1}{\gamma-1} \quad as \quad M_0 \to \infty \tag{12.124}$$

Also

$$\theta = \frac{2}{\beta}\frac{M_0^2\sin^2\beta - 1}{M_0^2(\gamma+2\cos 2\beta)+2}, \quad \Rightarrow \quad \frac{\beta}{\theta} = \frac{\gamma+1}{2} \quad as \quad M_0 \to \infty \tag{12.125}$$

since both θ and β are small.

The pressure coefficient is given by

$$C_p = \frac{4}{\gamma+1}\left(\sin^2\beta - \frac{1}{M_0^2}\right), \quad \Rightarrow \quad C_p = \frac{4}{\gamma+1}\sin^2\beta \quad as \quad M_0 \to \infty \tag{12.126}$$

In the additional limit $\gamma \to 1$, $\rho_1/\rho_0 \to \infty$, $\beta = \theta$, $C_p = 2\sin^2\theta$.

These results are expected to hold for high Mach numbers according to "Principle of Mach Number Independence".

As $M_0 \to \infty$ the layer between the shock and the body is of very high density and eventually the shock coincides with the surface. Busemann [63] noticed that for particles flowing round a curved surface, there must be a centrifugal pressure gradient, normal to the surface. The associated change in pressure is (see Eggers et al. [64])

$$\Delta p = \int_0^N \frac{\rho u^2}{R} dn \simeq \frac{\bar{u}}{\bar{R}}\int_0^N \rho u dn \tag{12.127}$$

where $\bar{u}$ and $\bar{R}$ are the mean values of the velocity and of radius of curvature of the thin layer of thickness N, see Fig. 12.5.

The mass flow rate past a section of a body of revolution is given by

$$\dot{m} = 2\pi r \int_0^N \rho u dn \simeq \pi r^2 \rho_0 U \tag{12.128}$$

Notice, $r_s \to r$. Hence

Fig. 12.5 Shock layer

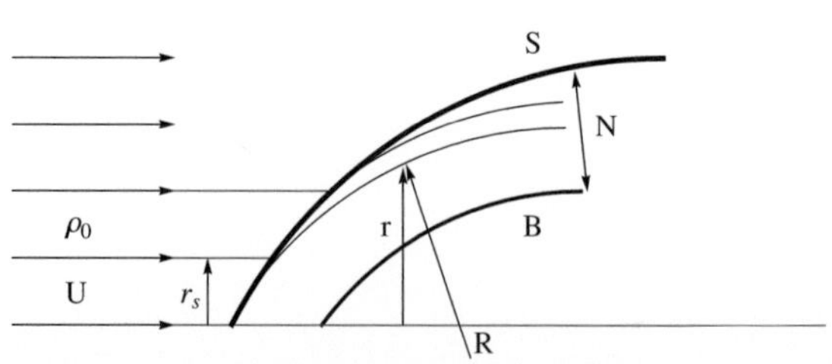

$$\Delta p = \frac{\bar{u}}{\bar{R}}\frac{r}{2}\rho_0 U^2, \quad and \quad C_p = \frac{r}{\bar{R}}\frac{\bar{u}}{U} \tag{12.129}$$

As $M_0 \to \infty$ and $\gamma \to 1$, a particle entering the shock with a tangential velocity $U \cos\theta$ retains that velocity, therefore the mean velocity $\bar{u}$ is

$$\bar{u} = \frac{1}{\pi r^2}\int_0^r 2\pi r U \cos\theta dr \tag{12.130}$$

The radius of curvature of the thin layer is governed by

$$\frac{r}{\bar{R}} = r \sin\theta \frac{d\theta}{dr} \tag{12.131}$$

Hence,

$$\Delta C_p = 2\frac{\sin\theta}{r}\frac{d\theta}{dr}\int_0^r \cos\theta r dr \tag{12.132}$$

Let $A = \pi r^2$, the cross-sectional area; The modified surface pressure coefficient given by Busemann is

$$C_p = 2\sin^2\theta + 2\sin\theta\frac{d\theta}{dA}\int_0^A \cos\theta dA \tag{12.133}$$

For a sphere, the above relation becomes

$$C_p = 2\left(1 - \frac{4}{3}\sin^2\theta\right) \tag{12.134}$$

Lighthill [65] suggested that the shock wave separates from the surface of the sphere at the point of vanishing C_p, $\theta = 60°$ measured from the forward stagnation point, Fig. 12.6.

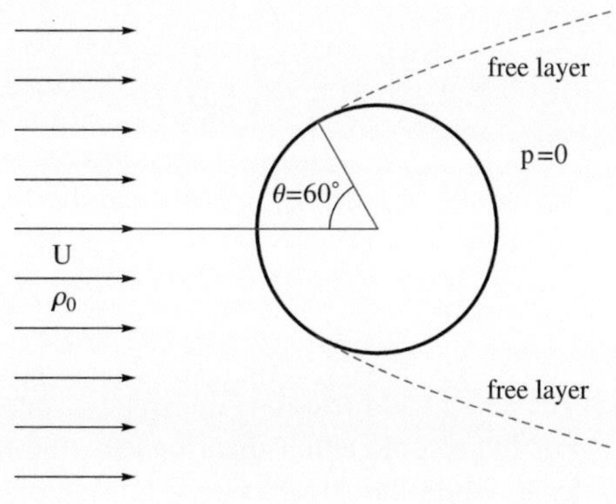

Fig. 12.6 Sphere in Hypersonic Flow

The shape of the shock wave may be determined from

$$0 = 2\sin^2\theta + 2\sin\theta \frac{d\theta}{dA}\int_0^A \cos\theta dA \tag{12.135}$$

It can be shown that the shock shape tends asymtotically to be $y \sim x^{1/3}$.
The modified Newton-Busemann formulas for slender bodies are

$$C_p = C_{p0}\left(\theta^2 + \frac{r}{2}\theta'\right), \quad \textit{for axisymmetric flow} \tag{12.136}$$

$$C_p = C_{p0}\left(\theta^2 + y\theta'\right), \quad \textit{for plane flow} \tag{12.137}$$

For cones and wedges, the Busemann corrections vanish and the Newtonian formulas are consistent with the shock relations as $M_0 \to \infty$ and $\gamma \to 1$.

$$C_{p,cone} = 2\frac{(\gamma+1)(\gamma+7)}{(\gamma+3)^2}\theta_c^2 \tag{12.138}$$

$$C_{p,wedge} = (\gamma+1)\theta_w^2 \tag{12.139}$$

12.2.13 Cole's Slender Body Theory of Newtonian Flow

The Newton-Busemann formulas can be obtained systematically from an asymptotic hypersonic small disturbance theory following Cole [66], who introduced the following parameters

$$\lambda = \frac{\gamma-1}{\gamma+1}, \quad \textit{(density ratio)} \tag{12.140}$$

$$H = \frac{1}{M_0^2\theta^2}, \quad \textit{(hypersonic similarity parameter)} \tag{12.141}$$

$$N = \frac{H}{\tau} = \frac{\gamma+1}{\gamma-1}\frac{1}{M_0^2\delta^2}, \quad \textit{(Newtonian flow parameter)} \tag{12.142}$$

The small parameter λ vanishes in the Newtonian limit $\gamma \to 1$ and

$$\gamma = \frac{1+\lambda}{1-\lambda} = 1 + 2\lambda + 2\lambda^2 + \cdots \tag{12.143}$$

The small parameter δ is the slope or maximum thickness ratio $\delta = e/c$.

For hypersonic small disturbances, the transverse perturbation velocity v is of order δ, while the streamwise perturbation velocity u is of order δ^2. The density perturbation is of order unity, while the perturbation in pressure is of order δ^2. These orders can be determined from shock wave relations or from Prandtl-Meyer expansion, provided $M_0\delta$ is not much less than unity.

Let the Newtonian flow parameter N be fixed as $\lambda \to 0$ and $H \to 0$. The first order approximate equations obtained from the hypersonic small disturbance equations for the pressure, density and the transverse velocity are ($\sigma = 0$ for two dimensional flow and $\sigma = 1$ for axial symmetry)

$$\frac{\partial}{\partial \hat{r}}(F\rho v) + \sigma \rho F' + F\frac{\partial \rho}{\partial x} = 0, \quad (continuity) \tag{12.144}$$

$$F'' = -\frac{1}{\rho}\frac{\partial p}{\partial \hat{r}}, \quad transverse\ momentum \tag{12.145}$$

$$\left(\frac{\partial}{\partial x} + v\frac{\partial}{\partial \hat{r}}\right)\left(\frac{p}{\rho}\right) = 0, \quad (entropy) \tag{12.146}$$

The boundary conditions are

$$v(x, 0) = 0, \quad at\ the\ body \tag{12.147}$$

$$\begin{cases} p(x, g) = -F'^2 \\ \rho(x, g) = F'^2/(N + F'^2) \\ v(x, g) = g' - F' - N/F' \end{cases}, \quad at\ the\ shock \tag{12.148}$$

In the above equations, $F(x)$ is the body shape and $S(x)$ is the shock shape, where x is a nondimensional coordinate using a characteristic length c, and $\hat{r} = r^*/\lambda$ and $r^* = r - F(x)$, r is nondimensional coordinate using characteristic length e. Notice $0 \geq r^* \geq G(x) = S - F$, and $g = \lambda G$.

Cole obtained general solution of the first order equations via introducing a stream function defined by

$$\frac{\partial \psi}{\partial x} = -F^\sigma \rho v, \quad \frac{\partial \psi}{\partial \hat{r}} = F^\sigma \rho \tag{12.149}$$

The continuity equation is satisfied automatically. The momentum equation reads

$$\frac{1}{F^\sigma}F''\frac{\partial \psi}{\partial \hat{r}} + \frac{\partial p}{\partial \hat{r}} = 0 \tag{12.150}$$

By integration from the surface $\hat{r} = 0$, one obtains

$$\frac{1}{F^\sigma}F''\psi + p = p_0(x) \tag{12.151}$$

where $\psi = 0$ at $\hat{r} = 0$.

To find the surface pressure distribution $p_0(x)$, ψ_{shock} is evaluated as follows

$$\frac{\partial \psi}{\partial x}|_{shock} = \frac{\partial \psi}{\partial x} + \frac{\partial \psi}{\partial \hat{r}} g' \quad (12.152)$$

Using the relation

$$\frac{\partial \psi}{\partial x}|_{shock} = F^{\sigma} F' \quad (12.153)$$

hence

$$\psi_{shock} = \frac{F^{\sigma+1}}{\sigma+1} = \psi(x, g) \quad (12.154)$$

using $\psi_{shock} = 0$ at $x = 0$.

Therefore

$$p_0(x) = F'^2 + \frac{1}{2} F F'', \quad \textit{for axial symmetry} \quad (12.155)$$

$$p_0(x) = F'^2 + F F'', \quad \textit{for planar flow} \quad (12.156)$$

In terms of the pressure coefficients

$$C_p(x) = \begin{cases} 2\delta^2 \left(F'^2 + \frac{1}{2} F F''\right), & \textit{for axial symmetry} \\ 2\delta^2 \left(F'^2 + F F''\right), & \textit{for planar flow} \end{cases} \quad (12.157)$$

Notice that the surface pressures are independent of the Newtonian parameter N, although the flow field and shock shape depend on N (i.e. the limiting values of the surface pressures as $N \to \infty$ should be a good approximation over a wide range of Mach numbers).

The entropy equation states that $p/\rho = K(\psi)$. From the shock relation $K = F'^2 + N$ on the shock, and since ψ_{shock} is given before, $K(\psi)$ can be found parametrically.

At $x = 0$, $\psi = 0$ and $K(0) = F'^2(0) + N$, so that

$$\rho_0(x) = \frac{p_0(x)}{F'(0) + N} \quad (12.158)$$

The surface temperature is found from the equation of state, where

$$\frac{p}{p_0} = 1 + \gamma M_0^2 \delta^2 \hat{p} = \frac{p_0}{N\lambda} \quad (12.159)$$

so that the surface temperature is constant

$$\frac{T}{T_0} = \frac{F'^2(0)}{N} + 1 = \frac{K(0)}{N} + \cdots \quad (12.160)$$

For cones, the first approximation is $F(x) = x$, $\delta = \tan\theta_c$ and the shock is $g(x) = Ax$. The shock conditions are

$$\begin{cases} p(x, Ax) = 1 \\ \rho(x, Ax) = 1/(N+1) \\ v(x, Ax) = A - (N+1) \end{cases} \tag{12.161}$$

The transverse momentum equation yields $\frac{\partial p}{\partial \hat{r}} = 0$, hence $p = 1,\ 0 \le \hat{r} \le Ax$.

From the entropy equation and the shock relations, it follows that the density is constant, $\rho = 1/(N+1)$.

The transverse component of the velocity can be found from the continuity equation

$$\frac{\partial x v}{\partial \hat{r}} = -1 \tag{12.162}$$

and, using the boundary condition $v = 0$ at $\hat{r} = 0$

$$v(x, \hat{r}) = -\frac{\hat{r}}{x} \tag{12.163}$$

This represents a flow filed whose streamlines are more inclined towards the cone surface as $\hat{r}$ increases.

The shock angle can be found from the boundary condition for v on the shock

$$-A = A - (N+1), \quad or \quad A = \frac{N+1}{2} \tag{12.164}$$

The shock shape is then

$$g(x) = \frac{N+1}{2}x \tag{12.165}$$

The pressure coefficient is

$$C_p = 2\delta^2 p = 2\tan^2\theta_c \tag{12.166}$$

and

$$\frac{\tan\theta_w}{\tan\theta_c} = 1 + \frac{N+1}{2}\lambda = \frac{3\gamma+1}{2(\gamma+1)} + \frac{1}{2}\frac{1}{M_0^2\delta^2} \tag{12.167}$$

The first order approximation is in good agreement with exact theory for $M_0 \tan\theta_c > 1$.

Cole obtained a second order approximation as well (for details see [66]) where

$$\frac{C_p}{\delta^2} = 2\left(1 + \frac{N+1}{4}\lambda\right) \tag{12.168}$$

12.2.14 Slender Body at Angle of Attack, Bow Shocks Around Blunt Bodies and Numerical Simulations of Hypersonic Inviscid Flows

Slender Body at Angle of Attack

Grimminger et al. [67], have derived general expressions for the lift and drag of inclined bodies of revolution. They assume that the streamlines remain parallel and leave the surface in a streamwise direction, see Fig. 12.7.

They used the Newtonian formula to calculate the pressure on the wetted part of the body.

Later, the Newtonian theory for slender bodies of revolution at small angle of attack was reported by Cole [66] and Cole and Brainerd [68], as a small perturbation on the solution at zero incidence. Linear equations are obtained with variable coefficients depending on the zero incidence solution. The results are

$$\frac{dC_L}{d\alpha} = \frac{4}{3}\left(1 + F'(1)\int_0^1 F(x)dx\right) \tag{12.169}$$

$$x_{c.p.} = 1 - \frac{4}{3}\frac{\int_0^1 F(x)dx}{\frac{dC_L}{d\alpha}} \tag{12.170}$$

where the body shape is given by $\hat{r} = \delta F(x)$, $F(1) = 1$, as stated before.

Hayes and Probstein [2] extended Cole's results for non slender bodies.

Cox and Crabtree [3] discuss the case of a cone of circular cross section and the appearance of what is called the vortical layer.

For a cone at zero incidence, the radial surfaces of constant entropy are shown in Fig. 12.8. In the same figure, the corresponding surfaces when $\alpha = 20°$ are shown indicating the existence of rotational flow.

Ferri [69, 71] and Ferri et al. [70] pointed out that the surface of the cone must have constant entropy and there exist a vortical singularity, where the entropy is multi-valued.

The thickness of the vortical layer is shown by Guiraud [72] to be

$$\epsilon_v \simeq \frac{\bar{\gamma}-1}{\bar{\gamma}+1} exp\left(-\frac{\bar{\gamma}+1}{\alpha(\bar{\gamma}-1)}\right) \tag{12.171}$$

where $\bar{\gamma}$ is a mean value of γ downstream of the shock.

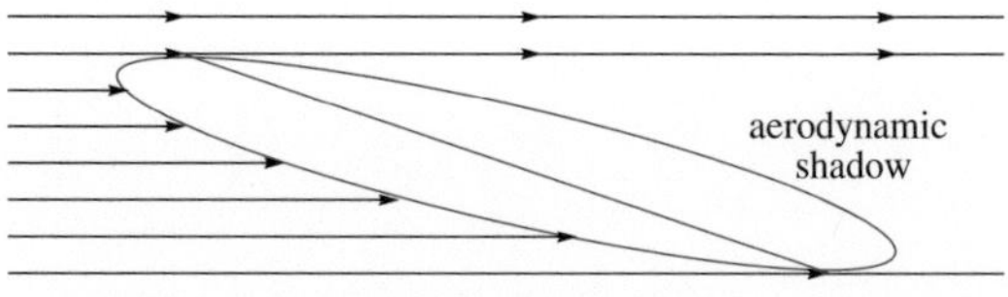

Fig. 12.7 Flow past body at incidence with Newtonian model

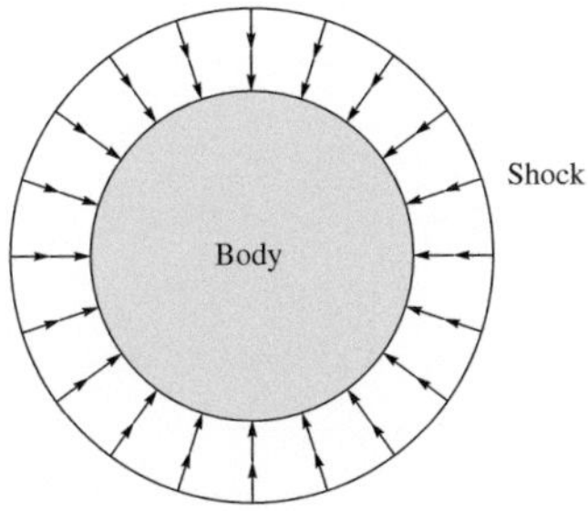

Fig. 12.8 Surfaces of constant entropy for flow past cone

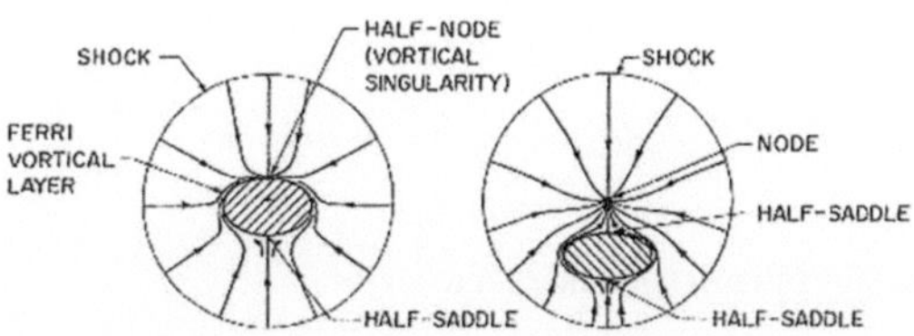

Fig. 12.9 Singularities in solutions for a cone (from Hypersonic Flow Theory, Vol. I, by Hayes and Probstein, with kind permission of Dover publication)

Based on Guiraud's analysis, the surface pressure distributions and the total forces and moment (relative to the nose) acting upon a body of revolution can be calculated in closed form. For slender bodies, Guiraud's results agree with Cole's.

Following Hayes and Probstein [2], singularities in solutions for an elliptic cone at small and large incidences are given in Fig. 12.9.

A topological law, based on a Poincaré theory, relates the number of elementary singular points according to the relation

$$N + F - S + O = 2 \tag{12.172}$$

where N is the number of nodes, F is the number of foci, S the number of saddle points and O the number of obstacles. For more details, see Ref. [74].

Over two decades, the problem of cones at small and large angles of attack has been studied by many authors, including Tsien [75], Lighthill [76], Young and Sisk [77], Stone [78, 79], Willet [80, 81], Holt [82], Cheng [83, 84], Sychev [85], Munson [86], Woods [87, 88], Melnick [89, 90]. Numerical simulations are discussed by Moretti [91], Dwyer [92] and Fletcher [93]. A comprehensive review can be found in the book by Bulakh [94].

Blunt Body Problems

There are two categories, the example for the first is cylinders or cones, while the second is a slender body with a blunt nose, see Fig. 12.10 from Ref. [2].

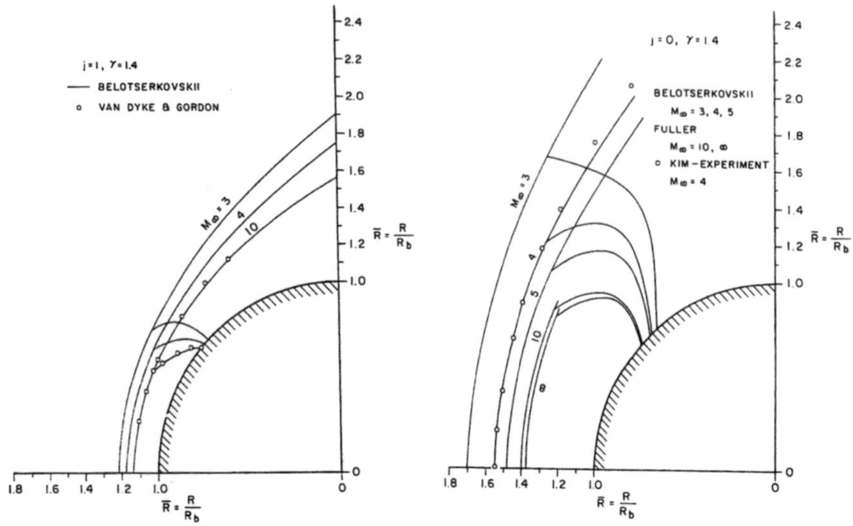

Fig. 12.10 Bow shock for sphere (*left*) and sphere (*right*)

For the bow shock stand-off distances for spheres and cylinders, approximate analytical relations are given by Hayes and Probstein [2]. For a sphere with a radius R_b and a shock of radius R_s, assuming constant density in the thin layer between the bow shock wave and the body surface

$$\Delta = R_S - R_b = \frac{\epsilon R_S}{1 + \sqrt{\frac{8}{3}\epsilon}} \tag{12.173}$$

where

$$\epsilon = \frac{\rho_0}{\rho_s} = \frac{\rho_0}{\rho_1} = \frac{(\gamma - 1)M_0^2 + 2}{(\gamma + 1)M_0^2} \tag{12.174}$$

Hence

$$\Delta = \epsilon R_s \left(1 - \sqrt{\frac{8}{3}\epsilon} + \frac{8}{3}\epsilon + O(\epsilon^{3/2})\right) \tag{12.175}$$

For the case of the circular cylinder

$$\Delta = R_S - R_b = \frac{1}{2}\epsilon R_s \left(\ln \frac{4}{3\epsilon} + \frac{\epsilon}{2}\left(\ln \frac{4}{3\epsilon} - 1\right) + O(\epsilon^3)\right) \tag{12.176}$$

For large M_0, $\epsilon \to \frac{\gamma-1}{\gamma+1}$, and for a sphere, the stand-off distance becomes

$$\Delta \simeq R_s \frac{\bar{\gamma} - 1}{\bar{\gamma} + 1} \tag{12.177}$$

Recent work on theoretical models for stand-off distance is discussed by Olivier in [95].

At the axis, the normal shock wave leads to the largest entropy jump and the entropy decreases as the shock angle approaches the Mach angle. Hence, according to Crocco's relation, vorticity is generated behind a curved shock.

For a blunt body at angle of attack, the entropy layer and the vortical layer will mix in the neighborhood of the body where the boundary layer is also growing.

The Flow field around a sphere at Mach 10 and Mach 50 are shown in Fig. 12.11.

The analogy of a bow shock on a blunt nosed body and unsteady one- dimensional explosion is sketched in Fig. 12.12.

These complicated flows have been studied over the years. In the series expansion methods, it is assumed that the shock shape is known. The conditions just downstream

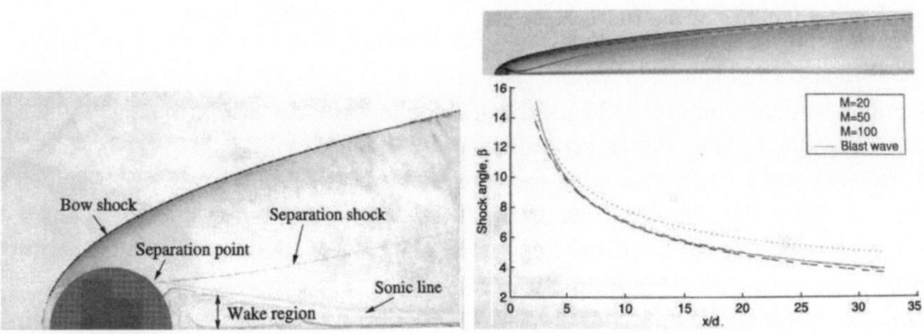

Fig. 12.11 Half-sphere at Mach 10 (*left*); Shock wave past a sphere at Mach 50 and comparison with blast wave analogy (*right*)

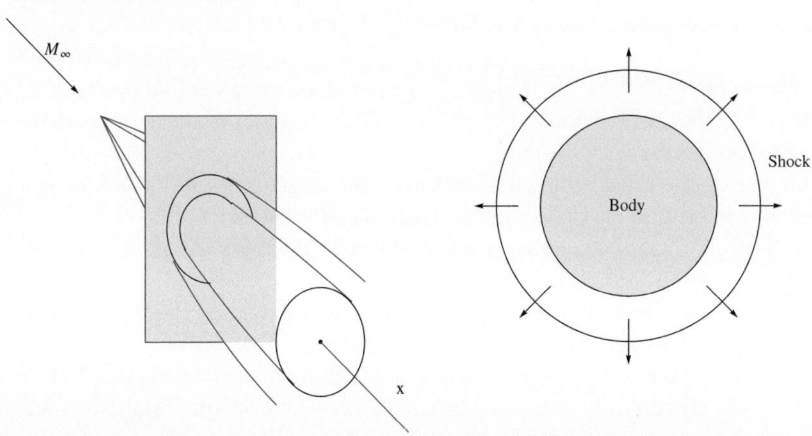

Fig. 12.12 Analogy of bow shock on a blunt nosed flat plate and unsteady one-dimensional explosion

of the shock can be obtained from the shock relations and the derivatives of the flow variables may be obtained using the governing equations. Taylor series in terms of the distance from the shock is used to describe the flow field between the shock and the body. Examples of this approach are given by Lin and Rubinov [96], Shen and Lin [97] and Cabannes [98, 99]. The convergence of the Taylor series, near the body is in general questionable as discussed by Van Dyke [100].

Numerical methods are more suitable for such a study as discussed by Hayes and Probstein [2].

Van Dyke [101] used an inverse method integrating the flow equations from the shock to the body by a marching technique, assuming the shock shape to be a conic section, see also Mangler [102]. Good results for stand-off distance for a sphere, in agreement with experimental measurements, are reported for different high Mach numbers. However, the problem is, in general, ill-posed as shown by Vaglio-Laurin [103], where almost indistinguishable shock shapes produce completely different body shapes. Special treatment is required to avoid numerical instabilities.

Another approach to solve the inverse problem was proposed by Garabedian and Lieberstein [104], based on the method of complex characteristics.

First the initial data for a known analytic shock curve are continued into a fictitious three-dimensional space which consists of the real value of one of the independent variables and the complex value of the other. The wave equation can be solved, in a stable manner, by a marching procedure, or by the method of characteristics.

However, any small change in the initial data in the real domain can result in a large change in the complex domain. Lin [105] noticed that the analytic continuation of the initial conditions from real values into the complex domain is itself the solution of an initial value problem for a Laplace equation. For simple closed form analytic shock shapes such as hyperbolas or parabolas for the hypersonic blunt body problem, the initial conditions are continued exactly into the complex domain avoiding any instabilities.

Moreover, Van Dyke [106], pointed out that the computational time of Garabedian and Lieberstein method was almost 200 times his straightforward marching procedure.

Inverse methods have been applied to asymmetric flows by Vaglio-Laurin and Ferri [107]. They treated bodies of revolution at small angles of attack by perturbation around a body at zero incidence.

Of course, the usual problem of interest is the direct problem, where the inverse methods can be used via trial and error. Hayes and Probstein [2], considered building up a catalog of solutions, a dangerous philosophy delaying the advances of direct methods.

In the late 50s, Dorodnitsyn proposed a direct method based on integral relations, see also Belotserkovskii and Chushkin [108]. See also Holt [109].

The basic idea is to solve a system of partial differential equations in a domain by dividing the domain into strips and obtain independent integral relations to obtain the solutions of the variables in each strip. Assuming a polynomial variation in one variable of the flow variables between the shock and the body and substituting into the equations of motion, a system of ordinary differential equations is generated and may be integrated numerically across the strips. An initial guess of the shock shape is assumed and the solution is obtained by iteration. Dorodnitsyn [110] has generalized his method, by multiplying the basic equations by a function to obtain the moments of the equations in an attempt to increase the accuracy of the calculations. Notice the analogy with the early boundary layer methods). Many variants exist and for more details see Holt [109].

Perhaps, more interesting are the applications of finite differences (and finite volumes) using relaxation techniques or unsteady approach to steady-state solutions of the governing equations.

Mitchell and McCall [111], calculated the supersonic flow over blunt bodies using the stream function/vorticity formulation where the shock wave was assumed known from experimental measurements. A serious problem was the breakdown of the relaxation procedure in the neighborhood of the sonic line. Hayes and Probstein [2], discussed how to carry the calculations through the sonic region and somewhat beyond the limiting characteristics and established consistency relations used as boundary conditions on the sonic line. These conditions may indicate how the shock shape should be updated to obtain an improved solution. (The supersonic flow field can be always obtained via the method of characteristics).

In such calculations, the shock is fitted. A more attractive approach is the shock capturing, where conservative Euler equations, in primitive variables, are solved explicitly to reach steady-state. There are two main schools, the first is based on explicit artificial viscosity as in the Lax-Wendroff scheme and the second is based on upwind schemes as in the Godunov method. Early applications to hypersonic flows of Lax-Wendroff type scheme was reported by Burstein and Rusanov using his version of the Lax scheme. Applications of Godunov' schemes are discussed in Holt's book [109]. Implicit methods to solve the unsteady Euler equations have been developed since then, see for example Richtmeyer and Morton [112]. For recent review of numerical methods for hypersonic inviscid flows see Pletcher et al. [113]. Notice, some of the early methods applied for hypersonic flows have been extended to deal with lower transonic flow regime, for example Garabedian's complex characteristics and the integral relation method as well as shock fitting and shock capturing techniques for the relaxation method of the stream function equation (see Emmons [114], Steger [115], Hafez and Lovell [116]), and for the unsteady Euler equations (see Hirsch [117], Van Leer [118], Roe [119] , Osher [120] for upwind schemes).

To conclude this section, we will discuss briefly the works of Chernyi and Cheng for the slender bodies with blunt noses.

In Chernyi's method [121], it is assumed that a plane piston starts to move from the nose with a velocity $V = U\theta$, normal to the flow direction, to generate a wedge or a cone of semi-angle θ (see also Cox and Crabtree [3]) for the governing integral equations expressing conservation of energy and momentum in the equivalent unsteady flow problem.

The method of Cheng [122], includes also the effect of the entropy layer.

Guiraud [123] treated directly the effect of slight bluntness. Yakura [124] used the inverse problem as discussed in Van Dyke [100]. In Fig. 12.13, in the direct problem,

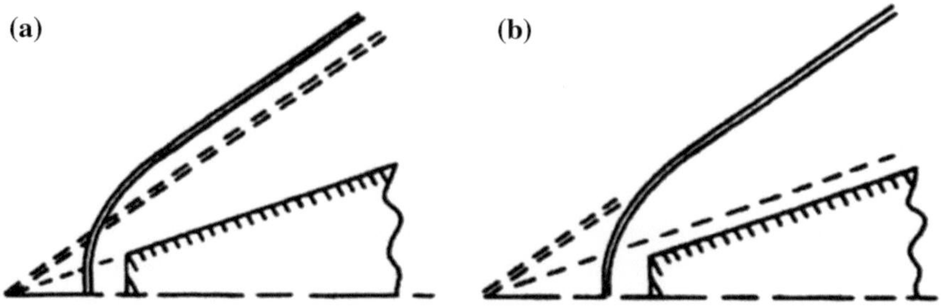

Fig. 12.13 Displacement effect of blunting. **a** Direct problem. **b** Inverse problem

Fig. 12.14 Blunt axisymmetric slender body producing a paraboloidal shock with infinite Mach number and $\gamma = 1.4$

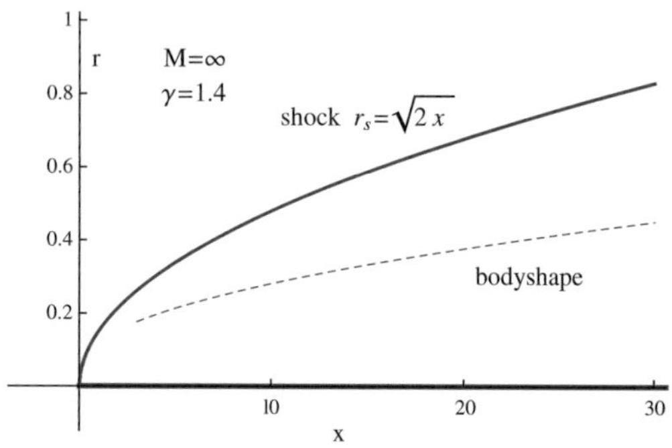

the shock wave is forced away from the body due to bluntness, while in the inverse problem the body must move instead.

Yakura has applied his method also to the axisymmetric problems of the blunted cone that supports a hyperbolidal shock wave and the body that produces a paraboloidal shock wave at $M = \infty$. Yakura's results are close to those of Sychev [125], see Fig. 12.14.

12.3 Viscous Hypersonic Flows

In the first part of this chapter, inviscid hypersonic flows are discussed, starting with hypersonic similarity parameters. High Mach number flows over plates, wedges and cones are treated with hypersonic small disturbance theory as well as the Newtonian flow theory. Numerical methods to solve the blunt body problems are also covered.

To account for viscous effects, boundary layers and viscous/inviscid interaction procedures will be covered next. The main feature of hypersonic boundary layer is the high temperature effects and aerodynamic heating, while the main feature of the interaction problem is the vorticity and entropy in the inviscid layer.

Other important factors, including chemical nonequilibrium with dissociation and ionization as well as low density and rarefied gas dynamics will not be covered here and the reader is referred to more specialized references (see Park [12]).

The discussion is based on the continuum model, where the mean free path of the molecules λ is much smaller than the characteristic length of the body, i.e. the Knudsen number is small (say less than 0.1), where

$$Kn = \frac{\lambda}{l} \tag{12.178}$$

From kinetic theory of gases, $Kn = 1.26\sqrt{\gamma}M_0/Re_l$. For $Kn > 10$, free molecular regime exists and Boltzmann equations must be used.

For the continuum model, the flow is governed by the Navier-Stokes equations. At a solid surface, no penetration and no slip boundary conditions are imposed. Either the temperature or the heat flux are also specified (the effect of slip velocity and jump in temperature at the boundary will not be covered). Moreover, we will assume perfect gas. In general, real gas is governed by $p = \rho RTZ$, where Z is the compressibility factor and γ, the ratio of specific heat under constant pressure and constant volume ($\gamma = C_p/C_v$), is different before and after the shock.

Another important factor is the dependence of the viscosity coefficient on temperature, and we will assume in this discussion a power law, namely

$$\frac{\mu}{\mu_0} = \left(\frac{T}{T_0}\right)^{\omega} \tag{12.179}$$

and for simplicity, let $\omega = 1$. Also, radiation effects will not be considered.

12.3.1 Navier-Stokes Equations

Writing the conservation laws of mass, momentum and energy including the viscous and heat conduction effects, in nondimensional form, reveals the following nondimensional parameters (see White [126]), γ, M_0, Re, Pr, St, where

$$M_0 = \frac{U}{a_0}, \quad Re = \frac{\rho_0 U l}{\mu_0}, \quad Pr = \frac{\mu C_p}{k}, \quad St = \frac{f\,l}{U} \tag{12.180}$$

In the above formulas, k is the heat conductivity which is assumed to be proportional to μ, and f is the characteristic frequency of unsteady motion. For reference quantities, a characteristic length of the body, l, free stream velocity U and density ρ_0 are used.

There is one more nondimensional parameter defined as $Da = t_c/t_m$. The Damkohler number Da is the ratio of the characteristic time to establish chemical equilibrium to the characteristic time of motion, $t_m = l/U$. Da is very small for

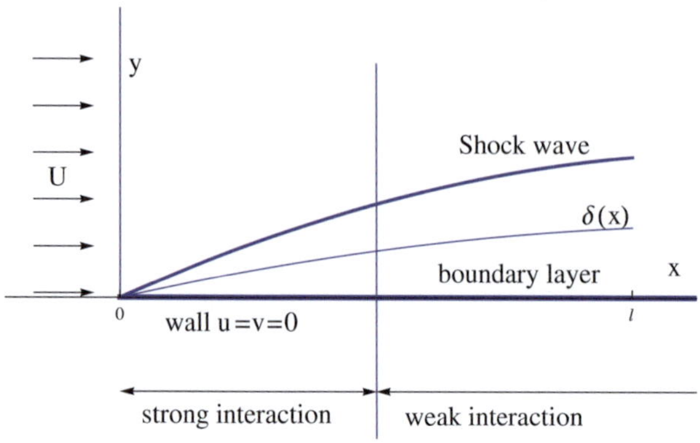

Fig. 12.15 Strong and weak interaction regimes

chemical equilibrium and very large for frozen chemistry, while it is of order one for finite rate chemistry.

Under all the assumptions mentioned above, consider a flat plate in hypersonic flow (but with finite, not even very high Mach number) and assume laminar boundary layer with two regimes, strong and weak interactions as shown in Fig. 12.15.

Across the shock wave, kinetic energy is transformed into enthalpy and the result is high internal energy flow. Since the total enthalpy is constant across the shock, the downstream state is related to the upstream, undisturbed state by

$$H = h_0 + \frac{1}{2}U^2 = h + \frac{1}{2}V^2 \tag{12.181}$$

For hypersonic flows, $V \ll U$ and $h_0 \ll h$, then $h \simeq U^2/2$.

Using the space shuttle as an example (Cousteix et al. [27]), let $U = 7.6$ km/s, the temperature for a perfect gas would be $T = 29{,}000$ K! In reality, some of the energy is stored in the form of chemical energy in the process of dissociation and for real gas $T = 6{,}000$ K instead. Even for $U = 2$ km/s, corresponding to $M_0 = 6$ (assuming $C_p \simeq 1$ KJ/kg · deg), the adiabatic compression leads to a high value $\Delta T \simeq 2{,}273$ K, where

$$C_p \Delta T = \frac{1}{2}U^2, \quad or \quad \frac{\Delta T}{T_0} = \frac{1}{2}\frac{U^2}{C_p T_0} = \frac{\gamma - 1}{2}M_0^2 \tag{12.182}$$

A considerable part of the heat is transported to the vehicle surface (if radiation is neglected).

In the flowing, the thermal boundary layer will be discussed for both isothermal and insulated (adiabatic) walls.

12.3.2 Laminar Boundary Layer in Weak Interaction Regime

For high Reynolds numbers, the viscous effects are confined to a layer next to the wall, of thickness $\delta \ll l$, and inside such a layer $v \ll u$. Based on order of magnitude analysis, the pressure does not vary across the boundary layer, hence the governing equations in two dimensions are (see Hayes and Probstein [127])

$$\frac{\partial \rho u}{\partial x} + \frac{\partial \rho v}{\partial y} = 0 \tag{12.183}$$

$$\frac{\partial \rho u^2}{\partial x} + \frac{\partial \rho u v}{\partial y} = -\frac{\partial p}{\partial x} + \frac{\partial}{\partial y}\left(\mu \frac{\partial u}{\partial y}\right) \tag{12.184}$$

$$\frac{\partial \rho u H}{\partial x} + \frac{\partial \rho v H}{\partial y} = \frac{\partial}{\partial y}\left(k \frac{\partial T}{\partial y}\right) + \frac{\partial}{\partial y}\left(\mu \frac{\partial u^2/2}{\partial y}\right) \tag{12.185}$$

where

$$p = \rho R T, \quad and \quad H = C_p T + \frac{1}{2} u^2 \tag{12.186}$$

The energy equation can also be written in the form

$$\rho C_p \left(u \frac{\partial T}{\partial x} + v \frac{\partial T}{\partial y}\right) = u \frac{\partial p}{\partial x} + \frac{\partial}{\partial y}\left(k \frac{\partial T}{\partial y}\right) + \mu \left(\frac{\partial u}{\partial y}\right)^2 \tag{12.187}$$

see Liepmann and Roshko [128], Shapiro [129] and Schlichting [130].

The pressure gradient is determined by the external flow

$$\frac{dp}{dx} = -\rho_e u_e \frac{du_e}{dx} = \rho_e C_p \frac{dT_e}{dx} \tag{12.188}$$

Since $\partial p/\partial y \simeq 0$, the temperature and density are related and satisfy the relation

$$\rho T = \rho_e T_e \tag{12.189}$$

The wall boundary conditions are $u = v = 0$. Either T_w or $\partial T/\partial y$ is specified at the wall. the thermal boundary layer is shown in Fig. 12.16.

Notice $(\partial T/\partial y)_{y=0} > 0$ for cold plate and $(\partial T/\partial y)_{y=0} < 0$ for hot plate.

The energy equation can be rewritten in terms of Pr (assuming C_p is constant)

$$\frac{\partial \rho u H}{\partial x} + \frac{\partial \rho v H}{\partial y} = \frac{\partial}{\partial y}\left(\mu \frac{\partial H}{\partial y}\right) + \frac{\partial}{\partial y}\left(\left(\frac{1}{Pr} - 1\right) \frac{\partial T}{\partial y}\right) \tag{12.190}$$

For $Pr = 1$, the above equation admits a solution $H = C_p T_s = C_p T + u^2/2 = const$ where the subscript s stands for stagnation.

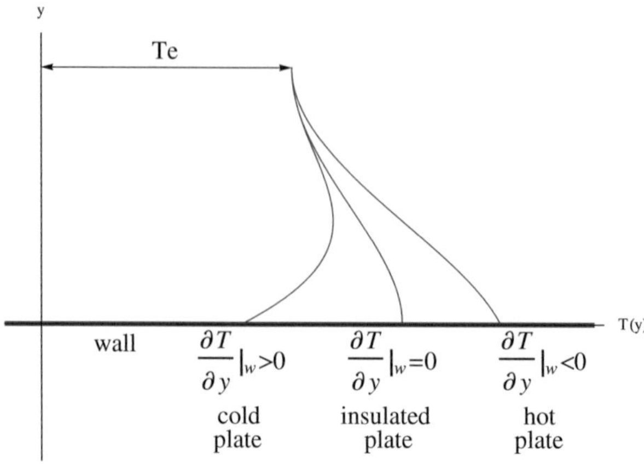

Fig. 12.16 Temperature distributions in boundary layer

Since $u = 0$ at the wall, it follows that the wall temperature is constant and equal to the stagnation temperature, i.e.

$$T_w = T_s = T + \frac{u^2}{2C_p} = T_e + \frac{u_e^2}{2C_p} \tag{12.191}$$

Moreover

$$\frac{\partial T}{\partial y} + u\frac{\partial u}{\partial y} = 0, \quad \Rightarrow \frac{\partial T}{\partial y}|_{y=0} = 0 \quad and \quad \frac{\partial T}{\partial u}|_{y=0} = 0 \tag{12.192}$$

The temperature profile is given by

$$T = T_e + \frac{1}{2C_p}\left(u_e^2 - u^2\right) \tag{12.193}$$

and

$$T_w = T_e\left(1 + \frac{\gamma - 1}{2}M_e^2\right) \tag{12.194}$$

Notice that the stagnation temperature can be constant throughout the boundary layer only if the heat transfer is balanced by the work of viscous stresses or

$$-k\frac{\partial T}{\partial y} = \mu\frac{\partial}{\partial y}\left(\frac{u^2}{2}\right), \quad or \quad \frac{\partial}{\partial y}\left(T + Pr\frac{u^2}{2C_p}\right) = 0 \tag{12.195}$$

therefore H is constant only when $Pr = 1$.

For the case of the flat plate, $dp/dx = 0$, a particular solution of the energy equation is $H = u$ since the energy equation reduces to the momentum equation. More generally, let $H = a + bu$. To find a and b, we impose the boundary conditions $u = 0$ and $T = T_w$ at $y = 0$ and $T = T_e$ and $u = u_e$ at the edge of the boundary layer. Thus the relation between u and T inside the boundary layer is given by

$$C_p T + \frac{1}{2}u^2 = C_p T_w + \left(C_p(T_e - T_w) + \frac{1}{2}u_e^2\right)\frac{u}{u_e} \tag{12.196}$$

Now, the heat transfer and skin friction at the wall are given by

$$q_w = k_w \frac{\partial T}{\partial y}|_w, \quad \tau_w = \mu_w \frac{\partial u}{\partial y}|_w$$

hence

$$\frac{q_w}{\tau_w} = \frac{k_w}{\mu_w}\frac{T_e}{u_e}\left(1 - \frac{T_w}{T_e} + \frac{\gamma - 1}{2}M_e^2\right) \tag{12.197}$$

Another form of the temperature equation is

$$\frac{T - T_w}{T_e} = \left(1 - \frac{T_w}{T_e}\right)\frac{u}{u_e} + \frac{\gamma - 1}{2}M_e^2 \frac{u}{u_e}\left(1 - \frac{u}{u_e}\right) \tag{12.198}$$

The direction of the heat flow can be known from the temperature gradient at the wall. Assuming $(\partial u/\partial y)_{y=0} > 0$ and from $\partial T/\partial u$ given by

$$\frac{u_e}{T_e}\left(\frac{dT}{du}\right)_w = 1 - \frac{T_w}{T_e} + \frac{u_e^2}{2C_p T_e} \tag{12.199}$$

therefore, if $(dT/du)_w < 0$, there is a flow of heat from the wall to the fluid, and if $(dT/du)_w > 0$, there is a flow of heat from the fluid to the wall ($Pr = 1$).

Now, it is useful to mention some definitions.

A film coefficient of heat transfer, f, is defined as

$$f = \frac{q_w}{T_w - T_{wa}} \tag{12.200}$$

and the Stanton number is defined as

$$C_H = \frac{f}{C_p \rho_0 U} = \frac{C_f}{2} \tag{12.201}$$

This is known as the Reynolds analogy. In terms of the Nusselt number

$$Nu = \frac{C_f}{2} Re_0 \tag{12.202}$$

or, in terms of wall properties

$$C_{Hw} = \frac{C_{fw}}{2} Re_w \tag{12.203}$$

If $Pr \neq 1$, it is conventional to define the adiabatic wall temperature as

$$T_a = T_e \left(1 + r\frac{\gamma - 1}{2} M_e^2\right) \tag{12.204}$$

For laminar flow, $r \simeq \sqrt{Pr}$ as demonstrated by Schlichting [130]. r is called the recovery factor and it represents the ratio of the frictional temperature increase of the plate to that due to adiabatic compression.

12.3.3 Solutions of Laminar Boundary Layer Equations at Hypersonic Speeds

Following Lees [131], Hayes and Probstein [1] and Cheng [132], the boundary layer equations are rewritten in terms of Howarth-Dorodnitsyn variables

$$\xi = \int_0^x c\frac{\rho}{\rho_e} dx, \quad \eta = \sqrt{\frac{Re}{\xi}} \int_0^y \frac{\rho}{\rho_e} dy \tag{12.205}$$

where $Re = \rho_e u_e l / \mu_e$ and $\mu/\mu_0 = cT/T_0$.

Assuming $Pr = 1$, one obtains

$$2\frac{d^3 f}{d\eta^3} + f\frac{d^2 f}{d\eta^2} = \epsilon h, \quad and \quad 2\frac{d^2 g}{d\eta^2} + f\frac{dg}{d\eta} = 0 \tag{12.206}$$

where $\epsilon = (\gamma - 1)/(\gamma + 1)$. f is a reduced streamfunction such that $df/d\eta = u/u_e$ and $g = (H - H_w)/(H_e - H_w)$. h consists of terms including $\int_0^x p dx$.

The boundary conditions are

$$f = \frac{df}{d\eta} = g = 0, \quad at\ \eta = 0 \tag{12.207}$$

$$\frac{df}{d\eta} = g = 1, \quad at\ \eta = \infty \tag{12.208}$$

Therefore, for small ϵ, there is a similar solution governed by Blasius equation and in this case, $g = df/d\eta$.

If terms of order ϵ are retained, solutions for heat transfer rate, skin friction and displacement thickness can be written explicitly in terms of pressure distribution. The results of this analysis, called local similarity, are given in terms of a nondimensional parameter χ which depends on M_0 and Re (which is called the hypersonic viscous interaction parameter)

$$\chi = M_0^3 \sqrt{\frac{c}{Re}} \tag{12.209}$$

The Stanton number and skin friction coefficients based on the edge conditions are

$$M_0^3 St \simeq 0.332\chi \left(\frac{p}{p_e}\right) \left(\int_0^x \frac{p}{p_e} dx\right)^{-1/2} \tag{12.210}$$

$$C_f \simeq 2St \tag{12.211}$$

$$M_0 \frac{\delta^*}{l} \simeq \epsilon \left(0.664 + 1.73 \frac{T_w}{T_e}\right) \chi \left(\int_0^x \frac{p}{p_e} dx\right) \left(\frac{p}{p_e}\right)^{-1} \tag{12.212}$$

Notice that $\delta^* \simeq \delta$.

Hayes and Probstein [1] reviewed similar solutions of boundary layers with chemical reactions including constant pressure solutions, stagnation point solutions, locally similar solutions with $Pr = 1$. Discussions of local similarity concepts can be found in [1] as well as in the work of Moore [133].

We will consider next the integral methods, where the profiles of the unknown functions are assumed and hence the partial differential equations are reduced to ordinary differential equations. The integrated momentum and energy equations give an estimate of the x-dependence of the boundary layer solutions and indicate when the local similarity concept breaks down.

Following Shapiro [129] and Schlichting [130], the integral momentum equation is given by

$$\frac{d}{dx} \int_0^\delta \left(\rho_e u_e^2 - \rho u^2\right) dy - u_e \frac{d}{dx} \int_0^\delta (\rho_e u_e - \rho u)\, dy = \tau_w \tag{12.213}$$

or, in terms of displacement thickness δ^* and momentum thickness θ, the above equation becomes

$$\frac{1}{\rho_e u_e^2} \frac{d}{dx} \left(\rho_e u_e^2 \theta\right) + \frac{\delta^*}{u_e} \frac{du_e}{dx} = \frac{\tau_w}{\rho_e u_e^2} \tag{12.214}$$

The integral energy equation is given by

$$\frac{d}{dx} \int_0^\delta \rho u H dy - H_e \frac{d}{dx} \int_0^\delta \rho u dy = q_w \tag{12.215}$$

or, in terms of the enthalpy thickness for $H_e = const.$

$$\frac{d\delta_H}{dx} = -\frac{q_w}{\rho_e u_e H_e} \tag{12.216}$$

where

$$\delta_H = \int_0^\delta \frac{\rho u}{\rho_e u_e}\left(1 - \frac{H}{H_e}\right) dy \tag{12.217}$$

To avoid assuming the profiles of unknown functions, numerical solutions of the boundary layer equations can be obtained using finite difference methods. Applications to hypersonic flows, with and without chemical reactions are reported by Blottner [134].

12.3.4 Weak Viscous/Inviscid Interactions

The standard boundary layer theory does not take into account normal gradients outside the boundary layer. A higher order boundary layer theory was developed by Van Dyke in [135]. The first order outer region equations are the standard Euler equations for inviscid flows, while the first order inner region equations are the standard boundary layer equations. The second order outer region equations are the linearized Euler equations, while the second order inner region equations are linear perturbation of the boundary layer equations with several source terms.

Because of linearity, the boundary conditions and the source terms can be split to different second order effects including curvature terms, displacement, entropy and total enthalpy gradients. Notice, the second order external flow is affected by the displacement of the first order boundary layer. The inner and outer solutions must match and a single composite formula valid in the whole domain can be obtained.

Recently, Cousteix developed a hypersonic interaction theory based on the defect approach [136]. (The defect approach was developed by Le Balleur for transonic flows). Cousteix used in the inner region the difference between the physical variables and the external solutions. In the outer region, the defect variables vanish and the equations for the outer flow region are identical to Van Dyke's. The equations in the inner region are different. For example, the pressure in the first order boundary layer varies and is equal to the local inviscid flow pressure.

In the defect approach, better matching is obtained between the inner and outer solutions than in Van Dyke's method, at the expense of calculating the inviscid flow on the boundary layer grid.

Successful applications of Euler/boundary layer coupling via defect approach were reported by Monnoyer [137] who demonstrated that proper accounting for entropy and vorticity variations in the inviscid flow outer region is an important aspect of hypersonic flow simulations.

12.3.5 Strong Viscous/Inviscid Interaction

Lee and Cheng [138] obtained a uniformly valid solution for hypersonic strong interaction problem including the second order boundary layer correction due to the uneven heating and external vorticity created by the curved leading edge shock wave. In general, there exist between the inviscid region and the boundary layer an intermediate transitional layer as demonstrated by Bush [139]. Near the leading edge, the problem is more complicated because of the slip velocity and the temperature jump in the wall boundary condition.

Cheng and Novack [140] introduced a set of composite equations, with proper boundary conditions, to solve such problems numerically. The equations are the conservation of mass, tangential momentum and energy. In the latter two equations, the second order derivatives in the tangential direction are ignored. An approximate version of the normal momentum equation is used as follows

$$\frac{\partial \rho u}{\partial x} + \frac{\partial \rho v}{\partial y} = 0 \tag{12.218}$$

$$\frac{\partial \rho u^2}{\partial x} + \frac{\partial \rho u v}{\partial y} = -\frac{\partial p}{\partial x} + \frac{\partial}{\partial y}\left(\mu \frac{\partial u}{\partial y}\right) \tag{12.219}$$

$$\frac{\partial \rho u v}{\partial x} + \frac{\partial \rho v^2}{\partial y} = -\frac{\partial p}{\partial y} + \frac{\partial}{\partial y}\left(\mu \frac{\partial v}{\partial y}\right) \tag{12.220}$$

$$\frac{\partial \rho u H}{\partial x} + \frac{\partial \rho v H}{\partial y} = \frac{\partial}{\partial y}\left(\frac{\mu}{Pr}\left[H - (1 - Pr)\frac{u^2}{2}\right]\right) \tag{12.221}$$

Cheng used a Crank-Nicolson scheme to obtain a numerical solution for attached flow over a flat plate, marching in the flow direction, on a relatively fine mesh ($\Delta x/l = 0.005$ and $\Delta y/\delta = 0.01$).

Cheng solve also blunt body problem with a similar set of equations in body fitted coordinates, where the transverse momentum equation is replaced by

$$\frac{\partial p}{\partial y} = -K_c \rho u^2 \tag{12.222}$$

where K_c is the longitudinal curvature of a reference surface.

Rudman and Rubin [141] and Rubin and Khosla [142] used similar formulations where they treated $\partial p/\partial x$ term as a higher order term.

It is instructive to study the type of this set of equations. Following Vigneron et al. [143] and also Ref. [35], the streamwise pressure gradient $\partial p/\partial x$ is multiplied by a coefficient ω.

For inviscid flows and $\omega = 1$, the equations are hyperbolic provided $u^2 = v^2 > a^2$ and space marching is feasible. For $0 \leq \omega \leq 1$, the equations are hyperbolic even in subsonic region provided

$$\omega \leq \frac{\gamma M_x^2}{1 + (\gamma - 1) M_x^2} \tag{12.223}$$

where $M_x = u/a$.

For viscous flows, ignoring the first derivative in y, the conditions that the equations are parabolic are

$$u > 0 \quad and \quad \omega \leq \frac{\gamma M_x^2}{1 + (\gamma - 1) M_x^2} \tag{12.224}$$

hence, the so called parabolized Navier-Stokes (PNS) equations can be marched only for attached flows and with a fraction of the streamwise pressure gradient. However, $\omega \to 0$ near the wall where $M_x = 0$. Lubard and Helliwell [144] showed that with backward difference approximation of $\partial p/\partial x$, there is a minimum Δx to be used for stability based of Fourier analysis and using finer meshes is not possible. See also Golovachov [39].

The question about the type of the PNS equations can be clearly seen in terms of velocity/vorticity formulation. Let

$$\rho u = \frac{\partial \psi}{\partial y}, \quad \rho v = -\frac{\partial \psi}{\partial x} \tag{12.225}$$

hence

$$\frac{\partial}{\partial x}\left(\frac{1}{\rho}\frac{\partial \psi}{\partial x}\right) + \frac{\partial}{\partial y}\left(\frac{1}{\rho}\frac{\partial \psi}{\partial y}\right) = -\omega \tag{12.226}$$

and

$$\frac{\partial}{\partial x}(\rho u \omega) + \frac{\partial}{\partial y}(\rho v \omega) + f(\rho) = \frac{\partial}{\partial y}\left(\mu \frac{\partial \omega}{\partial y}\right) \tag{12.227}$$

The vorticity equation can be marched since it is a convection/diffusion/reaction equation, however, the streamfunction equation is of mixed elliptic/hyperbolic type. In the outer inviscid supersonic flow, it is hyperbolic while near the solid surface, the speed is low and the equation is of elliptic type allowing for upstream influence and reversal flow. In boundary layer approximation, the streamline curvature term is ignored and the stream function equation becomes

$$\frac{\partial}{\partial y}\left(\frac{1}{\rho}\frac{\partial \psi}{\partial y}\right) = -\omega \tag{12.228}$$

and the system is of parabolic type. Therefore, it can be marched in space for attached flows. Couplings of Euler and parabolic equations are covered in Chap. 9.

To complete the formulation, the energy equation is solved for H or T and the pressure is calculated using the normal momentum equation in some form. The density is then calculated from the perfect gas law or its generalization.

It seems that marching the locally elliptic equation is a formidable task which appears in different forms in hypersonic flow research.

There are many attempts in the literature to use PNS with local separated regions using iterations locally. The obvious strategy is to use unsteady formulation to reach a steady state solution. Of course, such an option is more expensive.

12.3.6 Theoretical Developments

For the case of a flat plate, approximate analytical results are available for both weak and strong interaction (excluding the neighborhood of the leading edge).

Following Hayes and Probstein [1] and Cox and Crabtree [3], the displacement thickness for weak interaction on a flat plate is given by (as discussed before)

$$\delta^* \simeq \epsilon \left(0.664 + 1.73\frac{T_w}{T_e}\right) M_0^2 \sqrt{\frac{\nu_0 c}{U}}\sqrt{x} \tag{12.229}$$

Thus, the equivalent body slope due to the displacement effect is

$$\frac{d\delta^*}{dx} \simeq \frac{\epsilon}{2}\left(0.664 + 1.73\frac{T_w}{T_e}\right)\frac{M_0^2\sqrt{c}}{\sqrt{Re_x}} \tag{12.230}$$

where $Re_x = \rho_0 U x/\mu_0$.

On the other hand, the pressure ratio for oblique shocks is

$$\frac{p}{p_0} = 1 + \frac{\gamma(\gamma+1)}{4}K^2 + \gamma K^2\sqrt{\frac{(\gamma+1)^2}{4} + \frac{1}{K^2}} \tag{12.231}$$

where $K = M_0\, d\delta^*/dx$.

For $(M_0\, d\delta^*/dx)^2 \ll 1$, the pressure ratio over a wedge is obtained from

$$\frac{p}{p_0} = 1 + \gamma M_0 \frac{d\delta^*}{dx} = 1 + \frac{\gamma\epsilon}{2}\left(0.664 + 1.73\frac{T_w}{T_e}\right)\chi \tag{12.232}$$

where $\chi = M_0^3\sqrt{c}/\sqrt{Re_x}$.

For an insulated surface ($T_w/T_0 \to 1$) and a very cold wall ($T_w/T_0 \to 0$), the above results are reduced to

$$\frac{p}{p_0} = \begin{cases} 1 + 0.28\chi, & \textit{for insulated wall} \\ 1 + 0.078\chi, & \textit{for cold wall} \end{cases} \tag{12.233}$$

In the case of strong interaction, the pressure formula is based on strong oblique shock, i.e.

$$\frac{p}{p_0} = \frac{\gamma(\gamma+1)}{2}\left(M_0 \frac{d\delta^*}{dx}\right)^2 \tag{12.234}$$

It can be shown that for a consistent solution, the displacement thickness and induced pressure distribution must comply to the following behaviors, $\delta^* \sim x^{-3/4}$, $p \sim x^{-1/2}$.

Let $p/p_0 \sim kx^{-1/2}$, then

$$\delta^* \simeq \epsilon \left(0.664 + 1.73\frac{T_w}{T_e}\right) M_0^2 \sqrt{\frac{\nu_0 c}{U}} \sqrt{\frac{2}{k}} x^{3/4} \tag{12.235}$$

Hence

$$\left(\frac{d\delta^*}{dx}\right)^2 \simeq \frac{9}{8}\frac{\epsilon^2}{k}\left(0.664 + 1.73\frac{T_w}{T_e}\right)^2 M_0^4 \frac{\nu_0 c}{U} x^{-1/2} \tag{12.236}$$

and

$$\frac{p}{p_0} = \frac{\gamma(\gamma+1)}{2} M_0^2 \frac{9}{8}\frac{\epsilon^2}{k}\left(0.664 + 1.73\frac{T_w}{T_e}\right)^2 M_0^4 \frac{\nu_0 c}{U} x^{-1/2} = kx^{-1/2} \tag{12.237}$$

Therefore

$$k^2 = 9\frac{\gamma(\gamma+1)}{16}\epsilon^2 \left(0.664 + 1.73\frac{T_w}{T_e}\right)^2 M_0^6 \frac{\nu_0 c}{U} \tag{12.238}$$

and

$$\frac{p}{p_0} = \begin{cases} 0.53\chi, & \textit{for insulated wall} \\ 0.15\chi, & \textit{for cold wall} \end{cases} \tag{12.239}$$

The role of parameter χ is clear for both interaction regions. Comparison with experimental results are reasonable (see Hayes and Probstein [1]).

Later, a more comprehensive theory has been developed based on a triple deck structure, see reviews by Stewartson [145], Rothmayer and Smith [146], Sychev [147] and Brown et al. [148]. In this theory, upstream influence (suggested earlier by Lighthill [149]) as well as flow reversal and separation are studied in details.

12.4 Hypersonic Area Rule

An interesting "area rule", similar to transonic area rule, was introduced by Ladyzhenskii [150]. Following the discussion of Hayes and Probstein [2] and of Cheng [151], Ladyzhenskii showed that two slender bodies with blunt noses will have the same axial pressure distributions and the same drag under the following conditions:

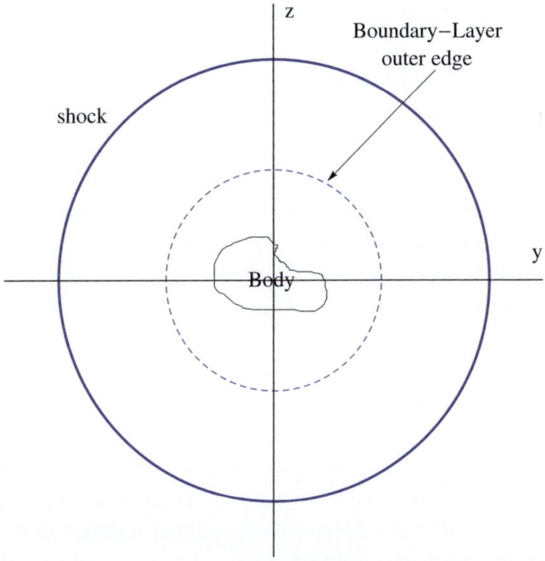

Fig. 12.17 A sketch in the transverse plane illustrating the concept of equivalent body of revolution

(i) the drag due to the blunt nose is the same for the two bodies,
(ii) the equivalent body will be within the entropy layer,
(iii) the cross section distribution is smooth.

Other variant requires the nose parts of equivalent bodies to have the same shape.

The analysis is based on the assumption that the transverse pressure gradients in the entropy layer may be neglected for slender bodies with blunt noses and asymmetric shapes or at small angle of attack. It follows that the outer field must be (almost) axisymmetric.

A stronger form was discussed by Cheng [151], where only a limited part of the entropy layer is affected by the asymmetric after body (see Fig. 12.17).

Ladyzhenskii treated another type of area rule based on the assumption that $\epsilon = (\gamma - 1)/(\gamma + 1) \ll 1$.

The rule is based on Chernyi's formulation for $M_0 = \infty$, where the pressure is constant across the entire shock layer. Experiments carried out by Krasovskii [152] to correlate the drag of equivalent bodies indicated that blunting can indeed lower the drag slightly (compared to sharp cones), while the yawed body data showed a reduction in lift as large as 60 % for $M_0 = 18$ and $Re = 10^6$.

Strictly speaking, the rule implies no lift and no moment. Indeed, slender bodies at hypersonic speeds suffer from a serious aerodynamic control problem.

12.5 Hypersonic Similitude

The inviscid hypersonic similitude was given first by Tsien [41] as discussed earlier. Based on the equivalence principle by Hayes [153], the similitude is applicable to general equations of state and three dimensional bodies. Hayes and Probstein [2], extended the similitude to take into account the interaction effect of the displacement thickness of the boundary layer and in this case, the total drag, including frictional drag, obeys the similarity law for the pressure drag. In viscous hypersonic similitude, the fluid is assumed however to be a perfect gas with additional assumptions on the dependence of the viscosity on temperature.

More recently, Viviand [154] studied the general similitude for the full Navier-Stokes equations for hypersonic flows involving high temperature real gas effects, whether in thermodynamic equilibrium or not, and showed that exact similitude is not possible in general. Viviand however introduced the concept of approximate similitude and applied it for simple models.

In the following the main important results of these studies are summarized.

A starting point is the Mach number Independence Principle by Oswatitsch [155]. In the case of a perfect gas, a limiting solution of the governing equations and boundary conditions is obtained as M_0 approaches infinity. For a general fluid, the free stream density ρ_0 and velocity U must be fixed and the quantities a_0, p_0, T_0 and h_0 approach zero. If M_0 is sufficiently large, the solution behind the bow shock wave becomes independent of M_0 and hence of p_0 and T_0. Hayes and Probstein [1] considered this principle is more than a similitude: "Two flows behind the bow shock waves, with different sufficiently high values of M_0, are not merely similar but are essentially identical".

The Mach Number Independence Principle, with ρ_0 and U fixed applies to real fluids, including viscosity, heat conduction, relaxation, diffusion and rarefied gas effects. (For a perfect gas with constant γ, the requirement that ρ_0 and U be fixed can be relaxed provided the gas is inviscid). Hayes and Probstein [1] introduced the "Hypersonic Boundary Layer Independence" where they showed that the principal part of a hypersonic boundary layer, with given pressure and wall temperature distributions and free stream total enthalpy is independent of the external Mach number distribution outside the boundary layer. (The dependence upon M_δ and ρ_δ appears only within a thin transitional layer at the outer edge of the boundary layer).

Again, the above principle is considered more than a similitude. The viscous hypersonic similitude was then developed taking into consideration the interaction between the displacement thickness distribution of the boundary layer and the external flow field. it is required that both the body and the displaced body are affinely related, hence

$$\frac{\delta^*}{\tau c} = f\left(\frac{x}{l}\right) \tag{12.240}$$

Hayes and Probstein [1] concluded that χ, the interaction parameter, must be invariant, where

$$\chi = M_0^3 \sqrt{\frac{c}{Re_0}} \tag{12.241}$$

with $c = \mu_w T_0/(\mu_0 T_w)$ and $Re_0 = \rho_0 U l/\mu_0$.

The result for the surface pressure distribution is given, for two-dimensional flows, by

$$\frac{C_p}{\tau^2} = F\left(\frac{x}{l}, K, \chi, \frac{h_w}{Hh}, \gamma, Pr\right) \tag{12.242}$$

The skin friction and heat transfer coefficient become

$$\frac{C_f}{\tau^3} = G\left(\frac{x}{l}, K, \chi, \frac{h_w}{Hh}, \gamma, Pr\right) \tag{12.243}$$

$$\frac{C_H}{\tau^3} = H\left(\frac{x}{l}, K, \chi, \frac{h_w}{Hh}, \gamma, Pr\right) \tag{12.244}$$

Similarly, the total force coefficients C_L/τ^2, C_D/τ^3 as well as the pitching moment coefficient obey similar laws. (Notice, it is necessary that the base pressure follow the similarity law for the base pressure drag to follow the similarity law of the drag).

Formulas for both weak and strong viscous interactions, in terms of the parameter χ, have been discussed earlier.

12.6 Hypersonic Vehicle Design

Townend [156] considered three applications:

(i) space launchers with reusable aerospace planes,
(ii) hypersonic airlines,
(iii) trans-atmospheric orbital transfer vehicles.

In all three applications, a central theme is integrating air-breathing propulsion into an aerodynamic design called "waverider", a concept introduced by Nonweiler [157]. The idea is to generate three dimensional lifting bodies in high Mach number flows utilizing the streamlines behind a known shock wave.

It is well known that the maximum lift to drag ratio, $(L/D)_{max}$, decreases as Mach number increases in the hypersonic regime, due to the strength of the shock wave. As an example is the dependence of L/D versus angel of attack for a flat plate, according to Newtonian flow theory. Accounting for laminar skin friction, $(L/D)_{max}$ is 6.5 for $M_0 = 10$ and $Re = 3.0\,10^6$. For a Boeing 707 cruising near Mach one, $(L/D)_{max} \simeq 20$. To generate higher L/D, waveriders are considered (see Anderson [9], Hayes and Probstein [2]).

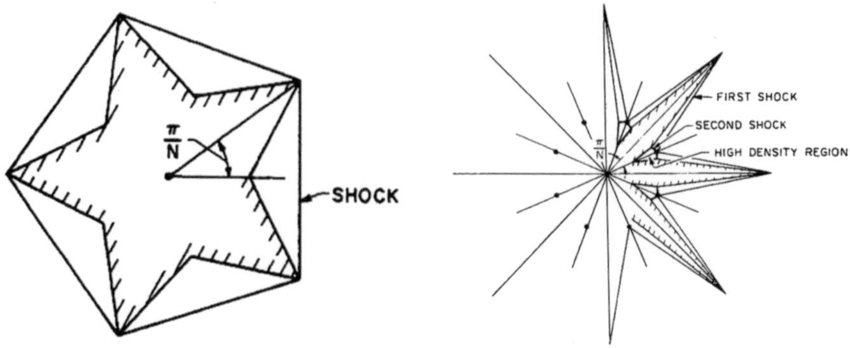

Fig. 12.18 Maikapar's star shaped body and shock waves (*left*); Gonor's star shaped body and associated shocks (*right*) (from Hypersonic Flow Theory, Vol. I, with kind permission of Dover publication, 2004)

Fig. 12.19 Nonweiler caret wing

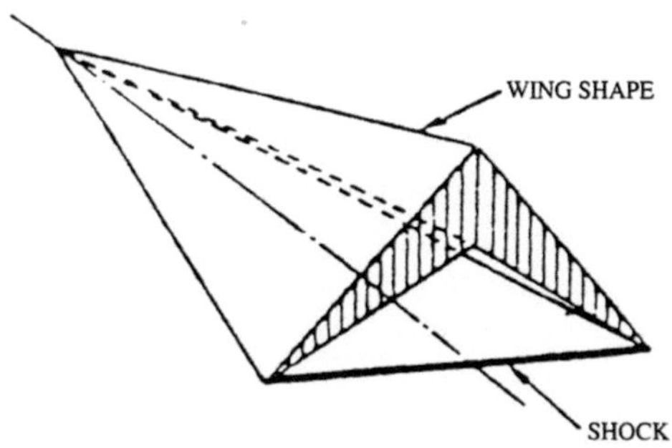

Inviscid flow solutions for star-shaped conical bodies using wedge solutions have been given by Maikapar [158] and Gonor [159]. See Fig. 12.18 for the cross section of the body and the shock system.

Such a design may have much less drag than a circular cone of the same cross section area.

Nonweiler [157] and Peckham [160] have proposed delta wings of inverted V-shape, which can be considered as pieces of Maikapar's shape.

The shock waves in Maikapar's solution are weak. In Gonor's solution, a straight weak shock is reflected from a plane of symmetry, in a regular manner and strikes the arm of the star.

An example of Nonweiler wing and shock known as "caret wing" is shown in Fig. 12.19.

The flow over a cone at zero angle of attack can be used to generate waverider shapes. as done by Jones et al. [161]. See Fig. 12.20 constructed by Anderson [162].

Surveys on waveriders are given by Townend [156], Eggers et al. [163], Kuchemann [18], Roe [164] and Koppenwwalter [165].

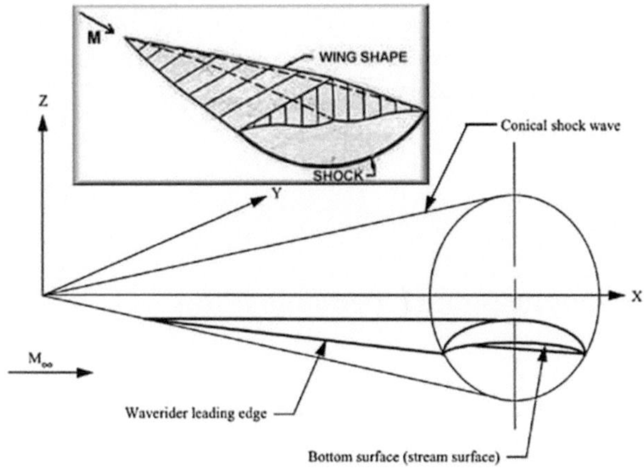

Fig. 12.20 Bodies supporting and containing parts of conical shock waves (from en.wikipedia.org)

Recently, Sobieczky et al. [166] introduced inverse aerodynamic design methods to obtain more general shapes of waveriders from given shock waves. In this approach, a flow field bounded by a shock wave, a stream surface defining a solid body contour and an exit surface are calculated. For 2D inviscid supersonic flow, marching in the flow direction or normal to it are both possible. For three dimensional flows, cross marching is an ill-posed problem. However, axisymmetric equations can be used to approximate (to second order) locally three dimensional flows. Location of the axis of this "osculating" axisymmetric flow depends on local flow curvature and velocity gradient. The calculations can be made in a 2D meridian plane by the method of characteristics, however occurrence of limit surfaces (with multi-valued solution) requires initial data modification. Alternatively, the authors solved Euler equations in an inverse approach, guided by characteristic domain of dependence.

Rasmussen and Stevens [167] were able to use calculus of variations to optimize the waverider shapes utilizing the available analytical solutions and the associated properties of inviscid flow. In this regard, the reader is referred to the book on optimum aerodynamic shapes edited by Miele [168].

Recently, Cole [169] designed optimal conical wings in hypersonic flows based on Euler codes.

Long [170] studied the off design condition performance of hypersonic waveriders using Euler codes.

In the above works, skin friction drag is not included, hence the inviscid L/D is not reliable since the waveriders have large wetted surface areas. Anderson [9] introduced a family of waveriders called viscous-optimized hypersonic waveriders using CFD code and numerical optimization techniques. Comparison with wind tunnel tests are satisfactory. Their designs differ considerably in shape, depending on laminar to turbulent transition models. Anderson [9] considered also nonequilibrium chemistry on waverider aerodynamics. Nonweiler [157] raised the concern whether the sharp leading edges of waveriders can survive the heat flux without using active cooling. With the available new materials, the solid body conductivity and radiative cooling can indeed limit the temperature to acceptable levels.

References

A-General References on Hypersonic Flows

1. Hayes, W., Probstein, R.: Hypersonic Flow Theory. Academic Press, New York (1959)
2. Hayes, W., Probstein, R.: Hypersonic Flow Theory, Vol. I: Inviscid Flows. Academic Press, New York (1966)
3. Cox, R.N., Crabtree, L.F.: Elements of Hypersonic Aerodynamics. The English University Press, London (1965)
4. Chernyi, G.G.: Introduction to Hypersonic Flow. Academic Press, New York (1961)
5. Truitt, R.W.: Hypersonic Aerodynamics. The Ronald Press Company, New York (1959)
6. Truitt, R.W.: Fundamentals of Aerodynamic Heating. The Ronald Press Company, New York (1960)
7. Dorrance, W.H.: Viscous Hypersonic Flow. McGraw Hill, New York (1962)
8. Zierep, J.: Theorie der Schallnahen und der Hyperschallströmungen, Verlag G. Braun, Karlsruhe, Germany (1966)
9. Anderson, J.: Hypersonic and High-Temperature Gas Dynamics. 2nd edn. AIAA, Reston, Virginia (2006)
10. Bertin, J.: Hypersonic Aerothermodynamics. AIAA. Reston, Virginia (1994)
11. Rasmussen, M.: Hypersonic Flow. Wiley, New York (1994)
12. Park, C.: Nonequilibrium Hypersonic Aerothermodynamics. Wiley, New York (1990)
13. Neiland, V.Ya., Bogohepov, V.V., Dudin, G.N., Lipatov, I.I.: Asymptotic Theory of Supersonic Viscous Gas Flows. Elsevier, Amsterdam (2008)
14. Lunev, V.: Real Gas Flows with High Velocities. CRC Press, Boca Raton, Florida (2010)
15. Hirschel, E.H.: Basics of Aerothermodynamics. AIAA, Reston, Virginia (2005)
16. Lees, L.: Hypersonic Flow. In: Proceedings of the 5th International Aeronautical Conference, AIAA, Los Angeles, (1955)
17. Cheng, H.K.: Recent advances in hypersonic flow research. AIAA J. **1**(2), 295–310 (1963)
18. Küchemann, D.: Hypersonic Aircrafts and their Aerodynamics Problems. Progress in Aeronautical Sciences, vol. 6. Pergamon Press (1965)
19. Collar, A.R., Tinkler, J. (eds.): Hypersonic Flow. Academic Press, New York (1960)
20. Riddle, F. (ed.): Hypersonic Flow Research. Academic Press, New York (1962)
21. Hall, G. (ed.): Fundamental Phenomena in Hypersonic Flow. Cornell University Press, New York (1966)
22. Holt, M. (ed.): Basic Developments in Fluid Dynamics, vol. I. Academic Press, New York (1968)
23. Loh, W.H.T. (ed.): Modern Developments in Gas Dynamics. Plenum Press, New York (1969)
24. Mikhailov, V.V., Neiland, V.Ya., Sychev, V.V.: The theory of viscous hypersonic flow. Annu. Rev. Fluid Mech. **3**, 371–396 (1971)
25. Cheng, H.K.: Perspectives on hypersonic viscous flow research. Annu. Rev. Fluid Mech. **25**, 455–485 (1993)
26. Cheng, H.K., Emanuel, G.: Perspective on hypersonic nonequilibrium flow. AIAA J. **33**(3), 385–400 (1995)
27. Cousteix, J., Arnal, D., Aupoix, B., Brazier, J. Ph. Lafon, A.: Shock layers and boundary layers in hypersonic flows. Prog. Aerosp. Sci. **30**, 95–212 (1994). see also Cousteix, J., Mauss, J.: Asymptotic Analysis and Boundary Layers. Springer, New York (2007)

28. Bertin, J., Cummings, R.: Fifty years of hypersonics: where we've been, where we're going. Prog. Aerosp. Sci. **39**, 511–536 (2003)
29. Bogdonoff, S. (ed.): Hypersonic Boundary Layers and Flow Fields. AGARD Conference Proceedings, No. 30 (1968)
30. Wendt, J.F., Pankhurst, R.C. (eds.): Aerodynamic Problems of Hypersonic Vehicles. AGARD Lecture Series, No. 42 (1972)
31. Wendt, J.F.: Hypersonic Aerothermodynamics. VKI Lecture Series, Von Karman Institute for Fluid Dynamics -1984-01. Belgium (1984)
32. Sacher, P.W. (ed.): Aerodynamics of Hypersonic Lifting Vehicles. AGARD Conference Proceedings, No. 428 (1987)
33. Bignell, P.R., Dujarric, C.: Theoretical and Experimental Methods in Hypersonic Flows. AGARD Conference Proceedings, No. 514 (1993)
34. Sarma, G.S.R.: Aerothermochemistry for Hypersonic Technology. VKI Lecture Series, Von Karman Institute for Fluid Dynamics -1995-04. Belgium (1995)
35. Bertin, J., Glowinski, R., Periaux, J. (eds.): Hypersonics, vol. I and II. Birkhäuser, Boston (1989)
36. Desideri, J., Glowinski, R., Periaux, J. (eds.): Hypersonic Flows for Reentry Problems. Springer, Berlin (1991)
37. Murthy, T.K.S. (ed.): Computational Methods in Hypersonic Aerodynamics. Kluwer Academic Publishers, Dordrecht (1991)
38. Rubin, S., Tannehill, J.: Parabolized/reduced Navier-Stokes computational techniques. Annu. Rev. Fluid Mech. **24**, 117–144 (1992)
39. Golovachov, Yu.: Numerical Simulation of Viscous Shock Layer Flows. Kluwer Academic Publishers, Dordrecht (1995)
40. Bertin, J., Periaux, J., Ballman, J. (eds.): Advances in Hypersonics: Modeling Hypersonic Flows. Springer, New York (2012)

B-References Cited in the Text

41. Tsien, H.S.: Similarity laws of hypersonic flows. J. Math. Phys. **25**, 247–251 (1946)
42. Linnell, R.D.: Two-dimensional airfoils in hypersonic flow. J. Aerosp. Sci. **16**, 22–30 (1949)
43. Dorrance, W.H.: Two dimensional airfoils in hypersonic flows. J. Aerosp. Sci. **19**, 593–600 (1952)
44. Ehret, D., Rossow, V., Stevens, V.: An Analysis of the Applicability of the Hypersonic Similarity Law to the Study of Flow About Bodies of Revolution at Zero Angle of Attack. NACA TN 2250 (1952)
45. Van Dyke, M.: A Study of Hypersonic Small-Disturbance Theory. NACA report no. 1194 (1954)
46. Goldsworthy, F.A.: Two-dimensional rotational flow at high mach number past thin aerofoils. Q. J. Mech. Appl. Math. **5**, 54–63 (1952)
47. Bird, G.A.: Some remarks on the foundation of piston theory and supersonic-hypersonic similarity. J. Aerosp. Sci. **25**, 138–139 (1958)
48. Hamaker, F.M., Wong, T.J.: The Similarity Law for Nonsteady Hypersonic Flows and Requirements for the Dynamical Similarity of Related Bodies in Free Flight, NACA TN No. 2631 (1952)
49. Lighthill, M.J.: Higher Approximations in General Theory of High Speed Aerodynamics. Princeton University Press, New Jersey (1954)
50. Kogan, A.: An application of crocco's stream function to the study of rotational supersonic flow past airfoils. Q. J. Mech. Appl. Math. **11**, 1–23 (1958)
51. Shen, S.F., Lin, C.C.: On the Attached Curved Shock in Front of a Sharp-Nosed Symmetric Body Placed in a Uniform Stream, NACA TN No. 2505 (1951)

52. Velesko, L.G., Grodzovskii, G.L., Krashchennikova, N.L : Table of flow parameters for power law bodies of revolution at hypersonic speeds. TsAGI Report (1956)
53. Sedov, L.: Propagation of strong blast waves. Prikl. Mat. Mekh. **10**, 241–250 (1946). See also, Similarity and Dimensional Methods in Mechanics. Academic Press, New York (1959)
54. Gusev, V.N.: Onunsteady self-similar motion of a gas displaced by a piston according to an exponential law. TsAGI Report (1957)
55. Grodzovskii, G.L.: Certain peculiarities of the flow around bodies at high supersonic velocities. Izv. Akad. Nauk SSSR, Otd. Tekhn. Nauk (1957)
56. Chernyi, G.G.: Hypersonic flow around a body by an ideal gas. Izv. Akad. Nauk. SSSR, Otd. Tekn. Nauk **6**, 77–85 (1957)
57. Stanyukovich, K.P.: Unsteady Motion of Continuous Media. Pergamon Press, London (1960)
58. Lees, L., Kubota, T.: Inviscid hypersonic flow over blunt-nosed slender bodies. J. Aerosp. Sci. **24**, 195–202 (1957)
59. Mirles, H.: Hypersonic flow over slender bodies associated with power-law shocks. Advances in Applied Mechanics, vol. VII, pp. 1–54. Academic Press, New York (1962)
60. Bechert, K.: *Ü*ber der Differentialgleichungen der Wellenausbreitungen in Gasen. Ann. Physik **39**, 357–372 (1941)
61. Guderley, G.: Starke Kugelige und Zylinderische Verdichtungsst*ö*sse in der N*ä*he des Kugelmittelpunktes bzw. der Zylinderasche. Luftfahrtforschung **19**, 302–312 (1942)
62. Kubota, T.: Inviscid flow over blunt nosed slender bodies. In: Proceedings of the 5th Heat Transfer and Fluid Mechanics Institute Conference, Pasadena (1957)
63. Busemann, A.: Aerodynamic lift at supersonic speeds. Luftfahrt-Forschung **12**(6), 210–220 (1935)
64. Eggers, A.J., Resknikoff, M.M., Dennis, D.H.: Bodies of revolution having minimum drag at high supersonic speeds. NACA Report No. 1306 (1957)
65. Lighthill, M.J.: Dynamics of a dissociated gas. J. Fluid Mech. **2**, 1–32 (1957)
66. Cole, J.D.: Newtonian flow theory for slender bodies. J. Aerosp. Sci. **24**, 448–455 (1957). See also Cole, J.D.: Note on the lift of slender nose shapes according to newtonian theory. J. Aerosp. Sci. **25**(6), 399 (1958)
67. Grimminger, G., Williams, E.P., Young, G.B.W.: Lift on inclined bodies of revolution in hypersonic flow. J. Aerosp. Sci. **17**, 675–690 (1950)
68. Cole, J.D., Brainerd, J.J.: Slender wings at high angles of attack in hypersonic flows. Hypersonic Flow Research, pp. 321–344. Academic Press, New York (1962)
69. Ferri, A.: Supersonic Flow Around Circular Cones at Angle of Attack. NACA Report No. 1045 (1951)
70. Ferri, A., Ness, N., Kaplita, T.: Supersonic flow over conical bodies without axial symmetry. J. Aerosp. Sci. **20**, 563–569 (1953)
71. Ferri, A.: Supersonic Flows With Shock Waves. General Theory of High Speed Aerodynamics. Princeton University Press, New Jersey (1955)
72. Guiraud, J.P.: Newtonian flow over a surface-theory and application. In: Collar, A.R., Tinkler, J. (eds.) Hypersonic Flow, pp. 253–296. Butterworths, London (1960)
73. Stocker, P.M., Mauger, F.E.: Supersonic flow past cones of general cross-section. J. Fluid Mech. **13**, 383–399 (1962)
74. Scheuing, R.A., Mead. H.R., Brook, J.W., Melnik, R.E., Hayes, W.D., Gray, K.E., Donaldson, C. Dur and Sullivan, R.D.: Theoretical prediction of pressures in hypersonic flow with special reference to configurations having attached leading-edge shock. ASD Technical Report No. 61–60, USAF, Wright Patterson Air Force Base, Ohio (1962)
75. Tsien, H.S.: Supersonic flow over inclined body of revolution. J. Aerosp. Sci. **5**, 480–483 (1938)
76. Lighthill, M.J.: Supersonic flow past slender pointed bodies of revolution at Yaw. Q. J. Mech. Appl. Math. **1**, 76–89 (1948)
77. Young, G., Sisk, C.: Supersonic flow around cones at large yaw. J. Aerosp. Sci. **19**, 111–119 (1952)
78. Stone, A.H.: On supersonic flow past a slightly yawing cone. J. Math. Phys. **27**, 67–81 (1948)

79. Stone, A.H.: On supersonic flow past a slightly yawing cone. Part II. J. Math. Phys. **27**, 220–233 (1948)
80. Willet, J.E.: Supersonic flow at the surface of a circular cone at angle of attack. J. Aerosp. Sci. **27**, 907–912 (1962)
81. Willet, J.E.: A note on the inviscid streamlines at the surface of a yawed cone in supersonic stream. J. Aerosp. Sci. **29**, 375–376 (1962)
82. Holt, M.: A vortical singularity in conical flow, Part 4. Q. J. Mech. Appl. Math. **VI**, 438–445 (1954)
83. Cheng, H.K.: On the structure of vortical layers in supersonic and hypersonic flows. J. Aerosp. Sci. **27**, 155–156 (1960)
84. Cheng, H.K.: Hypersonic flows past a yawed circular cone and other pointed bodies. J. Fluid Mech. **12**, 169–191 (1962)
85. Sychev, V.V.: Three-dimensional hypersonic gas flow past slender bodies at high angles of attack. J. Appl. Math. **24**, 296–306 (1960)
86. Munson, A.: The vortical layer on an inclined cone. J. Fluid Mech. **20**, 625–643 (1964)
87. Woods, B.A.: The flow close to the surface of a circular cone at incidence to a supersonic stream. Aerosp. Q. **XIII**, 115–128 (1962)
88. Woods, B.A.: The supersonic flow past a circular cone at incidence. Aeronautical research council R and M no. 3413. London (1965)
89. Melnik, R.: Newtonian entropy layer in the vicinity of a conical symmetry plane. AIAA J. **3**(3), 520–522 (1965)
90. Melnik, R.: Vortical singularities in conical flow. AIAA J. **5**(4), 631–637 (1967)
91. Moretti, G.: Inviscid flowfield about a pointed cone at an angle of attack. AIAA J. **5**(4), 789–791 (1967)
92. Dwyer, H.A.: Boundary layer on a hypersonic sharp cone at small angle of attack. AIAA J. **9**(2), 277–284 (1971)
93. Fletcher, C.: Vortical singularity behind a highly yawed cone. AIAA J. **13**(8), 1073–1078 (1975)
94. Bulakh, B.M.: Nonlinear Conical Flow. Delft University Press, Delft (1985)
95. Olivier, H.: A theoretical model for the shock stand-off distance in frozen equilibrium flows. J. Fluid Mech. **413**, 345–353 (2000)
96. Lin, C.C., Rubinov, S.I.: On the flow behind curved shocks. J. Math. Phys. **27**(2), 105–129 (1948)
97. Shen, S.F., Lin, C.C.: On the Attached Curved Shock in Front of a Sharp-Nosed Axially Symmetric Body Placed in a Uniform Stream. NACA TN No. 2505 (1951)
98. Cabannes, H.: Tables pour la détermination des ondes de choc détachées. Rech. Aéronaut. **49**, 11–15 (1956)
99. Cabannes, H.: Théorie des Ondes de Choc, Encyclopedia of Physics, vol. IX, III. Springer, Berlin (1960)
100. Van Dyke, M.: Perturbation Methods in Fluid Mechanics. Parabolic Press, Stanford (1975)
101. Van Dyke, M.: The supersonic blunt body problem-review and extension. J. Aerosp. Sci. **25**, 485–496 (1958)
102. Mangler, K.W.: The Calculation of the Flow Field Between a Blunt Body and Bow Wave. Hypersonic Flow. Butterworths Scientific Publications, London (1960)
103. Vaglio-Laurin, R.: On the PLK method and the supersonic blunt-body problem. J. Aerosp. Sci. **29**, 185–206 (1962)
104. Garabedian, P., Lieberstein, H.M.: On the numerical calculation of detached bow shock waves in hypersonic flow. J. Aerosp. Sci. **25**, 109–118 (1958)
105. Lin, C.C.: A note on Garabedian's paper "numerical construction of detached shock waves". J. Math. Phys. **36**(2), 206–209 (1957)
106. Van Dyke, M.: The blunt-body problem revisited. Fundamental Phenomena in Hypersonic Flow. Cornell University Press, New York (1966)
107. Vaglio-Laurin, R., Ferri, A.: Theoretical investigation of the flow field about blunt-nosed bodies in supersonic flight. J. Aerosp. Sci. **25**, 761–770 (1958)

108. Belotserkovskii, O., Chushkin, P.: Supersonic flow past blunt bodies. Arch. Mech. Stos. **14**, 461–490 (1962)
109. Holt, M.: Numerical Methods in Fluid Dynamics. Springer, Berlin (1984)
110. Dorodnitsyn, A.: Numerical methods in gas dynamics. Arch. Mech. Stos. **12**, 13–27 (1960)
111. Mitchell, A.R., McCall, F.: The rotational field behind a bow shock wave in axially symmetric flow using relaxation methods. Proc. R. Soc. Edinb. Sect. A **53**, 371–380 (1952)
112. Richtmeyer, R., Morton, W.: Difference Methods for Initial-Value Problems. Wiley, New York (1967)
113. Pletcher, R., Tannehill, J., Anderson, D.: Computational Fluid Mechanics and Heat Transfer. CRC Press, Boca Raton (2013)
114. Emmons, H.W.: The Flow of a Compressible Fluid Past a Symmetrical Airfoil in a Wind Tunnel and in Free Air. NACA TN No. 1746 (1948)
115. Steger, J.L.: Application of cyclic relaxation procedures to transonic flow fields. Ph.D. Dissertation, Iowa State University (1969)
116. Hafez, M., Lovell, D.: Numerical solution of transonic stream function equation. AIAA J. **21**(3), 327–335 (1983)
117. Hirsch, C.: Numerical Computation of Internal and External Flows, vol. I and II. Wiley, New York (1990)
118. Van Leer, B.: Progress in multi-dimensional upwind differencing. Lect. Notes Phys. **414**, 1–26 (1990)
119. Roe, P.: Approximate Riemann solvers, parameter vectors and difference schemes. J. Comput. Phys. **43**, 357–372 (1981)
120. Osher, S.: Riemann solvers, entropy conditions and difference approximations. SIAM J. Numer. Anal. **21**, 217–235 (1984)
121. Chernyi, G.G.: The effect of slight blunting of the leading edge of a profile on flow at high supersonic speeds. Izv. Akad. nauk sssr, otd. tekhn. nauk no, 4 (1958)
122. Cheng, H.K.: Hypersonic Flow with Combined Leading-Edge Bluntness and Boundary Layer Displacement Effect. Cornell Aero. Lab., Report, No. AF-1285-A-4 (1960)
123. Guiraud, J.P.: Écoulement hypersonique d'un fluide parfait sur une plaque plane comportant un bord d'Attaque d'Épaisseur finie. C. R. Acad. Sci., Paris. **246**, 2842–2845 (1958)
124. Yakura, J.K.: Theory of entropy layers and nose bluntness in hypersonic flow. Hypersonic Flow Research. Academic Press, New York (1962)
125. Sychev, V.V.: On the theory of hypersonic gas flow with a power law shock wave. J. Appl. Math. Mech. **24**, 756–764 (1960)
126. White, F.: Viscous Fluid Flow. McGraw Hill, New York (1974)
127. Hayes, W., Probstein, R.: Viscous hypersonic similitude. J. Aerosp. Sci. **26**, 815–824 (1959)
128. Liepmann, H., Roshko, A.: Elements of Gas Dynamics. Wiley, New York (1957)
129. Shapiro, A.: The Dynamics and Thermodynamics of Compressible Fluid Flow. The Ronald Company, New York (1954)
130. Schlichting, H.: Boundary Layer Theory. McGraw Hill, New York (1979)
131. Lees, L.: On the boundary layer equations in hypersonic flow and their approximate solutions. J. Aerosp. Sci. **20**, 143–145 (1953)
132. Cheng, H.K.: Investigations Related to Hypersonic Flow and Boundary Layer Phenomena. Cornell Aero. Lab. Report, No. AF-1180-A-1 (1957)
133. Moore, F.K.: On the local Flat Plate Similarity in the Hypersonic Boundary Layer. Cornell Aero. Lab. Report, No. AF-1285-A-2 (1960)
134. Blottner, F.: Computational Techniques for Boundary Layers. AGARD lecture Series No. 73. VKI, Belgium (1975)
135. Van Dyke, M.: Higher approximations in boundary-layer theory. J. Fluid Mech. **19**, 145–159 (1964)
136. Le Balleur, J.Cl.: Numerical flow calculation and viscous-inviscid interaction techniques. Pineridge Press, U.K, Viscous Flows (1984)
137. Monnoyer, F.: Hypersonic boundary layer flows, Flows at large reynolds numbers. Computational Mechanics Publication, Southhampton, U.K (1997)

138. Lee, R., Cheng, H.K.: On the outer-edge problem of a hypersonic boundary layer. J. Fluid Mech. **38**, 161–179 (1969)
139. Bush, W.: Hypersonic strong-interaction similarity solutions for flow past a flat plate. J. Fluid Mech. **25**, 51–64 (1966). See also vol. 29, 349–359
140. Novack, B.B., Cheng, H.K.: Numerical Analysis and Modeling of Slip Flows at Very High Mach Numbers. Lecture Notes in Physics, vol. 8. Springer, Berlin (1970)
141. Rudman, R., Rubin, S.: Hypersonic viscous flow over slender bodies with sharp leading edges. AIAA J. **6**, 1883–1889 (1968). Also AIAA Paper 68–3
142. Rubin, S., Koshla, P.: A review of reduced navier-stokes computations for compressible viscous flows. J. Comput. Syst. Eng. **1**, 549–562 (1990)
143. Vigneron, Y.C., Rakich, J.V., Tannehill, J.: Calculation of Supersonic Viscous Flow Over Delta Wings with Sharp Subsonic leading Edges, AIAA Paper 78–1137 (1978). Also NASA TM-78500
144. Lubard, S.C., Helliwell, W.S.: Calculation of the flow on a cone at high angle of attack. AIAA J. **12**, 965–974 (1974). Also AIAA Paper 75–149
145. Stewartson, K.: Multistructured boundary layers on flat plates and related bodies. Adv. Appl. Mech. **14**, 145–239 (1974)
146. Rothmayer, A.P., Smith, F.T.: High Reynolds number asymptotic theories. Handbook of Fluid Dynamics. CRC Press, Boca Raton (1998)
147. Sychev, V.V.: On laminar separation. Fluid Dyn. **7**(3), 407–417 (1972)
148. Brown, S., Cheng, H.K., Lee, C.J.: Inviscid-viscous interaction on triple-deck scales in a hypersonic flow with strong wall cooling. J. Fluid Mech. **220**, 309–337 (1990)
149. Lighthill, M.J.: On boundary layer and upstream influence. Proc. R. Soc. Lond. Ser. A, 478–507 (1953)
150. Ladyzhenskii, M.D.: Hypersonic area rule. AIAA J. **1**(11), 2696–2698 (1963)
151. Cheng, H.K.: Viscous and Inviscid Slender-Body Problems of Hypersonic Flow. University of Southern California, California, Report, USC-1408 (1969)
152. Krasovskii, V.M.: Experimental study of hypersonic helium flow over blunt-nosed bodies. Inzh. Zh. **5**(2), 249–253 (1965)
153. Hayes, W.: On hypersonic similitude. Q. Appl. Math. **9**(1), 105–106 (1947)
154. Viviand, H.: Hypersonic flows for reentry problems. Similitude in Hypersonic Aerodynamics. Springer, Berlin (1991)
155. Oswatitsch, K.: Similarity and equivalence in compressible flow. Advances in Applied Mechancis, vol. VI, pp. 153–271. Academic Press, New York (1960)
156. Townend, L.H.: Lectures on Waveriders. VKI Lecture Series 1984–01. Brussels, Belgium (1984)
157. Nonweiler, T.R.F.: Delta wings of shapes amenable to exact shock wave theory. J. R. Aerosp. Soc. Ser. A **67**(39), (1963)
158. Maikapar, G.I.: On the wave drag of nonaxisymmetric bodies at supersonic speeds. J. Appl. Math. Mech. **23**, 528–531 (1959)
159. Gonor, A.L.: Exact solution of the problem of supersonic flow of gas past some three-dimensional bodies. J. Appl. Math. Mech. **28**, 1178–1181 (1964)
160. Peckham, D.H.: On Three-Dimensional Bodies of Delta Planform Which Can Support Plane Attached Shock Waves. ARC CP 640. HMSO, London (1962)
161. Jones, J.G., Moore, K.C., Pike, J., Roe, P.: A method for designing lifting configurations for high supersonic speeds using axisymmetric flow fields. Ingenieur-Archiv. **37**(1), 56–72 (1968)
162. Anderson, J., et al.: Hypersonic waveriders for high altitude applications, AIAA Paper 91–0530 (1991)
163. Eggers, A.J., Resnikoff, M.M., Dennis, D.H.: Bodies of revolution having minimum drag at high supersonic air speeds. NACA Report No. 1306 (1957)
164. Roe, P.: Aerodynamic Problems of Hypersonic Vehicles. Lecture 1–5, AGARD-LS, 42, vol. 1 (1972)

165. Koppenwalter, G.: Hypersonic aerothermodynamics. Fundamentals of Hypersonics: Aerodynamics and Heat transfer. VKI Lecture Series 1984–01. (1984)
166. Sobieczky, H., et al.: Hypersonic waverider design from given shock wave. first int waverider symposium, University of Maryland, (1990)
167. Rasumussen, M.L., Stevens, D.: On waverider shapes applied to aerospace plane forebody configurations, AIAA Paper 87–2550 (1987)
168. Miele, A.: Theory of Optimum Aerodynamic Shapes. Academic Press, New York (1965)
169. Cole, J.D.: Frontiers of Computational Fluid Dynamics. Optimal Hypersonic Conical Wings. World Scientific, Singapore (1998)
170. Long, L.N.: The Off-Design Performance of Hypersonic Waveriders. AGARD Conference Proceedings No. 428 (1987)

Chapter 13
Flow Analogies

In this chapter we will discuss some experimental techniques for flow simulation and visualization, including the Hele-Shaw flow [1], shallow water, hydraulic analogy, electric analogy and analog computers. In each case, the theory and the limitation will be studied.

13.1 Hele-Shaw Flows

Consider a slow motion of fluid between two parallel plates separated by a small distance 2h. If a cylindrical body of arbitrary cross-section is inserted between the two plates at right angle, from wall to wall, the streamline pattern will be the same as that of potential flow around the same shape!

The creeping flow in general is dominated by viscous effects and inertia terms can be neglected leading to the Stokes equations for incompressible flows

$$\nabla p = \mu \nabla^2 \mathbf{V} \quad and \quad \nabla . \mathbf{V} = 0 \tag{13.1}$$

The no slip and no penetration boundary conditions on the two walls are satisfied with $\mathbf{V} = 0$.

Taking the divergence of the gradient of pressure yields the Laplacian equation, namely

$$\nabla . (\nabla p) = \nabla^2 p = \mu \nabla . \left(\nabla^2 \mathbf{V} \right) = \mu \nabla^2 (\nabla . \mathbf{V}) = 0 \tag{13.2}$$

Hence the pressure $p(x, y, z)$ is a potential function.

Now, taking the rotational of the equation yields

$$\nabla \wedge (\nabla p) = \mathbf{0} = \mu \nabla^2 (\nabla \wedge \mathbf{V}) = \mu \nabla^2 \omega \tag{13.3}$$

J.J. Chattot and M.M. Hafez, *Theoretical and Applied Aerodynamics*,
DOI 10.1007/978-94-017-9825-9_13

Hence, the $z-$component of the rotational vector satisfies

$$\nabla^2\omega_z = \nabla^2\left(\frac{\partial v}{\partial x} - \frac{\partial u}{\partial y}\right) = 0 \tag{13.4}$$

Moreover, for two dimensional flows, a stream function exists such that $u = \psi_y$ and $v = -\psi_x$, which satisfies

$$\nabla^2\psi = -\frac{\partial v}{\partial x} + \frac{\partial u}{\partial y} = -\omega_z \tag{13.5}$$

Therefore, the governing equation for ψ is

$$\nabla^4\psi = 0 \tag{13.6}$$

the biharmonic equation.

Next, for a thin film of fluid, such that $h \ll l$, where l is a characteristic length (i.e. cylinder diameter), the above equations are further simplified, because of the strong gradient of the velocity in the normal $z-$direction. The governing equations become

$$\frac{\partial p}{\partial x} = \mu\frac{\partial^2 u}{\partial z^2}, \quad \frac{\partial p}{\partial y} = \mu\frac{\partial^2 v}{\partial z^2}, \quad \frac{\partial p}{\partial z} = \mu\frac{\partial^2 w}{\partial z^2} \tag{13.7}$$

and

$$\frac{\partial u}{\partial x} + \frac{\partial v}{\partial y} + \frac{\partial w}{\partial z} = 0 \tag{13.8}$$

with the condition that

$$Re\left(\frac{h}{l}\right)^2 \ll 1 \tag{13.9}$$

Since w is much smaller than the (u, v) components, $\partial p/\partial z$ is much smaller that the other components of the pressure gradient. Hence, to a first approximation $p = p(x, y)$. The u and v components are obtained via integration in the $z-$direction and applying the no slip conditions at both $z = \pm h$ gives

$$u = -\frac{1}{2\mu}\frac{\partial p}{\partial x}(h^2 - z^2) \tag{13.10}$$

$$v = -\frac{1}{2\mu}\frac{\partial p}{\partial y}(h^2 - z^2) \tag{13.11}$$

Notice that the ratio v/u does not depend on z, i.e. the direction of the flow and the streamline patterns are independent of z. Notice also, the stress tensor

$$\tau_{ij} = -p\delta_{ij} + \mu\left(\frac{\partial u_i}{\partial x_j} + \frac{\partial u_j}{\partial x_i}\right) \tag{13.12}$$

is simplified to be

$$\tau_{ij} = -p\delta_{ij} \tag{13.13}$$

since the pressure magnitude is $p = O\left(\mu U l/h^2\right)$ and the tangential stresses are of order $O\left(\mu U/h\right)$. Thus, in a thin film, the normal stresses are dominant. At the cylinder surface, with this model, only the tangency (no penetration) condition is satisfied and the no slip condition is not satisfied.

From the above results for u and v, upon elimination of the pressure, one obtains

$$\frac{\partial v}{\partial x} = \frac{\partial u}{\partial y} \tag{13.14}$$

Thus, the flow past a cylinder, at any given z, correspond to a 2-D irrotational flow past that cylinder.

Notice the average velocity components are

$$\bar{u} = \frac{1}{h}\int_0^h u\,dz = -\frac{h^2}{3\mu}\frac{\partial p}{\partial x} \tag{13.15}$$

$$\bar{v} = \frac{1}{h}\int_0^h v\,dz = -\frac{h^2}{3\mu}\frac{\partial p}{\partial y} \tag{13.16}$$

hence

$$\bar{\mathbf{V}} = -\frac{h^2}{3\mu}\nabla p \tag{13.17}$$

where $\bar{\mathbf{V}}$ and p depend only on x and y.

Therefore, a two-dimensional potential flow exists for which the potential function is proportional to the pressure. The above relation is of the same form as Darcy's law for flows in porous media.

Applications in different fields can be found in a recent book by Gustafsson and Vasilev [2] on "Conformal and Potential Analysis in Hele-Shaw Cells."

Hele-Shaw in 1898 used this analogy to obtain experimental patterns of streamlines in potential flow about arbitrary 2-D bodies, by introducing coloring matter at a few points at the inlet section, for visual demonstration.

There are, however, some limitations. The circulation Γ around any closed curve C lying in a (x, y)-plane, whether enclosing the cylinder or not, must be zero. This is clear from the definition

$$\Gamma = \oint u\,dx + v\,dy = -\frac{1}{2\mu}(h^2 - z^2)\oint \frac{\partial p}{\partial x}dx + \frac{\partial p}{\partial y}dy = -\frac{1}{2\mu}(h^2 - z^2)\oint \frac{\partial p}{\partial s}ds = 0 \tag{13.18}$$

since p is a single valued function of x and y.

Visualizations of potential flows in simply connected domains, for example, flow in a diffuser, are possible.

Ironically, potential streamlines are visualized using viscous dominated flows. beautiful pictures of Hele-Shaw flows past a circle, and inclined plate, a Rankine half-body, inclined airfoil, a rectangular block on a plate, uniform flow normal to a plate, can be found in Van Dyke's Album of Fluid Motion [3]. Other pictures for flow past an ellipse and past symmetrical Joukowski airfoil at zero angle of attack can be found in Visualized Flow, compiled by the Japan Society of Mechanical Engineers. For more details of the theory, see Schlichting [4], Batchelor [5] and Acheson [6]. Two examples of Hele-Shaw flows are presented, the flow past a flat plate at high incidence and the flow past a rectangular block. In Fig. 13.1, the flat plate produces a flow that is antisymmetrical with respect to the plate center. Hence, the Kutta-Joukowski condition at the trailing edge is not satisfied and there is no circulation.

Fig. 13.1 NASA/Ames reeves electronic analog computer (1949) (from http://www.nasa.gov/centers/ames/multimedia/images/2010/iotw/reeves.html)

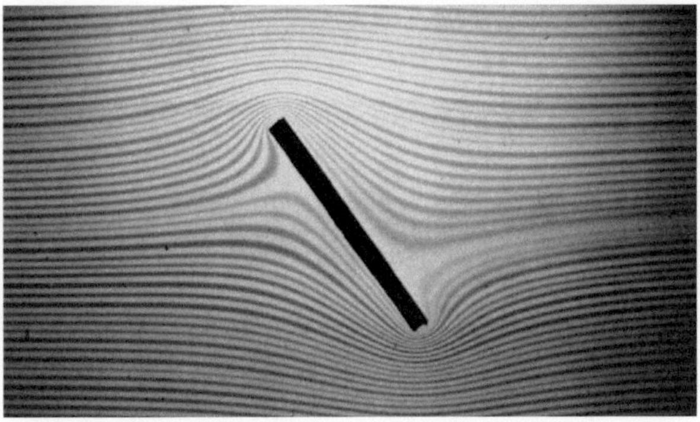

Fig. 13.2 Hele-Shaw flow past a flat plate (from en.wikipedia.org)

The flow past a rectangular block is shown in Fig. 13.2. No visible separation occurs with the creeping flow regime.

Hele-Shaw flow and pattern formation in a time-dependent gap is studied in Ref. [3].

13.2 Hydraulic Analogy

The equations of unsteady two dimensional compressible inviscid flows are analogous to the shallow water equations where the height of the water corresponds to the density of the compressible fluid. To show such analogy, we will start with the isentropic Euler equations in conservative form

$$\frac{\partial \rho}{\partial t} + \frac{\partial \rho u}{\partial x} + \frac{\partial \rho v}{\partial y} = 0 \tag{13.19}$$

$$\frac{\partial \rho u}{\partial t} + \frac{\partial \rho u^2}{\partial x} + \frac{\partial \rho uv}{\partial y} = -\frac{\partial p}{\partial x} \tag{13.20}$$

$$\frac{\partial \rho v}{\partial t} + \frac{\partial \rho uv}{\partial x} + \frac{\partial \rho v^2}{\partial y} = -\frac{\partial p}{\partial y} \tag{13.21}$$

$$\frac{p}{p_0} = \left(\frac{\rho}{\rho_0}\right)^{\gamma} \tag{13.22}$$

The above equations imply conservation of mass, momentum and entropy. Energy is not conserved in general. Only for smooth flow, energy will be conserved.

Next, consider a thin layer of water over a flat smooth plate, where the height of the water is given by $h(x, y, t)$.

The conservation of mass requires

$$\frac{\partial \rho_w h}{\partial t} + \frac{\partial \rho_w h u}{\partial x} + \frac{\partial \rho_w h v}{\partial y} = 0 \tag{13.23}$$

Since $\rho_w = const$, it cancels out and the equation becomes similar to the compressible flow continuity equation, where ρ is replaced by h.

The momentum equations for the column of water with height h is given, ignoring the viscous effects, by

$$\frac{\partial \rho_w h u}{\partial t} + \frac{\partial \rho_w h u^2}{\partial x} + \frac{\partial \rho_w h u v}{\partial y} = -\rho_w g h \frac{\partial h}{\partial x} = -\frac{1}{2}\rho_w g \frac{\partial h^2}{\partial x} \tag{13.24}$$

$$\frac{\partial \rho_w h u v}{\partial t} + \frac{\partial \rho_w h u v}{\partial x} + \frac{\partial \rho_w h v^2}{\partial y} = -\rho_w g h \frac{\partial h}{\partial y} = -\frac{1}{2}\rho_w g \frac{\partial h^2}{\partial y} \tag{13.25}$$

Once again, ρ_w cancels out.

Assuming small variations of the water surface and ignoring the vertical acceleration, the pressure is governed by the hydrostatic equation

$$p = p_{atm} + \rho_w g(h - z), \quad 0 \le z \le h \tag{13.26}$$

where z is the elevation of a point above the flat plate.

Hence

$$\frac{\partial p}{\partial x} = \rho_w g \frac{\partial h}{\partial x}, \quad \frac{\partial p}{\partial y} = \rho_w g \frac{\partial h}{\partial y} \tag{13.27}$$

On the other hand, from isentropic relation of compressible flows,

$$\frac{\partial p}{\partial x} = \frac{\partial p}{\partial \rho}\frac{\partial \rho}{\partial x} = a^2 \frac{\partial \rho}{\partial x}, \quad \frac{\partial p}{\partial y} = \frac{\partial p}{\partial \rho}\frac{\partial \rho}{\partial y} = a^2 \frac{\partial \rho}{\partial y} \tag{13.28}$$

where $a^2 = \gamma RT$. Hence the analogy holds if a^2 corresponds to gh. In this case, the Mach number $M = V/a$ corresponds to the Froude number $F = V/\sqrt{gh}$ of the shallow water waves where $\sqrt{gh}$ is the speed of surface waves due to small disturbances.

Notice the pressure of compressible flow corresponds to h^2, hence the analogy holds only for $\gamma = 2$. In this case the temperature of the compressible flow corresponds to h.

To summarize these relations, the different non dimensional quantities for compressible and shallow water flows are given in Table 13.1, where subscript r stands for reference quantities

Table 13.1 Analog quantities for compressible flow and shallow water models

Compressible flow	Shallow water
ρ/ρ_r	h/h_r
p/p_r	$(h/h_r)^2$
T/T_r	h/h_r
a/a_r	$\sqrt{h/h_r}$
M	F

There are several special cases of interest including

(i) Quasi-one dimensional unsteady flows in nozzles,
(ii) Two dimensional steady external and internal flows,
(iii) irrotational flows.

The above derivation shows that the analogy is valid for rotational flows where vorticity may be generated due to the variation of the total enthalpy according to Crocco's relation. One can consider the special case where vorticity vanishes and a potential flow exists for both steady as well as unsteady phenomena. In all cases, the flow is assumed isentropic. In the following, hydraulic jumps (corresponding to shock relations of compressible flows) will be studied.

The propagation of sound corresponds to the propagation of an infinitesimal gravity wave, since

$$\frac{D\mathbf{u}}{Dt} + a^2 \frac{\nabla \rho}{\rho} = 0 \tag{13.29}$$

corresponds to

$$\frac{D\mathbf{u}}{Dt} + gh \frac{\nabla h}{h} = 0 \tag{13.30}$$

with $a^2 = gh$.

Also, the linearization of the shallow water equation yields the classical wave equation

$$\frac{\partial^2 h}{\partial t^2} - gh_0 \nabla^2 h = 0 \tag{13.31}$$

where $a_0 = \sqrt{gh_0}$.

Real time observation is possible since $g = 9.81\,\text{m/s}^2$ and for $h = 0.01\,\text{m}$, $a = 0.3\,\text{m/s}$. In fact, shallow water experiment can be performed at the kitchen sink!

For smooth steady flows, one can obtain the isentropic relation and corresponding shallow water equations. For example

$$\frac{p}{p_0} = \left(1 + \frac{\gamma - 1}{2} M^2\right)^{\frac{\gamma}{\gamma-1}} \quad and \quad \frac{h_0}{h} = 1 + \frac{1}{2} F^2 \tag{13.32}$$

Supersonic ($M > 1$) and subsonic ($M < 1$) flows correspond to supercritical ($F > 1$) and subcritical ($F < 1$) flows.

The analogs of shock waves are called hydraulic jumps. Moving with the discontinuity and considering the relative normal velocity, one can derive the following steady relations

$$u_1 h_1 = u_2 h_2 = m \tag{13.33}$$

To evaluate the momentum balance, a control volume is used as shown in Fig. 13.3. Friction on the bottom surface is neglected.

The net acting force per unit span in the x-direction due to pressure is

$$\int_0^{h1} \rho_w g(h_1 - z)\, dz - \int_0^{h2} \rho_w g(h_2 - z)\, dz = \frac{1}{2}\rho_w g \left(h_1^2 - h_2^2\right) \tag{13.34}$$

Hence, the rate of change of x-momentum reads

$$\rho_w u_2^2 h_2 - \rho_w u_1^2 h_1 = \frac{1}{2}\rho_w g \left(h_1^2 - h_2^2\right) \tag{13.35}$$

or

$$u_1^2 h_1 + \frac{1}{2} g h_1^2 = u_2^2 h_2 + \frac{1}{2} g h_2^2 = n \tag{13.36}$$

The above equations can be solved for $(u, h) = (u_2, h_2)$ given (u_1, h_1) as follows

$$uh = m, \quad and \quad u^2 h + \frac{1}{2} g h^2 = n \tag{13.37}$$

Fig. 13.3 Hele-Shaw flow past a rectangular block (from en.wikipedia.org)

hence

$$\frac{(uh)^2}{h} + \frac{1}{2}gh^2 = n \tag{13.38}$$

or

$$m^2 + \frac{1}{2}gh^3 - nh = 0 \tag{13.39}$$

Since

$$m^2 + \frac{1}{2}gh_1^3 - nh_1 = 0 \tag{13.40}$$

therefore

$$\frac{1}{2}g\left(h^3 - h_1^3\right) - n\left(h - h_1\right) = 0 \tag{13.41}$$

The trivial root is $h = h_1$, i.e. no jump. Other roots satisfy the relation

$$\frac{1}{2}gh^2 + \frac{1}{2}gh_1h + \frac{1}{2}gh_1^2 - n = 0 \tag{13.42}$$

A solution of this quadratic equation is given by

$$\frac{h}{h_1} = -\frac{1}{2} + \sqrt{\frac{1}{4} + 2F_1^2} \tag{13.43}$$

for $F_1 \geq 1, h/h_1 \geq 1$.

The negative root is not physical.

Notice, from conservation of mass,

$$\frac{F_1}{F_2} = \left(\frac{h_2}{h_1}\right)^{\frac{3}{2}} \tag{13.44}$$

The solution with $F_1 \leq 1$ and $h/h_1 \leq 1$ is excluded. It corresponds to an expansion shock. To see that analogy, consider the specific energy (see White [7])

$$E = gh + \frac{1}{2}u^2 = gh + \frac{1}{2}\frac{m^2}{h^2} \tag{13.45}$$

For a given m (discharge), there are two possible states for the same E, see the sketch in Fig. 13.4 and their relation is plotted in Fig. 13.5.

There is a minimum value of E at a certain value of h called the *critical value*. Setting $dE/dh = 0$, E_{min} occurs at $g - m^2/h^3 = 0$, or $h_c = \left(m^2/g\right)^{1/3}$ and $E_{min} = gh_c + m^2/(2h_c^2)$.

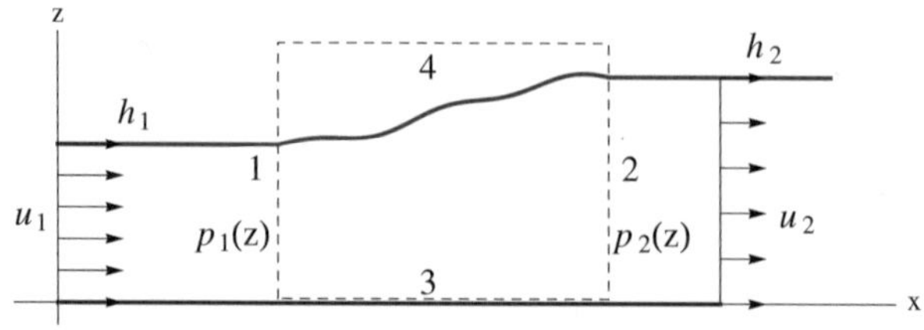

Fig. 13.4 Hydraulic jump: Control volume for momentum calculation

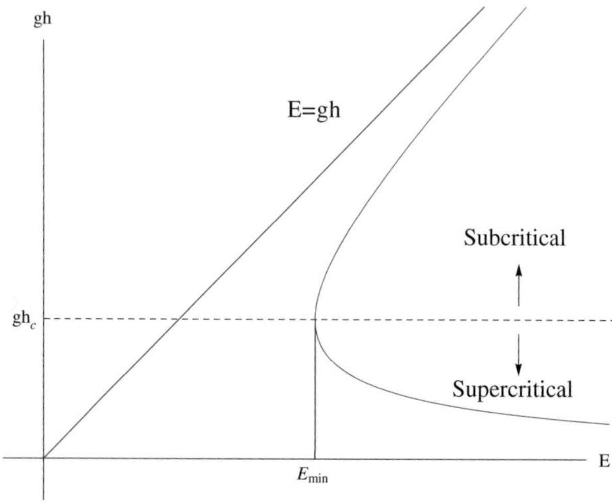

Fig. 13.5 Energy versus height

At the critical depth, $m^2 = gh_c^3 = c_0^2 h_c^2$, where $c_0 = \sqrt{gh_c}$ and $F = 1$.

For $E < E_{min}$, no solution exists and for $E > E_{min}$ two solutions are possible with $h > h_c$ and $u < c_0$ as well as $h < h_c$ and $u > c_0$. The analogy with Mach waves is clear with the angle of the wave given by

$$\sin \mu = \frac{c_0}{u} \tag{13.46}$$

Now, across the jump

$$E_1 - E_2 = g\frac{(h_2 - h_1)^3}{4h_1 h_2} \tag{13.47}$$

This equation shows that dissipation loss is positive only if $h_2 > h_1$, a requirement implied by the second law of thermodynamics.

A more consistent analysis would be based on the total enthalpy (and not just potential and kinetic energy)

$$\frac{H}{u_r^2} = \frac{p/p_r}{(\gamma-1)\rho/\rho_r}\frac{1}{M_r^2} + \frac{1}{2}\left(\frac{u}{u_r}\right)^2 \tag{13.48}$$

The corresponding quantity for shallow water is

$$\frac{H}{u_r^2} = \frac{h}{h_r}\frac{1}{F^2} + \frac{1}{2}\left(\frac{u}{u_r}\right)^2 \tag{13.49}$$

The total enthalpy should decrease for steady isentropic Euler shocks (see [8]) and hence expansion shocks are excluded and the same is true for hydraulic jumps.

Again, from the conservation of mass across the hydraulic jump

$$u_1 h_1 = u_2 h_2 = m \tag{13.50}$$

and from the momentum balance

$$u_1^2 h_1 + \frac{1}{2} g h_1^2 = u_2^2 h_2 + \frac{1}{2} g h_2^2 = n \tag{13.51}$$

one can show, upon elimination of h that

$$m u_1 + \frac{1}{2} g \frac{m^2}{u_1^2} = m u_2 + \frac{1}{2} g \frac{m^2}{u_2^2}, \quad \Rightarrow \quad u_2 - u_1 + \frac{1}{2} g m \left(\frac{1}{u_2^2} - \frac{1}{u_1^2}\right) = 0 \tag{13.52}$$

Introducing the jump notation $[a] = a_2 - a_1$, one can write

$$< u > + \frac{1}{2} g m \left\langle \frac{1}{u^2} \right\rangle = 0 \tag{13.53}$$

But, using the identity $< a^2 > = \bar{a} < a >$, where $\bar{a} = (a_1 + a_2)/2$, yields

$$< u > + g m \frac{\bar{1}}{u} \left\langle \frac{1}{u} \right\rangle = < u > + g \bar{h} \left\langle \frac{1}{u} \right\rangle = 0 \tag{13.54}$$

Hence

$$\frac{< u >}{\left\langle \frac{1}{u} \right\rangle} = -u_1 u_2 = -g\bar{h} \tag{13.55}$$

The latter is analogous to the Prandtl relation for normal shock of compressible fluid flow.

Hydraulic analogy can provide experimental demonstration of compressible flow phenomena using a device called water table, which consists of an inexpensive sheet of glass with suitable sides. Either a model is moved in a thin layer of water or the model is fixed and the water flows around it. The analogy is particularly useful for transonic flows including shocks. Examples are:

(i) Transonic flows in convergent/divergent nozzle at design and off-design conditions,
(ii) Steady two dimensional supersonic flows over airfoils (flat plate, parabolic or diamond airfoils). Compression shocks and expansion fans are easily identified. Attached and detached shocks as well as fish tail shocks can be observed, including wave reflected from the walls,
(iii) Unsteady flows, including accelerated and decelerated as well as oscillating airfoils.

There are limitations, of course. Strong shocks are not well represented. The assumptions that vertical accelerations are small, the viscous force exerted by the horizontal bottom is negligible and the vertical variation of the velocity can be neglected in the continuity equation are questionable if quantitative results are required. In general, separated flows are not representative since the Reynolds numbers are very different in water and compressible flows. For more details, the reader is referred to Oswatitsch [9], Loh [10], Thompson [11] and White [7].

13.3 Electric Analogy

This analogy is based on the relation between the equation governing the electrical potential in a conductor and the velocity potential of fluid flows. In the absence of interior sources within the conductor, the electric potential ϕ is governed by

$$\nabla . (\sigma \nabla \phi) = 0 \tag{13.56}$$

where σ is the conductivity.

For homogeneous conductors with constant conductivity, the above equation reduces to the Laplacian

$$\nabla^2 \phi = 0 \tag{13.57}$$

which is the same equation for the velocity potential of inviscid, incompressible fluid flow, even for unsteady problems (for example oscillating wing).

For two-dimensional models, it is possible to represent variable conductivity by a model having variable thickness. In this case, the electric potential is governed by

$$\frac{\partial}{\partial x}\left(\sigma h \frac{\partial \phi}{\partial x}\right) + \frac{\partial}{\partial y}\left(\sigma h \frac{\partial \phi}{\partial y}\right) = 0 \tag{13.58}$$

Moreover, one can introduce a stream function ψ, such that

$$\frac{\partial \phi}{\partial x} = \frac{1}{\sigma h}\frac{\partial \psi}{\partial y}, \quad and \quad \frac{\partial \phi}{\partial y} = -\frac{1}{\sigma h}\frac{\partial \psi}{\partial x} \tag{13.59}$$

The equation for ψ is given by

$$\frac{\partial}{\partial x}\left(\frac{1}{\sigma h}\frac{\partial \phi}{\partial x}\right) + \frac{\partial}{\partial y}\left(\frac{1}{\sigma h}\frac{\partial \phi}{\partial y}\right) = 0 \tag{13.60}$$

The Dirichlet and Neumann boundary conditions are simple to reproduce in the electrical model. For $\phi = const.$, the boundary is covered by an electrode which is raised to the potential level specified. For $\partial\phi/\partial n = 0$, the boundary is insulated (i.e. no electric current crosses the boundary).

The conducting media can be a weak electrolyte contained in a tank leading to "electrolytic basin" technique metallic plates or special conducting papers are also used.

Notice, for compressible (subsonic) flow studies, the height distribution must be adjusted iteratively to be related to the density distribution Which is unknown a priori).

For details of the apparatus and the experimental techniques, see Malavard [12]. In the following, some applications are discussed.

13.3.1 Analog Representation of Circulation Around Lifting Airfoils

In this case, the velocity potential is not single-valued function. The analogy in which it is identified with an electric potential is not simple since cuts must be provided in the conductor in order to make the electric potential single-valued.

Instead, the stream function of the fluid flow, which is continuous, will be identified with an electric potential. However, to give the circulation around the body a prescribed value, the model must function as an extra electrode. it must be connected through an adjustable resistance, R, to one of the poles (chosen to give the correct sign of Γ). It is possible, however, to establish a "cut" in the electric field, which can be chosen along an equipotential line, extending outward from the surface of the body. This cut is realized in the electric tank by two auxiliary electrodes separated by a thin insulating strip. The potential difference between theses two electrodes corresponds to the value of the circulation (the flux crossing the cut must be conservative).

Studies of interaction problems using combination of airfoil sections are possible. Also, airfoils with flaps and hinged leading edges as well as slots have been considered. Since conformal mappings satisfy the Cauchy-Riemann equations, it is feasible to construct electric analog of conformal transformation of an airfoil (or any closed region) onto a circle.

Analog study of airfoil cascades is based on representing the periodicity of the field.

13.3.2 Analog Study of Flows Around Bodies of Revolution

The potential and stream function equations in a meridian plane satisfy the following equations

$$\frac{\partial}{\partial x}\left(\rho r \frac{\partial \phi}{\partial x}\right) + \frac{\partial}{\partial r}\left(\rho r \frac{\partial \phi}{\partial r}\right) = 0 \tag{13.61}$$

and

$$\frac{\partial}{\partial x}\left(\frac{1}{\rho r}\frac{\partial \psi}{\partial x}\right) + \frac{\partial}{\partial r}\left(\frac{1}{\rho r}\frac{\partial \psi}{\partial r}\right) = 0 \tag{13.62}$$

These equations can be compared to the equations governing the electrical potential in a conducting medium of varying thickness. The depth of the conducting material will be proportional or inversely proportional to ρr, depending on whether ϕ or ψ is simulated. In the first case, inclined tank can be used to simulate flow around streamlined bodies of revolution, convergent wind tunnel ducts and axially symmetric air intakes. For the representation of the stream function, a tank with hyperbolic bottom can be constructed where the depth varies as $1/y$, see Ref. [12].

13.3.3 Hodograph Tank

In the hodograph plane, the velocity potential ϕ and the stream function ψ, of compressible fluid flow are related by the following equations

$$\frac{\partial \phi}{\partial q} = -\frac{\rho_0}{\rho q}(1 - M^2)\frac{\partial \psi}{\partial \theta} \tag{13.63}$$

$$\frac{\partial \phi}{\partial \theta} = q\frac{\rho_0}{\rho}\frac{\partial \psi}{\partial q} \tag{13.64}$$

where the independent variables q and θ are the modulus and argument of the velocity, M is the local Mach number and ρ_0 is the density for $q = 0$. The potential differential equations for ϕ and ψ are given by

$$\frac{\partial}{\partial q}\left(\frac{1}{\frac{\rho_0}{\rho q}(1 - M^2)}\frac{\partial \phi}{\partial q}\right) + \frac{\partial}{\partial \theta}\left(\frac{\rho}{\rho_0 q}\frac{\partial \phi}{\partial \theta}\right) = 0 \tag{13.65}$$

and

$$\frac{\partial}{\partial q}\left(q\frac{\rho_0}{\rho}\frac{\partial \psi}{\partial q}\right) + \frac{\partial}{\partial \theta}\left(\frac{\rho_0}{\rho q}(1 - M^2)\frac{\partial \psi}{\partial \theta}\right) = 0 \tag{13.66}$$

these equations may be compared to the equation of the electric potential in a conductor, in terms of polar coordinates

$$\frac{\partial}{\partial r}\left(hr\frac{\partial \phi}{\partial r}\right)+\frac{\partial}{\partial \theta}\left(\frac{h}{r}\frac{\partial \phi}{\partial \theta}\right)=0 \tag{13.67}$$

It is possible to establish the correspondence with $r = r(q)$ and $h = h(q)$.

The electric tank may be constructed in a circular form, where the center corresponds to the origin of the (q, θ)-plane. The tank radius is limited to a value corresponding to M slightly less than one (for $Mach = 1$, the depth is infinite). The image of uniform flow is given by a singularity placed at a point, the distance of which from the origin corresponds to the magnitude of the velocity of the uniform stream at infinity. The singularity is represented in the tank by two electrodes, properly chosen.

To determine the form of the wing section which corresponds to the hodograph chosen, the following relations are used

$$dx = \frac{\cos\theta}{q}d\phi, \quad dy = \frac{\sin\theta}{q}d\phi \tag{13.68}$$

where x and y are obtained via simple integration.

13.3.4 Analog Study of Supersonic Conical Flows

The governing equation of the perturbation potential for linearized three-dimensional supersonic flow is given by

$$\beta^2\frac{\partial^2\varphi}{\partial x^2}=\frac{\partial^2\varphi}{\partial y^2}+\frac{\partial^2\varphi}{\partial z^2} \tag{13.69}$$

where $\beta^2 = M^2 - 1$.

When the flow is conical, φ depends only on two variables as follows. Let

$$y = r\cos\theta, \quad z = r\sin\theta, \quad x = \beta r X \tag{13.70}$$

then, any of the three velocity components, u, v, w depend only on θ and X. For $X = \cos\xi$, a Laplace equation is obtained for the flow interior to the mach cone

$$\frac{\partial^2 u}{\partial \theta^2}+\frac{\partial^2 u}{\partial \xi^2}=0, \quad X < 1 \tag{13.71}$$

For $X > 1$ (exterior of the mach cone), with $X = \cosh\xi$, the wave equation is obtained

$$\frac{\partial^2 u}{\partial \theta^2} - \frac{\partial^2 u}{\partial \xi^2} = 0, \quad X > 1 \tag{13.72}$$

For more details, see Stewart [13].

To study the flow inside the Mach cone, the complex variable $Z = \varrho e^{i\theta}$, where

$$\frac{x}{\beta r} = X = \frac{1 + \varrho^2}{2\varrho} \tag{13.73}$$

Either u, v, w may be identified with the electrical potential in a plane circular tank.

Germain [14] analyzed in detail the analog representation for a flat plate inside the Mach cone. he also studied wings with different cross sections and with dihedral angles. The study of wings in supersonic flow regime, steady and unsteady, was carried out by Enselme [15].

13.4 Analog Study of Three-Dimensional Flows

In general, the study of lifting surface problems can be performed with the help of electric tanks.

The low and high aspect ratio (AR) wings are of special interest since the formulation will reduce to two-dimensional problems.

In the case of slender wings and slender bodies at angle of attack, the flow can be decomposed into axial and normal components. The cross flow at the end section can be easily studied by electric analogy (as a special case of two-dimensional flows with $\alpha = 90°$) using polar coordinates.

On the other hand, the lifting line theory for high AR wings can be studied using the Trefftz plane. The Prandtl integro-differential equation is given by

$$\Gamma(y) = \pi U c(y) \left(\alpha + \frac{w_w}{U} \right) \tag{13.74}$$

where the downwash at the lifting line is given by

$$w_w = -\frac{1}{4\pi} \int_{-b}^{b} \frac{\Gamma'(\eta) d\eta}{y - \eta} \tag{13.75}$$

Notice that $\Gamma(-b) = \Gamma(b) = 0$ and $\int_{-b}^{b} d\Gamma(\eta) = 0$. A potential function can be introduced in the Trefftz plane such that $\Gamma(y) = 2\varphi$ and $w = -\frac{1}{2}\frac{\partial \varphi}{\partial z}$.

Now, φ is zero outside the span $(-b, b)$, and in between

$$\varphi = \frac{1}{2} \pi U c(y) \left(\alpha + \frac{1}{2U} \frac{\partial \varphi}{\partial z} \right) \tag{13.76}$$

Also, φ need only to be considered in a half-plane since $\varphi(y, -z) = -\varphi(y, z)$.

The boundary conditions along the axis can be simulated using a distribution of electrodes to impose the above mixed boundary condition as well as the Dirichlet condition, in a plane tank of large dimensions.

Nonlinear local lift coefficient in the neighborhood of maximum lift can also be considered using successive approximations.

Moreover, wind tunnel wall effects can be readily shown using $\partial\varphi/\partial n = 0$ on solid walls and $\varphi = const.$ in open section, where the pressure is constant across the wall boundaries.

Analog study of propeller's performance following Goldstein's analysis [16] is also possible. The potential of the velocity induced by the trailing vortex sheets satisfies the equation

$$\frac{1}{r}\frac{\partial}{\partial r}\left(r\frac{\partial \phi}{\partial r}\right) + \frac{1}{r^2}\frac{\partial^2 \phi}{\partial \theta^2} + \frac{\partial^2 \phi}{\partial z^2} = 0 \tag{13.77}$$

where the z-axis coincide with the propeller axis. If the propeller has p blades, the trailing vortex system is composed of p helicoids of geometric pitch $h = 2\pi w/\omega$, where w is the translation velocity along the z-axis and ω is the angular speed of rotation.

The equation of these helicoidal surfaces is defined by

$$\theta - \frac{z}{\lambda} = 2\pi\frac{k}{p}, \quad k = 0, 1, \ldots p-1, \quad 0 \leq r \leq R \tag{13.78}$$

where $\lambda = w/(\omega R)$, R is the radius of the propeller. The potential equation can be written in terms of $\xi = r/R$ and $\zeta = \theta - z/x$ as follows

$$\frac{\partial}{\partial \xi}\left(\xi\frac{\partial \phi}{\partial \xi}\right) + \left(\frac{1}{\lambda^2} + \frac{1}{\xi^2}\right)\frac{\partial^2 \phi}{\partial \zeta^2} = 0 \tag{13.79}$$

Moreover, the periodicity of the field with respect to ξ (period $2\pi/p$) and the antisymmetry of ϕ with respect to a cut, allow the domain for ϕ to be reduced to a sector $0 \leq \zeta \leq \pi/p$.

The boundary conditions are

$$\phi = 0 \quad at\,\zeta = 0,\ \xi > 1 \tag{13.80}$$

$$\phi = 0 \quad at\,\zeta = \pi/p,\ \xi > 1 \tag{13.81}$$

$$\phi(\xi) = \frac{\Gamma(\xi)}{2}, \quad \zeta = 0,\ 0 \leq \xi \leq 1 \tag{13.82}$$

The circulation is calculated as in the lifting line problem assuming that the airfoil section ξ of the blade has the same properties as that of infinite aspect ratio with a wind velocity $\sqrt{w^2 + (\omega x)^2}$, at an effective angle of attack $(\alpha - \delta\alpha)$, where α is the geometric angle of attack and $\delta\alpha$ is the induced angle.

Imposing the periodicity condition is of interest also in the study of cascade problems.

13.5 Sobieczky's Rheograph-Transformations

Sobieczky [17] used Busemann normalized hodograph plane, where ϕ and ψ (potential and stream function) can be written as Beltrami equations:

$$\frac{\partial \phi}{\partial q} = \pm K(q) \frac{\partial \psi}{\partial \theta} \tag{13.83}$$

$$\frac{\partial \phi}{\partial \theta} = K(q) \frac{\partial \psi}{\partial q} \tag{13.84}$$

where q (and K) are functions of the local Mach number. The above linear system for generalized potentials, is hyperbolic for $M \geq 1$ $(q(M) \geq 0)$ and elliptic for $M \leq 1$ $(q(M) \leq 0)$.

For transonic plane flow, the values of ϕ and ψ for the subsonic and supersonic parts are connected along the transformed sonic line, the axis $q = 0$.

Let, $\phi, \psi(q = 0, \theta) = \phi^*, \psi^*(\theta)$. For the elliptic region, ϕ^* and ψ^* are used to complete the formulation of an elliptic boundary value problem, and for the hyperbolic domain, ϕ^* and ψ^* are initial values which describe the solution of the above equations in a triangular region of dependence. It should be mentioned that Sobieczky's formulation of transonic boundary value problem is different from the well known Tricomi problem, since there is no prescribed boundary in the hyperbolic region.

In the elliptic domain $(q \leq 0)$, the elliptic system does not change its nature by conformal mappings. Sobieczky introduced several mappings called rheograph-transformations to facilitate the solution of the equations in the elliptic domain. He used electric analogy and conducting papers which have to be inhomogenized to vary the local conductivity to simulate the coefficient K in the governing equations.

On the other hand, the solution of the hyperbolic equations can be found by series expansion or by the method of characteristics, where the computation starts from the given sonic line.

In passing, for near sonic flows, where the mach number is close to one in the whole flow field, Sobieczky replaced $q(M)$ and $K(M)$ by the first term of their series expansions around $M = 1$. The governing equations take now the form of generalized axially symmetric potential equations treated by Weinstein, thus Sobieczky obtained

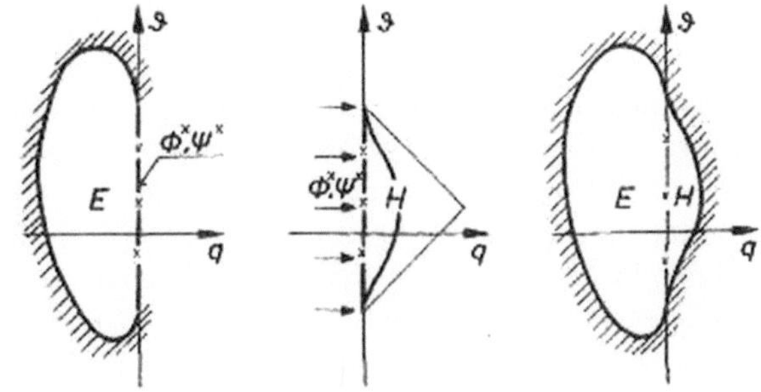

Fig. 13.6 Formulation of a transonic boundary value problem (from IUTAM Symposium Transonicum II, 1976, Springer Verlag, with kind permission of Springer)

exact solutions in closed form which are identical to the solutions of the Tricomi equation.

The electric analogy together with the method of characteristics were used successfully to design supercritical wing sections as well as to design transonic cascades. For a given upstream velocity, the elliptic boundary value problem is defined and an analog model can be produced. Continuation of the analog model beyond the sonic line, using numerical methods, allows the calculation of the dependent supersonic flow region and hence the airfoil contour, see Fig. 13.6.

Later, Sobieczky introduced the method of fictitious gas, where the density is modified in the supersonic region, and the flow field is solved to produce the sonic line. The elliptic flow field with the sonic line are used to construct shock free airfoil via continuation based on the method of characteristics and the actual density in the supersonic region, to produce the contour of the airfoil adjacent to the supersonic flow. See Ref. [18] for more details (Notice that the direct extension to three dimensional flows over a wing leads to unstable calculations) (Fig. 13.6).

13.6 Electronic Analog Computers: Networks Versus Tanks

The analogy between an electrical system and a mechanical system is well known. For example, a mass-spring damper system is equivalent to a circuit of resistors, capacitors and inductors, governed by the same second order ordinary differential equation and can be solved on an analog computer using summation and integration elements with continuous real time simulation and low power requirement. Systems of two degrees can be simulated via solving two coupled differential equations and coupling partial circuits. Accuracy is limited however, due to the noise in the circuits.

Analog computers were used in aeronautical industry in the fifties and sixties (before digital computers took over). Evolutionary partial differential equations can be reduced to systems of ordinary differential equations (continuous in time) via method of lines, where space derivatives are approximated with finite differences. Examples of heat and wave equations can be solved using analog computers with real time simulation. Stiff and nonlinear equations can also be treated using hybrid computers (see Ref. [19]). New ideas in this field are promising, for example the neural network systems [20].

13.7 Concluding Remarks

In this chapter, we discussed flow analogies including the Hele-Shaw cell, hydraulic analogy and electric analogy. Analog computer is also briefly covered. Other analogies exist. For example flowing foam, (a foam is a mixture of gas with a liquid at constant mass ratio).

Let μ be the mass concentration of the gas and $1 - \mu$ that of the liquid, then the specific volume of the foam is

$$\frac{1}{\bar{\rho}} = \frac{\mu}{\rho} + \frac{1-\mu}{\sigma} \tag{13.85}$$

where ρ and σ are the densities of the gas and the liquid component, respectively.

Assuming the expansion of the gas in the bubbles to be isentropic and $\mu \gg \rho/\sigma$, hence

$$\bar{\rho} = \frac{\rho}{\mu}, \quad and \quad \frac{\bar{\rho}}{\bar{\rho}_0} = \left(\frac{p}{p_0}\right)^{\frac{1}{\gamma}} \tag{13.86}$$

The speed of sound $\bar{c}^2 = dp/d\bar{\rho} = \mu dp/d\rho = \mu c^2$, and from Bernoulli's law, the velocity is given by

$$\frac{1}{2}\bar{q}^2 = \int_{p_0}^{p} \frac{dp}{\bar{\rho}} = \mu \int_{p_0}^{p} \frac{dp}{\rho} = \mu \frac{1}{2} q^2 \tag{13.87}$$

Hence,

$$\bar{M} = \frac{\bar{q}}{\bar{c}} = \frac{q}{c} = M \tag{13.88}$$

Flowing foams and cavitating liquids are closely related. For more details see Oswatitsch [9].

In this regard, it is of historical interest that the analog solution of the torsion problem of a bar was proposed by L. Prandtl, and used by G.I. Taylor, two giants of fluid mechanics. An opening in the plane that has the same shape as the cross-section of a torsion bar covered with homogeneous elastic membrane, such as a soap film, under pressure on one side will have a lateral displacement satisfying the same equation as the Prandtl torsion stress function. Prandtl's membrane analogy is discussed in many strength of materials books, see for example [21].

References

1. Hele-Shaw, H.J.S.: Investigation of the nature of the surface resistance of water and of stream-line motion under certain experimental conditions. Trans. Inst. Nav. Archit. **40**, 25 (1898)

2. Gustafsson, B., Vasiliev, A.: Conformal and Potential Analysis in Hele-Shaw Cells. Birkhäuser Verlag, Basel, Switzerland (2006)
3. Van Dyke, M.: Album of Fluid Motion. Parabolic Press, Stanford (1975)
4. Schlichting, H.: Boundary Layer Theory. McGraw Hill, New York (1960)
5. Batchelor, G.K.: An Introduction to Fluid Dynamics. Cambridge University Press, Cambridge (1967)
6. Acheson, D.J.: Elementary Fluid Dynamics. Oxford University Press, U. K. (1990)
7. White, F.M.: Fluid Mechanics. Wiley, New York (1986)
8. Rajaratnam, N.: Hydraulic Jumps. Advances in Hydroscience, vol. 4. Academic Press, New York (1967)
9. Oswatitsch, K.: Gas Dynamics. Academic Press, New York (1956)
10. Loh, W. (ed.): Modern Developments in Gas Dynamics. Plenum Press, New York (1969)
11. Thompson, P.: Compressible Fluid Dynamics. McGraw Hill, New York (1972)
12. Malavard, L.C.: The use of photoelectric analogies in aerodynamics. AGARDograph No. 18 (1956)
13. Stewart, A.J.: The lift of the delta wing at supersonic speeds. Q. Appl. Math. **IV**(3), 246–254 (1964)
14. Germain, P.: La Théorie des Mouvements Coniques et ses Applications à l' Aérodynamique, Utilisation des Analogies Eléctriques. Publication O.N.E.R.A. No.34 (1949)
15. Enselme, M.: Contribution de l'analogie électrique au calcul des caractéristiques aérodynamiques d'un avion en écoulement supersonique stationnaire ou instationnaire. O.N.E.R.A. Paris (1967)
16. Goldstein, S.: On the vortex theory of screw propellers. Proc. R. Soc. Lond. Ser. A **123**, 440–465 (1929)
17. Sobieczky, H.: Application of generalized potentials to plane transonic low. IUTAM Symposium Transonicum II. Springer, Berlin
18. Sobieczky, H.: Rheograph Transformation and Continuation Methods, Von Karman Institute Fluid Mech. Lecture Series 1980-4 (1980)
19. Cowan, G.: A VLSI analog computer/math coprocessor for a digital computer. Ph.D. Dissertation, Columbia University (2005)
20. Ratier, N.: Analog Computing of Partial Differential Equations. 6th International Conference on Sciences of Electronics, Technologies of Information and Telecommunications, Sousse, Tunisia (2012)
21. Boresi, A.P., Schmidt, R.J., Sidebottom, O.M.: Advanced Mechanics of Materials. Wiley, New York (1993)

Part III
Problems and Solutions

Chapter 14
Problems

14.1 Problem 1

14.1.1 Thin Airfoil Theory (2-D Inviscid Flow)

14.1.1.1 Quiz—Answer by Yes/no

- in subsonic flow, does the aerodynamic coefficient given below depend upon the small parameters:

Aerodynamic coefficient	e/c	d/c	α
C_l			
$C_{m,o}$			
$C_{m,ac}$			
C_d			

- in supersonic flow, does the aerodynamic coefficient given below depend upon the small parameters:

Aerodynamic coefficient	e/c	d/c	α
C_l			
$C_{m,o}$			
$C_{m,ac}$			
C_d			

14.1.1.2 Lift Coefficient

A thin profile with parabolic camber $\frac{d}{c} = 0.02$ is set at angle of attack $\alpha = 5°$. Find the lift coefficient C_l in incompressible flow. Find the lift coefficient in compressible flow at $M_0 = 0.7$. Hint: use linear correction method.

J.J. Chattot and M.M. Hafez, *Theoretical and Applied Aerodynamics*,
DOI 10.1007/978-94-017-9825-9_14

14.1.1.3 Drag Coefficient

If the drag coefficient of the same profile at $\alpha = 0$ and $M_0 = 1.4$ is $(C_d)_{M_0=1.4} = 0.02$, find the drag at $M_0 = 2$. (hint: look at the definition of $(C_d)_{\alpha=0}$). What will the drag be at $M_0 = 2$. if $\alpha = 5°$? What is the drag coefficient at $M_0 = 0.7$?

14.1.1.4 Equilibrium About an Axis

Give the definition of the aerodynamic moment coefficient $C_{m,o}$ in terms of the dimensional quantities.

If the profile is attached to a vertical axis located at $x = 0$ (nose of profile), write the equation of equilibrium that will give the equilibrium angle of attack α_{eq}. Find α_{eq}, and sketch the profile at equilibrium. Is the equilibrium stable?

14.1.2 Lifting Line Theory

14.1.2.1 Flow Model

Explain, in simple terms, how an inviscid flow model, Prandtl Lifting Line theory, can account for a drag in the flow past a finite wing. Use arguments such as Joukowski Lift theorem with a sketch, or energy equation.

Give the expressions for the lift coefficient C_L and induced drag coefficient C_{Di} of an arbitrary wing, in terms of the Fourier coefficients $A_1, A_2, ..., A_n$. The area of reference, $A_{ref} = S$ is the wing area.

14.1.2.2 Non Singular Lift Distribution

To lower the risks of tip vortex encounter in airport traffic associated with intense vorticity ($\frac{d\Gamma}{dy}\left(\pm\frac{b}{2}\right) = \pm\infty$), the new generation of airplanes is required to have weak tip vortices. This is achieved by enforcing a Γ distribution such that $\frac{d\Gamma}{dy}\left(\pm\frac{b}{2}\right) = 0$.

For a given $C_L \neq 0$, a distribution that satisfies this condition and has a low induced drag is a combination of modes 1 and 3:

$$\begin{cases} \Gamma\left[y(t)\right] = 2Ub\left(A_1 \sin t + A_3 \sin 3t\right) \\ y(t) = -\frac{b}{2}\cos t \end{cases}, \quad 0 \leq t \leq \pi$$

Find the relation between A_1 and A_3 that satisfies $\frac{d\Gamma}{dy}\left(-\frac{b}{2}\right) = 0$ (by symmetry, the other condition will be satisfied).

Make a graph of the corresponding distribution $\Gamma(y)$ or $\Gamma(t)$.

14.1.2.3 Drag Penalty

For a given C_L (at take-off), find the penalty in induced drag for a given aspect ratio AR compared to the elliptic loading. Work the answer formally (do not plug in numbers).

Find the efficiency factor $e = (1 + \delta)^{-1}$.

14.1.3 Equilibrium of the Glider (3-D Incompressible Flow)

14.1.3.1 Glider Effective Aspect Ratio

As a glider (engine off), the glider has the following lift and moment coefficients in terms of the geometric angle of attack α (rd) and tail setting angle t_t (rd):

$$\begin{cases} C_L(\alpha, t_t) = 5.3\alpha + 1.56 + 0.5t_t \\ C_{M,0}(\alpha, t_t) = -1.7\alpha - 0.43 - 0.45t_t \end{cases}$$

Find the effective aspect ratio AR of the glider (wing+tail), assuming ideal loading.

14.1.3.2 Equilibrium Angle of Attack

Find $\alpha_{eq}(t_t)$ for the glider, given that $\frac{x_{cg}}{l_{ref}} = 0.29$. Hint: the equilibrium of the moment corresponds to

$$C_{M,0}(\alpha_{eq}, t_t) + \frac{x_{cg}}{l_{ref}} C_L(\alpha_{eq}, t_t) = 0$$

Find the angle of attack α_{eq} and the corresponding setting angle of the tail t_t for $C_L = 1.9$.

14.1.3.3 Landing Speed

At sea level, $\rho = 1.2\,\text{kg/m}^3$, $g = 9.81\,\text{ms}^{-2}$, find the landing speed, given that the total mass $m = 15\,\text{kg}$, the maximum lift coefficient $(C_L)_{max} = 1.9$ corresponding to the reference area $A_{ref} = 0.68\,\text{m}^2$. Hint: use the equilibrium equation in the direction perpendicular to the flight path.

14.2 Problem 2

14.2.1 Thin Airfoil Theory (2-D Inviscid Flow)

14.2.1.1 Incompressible Flow ($M_0 = 0$)

A thin airfoil with parabolic camber line $d(x) = 4d\frac{x}{c}(1 - \frac{x}{c})$ is moving with velocity U in a uniform atmosphere of density ρ. The chord of the airfoil is c.

Define the lift and moment coefficients, C_l and $C_{m,o}$ in terms of the dimensional quantities $\rho,\ U,\ c,\ L,\ M_{,o}$, where L and $M_{,o}$ are the lift (N/m) and the moment (Nm/m) per unit span.

Use thin airfoil theory results to express the C_l and $C_{m,o}$ in terms of α, the incidence angle (rd) and $\frac{d}{c}$ the relative camber.

Given $\frac{d}{c} = 0.086$ find the angles of incidence in degree for which $C_l = 0.5,\ C_l = 2.0$ and calculate the corresponding moments $C_{m,o}$.

Find the location of the center of pressure $\frac{x_{cp}}{c}$ in both cases.

Show on a sketch the position of the center of pressure relative to the aerodynamic center.

14.2.1.2 Supersonic Linearized Theory ($M_0 > 1$)

The same airfoil as in 1.1 is moving at supersonic speed.

Give the expression for the lift coefficient $C_l(\alpha)$ in supersonic flow.

Give the expression for the moment coefficient $C_{m,o}(\alpha)$ in supersonic flow and calculate $(C_{m,o})_{\alpha=0}$.

Express $\frac{x_{cp}}{c}$ in terms of α.

14.2.2 Lifting Line Theory (3-D Inviscid Flow)

14.2.2.1 Flow Model

Which velocity components are continuous across the vortex sheet and why?

Sketch on your own sheet the velocity vectors at points $A - E$ located near the vortex sheet in the Trefftz plane. See Fig. 14.1.

14.2.2.2 Improved Lift of a Rectangular Wing

The maximum lift of a rectangular wing is obtained when $\Gamma(y) = \Gamma_{max}$. However, this solution corresponds to two vortices of strength Γ_{max} and $-\Gamma_{max}$ shed by the left and right wing tips respectively, for which, as noted by Prandtl, the induced drag

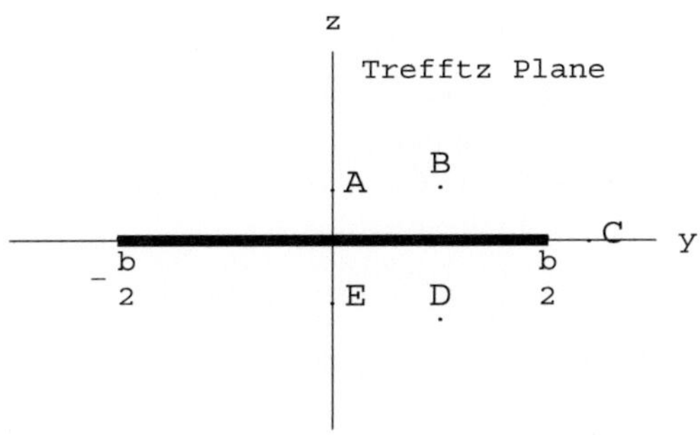

Fig. 14.1 Sketch of vortex sheet in the Trefftz plane

is infinite. A more desirable result is obtained by combining a few modes to improve the lift. Consider the following combination where only the odd modes, which are symmetric in y are used:

$$\begin{cases} \Gamma\left[y\left(t\right)\right] = 2UbA_1\left(\sin t + \frac{1}{9}\sin 3t\right) \\ y\left(t\right) = -\frac{b}{2}\cos t \end{cases}, \; 0 \le t \le \pi$$

Note that $\frac{d\Gamma}{dt} = \frac{8UbA_1}{3}\cos^3 t$.

Find the value A_1 such that the maximum circulation is equal to Γ_{max}, and sketch the distributions of circulation for the high lift wing and the elliptic loading.

Find the total lift C_L in terms of aspect ratio AR, Γ_{max}, U and b. Compare with the maximum lift for the elliptic loading.

Find the downwash distribution $w_T(t)$ for the high lift wing and sketch it with the downwash for the elliptic loading.

14.2.2.3 Drag Penalty

Compute the induced drag of the high lift wing and compare it with the induced drag for the elliptic loading.

14.2.3 Airplane Longitudinal Equilibrium

14.2.3.1 Global Coefficients

The airplane has the following lift and moment coefficients in terms of the geometric angle of attack α (rd) and tail setting angle t_t (rd):

$$\begin{cases} C_L(\alpha, t_t) = 3.88\alpha + 0.5 + 0.481t_t \\ C_{M,0}(\alpha, t_t) = -1.31\alpha - 0.124 - 0.452t_t \end{cases}$$

Find the location $\frac{x_{ac}}{l_{ref}}$ of the aerodynamic center.

Find the location $\frac{x_{cg}}{l_{ref}}$ of the center of gravity, given a 4 % static margin of stability.

Give the definition of the aerodynamic center.

Derive the expression for the moment about the aerodynamic center, $C_{M,ac}(t_t)$ as a function of the tail setting angle.

14.2.3.2 Take-Off Conditions

Using the transfer of moment formula, derive the moment about the center of gravity, $C_{M,cg}(\alpha, t_t)$.

Find the equilibrium conditions, $\alpha_{eq}(t_t)$ and $C_{Leq}(t_t)$ for the airplane.

If the airplane take-off lift coefficient is $(C_L)_{t-o} = 1.44$, find the tail setting angle that will be needed, and the corresponding value of $(\alpha_{eq})_{t-o}$.

14.3 Problem 3

14.3.1 Thin Airfoil Theory (2-D Inviscid Flow)

14.3.1.1 Incompressible Flow ($M_0 = 0$)

A thin airfoil with parabolic camber line $d(x) = 4d\frac{x}{c}(1 - \frac{x}{c})$ is moving with velocity U in a uniform atmosphere of density ρ. The chord of the airfoil is c, $\frac{d}{c} = 0.086$.

Give the expression of the Fourier coefficients $A_0, A_1, A_2, \ldots, A_n$, for this airfoil.

What is the angle of adaptation or ideal angle of attack? Sketch the corresponding flow. How is the flow at the leading edge?

Calculate the incidence of zero lift, α_0 in deg. Show that the moments $C_{m,o} = C_{m,a.c.}$ in this case.

Estimate the upper limit of the weight (in N) that a wing of chord $c = 0.368$ m and span $b = 4.877$ m could lift at $C_l = 2$ if the 2-D solution were applicable at take-off velocity $V = 10.5$ m/s (use $\rho = 1.225$ kg/m^3). What would the take-off mass be in kg?

14.3.1.2 Supersonic Linearized Theory ($M_0 > 1$)

A biconvex profile of equation $z = 2e\frac{x}{c}(1 - \frac{x}{c})$ equips a fin moving through the air at $M_0 = 2$ and $\alpha = 1°$.

Calculate the wave drag coefficient $(C_d)_{\alpha=0}$ and give the formula for $C_l(\alpha)$ and $C_d(\alpha)$ in terms of α and calculate the values for the given incidence and for a 10 % thickness ratio $\frac{e}{c} = 0.1$.

Estimate the upper limit of the lift and drag forces (in N) on the fin if $c = 0.1$ m, $b = 0.2$ m and $V = 633$ m/s at $z = 6000$ m altitude ($\rho_{air} = 0.657$ kg/m^3).

14.3.2 Lifting Line Theory (3-D Inviscid Flow)

14.3.2.1 Induced Drag in Cruise

Give the formula for the induced drag D_i in (N), in terms of the wing efficiency e, the dynamic pressure $\frac{1}{2}\rho V^2$, and the other geometric characteristics of the wing.

Give the equilibrium equation along the vertical axis during horizontal cruise in dimensional form (neglect the lift of the tail).

Eliminate C_L between the two results and rewrite D_i.

Application: calculate D_i for $V = 20$ m/s, M = 24.7 kg and $b = 4.877$ m, $e = 0.9$ (use $\rho = 1.225$ kg/m^3).

14.3.2.2 Turn

In order to turn, the airplane must roll about its longitudinal axis. This can be achieved by using the ailerons, small flaps located along the trailing edge of the wing.

Assume that the wing loading in cruise is given by:

$$\begin{cases} \Gamma\left[y\left(t\right)\right] = 2UbA_1\left(\sin t + \frac{1}{9}\sin 3t\right) \\ y\left(t\right) = -\frac{b}{2}\cos t \end{cases}, \quad 0 \le t \le \pi$$

What is the induced drag coefficient during cruise in terms of A_1?

The turn will require to add the antisymmetric mode $A_2 \sin 2t$. Sketch mode 2.

14.3.2.3 Drag Penalty

Find the drag penalty associated with the turn by calculating $\frac{\Delta C_{Di}}{C_{Di}}$, given that $A_2 = A_1/3$.

14.3.3 Glider Equilibrium

Away from high lift, a linear model of the glider in terms of the angle of incidence α (deg) and tail setting angle t_t (deg) is given by:

$$\begin{cases} C_L(\alpha, t_t) = 0.1\alpha + 0.01t_t + 0.7 \\ C_{M,0}(\alpha, t_t) = -0.02\alpha - 0.01t_t - 0.13 \end{cases}$$

14.3.3.1 Definition

Give the definition of the aerodynamic center.

14.3.3.2 Aerodynamic Center

Find the location x_{ac} of the aerodynamic center, given that the reference length is $l_{ref} = 1.5\,\text{m}$.

14.3.3.3 Moment at Aerodynamic Center

Derive $C_{M,ac}$.

14.3.3.4 Moment at Center of Gravity

Derive $C_{M,cg}$ given a 4 % positive static margin.

Is the airplane statically stable? Explain.

14.3.3.5 Equilibrium

Derive the equilibrium incidence $\alpha(t_t)$.

14.4 Problem 4

14.4.1 2-D Inviscid, Linearized, Thin Airfoil Theories

14.4.1.1 Incompressible Flow ($M_0 = 0$)

Thickness Effect

In Thin Airfoil theory, what type of singularity distribution is used to represent thickness?

Which forces and moment coefficients are not zero in the symmetric problem associated with thickness, C_l, $C_{m,o}$, and C_d?

Is the pressure coefficient C_p affected by thickness? Explain.

Camber Effect

The Selig 1223 has a mean camber line which is parabolic to a good approximation, and a relative camber $d_m = d/c = 0.086$.

Write the expressions of C_l and $C_{m,o}$ in terms of α and d_m (Use results derived in class).

Find the angle of zero lift, $(\alpha)_{Cl=0}$.

What are the values of the following coefficients in the linear expressions of the lift and moment curves:

$$\frac{dC_l}{d\alpha}, \quad C_{l0}, \quad \frac{dC_{m,o}}{d\alpha}, \quad C_{m,o0}$$

Sketch the flow past the mean camber line at $\alpha = 0$. Explain.

Equilibrium About an Axis

If a hinge is placed at the nose of the Selig 1223 and the profile is allowed to rotate freely about it, what would the equilibrium angle be in deg (neglect weight)?

Is the equilibrium stable? Why?

Where is the center of pressure at equilibrium?

Sketch the free-body diagram of the profile at equilibrium, including the reaction, C_r of the hinge.

14.4.1.2 Supersonic Flow ($M_0 > 1$, $\beta = \sqrt{M_0^2 - 1}$)

A thin airfoil with parabolic camber line $d(x) = 4d_m x(1-x/c)$ is moving with Mach number M_0 in a uniform atmosphere. The chord of the airfoil is c, $d_m = d/c = 0.086$.

Pressure Distributions

Plot $-C_p^+$ and $-C_p^-$ versus x for this airfoil at $\alpha = 0$.

What is the corresponding lift coefficient C_l?

Lift and Moment Coefficients

Calculate the zero incidence moment coefficient $\left(C_{m,o}\right)_{\alpha=0}$.

Give the values of the coefficients C_l and $C_{m,o}$ for the general case $\alpha \neq 0$.

Equilibrium About an Axis

Calculate the moment $C_{m,D}$ about an arbitrary point D along the chord and if the profile is allowed to rotate freely about point D, show that the equilibrium will be stable provided $x_D/c < 1/2$ (neglect weight).

Give the equilibrium incidence, α_{eq} in terms of x_D.

Does α_{eq} depend on Mach number?

14.4.2 Prandtl Lifting Line Theory

14.4.2.1 Vortex Sheet

Describe briefly the vortex sheet behind a finite wing with sharp trailing edge and explain whether the velocity components (u, v, w) have jumps and why.

14.4.2.2 Designing for Tip Vortices

In general, the wing loading is such that $\Gamma' \to \mp\infty$ as $y \to \pm b/2$, an indication of strong tip vortices.

Combine modes 1 and 3 to satisfy $\Gamma'(\pm b/2) = 0$, given that the total lift is $(C_L)_{t-o}$ at take-off.

Find A_1 and A_3 in terms of $(C_L)_{t-o}$ and wing aspect ratio AR.

Sketch the corresponding distributions of circulation and downwash (Use $\sin 3t = \sin t\ (4\cos^2 t - 1)$).

Show that there is upwash near the wing tips.

14.4.2.3 Induced Drag

Calculate the induced drag and the gain/loss compared to the elliptic loading.

14.4.3 Equilibrium of the Aggie Micro Flyer

The AMF has a span $b_m = 1\,\text{m}$ and a constant chord $c_{xm} = 0.192\,\text{m}$.

14.4.3.1 Main Wing

What is the wing aspect ratio, AR_m?

Using the viscous polar of the Selig 1223 at Re = 200,000 with the "prandtline.f" code, it was found that the wing has a lift slope $dC_{Lm}/d\alpha = 4.8701$. Compare this result with the theoretical formula $dC_{Lm}/d\alpha = 2\pi/(1 + 2/AR_m)$.

14.4.3.2 Lift Curves

The linearized lift curve at low incidences for the main wing is: $C_{Lm} = 4.8701\alpha + 0.6997$, and for the tail: $C_{Lt} = 1.6338\alpha + 3.4529t_t - 0.26135$, α and t_t in *radian*.

Give the general formula for the lift $C_L(\alpha, t_t)$ of the combination of the wing and tail of areas A_m and A_t, and reference area $A_{ref} = A_m + A_t$.

Application: check that with $A_m = 0.192\,\text{m}^2$, $A_t = 0.1525\,\text{m}^2$, $C_L(\alpha, t_t) = 3.4374\alpha + 1.5286t_t + 0.2742$.

14.4.3.3 Moment Curve

The moment coefficient for the complete airplane is given by: $C_{M,o}(\alpha, t_t) = -1.378\alpha - 1.268t_t - 0.03492$.

Write the equilibrium equation for the aerodynamic moment (or equivalently write $C_{M,cg} = 0$), given that $x_{cg}/l_{ref} = 0.3064$. Neglect moment contribution from the engine.

Solve for $\alpha_{eq}(t_t)$.

Application: The maximum speed corresponding to full throttle and horizontal flight corresponds to $t_t = 1.63°$. Find the angle of incidence α_{eq} in *deg*.

14.4.3.4 Take-Off Conditions

Using the same linearized lift curve, find the take-off incidence, $(\alpha_{eq})_{t-o}$ in *deg* corresponding to $C_{Lm} = 1.9$.

Show that at take-off, the tail has positive lift (Hint: calculate the tail setting angle $(t_t)_{t-o}$ at take-off).

14.4.3.5 Extra Credit

Sketch the forces and forces locations at take-off with respect to the aerodynamic centers of the main wing ($x_{acm}/l_{ref} = 0.27$), of the tail ($x_{act}/l_{ref} = 0.83$), of the complete configuration (x_{ac}/l_{ref}) and with respect to the center of gravity, given that there is an 8 % static margin.

14.5 Problem 5

14.5.1 2-D Inviscid, Linearized, Thin Airfoil Theories

14.5.1.1 Incompressible Flow ($M_0 = 0$)

Definition

In 2-D, define the following aerodynamic coefficients C_l, $C_{m,o}$, and C_d in terms of the forces and moment per unit span (L', M'_o, D'), the density (ρ), incoming flow velocity (U) and chord (c). Give the expressions for $C_l(\alpha)$, $C_{m,o}(\alpha)$ and C_d for a symmetric profile.

Suction Force

Consider a symmetric profile ($d(x) = 0$). The lifting problem corresponds to the flow past a flat plate at incidence. Sketch the flat plate, the force due to pressure integration and the suction force.

Calculate the suction force coefficient $C_s = F'_s/(\frac{1}{2}\rho U^2 c)$, where F'_s is the suction force per unit span, from the C_l expression.

In what sense does the thickness distribution contribute to recuperating the suction force? Explain.

Center of Pressure

Give the definition of the center of pressure.

Find the location of the center of pressure for a symmetric profile.

Calculate $C_{m,ac}$ for a symmetric profile.

Verify your result for $C_{m,o}(\alpha)$ using the change of moment formula, from the aerodynamic center to the profile leading edge.

14.5.1.2 Supersonic Flow ($M_0 > 1$, $\beta = \sqrt{M_0^2 - 1}$)

A thin double wedge airfoil equips the fins of a missile cruising at Mach number $M_0 > 1$ in a uniform atmosphere. The chord of the airfoil is c. The profile camber and thickness are:

$$\begin{cases} d(x) = 0 \\ e(x) = \begin{cases} 2\theta x, & 0 \le x \le c/2 \\ 2\theta(c - x), & c/2 \le x \le c \end{cases} \end{cases}$$

Fig. 14.2 Double wedge in supersonic flow at zero incidence

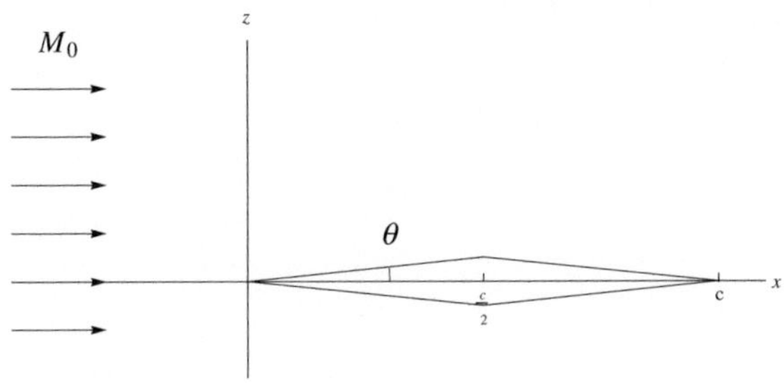

with $z^{\pm}(x) = d(x) \pm e(x)/2 + \alpha(c - x)$. See Fig. 14.2.

Pressure Distributions

Plot $-C_p^+$ and $-C_p^-$ versus x for this airfoil at $\alpha = 0$.

Lift Coefficient

Give the lift coefficient $C_l(\alpha)$ for this airfoil in supersonic flow.

Drag Coefficient

Calculate $(C_d)_{\alpha=0}$.

Give the value of the coefficient C_d for the general case $\alpha \neq 0$.

Moment Coefficient

Calculate the zero incidence moment coefficient $\left(C_{m,o}\right)_{\alpha=0}$.

Give the values of the coefficient $C_{m,o}$ for the general case $\alpha \neq 0$.

Maximum Finess

Form the expression of the inverse of the finess, $1/f = C_d/C_l$ and find the value of α that maximizes f (minimizes $1/f$).

Sketch the profile at the maximum finess incidence and indicate on your drawing the remarkable waves (shocks, expansions).

14.5.2 Prandtl Lifting Line Theory

14.5.2.1 Origin of Induced Drag

Explain how an inviscid flow (no friction) past a finite wing, can produce a drag. (Hint: you can discuss an energy balance).

14.5.2.2 Maximum Lift of a Wing

The maximum lift of a wing would ideally correspond to a constant circulation along the span, $\Gamma(y) = \Gamma_{max}$, where Γ_{max} is the maximum circulation that a profile can sustain, given the flow conditions (Reynolds number, etc.). Such a distribution is given by:

$$\Gamma\left[y(t)\right] = 2Ub \sum_{p=0}^{\infty} \frac{2\Gamma_{max}}{\pi Ub(2p+1)} \sin(2p+1)t$$

where $y(t) = -\frac{b}{2}\cos t,\ 0 \leq t \leq \pi$. It corresponds to a single, finite strength horseshoe vortex trailing the wing from the wing tips. The problem with this model, as found out by Prandtl, is that the induced drag is infinite. Since this is not feasible, we are going to consider only a few terms of the infinite series.

Consider modes 1 and 3 only ($p = 0, 1$).

Give the values of A_1 and A_3.

Sketch the corresponding distribution of circulation and downwash (Use $\sin 3t = \sin t\ (3 - 4\sin^2 t)$).

Show that there is no upwash.

14.5.2.3 Induced Drag

Calculate the induced drag and the gain/loss compared to the elliptic loading.

Show that the induced drag due to two finite strength vortices is infinite.

14.5.3 Equilibrium of the Aggie Micro Flyer

The AMF III has a span $b_m = 2.1\,\text{m}$ and a constant chord $c_{xm} = 0.263\,\text{m}$. The corresponding tail is defined as $b_t = 0.527\,\text{m}$ with constant chord $c_{xt} = 0.176\,\text{m}$, using the rapid prototyping code, which also provides the take-off velocity $V_{to} = 13.7\,\text{m/s}$ and a maximum take-off mass $M = 15.5\,\text{kg}$.

14.5.3.1 Airplane Lift and Moment Curves

The equilibrium code calculates the lift and moment coefficients for the complete configuration at low incidences to be:

$$\begin{cases} C_L(\alpha, t_t) = 4.47\alpha + 0.37t_t + 1.11 \\ C_{M,o}(\alpha, t_t) = -1.30\alpha - 0.33t_t - 0.28 \end{cases}$$

where α is the geometric incidence (in radians, measured from the fuselage axis) and t_t is the tail setting angle (in radians). $C_{M,o}$ is the aerodynamic moment about the origin of the coordinate system (located at the nose O).

What is the effective aspect ratio AR for the complete configuration? Compare it with the main wing aspect ratio AR_m. Explain the difference. (Hint: look at how the total lift coefficient is calculated).

Find the location of the aerodynamic center x_{ac} given that $l_{ref} = 2.062\,\text{m}$.

14.5.3.2 Equilibrium Incidence

The center of gravity is located at $x_{cg}/l_{ref} = 0.21$.

Find the static margin in % of l_{ref}.

Derive the formula for the equilibrium incidence $\alpha(t_t)$. Check your result carefully as other results depend upon it.

14.5.3.3 Take-Off Conditions

The take-off lift coefficient for the complete configuration is $C_{L,to} = 2.0$.

Find the tail setting angle and the incidence at take-off.

The tail lift curve is given by

$$C_{Lt} = 2.49\alpha + 2.61t_t - 0.36$$

Find the tail lift coefficient at take-off.

Find the lift force (in N) on the tail at take-off. Take $\rho = 1.225\,\text{kg/m}^3$.

Is it up or down?

14.5.3.4 Extra Credit

Sketch qualitatively the forces (main wing and tail lifts, weight) and forces locations at take-off with respect to the aerodynamic centers of the main wing ($x_{acm}/l_{ref} = 0.21$), of the tail ($x_{act}/l_{ref} = 0.85$), of the complete configuration (x_{ac}/l_{ref}) and with respect to the center of gravity.

14.6 Problem 6

14.6.1 2-D Inviscid, Linearized, Thin Airfoil Theories

14.6.1.1 Incompressible Flow ($M_0 = 0$)

Airfoil Design

Consider a thin airfoil of chord c with *parabolic camberline*. The design requirements are the following:

- the take-off lift coefficient is $(C_l)_{t-o} = 1.8$
- the location of the center of pressure at take-off is $\left(\frac{x_{cp}}{c}\right)_{t-o} = 0.4$

The airfoil geometry will be completely determined by the airfoil Fourier coefficients.

First, write C_l in terms of the Fourier coefficients.

Second, express the center of pressure location in terms of the Fourier coefficients.

Write as a system and solve for A_0 and A_1. Check your solution carefully.

Find the take-off incidence.

Find the profile relative camber.

Can you think of a profile that can fulfill these requirements?

Lift Curve

Give the value of the lift slope $\frac{dC_l}{d\alpha}$.

Give the value of the C_{l0} corresponding to $\alpha = 0$.

Sketch the lift curve on a graph $C_l(\alpha) = \frac{dC_l}{d\alpha}\alpha + C_{l0}$.

Equilibrium with Tail

If the center of gravity of the wing+tail configuration is located at $\frac{x_{cg}}{c} = 0.5$ at take-off, what sign do you expect the tail lift coefficient to be, $C_l >=< 0$?

Sketch this situation.

14.6.1.2 Supersonic Flow ($M_0 > 1$, $\beta = \sqrt{M_0^2 - 1}$)

A flat plate equips the fins of a missile cruising at Mach number $M_0 > 1$ in a uniform atmosphere. The chord of the airfoil is c.

Pressure Distribution and Global Coefficients

Plot $-C_p^+$ and $-C_p^-$ versus x for this airfoil at $\alpha > 0$.

Write the formula for the lift coefficient $C_l(\alpha)$, moment coefficient $C_{m,o}(\alpha)$ and drag coefficient $C_d(\alpha)$ according to inviscid, linearized supersonic flow theory.

Maximum Finess and Flow Features

Given a *viscous drag* coefficient C_{d0} independent of incidence, form the expression of the inverse of the finess, $1/f = C_d/C_l$ where C_d now includes the viscous drag and find the value of α that maximizes f (minimizes $1/f$).

Sketch the profile at incidence and indicate on your drawing the remarkable waves (shocks, expansions).

14.6.2 Prandtl Lifting Line Theory

14.6.2.1 Downwash Evolution

For a large aspect ratio (AR) wing without sweep (straight lifting line) how does the downwash vary with distance from the lifting line from upstream infinity to downstream infinity near the x-axis and in particular what is the relationship between $w_w(y)$ and $w_T(y)$, $-\frac{b}{2} < y < \frac{b}{2}$, where the former is the downwash at the lifting line and the latter the downwash in the Trefftz plane, far downstream. Explain with a simple argument.

14.6.2.2 Wing with Ideal Loading

Consider an aircraft wing with ideal loading (minimum induced drag) at a cruise lift coefficient $(C_L)_{cruise}$.

Calculate the maximum value of the circulation (root circulation) Γ_0 and the downwash at the lifting line $w_w(y)$.

14.6.2.3 Induced Downwash in Manoeuvre

The aircraft needs to roll to initiate a left turn. Given that the rolling moment about the aircraft axis is given in terms of the Fourier coefficients by

$$C_{M,0x} = -\frac{\pi}{4} AR\, A_2$$

sketch the circulation distribution and corresponding downwash that will allow such a manoeuvre. Hint: use the simplest Fourier distribution that needs to be added to the ideal loading.

14.6.2.4 Induced Drag in Manoeuvre

Calculate the induced drag and the gain/loss compared to the elliptic loading, assuming that the $(C_L)_{cruise}$ remains unchanged and that $A_2 = -A_1/10$.

14.6.3 Equilibrium of the Aggie Micro Flyer

The best design for the AMF IV has a span $b_m = 2.54$ m and a constant chord $c_{xm} = 0.279$ m. The corresponding tail is defined as $b_t = 1.0$ m with constant chord $c_{xt} = 0.254$ m. The equilibrium code provides the take-off velocity $V_{to} = 12.5\ m/s$ and a maximum take-off mass $M = 19$ kg.

14.6.3.1 Airplane Lift and Moment Curves

The equilibrium code calculates the lift and moment coefficients for the complete configuration at low incidences to be:

$$\begin{cases} C_L(\alpha, t_t) = 4.391\alpha + 0.757t_t + 0.928 \\ C_{M,o}(\alpha, t_t) = -1.733\alpha - 0.686t_t - 0.267 \end{cases}$$

where α is the geometric incidence (in radians, measured from the fuselage axis) and t_t is the tail setting angle (in radians). $C_{M,o}$ is the aerodynamic moment about the origin of the coordinate system (located at the nose O). We will use this linear model, even for take-off conditions.

Give the definition of the *aerodynamic center*.

Find the location of the aerodynamic center x_{ac} in m given that $l_{ref} = 1.6$ m.

The center of gravity is located at $x_{cg} = 0.503$ m. Find the *static margin SM* in % of l_{ref}.

14.6.3.2 Take-Off Conditions

Write the longitudinal equilibrium equation for the moment. How do you interpret this equation?

The main wing lift coefficient is given by

$$C_{Lm} = 4.927\alpha + 1.388$$

At take-off the lift coefficient for the main wing is $(C_{Lm})_{t-o} = 2.7$.
Find the take-off incidence $(\alpha)_{t-o}$.
Find the tail setting angle at take-off.
Find the location of the *center of pressure* in m at take-off.
Is the lift force on the tail up or down at take-off.

14.7 Problem 7

14.7.1 2-D Inviscid, Linearized, Thin Airfoil Theories

14.7.1.1 Incompressible Flow ($M_0 = 0$)

Cambered Plate Geometry

Consider a thin cambered plate of chord c of equation

$$d(x) = Ac\frac{x}{c}\left(1 - 2\frac{x}{c}\right)\left(1 - \frac{x}{c}\right)$$

Calculate $d'(x)$ to help you with the graph and make a plot of the plate (Hint: expand $d(x)$ before taking the derivative, A is not determined at this point). Verify your result as much of the rest depends on it.

Fourier Coefficients

Find all the Fourier coefficients $A_0, A_1, A_2, ..., A_n$ for this thin cambered plate and give the incidence of adaptation α_{adapt}. Use the identity $\cos^2 t = \frac{1}{2}(1 + \cos 2t)$.

Sketch the flow at the incidence of adaptation, showing in particular the streamlines near the leading and trailing edges.

Definition of Aerodynamic Center

Give the definition of the aerodynamic center.

Aerodynamic Coefficients

Give the aerodynamic coefficients $C_l(\alpha)$, $C_{m,o}(\alpha)$ and $C_{m,a.c.}$.

Static Equilibrium About an Axis

If an axis is located at the mid-chord, $\frac{x}{c} = \frac{1}{2}$, find the equilibrium angle α_{eq} if there are no forces other than the aerodynamic forces. Is the equilibrium stable, unstable, neutral?

14.7.1.2 Supersonic Flow ($M_0 > 1,\ \beta = \sqrt{M_0^2 - 1}$)

The same cambered plate equips the fins of a missile cruising at Mach number $M_0 > 1$ in a uniform atmosphere.

Pressure Distribution and Flow Features

Calculate and plot $-C_p^+$ and $-C_p^-$ versus x for this airfoil at $\alpha = 0$. Sketch the flow at $\alpha = 0$ (shocks, characteristic lines, expansion shocks).

Static Equilibrium About an Axis

If an axis is located at the mid-chord, $\frac{x}{c} = \frac{1}{2}$, find the equilibrium angle α_{eq} if there are no forces other than the aerodynamic forces. Is the equilibrium stable, unstable, neutral? (Hint: calculate $C_{m,\frac{1}{2}}$).

14.7.2 Prandtl Lifting Line Theory

14.7.2.1 Vortex Sheet Characteristics

Explain why the perturbation velocity components u and w are continuous across the vortex sheet behind a finite wing, i.e. $< u >=< w >= 0$.

How do you explain the existence of induced drag for a wing in incompressible, inviscid flow?

14.7.2.2 Circulation Representation

The circulation is represented by a Fourier series

$$\begin{cases} \Gamma[y(t)] = 2Ub \sum_{n=1}^{\infty} A_n \sin nt \\ y(t) = -\frac{b}{2} \cos t,\ 0 \leq t \leq \pi \end{cases}$$

Sketch the first three modes with unit coefficients $A_1 = A_2 = A_3 = 1$.

14.7.2.3 Efficiency Factor

Give the definition of the Efficiency Factor e in terms of the Fourier coefficients.

The AMAT09 wing has been designed with a rectangular planform, $b = 3.1$ m, $c = 0.55$ m, and is equipped with the SS1707-0723 double-element airfoil with high lift $(C_L)_{max} = 2.7$. Prandtl Lifting Line theory indicates that the Efficiency Factor is $e = 0.95$ for all phases of flight (take-off, top speed, power-off descent). Find all the coefficients $A_1, A_2, ..., A_n$ (not to be confused with their 2-D counterparts) for the following three phases of flight (symmetrical)

- take-off: $\alpha = 18°$, $C_L = 2.5$
- top speed: $\alpha = -3°$, $C_L = 1.0$
- power-off descent: $\alpha = 4°$, $C_L = 1.6$.

Here we assume that the rectangular wing is the "simplest" wing with $e = 0.95$, i.e., with the least number of non-zero coefficients.

14.7.3 Equilibrium of the AMAT09

The best design for the AMAT09 has a rectangular main wing with span $b_m = 3.1$ m and constant chord $c_{xm} = 0.55$ m. The lifting tail is also defined with a 33% moving flap. The equilibrium code provides the aircraft characteristics and a maximum take-off mass $M = 30$ kg.

14.7.3.1 Airplane Aerodynamic Center and Static Margin

The equilibrium code calculates the lift and moment coefficients for the complete configuration at low incidences to be:

$$\begin{cases} C_L(\alpha, t_t) = 3.912\alpha + 0.626t_t + 0.844 \\ C_{M,o}(\alpha, t_t) = -1.124\alpha - 0.625t_t - 0.138 \end{cases}$$

where α is the geometric incidence (in radians, measured from the fuselage axis) and t_t is the tail setting angle (in radians). $C_{M,o}$ is the aerodynamic moment about the origin of the coordinate system (located at the nose O). We will use this linear model, even for take-off conditions.

The center of gravity is located at $x_{cg}/l_{ref} = 0.227$. Find the *aerodynamic center* and the *static margin* SM in % of l_{ref}.

14.7.3.2 Equilibrium Condition and Static Stability

Derive the moment coefficient at the center of gravity, $C_{M,c.g.}(\alpha, t_t)$ and write the condition for equilibrium. Verify your result as the rest depends on it. Is the equilibrium stable?

Solve for $\alpha(t_t)$.

14.7.3.3 Top Speed

The top speed is obtained for $t_t = 7.6°$. Find α_{eq} at top speed.

Find the top speed, given that $A_{ref} = 2.225\,\text{m}^2$, $\rho = 1.2\,\text{kg/m}^3$ (Hint: use the equilibrium equation for horizontal flight).

14.8 Problem 8

14.8.1 2-D Inviscid, Linearized, Thin Airfoil Theories

14.8.1.1 Incompressible Flow ($M_0 = 0$)

Profile Geometry

Consider a half-double wedge profile of chord c of equation

$$\begin{cases} f^+(x) = d(x) + \frac{1}{2}e(x) = \Theta x, \ 0 \le x \le \frac{c}{2} \\ f^+(x) = d(x) + \frac{1}{2}e(x) = \Theta(c - x), \ \frac{c}{2} \le x \le c \end{cases}, \quad f^-(x) = d(x) - \frac{1}{2}e(x) = 0, \ 0 \le x \le c$$

where Θ is the wedge angle, see Fig. 14.3.

Calculate the distributions of camber $d(x)$ and thickness $e(x)$ for this profile. Check your result.

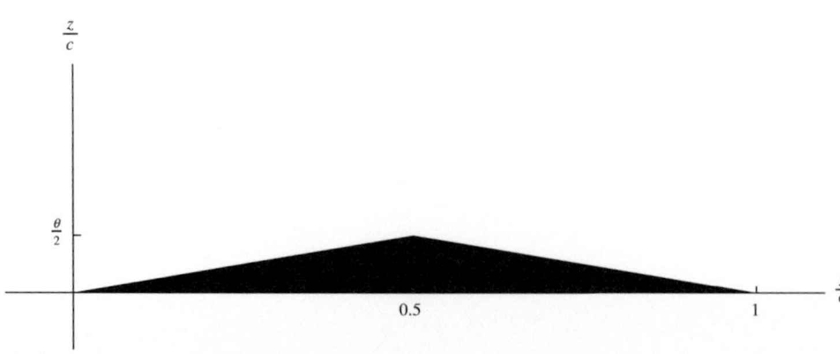

Fig. 14.3 Half double-wedge geometry

Fourier Coefficients

Give the expressions of the Fourier coefficients A_0 and A_n in the expansion of the vorticity for an arbitrary profile.

Calculate the Fourier coefficients $A_0,\ A_1$ and A_2 for this thin profile (Hint: you need to split the integral into two pieces $\int_0^{\frac{c}{2}} + \int_{\frac{c}{2}}^{c}$).

Give the incidence of adaptation α_{adapt}.

Sketch the flow at the incidence of adaptation, showing in particular the streamlines near the leading and trailing edges.

Definition of Aerodynamic Center

Give the definition of the aerodynamic center.

Aerodynamic Coefficients

Give the aerodynamic coefficients $C_l(\alpha)$ and $C_{m,o}(\alpha)$.

14.8.1.2 Supersonic Flow ($M_0 > 1,\ \beta = \sqrt{M_0^2 - 1}$)

The same profile equips the wing of an airplane cruising at Mach number $M_0 > 1$ in a uniform atmosphere.

Pressure Distribution and Flow Features

Calculate and plot $-C_p^+$ and $-C_p^-$ versus x for this airfoil at $\alpha = 0$. Sketch the flow at $\alpha = 0$ (shocks, characteristic lines, expansion shocks).

Aerodynamic Coefficients

At $\alpha = 0$, calculate the drag and moment coefficients $(C_d)_{\alpha=0}$ and $\left(C_{m,o}\right)_{\alpha=0}$.

Static Equilibrium About an Axis

If an axis is located at the leading edge, $x = 0$, find the equilibrium angle α_{eq} if there are no forces other than the aerodynamic forces. Is the equilibrium stable, unstable, neutral? (Hint: calculate $C_{m,o}(\alpha)$).

14.8.2 Prandtl Lifting Line Theory

14.8.2.1 Circulation Representation

A paper airplane wing is made with manila folder by cutting an ellipse of large aspect ratio (b, c_0). Then the wing is given some small camber such that the relative camber is constant $(d/c = const)$ along the span, and there is no twist, $t(y) = 0$. The circulation for an arbitrary wing is represented by a Fourier series

$$\begin{cases} \Gamma[y(\theta)] = 2Ub\sum_{n=1}^{\infty} A_n \sin n\theta \\ y(\theta) = -\frac{b}{2}\cos\theta,\ 0 \le \theta \le \pi \end{cases}$$

The chord distribution of the wing can be expressed as $c[y(\theta)] = c_0 sin\theta$ with the above change of variable.

Show that the circulation is given by the first mode only (Hint: use Prandtl Integro-differential equation to prove it).

What is the relationship between the root circulation Γ_0, the incidence α, the constant induced incidence α_i and the relative camber d/c?

14.8.2.2 Ideal Angle of Attack

From the previous result, calculate the lift coefficient, $C_L = C_l$ that corresponds to the geometric incidence, α_{ideal}, that will make the effective incidence zero, i.e. $\alpha_{eff} = \alpha_{ideal} + \alpha_i = 0$. (Hint: first find Γ_0 in this case and use the relationship between Γ_0 and C_l to eliminate Γ_0).

Find the corresponding value of α_{ideal} in term of the wing aspect ratio, AR, and d/c, given that the lift of this ideal wing is

$$C_L = \frac{2\pi}{1 + \frac{2}{AR}}\left(\alpha + 2\frac{d}{c}\right)$$

Make a sketch of the flow in a cross section of the wing (for example the wing root).

14.8.2.3 Including Twist

The paper wing is actually slightly warped with a linear twist $t(y) = -2t_x y/\mathrm{b}$, where t_x is a small positive number that represents the tip twist.

Show that this linear twist is represented by the second mode in the Fourier series. Use Prandtl Integro-differential equation to find A_2 in terms of t_x. (Hint: use the identity $sin(2\theta) = 2sin\theta cos\theta$).

14.8.2.4 Efficiency Factor

The warped wing loading is represented by the first two modes in the Fourier series.

Given that $A_2 = A_1/10$, calculate the induced drag of the warped wing and compare it with that of the ideal (untwisted) wing.

What is the percentage of increase of the induced drag?

Calculate the efficiency factor e.

14.8.3 Equilibrium of the AMAT10

The AMAT10 has a rectangular main wing with span $b_m = 3.6$ m and constant chord $c_{xm} = 0.35$ m. The tail is also defined with a 33% moving flap. The equilibrium code provides the aircraft characteristics and a maximum take-off mass $M = 24$ kg.

14.8.3.1 Airplane Aerodynamic Center and Static Margin

The equilibrium code calculates the lift and moment coefficients for the complete configuration at low incidences to be:

$$\begin{cases} C_L(\alpha, t_f) = 4.479\alpha + 0.808t_f + 0.9314 \\ C_{M,o}(\alpha, t_f) = -1.469\alpha - 0.7479t_f - 0.1565 \end{cases}$$

where α is the geometric incidence (in radians, measured from the fuselage axis) and t_f is the tail flap setting angle (in radians). $C_{M,o}$ is the aerodynamic moment about the origin of the coordinate system (located at the nose O). We will use this linear model for simplicity.

The center of gravity is located at $x_{cg}/l_{ref} = 0.268$.

Find the *aerodynamic center* and the *static margin* SM in % of l_{ref}.

14.8.3.2 Equilibrium Condition and Static Stability

Derive the moment coefficient at the center of gravity, $C_{M,c.g.}(\alpha, t_f)$ and write the condition for equilibrium. Verify your result as the rest depends on it. Is the equilibrium stable?

Solve for $\alpha(t_f)$.

14.8.3.3 Take-Off Conditions

The take-off speed of $U = 11.49$ m/s is obtained for $t_f = 7.13°$. Find α_{eq} at take-off.

Find the lift coefficient of the tail at take-off : the tail aerodynamic lift curve is given by

$$C_{Lt} = 3.032\alpha + 2.886t_f - 0.3259$$

Calculate the force on the tail in (N), given that the tail reference area is $A_t = 0.49\,\text{m}^2$, $\rho = 1.2\,\text{kg/m}^3$.

Is the force up or down?

14.9 Problem 9

14.9.1 2-D Inviscid, Linearized, Thin Airfoil Theories

14.9.1.1 Incompressible Flow ($M_0 = 0$)

Aerodynamic Center

Give the definition of the aerodynamic center.

Where is the aerodynamic center located for thin airfoils at low speeds?

Second Mode only Airfoil

Consider a thin cambered plate such that the vorticity distribution is given by the second mode with $A_2 > 0$ as

$$\begin{cases} \Gamma'[x(t)] = 2U\left\{A_0\frac{1+\cos t}{\sin t} + A_2\cos 2t\right\} \\ x(t) = \frac{c}{2}(1-\cos t),\ \ 0 \le t \le \pi \end{cases}$$

Use the formula for $d'[x(t)]$ to find A_0 and find the *incidence of adaptation* for this airfoil (Hint: use the identity $\cos 2t = 2\cos^2 t - 1$, and integrate in x: $d'(x)dx$ from zero to c; or in t: $d'[x(t)]\frac{dx}{dt}dt$ from zero to π.)

Eliminate A_0 and sketch the slope of the cambered plate along with the profile itself.

Aerodynamic Coefficients

Give the expression of $C_l(\alpha)$, $C_{m,o}(\alpha)$, $C_{m,a.c.}$ and C_d for this airfoil.

Moment About an Axis

Calculate the aerodynamic moment about the mid-chord, $C_{m,c/2}$ (Hint: use the change of moment formula.)

If the profile is allowed to rotate without friction about an axis located at mid-chord, find the equilibrium incidence, α_{eq}, if only aerodynamic forces and moment are present.

Is the equilibrium stable (Answer by Yes or No)?

14.9.1.2 Supersonic Flow ($M_0 > 1$, $\beta = \sqrt{M_0^2 - 1}$)

Consider the cubic plate of equation

$$d(x) = \frac{4}{3}Ac\frac{x}{c}\left(1 - 2\frac{x}{c}\right)\left(1 - \frac{x}{c}\right), \quad A > 0$$

The slope is given by

$$d'(x) = \frac{4}{3}A\left(1 - 6\frac{x}{c} + 6\frac{x^2}{c^2}\right)$$

This plate equips the fins of a supersonic rocket.

Pressure Distribution and Flow Features

Calculate and plot $-C_p^+$ and $-C_p^-$ versus x for this airfoil at $\alpha = 0$. Sketch the flow at $\alpha = 0$ (shocks, characteristic lines, expansion shocks).

Moment Coefficient

At $\alpha = 0$, calculate the moment coefficient $\left(C_{m,o}\right)_{\alpha=0}$ and give the expression of $C_{m,o}(\alpha)$.

Use the change of moment formula to evaluate $C_{m,c/2}$ and discuss whether or not there is an equilibrium about a mid-chord axis and why. How would you qualify this situation: stable, unstable, neutral?

14.9.2 Prandtl Lifting Line Theory

The WWII Spitfire was designed with an elliptic wing of span $b = 11.2\,\text{m}$ and wing area $S = 22.5\,\text{m}^2$. It is equipped with a NACA2209.4 profile of 2 % relative camber $(d/c = const = 0.02)$. The top speed in cruise is $V = 170\,\text{m/s}$ (378 mph) with a take-off mass of $M = 3{,}000\,\text{kg}$.

14.9.2.1 Vortex Sheet

Explain succinctly the key features of the vortex sheet (physically and mathematically) and the effect it has on the flow past a large aspect ratio wing (lifting line).

14.9.2.2 Lift Coefficient

Given the above data, calculate the lift coefficient $C_L$at top speed in cruise (take $\rho = 1.2\,\text{kg/m}^3$ and $g = 9.81\,\text{m/s}^2$).

14.9.2.3 Elliptic Loading

Assuming an elliptic planform, a constant relative camber and zero twist, calculate the induced drag C_{Di} and the first mode amplitude A_1 in the Fourier Series expansion of the circulation

$$\begin{cases} \Gamma[y(\theta)] = 2Ub\sum_{n=1}^{\infty} A_n \sin n\theta \\ y(\theta) = -\frac{b}{2}\cos\theta,\ 0 \le \theta \le \pi \end{cases}$$

Write the equation for the lift coefficient C_L in terms of aspect ratio AR, geometric incidence α and relative camber d/c.

Find α, the geometric incidence in cruise.

14.9.2.4 Added Twist

The designer of the Spitfire added "washout" ($-2.5°$ of twist between root and tip) so that the local lift coefficient is larger at the root than at the tip, to avoid tip stall. This corresponds to adding *only* the third mode with an amplitude $A_3 = -0.002$. $(A_2 = A_4 = A_5 = ...A_n = 0,\ n \ge 4)$.

Calculate the induced drag $(C_{Di})_{washout}$ of the wing with twist.

What is the percent change in induced drag?

What is the corresponding Efficiency factor e?

14.9.3 Equilibrium of the AMAT11

The AMAT11 has a rectangular main wing with span $b_m = 2.1$ m and constant chord $c_{xm} = 0.3$ m. The tail is also rectangular with span $b_t = 1.0$ m and chord $c_{xt} = 0.3$ m. The equilibrium code provides the aircraft aerodynamic characteristics and a maximum take-off mass $M = 19$ kg. The reference area is $A_{ref} = A_m + A_t$.

14.9.3.1 Aspect Ratios of Lifting Elements—Global Lift Slope

Calculate the aspect ratio of the wing, AR_m and of the tail, AR_t.

Given that

$$\frac{dC_{Lm}}{d\alpha} = 4.675, \quad \frac{dC_{Lt}}{d\alpha} = 2.696$$

give the global lift slope $dC_L/d\alpha$ for the wing+tail configuration.

14.9.3.2 Airplane Center of Gravity

The aerodynamic center is located at $x_{a.c.}/l_{ref} = 0.3$.

Find the center of gravity $x_{c.g.}/l_{ref}$, given a 6 % *static margin* (SM).

14.9.3.3 Equilibrium Condition and Static Stability

The equilibrium code calculates the linear model for lift and moment coefficients for the complete configuration, at low incidences, to be:

$$\begin{cases} C_L(\alpha, t_f) = 3.955\alpha + 0.984t_f + 0.712 \\ C_{M,o}(\alpha, t_f) = -1.188\alpha - 0.907t_f - 0.008 \end{cases}$$

Derive the moment coefficient at the center of gravity, $C_{M,c.g.}(\alpha, t_f)$ and write the condition for equilibrium. Verify your result as the rest depends on it. Is the equilibrium stable?

Solve for $\alpha_{eq}(t_f)$.

14.9.3.4 Take-Off Conditions

The take-off speed of $U = 13.11$ m/s is obtained for $t_f = 9.2°$. Find α_{eq} at take-off.

Find the lift coefficient of the tail at take-off : the tail aerodynamic lift curve is given by

$$C_{Lt} = 2.696\alpha + 2.705t_f - 0.347$$

Calculate the force on the tail in (N), given that $\rho = 1.2\,\text{kg/m}^3$.
Is the force up or down?

14.10 Problem 10

14.10.1 2-D Inviscid, Linearized, Thin Airfoil Theories

14.10.1.1 Incompressible Flow ($M_0 = 0$)

Profile Camber Estimation

A wing profile lift curve, calculated with a numerical method, gives a value of the lift coefficient to be $(C_l)_{profile} = 2.1878$ at $\alpha = 4°$. Using the result of thin airfoil theory, find all the Fourier coefficients of a thin parabolic plate equivalent to this airfoil. Estimate the relative camber d_m/c of the thin parabolic plate. Check your result as the rest depends on it.

Take-Off Incidence

For this equivalent thin parabolic plate, find the incidence in $deg.$ for which the lift coefficient is $C_l = 2.5$.

Nose Pitching Moment Coefficient

Give the pitching moment coefficient $C_{m,o}(\alpha)$ for the parabolic plate. Predict the nose pitching moment coefficient of the wing profile $\left(C_{m,o}\right)_{profile}$ at $\alpha = 4°$. Compare with the calculated result of -0.935.

Aerodynamic Center Pitching Moment Coefficient

Give the definition of the aerodynamic center. Give the pitching moment coefficient about the aerodynamic center $C_{m,a.c.}$.

14.10.1.2 Supersonic Flow ($M_0 > 1$, $\beta = \sqrt{M_0^2 - 1}$)

Consider the thin parabolic plate of equation

$$d(x) = 4d_m \frac{x}{c}\left(1 - \frac{x}{c}\right)$$

The slope is given by

$$d'(x) = 4\frac{d_m}{c}\left(1 - 2\frac{x}{c}\right)$$

where the relative camber is given to be $d_m/c = 0.14$.

Lift Coefficient

Give the lift coefficient $C_l(\alpha)$ for this airfoil.

Drag Coefficient

Calculate the drag coefficient $(C_d)_{\alpha=0}$ and give the expression of $C_d(\alpha)$ for this airfoil.

Pitching Moment Coefficient

If the profile is allowed to rotate freely about an axis placed at the quarter-chord, find the equilibrium incidence α_{eq} in deg. (Hint: Use the change of moment formula to evaluate $C_{m,c/4}$). How would you qualify the equilibrium situation: stable, unstable, neutral?

14.10.2 Prandtl Lifting Line Theory

14.10.2.1 Lifting Line Theory

In Lifting Line theory, explain the origin of downwash.

In terms of drag, what is the difference between downwash and upwash?

Is there a net drag benefit to design a wing with upwash?

14.10.2.2 Design of a Wing

The circulation distribution is represented by the Fourier series

$$\begin{cases} \Gamma[y(t)] = 2Ub\sum_{n=1}^{\infty} A_n \sin nt \\ y(t) = -\frac{b}{2}\cos t,\ 0 \le t \le \pi \end{cases}$$

Design the simplest wing (meaning the wing with the least number of non-zero Fourier coefficients) with zero strength tip vortices, in order to minimize risks for airplanes caught in the wake of the wing. To achieve this goal, combine modes 1 and 3 in such a way that

$$\frac{d\Gamma(y)}{dy} = 0, \quad at\ y = \pm\frac{b}{2}$$

Note: you can impose

$$\frac{d\Gamma[y(t)]}{dt} = 0, \quad at\ t = 0\ and\ \pi$$

to find a relation between A_1 and A_3(verify your result as the rest depends on it), then verify that

$$\frac{d\Gamma(y)}{dy} = \frac{d\Gamma[y(t)]}{dt} / \frac{dy(t)}{dt} = 0, \quad at\ t = 0\ and\ \pi$$

Use the following identity: $\cos 3t = \cos t\left(4\cos^2 t - 3\right)$.

Sketch the circulation distribution $\Gamma[y(t)]$ in terms of y or t. Use the following identity: $\sin 3t = \sin t\left(4\cos^2 t - 1\right)$.

14.10.2.3 Induced Velocity

From the result derived in class

$$w_w[y(t)] = -U\sum_{n=1}^{\infty} nA_n \frac{\sin nt}{\sin t}$$

find the distribution of induced velocity.

Sketch the distribution $w_w[y(t)]$ in terms of y or t.

Show that the wing experiences upwash.

14.10.2.4 Induced Drag

Calculate the induced drag of the wing without tip vortices and compare with the elliptic loading.

What is the percent change in induced drag?

What is the corresponding efficiency factor e?

14.10.3 Equilibrium of the Glider

The AMAT12 has a rectangular main wing with span $b_m = 2.1$ m and constant chord $c_{xm} = 0.3$ m. The tail is also rectangular with span $b_t = 1.0$ m and chord $c_{xt} = 0.338$ m. The equilibrium code provides the aircraft aerodynamic characteristics and a maximum take-off mass $M = 18$ kg. The reference area is $A_{ref} = A_m + A_t$. The reference length is $l_{ref} = 2.12$ m. The engine is turned off. The global aerodynamic lift and nose pitching moment coefficients are found to be

$$\begin{cases} C_L(\alpha, t_t) = 3.853\alpha + 0.902t_t + 0.716 \\ C_{M,o}(\alpha, t_t) = -1.336\alpha - 0.894t_t - 0.06 \end{cases}$$

where α represents the geometric incidence and t_t the tail setting angle in rd.

14.10.3.1 Equilibrium About the Center of Gravity

Calculate the pitching moment about the center of gravity $C_{M,c.g.}(\alpha)$, given that the center of gravity is located at $x_{c.g.}/l_{ref} = 0.2868$. (Hint: use the change of moment formula). Verify your result as the rest depends on it.

Is the airplane statically stable?

If the aerodynamic center of the glider is at $x_{a.c.} = 0.735$ m, give the static margin SM in %.

14.10.3.2 Equilibrium Incidence

Write the equilibrium condition and solve the equation for $\alpha_{eq}(t_t)$. Check your algebra by substituting back into the equation.

14.10.3.3 Trimming for Maximum Distance

The equilibrium code calculates the tail setting angle for maximum $C_L/C_D = 9.3$ to be $t_t = 12°$.

Find the corresponding values of C_L and C_D, as well as the descent angle β.

Calculate the lift forces (in N) on the main wing and the tail, given that the velocity is $U = 17.3$ m/s and the air density $\rho = 1.1214$ kg/m^3. The wing and tail lift curves are given respectively by

$$\begin{cases} C_{Lm}(\alpha) = 4.664\alpha + 1.312 \\ C_{Lt}(\alpha, t_t) = 2.342\alpha + 2.583t_t - 0.396 \end{cases}$$

Chapter 15
Solutions to Problems

15.1 Solution to Problem 1

15.1.1 Thin Airfoil Theory (2-D Inviscid Flow)

15.1.1.1 Quiz

In subsonic flow, the aerodynamic coefficient dependency upon the small parameters is given in Table 15.1. In supersonic flow, the aerodynamic coefficient dependency upon the small parameters is given in Table 15.2.

15.1.1.2 Lift Coefficient

The lift coefficient C_l in incompressible flow (low speed) is given by

$$(C_l(\alpha))_{M_0=0} = 2\pi\left(\alpha + 2\frac{d}{c}\right) = 6.28(0.0873 + .04) = 0.8$$

At $M_0 = 0.7$, using Prandtl-Glauert correction one finds

$$(C_l)_{M_0} = \frac{(C_l)_{M_0=0}}{\sqrt{1 - M_0^2}} = \frac{0.8}{0.714} = 1.12$$

15.1.1.3 Drag Coefficient

If the drag coefficient of the same profile at $\alpha = 0$ and $M_0 = 1.4$ is $(C_d)_{M_0=1.4} = 0.02$, the drag at $M_0 = 2.$ is given by

J.J. Chattot and M.M. Hafez, *Theoretical and Applied Aerodynamics*,
DOI 10.1007/978-94-017-9825-9_15

Table 15.1 Subsonic flow

Aerodynamic coefficient	e/c	d/c	α
C_l	n	y	y
$C_{m,o}$	n	y	y
$C_{m,ac}$	n	y	n
C_d	n	n	n

Table 15.2 Supersonic flow

Aerodynamic coefficient	e/c	d/c	α
C_l	n	n	y
$C_{m,o}$	n	y	y
$C_{m,ac}$	n	y	n
C_d	y	y	y

$$(C_d)_{M_0=2} = \frac{(\beta)_{M_0=1.4}}{(\beta)_{M_0=2}}(C_d)_{M_0=1.4} = \frac{0.9798}{1.732}0.02 = 0.0113$$

At $M_0 = 2$. if $\alpha = 5°$ then

$$C_d(\alpha) = (C_d)_{\alpha=0} + 4\frac{\alpha^2}{\beta} = 0.0113 + 4\frac{0.0873^2}{1.732} = 0.289$$

The drag coefficient at $M_0 = 0.7$ is

$$(C_d)_{M_0=0.7} = 0$$

since this is a 2-D, inviscid, subsonic flow.

15.1.1.4 Equilibrium About an Axis

The aerodynamic moment coefficient $C_{m,o}$ is defined as

$$C_{m,o} = \frac{M,o}{\frac{1}{2}\rho U^2 c^2}$$

If the profile is attached to a vertical axis located at $x = 0$ (nose of profile), the equation of equilibrium that will give the equilibrium angle of attack α_{eq} is given by the equation of the moment coefficient

$$C_{m,o}(\alpha_{eq}) = 0 = -\frac{\pi}{2}\left(\alpha_{eq} + 4\frac{d}{c}\right)$$

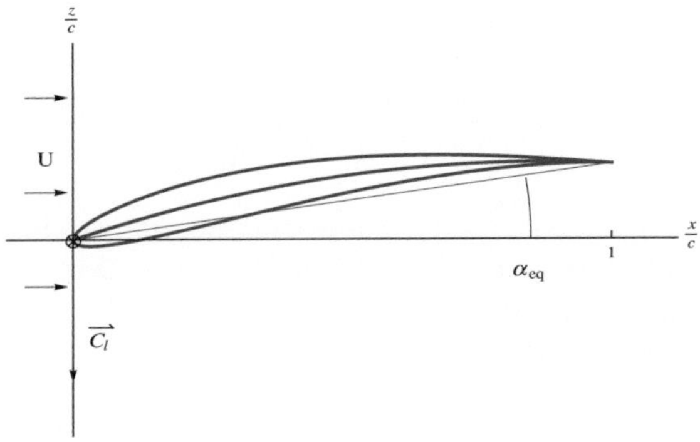

Fig. 15.1 Equilibrium position for the profile

α_{eq} is found to be

$$\alpha_{eq} = -4\frac{d}{c} = -0.08$$

See Fig. 15.1.

The equilibrium is stable because $\frac{dC_{m,o}}{d\alpha} < 0$.

15.1.2 Lifting Line Theory

15.1.2.1 Flow Model

The vortex sheet that is shed by the finite wing is composed of vortex filaments parallel to the x-axis, that induce two velocity components, v and w in its proximity, that vanish at infinity upstream, but not at infinity downstream (Trefftz plane). The kinetic energy of the flow increases along a stream tube. An energy balance indicates that work must be done, for the kinetic energy to increase. This work corresponds to the work of the thrust that is needed to balance the induced drag, in order to maintain the motion. The vortex sheets is a trace of the passage of the wing, that does not disappear in inviscid flow.

The lift and induced drag coefficients are given by

$$\begin{cases} C_L = \pi A R A_1 \\ C_{Di} = \pi A R \left(A_1^2 + 2A_2^2 + \cdots + nA_n^2 + \cdots\right) \end{cases}$$

15.1.2.2 Non Singular Lift Distribution

The combination of modes 1 and 3 reads

$$\begin{cases} \Gamma\left[y(t)\right] = 2Ub\left(A_1 \sin t + A_3 \sin 3t\right) \\ y(t) = -\frac{b}{2}\cos t \end{cases}, \quad 0 \le t \le \pi$$

Taking the derivative with respect to y gives

$$\frac{d\Gamma}{dy} = \frac{d\Gamma}{dt}\frac{dt}{dy} = \frac{\frac{d\Gamma}{dt}}{\frac{dy}{dt}} = 4U\frac{A_1 \cos t + 3A_3 \cos 3t}{\sin t}$$

A necessary condition is that the numerator go to zero when $t \to 0$. This will be satisfied if $A_3 = -\frac{A_1}{3}$. It is possible to verify that, with this condition, the derivative will go to zero, $\frac{d\Gamma}{dy} \approx 16UA_1 t,\ as\ t \to 0$.

See Fig. 15.2.

15.1.2.3 Drag Penalty

For a given C_L (at take-off), the coefficient A_1 is fixed since $A_1 = \frac{C_L}{\pi AR}$.

The induced drag of the wing is given by

$$C_{Di} = \pi ARA_1^2\left(1 + \cdots + n\frac{A_n^2}{A_1^2} + \cdots\right) = \frac{C_L^2}{\pi AR}\left(1 + 3\frac{A_3^2}{A_1^2}\right) = \frac{4}{3}\frac{C_L^2}{\pi AR}$$

The induced drag has been increased by 33 %, which is not negligible.

The efficiency factor $e = (1+\delta)^{-1} = \frac{3}{4} = 0.75$.

15.1.3 Equilibrium of the Glider (3-D Incompressible Flow)

15.1.3.1 Glider Effective Aspect Ratio

As a glider (engine off), the glider has the following lift and moment coefficients in terms of the geometric angle of attack α (rd) and tail setting angle t_t (rd):

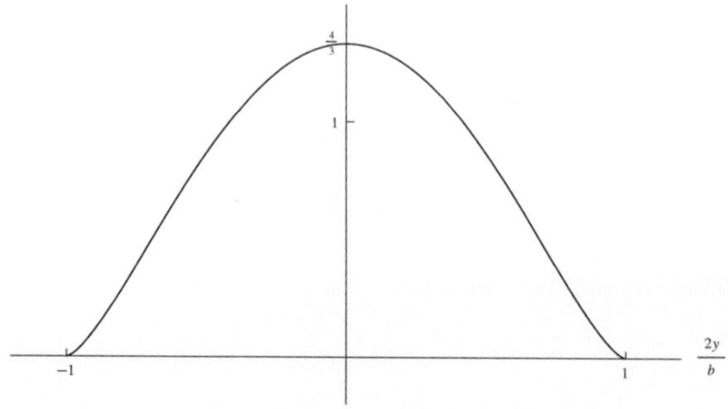

Fig. 15.2 Distribution of circulation

$$\begin{cases} C_L(\alpha, t_t) = 5.3\alpha + 1.56 + 0.5t_t \\ C_{M,o}(\alpha, t_t) = -1.7\alpha - 0.43 - 0.45t_t \end{cases}$$

The effective aspect ratio AR of the glider (wing+tail), assuming ideal loading is given by the lift slope

$$\frac{dC_L}{d\alpha} = \frac{2\pi}{1 + \frac{2}{AR}} = 5.3$$

Solving for AR gives

$$AR = \frac{2}{\frac{2\pi}{5.3} - 1} = 10.8.$$

15.1.3.2 Equilibrium Angle of Attack

The equilibrium of the moment corresponds to

$$C_{M,o}(\alpha_{eq}, t_t) + \frac{x_{cg}}{l_{ref}} C_L(\alpha_{eq}, t_t) = 0$$

When the lift and moment coefficients functions are replaced in the above equation, taking into account $\frac{x_{cg}}{l_{ref}} = 0.29$, the solution for $\alpha_{eq}(t_t)$ reads

$$\alpha_{eq} = -\frac{C_{M,o0} + \frac{x_{cg}}{l_{ref}} C_{L0}}{\frac{dC_{M,o}}{d\alpha} + \frac{x_{cg}}{l_{ref}} \frac{dC_L}{d\alpha}} = -\frac{-0.43 - 0.45t_t + 0.29(1.56 + 0.5t_t)}{-1.7 + 0.29\, 5.3} = 0.137 - 1.87t_t$$

Plugging this result in the lift coefficient gives

$$C_L(t_t) = 5.3(0.137 - 1.87t_t) + 1.56 + 0.5t_t = 2.286 - 9.411t_t$$

which for $C_L = 1.9$ results in $t_t = 0.041\,\text{rd} = 2.35°$.

The corresponding angle of attack is $\alpha_{eq} = 0.06\,\text{rd} = 3.46°$.

15.1.3.3 Landing Speed

At sea level, $\rho = 1.2\,\text{kg/m}^3$, $g = 9.81\,\text{ms}^{-2}$, given that the total mass $m = 15\,\text{kg}$, the maximum lift coefficient $(C_L)_{max} = 1.9$ corresponding to the reference area $A_{ref} = 0.68\,\text{m}^2$, the landing speed corresponds to the equilibrium equation for the force in the direction normal to the trajectory. i.e.

$$u_{eq} = \sqrt{\frac{2mg}{\rho A_{ref}\,(C_L)_{max}}} = 13.8\,\text{m/s}$$

15.2 Solution to Problem 2

15.2.1 Thin Airfoil Theory (2-D Inviscid Flow)

15.2.1.1 Incompressible Flow ($M_0 = 0$)

By definition:

$$C_l = \frac{L}{\frac{1}{2}\rho U^2 c}, \; C_{m,o} = \frac{M_{,o}}{\frac{1}{2}\rho U^2 c^2}$$

As seen in class

$$C_l = 2\pi\left(\alpha + 2\frac{d}{c}\right), \; C_{m,o} = -\frac{\pi}{2}\left(\alpha + 4\frac{d}{c}\right)$$

For a given value of camber and C_l, the lift coefficient relation can be solved for the angle of incidence as

$$(\alpha)_{C_l=0.5} = \frac{C_l}{2\pi} - 2\frac{d}{c} = -0.0924\,\text{rd} = -5.3°$$

$$(\alpha)_{C_l=2.0} = \frac{C_l}{2\pi} - 2\frac{d}{c} = 0.1463\,\text{rd} = 8.4°$$

The corresponding moments are

$$(C_{m,o})_{C_l=0.5} = -0.395$$

$$(C_{m,o})_{C_l=2.0} = -0.77$$

The center of pressure is found from $\frac{x_{cp}}{c} = -\frac{C_{m,o}}{C_l}$, that is

$$(\frac{x_{cp}}{c})_{C_l=0.5} = 0.79$$

$$(\frac{x_{cp}}{c})_{C_l=2.0} = 0.385$$

See Fig. 15.3.

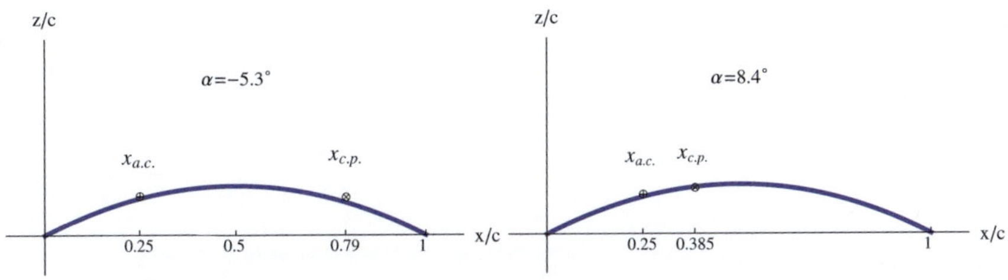

Fig. 15.3 Relative position of aerodynamic center and center of pressure

15.2.1.2 Supersonic Linearized Flow ($M_0 > 1$)

The lift coefficient C_l in supersonic flow is given by

$C_l(\alpha) = 4\frac{\alpha}{\beta}$, where $\beta = \sqrt{M_0^2 - 1}$.

The moment coefficient is

$$C_{m,o}(\alpha) = (C_{m,o})_{\alpha=0} - 2\frac{\alpha}{\beta}$$

where $(C_{m,o})_{\alpha=0} = \frac{2}{\beta}\int_0^c (f'^+(x) + f'^-(x))\frac{x}{c}\frac{dx}{c}$.

The thickness distribution has no effect on the moment which reduces to the camber contribution

$$(C_{m,o})_{\alpha=0} = \frac{4}{\beta}\int_0^c d'(x)\frac{x}{c}\frac{dx}{c} = \frac{16}{\beta}\frac{d}{c}\int_0^1 (\xi - 2\xi^2)d\xi = -\frac{8}{3\beta}\frac{d}{c}$$

The moment reads $C_{m,o}(\alpha) = -\frac{8}{3\beta}\frac{d}{c} - 2\frac{\alpha}{\beta}$.

The center of pressure is given by

$$\frac{x_{cp}}{c} = \frac{1}{2} - \frac{\beta(C_{m,o})_{\alpha=0}}{4\alpha} = \frac{1}{2} + \frac{2}{3}\frac{d}{c}\frac{1}{\alpha}$$

15.2.2 *Lifting Line Theory (3-D Inviscid Flow)*

15.2.2.1 Flow Model

The vortex sheet is a stream surface, wetted on both sides by the fluid, hence the tangency condition imposes that the w-component of the perturbation be continuous. Across the vortex sheet the pressure is continuous, hence the u-component of the perturbation is continuous ($C_p = -2u/U$).

Velocity vectors near the vortex sheet: see Fig. 15.4.

Fig. 15.4 Induced velocities near the vortex sheet

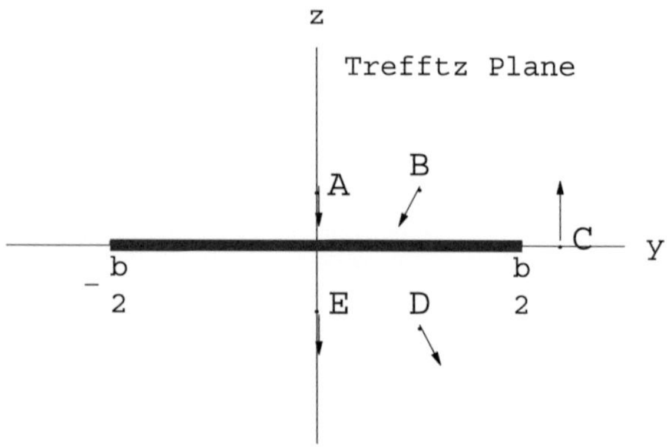

15.2.2.2 Improved Lift of a Rectangular Wing

The combination of modes 1 and 3 is represented in Fig. 15.5.

The total lift of a wing is given solely in terms of A_1 and can be written as $C_L = \pi AR\, A_1$. The maximum circulation is obtained for $t = \frac{\pi}{2}$ and is $\Gamma_{max} = \frac{16}{9} U b A_1$. Hence $A_1 = \frac{9\Gamma_{max}}{16Ub}$ and the corresponding lift is $C_L = \pi AR \frac{9\Gamma_{max}}{16Ub} = \frac{9}{8}(C_L)_{elliptic}$, a 12.5 % increase for the high lift rectangular wing.

As seen in class, the downwash for the high lift wing is given by

$$w_T[y(t)] = -2U\left(A_1 + 3A_3 \frac{\sin 3t}{\sin t}\right) = -\frac{4}{3} U A_1 (1 + 2\cos^2 t)$$

$$= -\frac{3}{4}\frac{\Gamma_{max}}{b}(1 + 2\cos^2 t) = -\frac{3}{4}\frac{\Gamma_{max}}{b}\left(1 + 2\left(\frac{2y}{b}\right)^2\right)$$

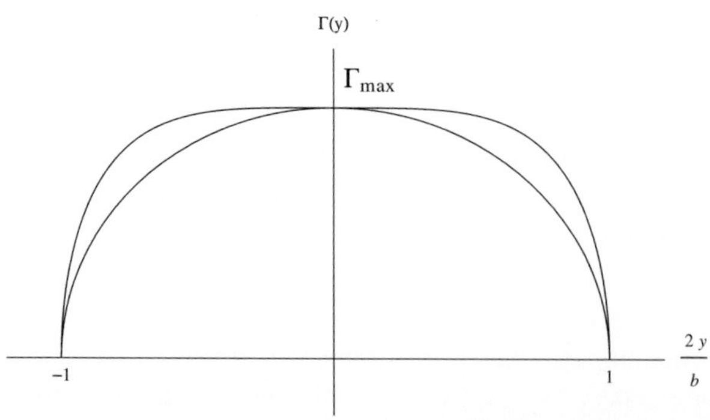

Fig. 15.5 Combined 1 and 3 modes

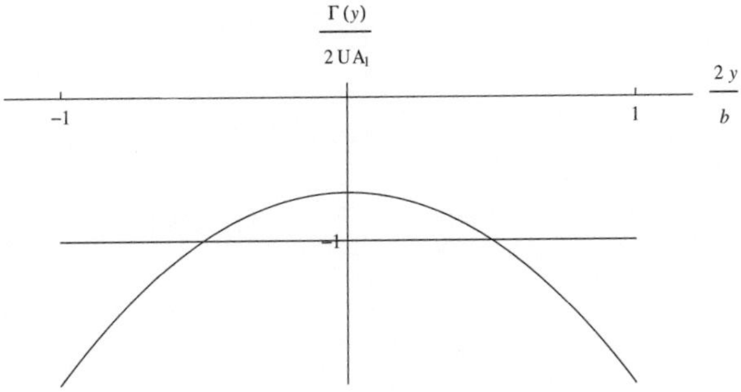

Fig. 15.6 Downwash for the high lift and elliptic-loaded wing

The distribution of downwash is parabolic and is compared with that of the elliptic wing $w_T[y(t)]_{elliptic} = -\frac{\Gamma_{max}}{b}$. See Fig. 15.6.

15.2.2.3 Drag Penalty

The induced drag of the wing is given by

$$C_{Di} = \pi AR\, A_1^2\left(1 + \cdots + n\frac{A_n^2}{A_1^2} + \cdots\right) = \pi AR\, A_1^2\left(1 + 3\left(\frac{1}{9}\right)^2\right) = \frac{28}{27}(C_{Di})_{elliptic}$$

The drag penalty is a 3.7 % increase in induced drag.

15.2.3 Airplane Longitudinal Equilibrium

15.2.3.1 Global Coefficients

Aerobrick 2003 has the following lift and moment coefficients in terms of the geometric angle of attack α (rd) and tail setting angle t_t (rd):

$$\begin{cases} C_L(\alpha, t_t) = 3.88\alpha + 0.5 + 0.481t_t \\ C_{M,o}(\alpha, t_t) = -1.31\alpha - 0.124 - 0.452t_t \end{cases}$$

The aerodynamic center is given by $\frac{x_{ac}}{l_{ref}} = -\frac{\frac{dC_{M,o}}{d\alpha}}{\frac{dC_L}{d\alpha}} = \frac{1.31}{3.88} = 0.338$.

To satisfy the 4 % static margin, the center of gravity must be located in front of the aerodynamic center with

$$\frac{x_{cg}}{l_{ref}} = \frac{x_{ac}}{l_{ref}} - 0.04 = 0.298$$

Definition: the aerodynamic center is the point about which the moment is independent of the incidence angle.

It is clear from the above calculation of $\frac{x_{ac}}{l_{ref}}$ that $C_{M,o}(\alpha) + \frac{x_{ac}}{l_{ref}} C_L(\alpha)$ is independent of α and reads

$$C_{M,ac}(t_t) = 0.045 - 0.289 t_t$$

15.2.3.2 Take-Off Conditions

The moment at the center of gravity is given by

$$C_{M,cg}(\alpha, t_t) = C_{M,o}(\alpha, t_t) + \frac{x_{cg}}{l_{ref}} C_L(\alpha, t_t) = -0.154\alpha + 0.025 - 0.309 t_t$$

At equilibrium the moment is zero, hence, solving for the equilibrium incidence:

$$\alpha_{eq}(t_t) = 0.162 - 2.0 t_t$$

The equilibrium lift is then

$$C_{Leq}(t_t) = 1.129 - 7.28 t_t$$

At take-off, the airplane lift coefficient is $C_L = 1.44 = 1.129 - 7.28 (t_t)_{t-o}$, which can be solved to give

$$(t_t)_{t-o} = -0.0427 = -2.45^\circ$$

At take-off the angle of attack will be

$$(\alpha_{eq})_{t-o} = 0.2474 = 14.17^\circ$$

15.3 Solution to Problem 3

15.3.1 Thin Airfoil Theory (2-D Inviscid Flow)

15.3.1.1 Incompressible Flow ($M_0 = 0$)

As seen in class, the coefficients for a parabolic camberline are

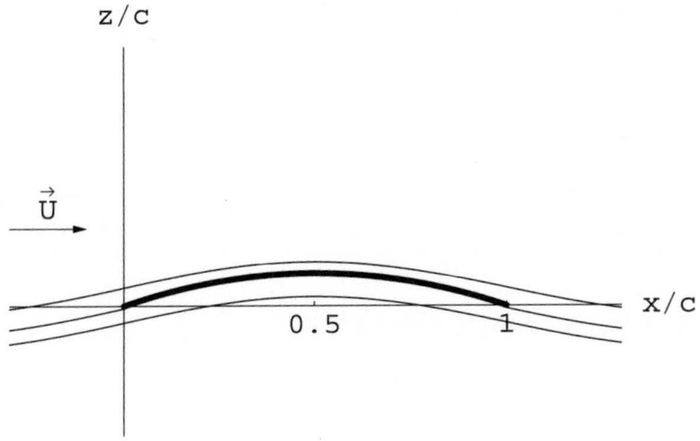

Fig. 15.7 Thin parabolic plate at incidence of adaptation

$$A_0 = \alpha$$

$$A_1 = \frac{4d}{c}$$

$$A_2 = \cdots = A_n = 0, \; n \geq 2$$

The angle of adaptation or ideal angle of attack is $\alpha = 0$ since $A_0 = \alpha$. The flow satisfies two Kutta-Joukowski conditions. The leading edge flow has finite velocities. See Fig. 15.7.

The lift coefficient is given by $C_l = 2\pi(A_0 + A_1/2) = 2\pi(\alpha + \frac{2d}{c})$. The incidence of zero lift is thus $\alpha_0 = -\frac{2d}{c} = -0.172\,\text{rd} = -9.85°$.

The moment $C_{m,ac}$ is given by $C_{m,ac} = C_{m,o} + \frac{x_{ac}}{c} C_l = C_{m,o}$ in this case. Indeed: $C_{m,o} = -\frac{\pi}{2}(A_0 + A_1 - A_2/2) = -\pi \frac{d}{c}$ which is the value of $C_{m,ac}$ for this profile.

The upper limit of the weight (in N) that a wing of chord $c = 0.368\,\text{m}$ and span $b = 4.877\,\text{m}$ could lift at $C_l = 2$ corresponds to the 2-D solution. The lift per unit span is given by

$$L' = \frac{1}{2}\rho V^2 c\, C_l$$

hence the total lift would be

$$L = \frac{1}{2}\rho V^2 bc\, C_l = 242.4\; N$$

The corresponding mass at take-off would be $M = 24.7\,\text{kg}$.

15.3.1.2 Supersonic Linearized Theory ($M_0 > 1$)

The wave drag coefficient $(C_d)_{\alpha=0}$ is given by

$$(C_d)_{\alpha=0} = \frac{2}{\beta}\int_0^c \left[\left(f'^{+}(x)\right)^2 + \left(f'^{-}(x)\right)^2\right]\frac{dx}{c} = \frac{16}{3\beta}\left(\frac{e}{c}\right)^2$$

as seen in class.

Hence the lift and wave drag coefficients are:

$$C_l(\alpha) = \frac{4\alpha}{\beta} = 0.0403$$

$$C_d(\alpha) = \frac{16}{3\beta}\left(\frac{e}{c}\right)^2 + \frac{4\alpha^2}{\beta} = 0.0315$$

The lift and wave drag upper bounds for the fin are

$$L = \frac{1}{2}\rho V^2 bc\ C_l = 106\ N$$

$$D = \frac{1}{2}\rho V^2 bc\ C_d = 82.9\ N$$

15.3.2 *Lifting Line Theory (3-D Inviscid Flow)*

15.3.2.1 Induced Drag in Cruise

The induced drag D_i in (N) is given by

$$D_i = \frac{1}{2}\rho V^2 S \frac{C_L^2}{\pi e AR}$$

The equilibrium equation along the vertical axis during horizontal cruise in dimensional form (neglecting the lift of the tail) reads:

$$L = \frac{1}{2}\rho V^2 S\ C_L = W$$

Elimination of C_L yields:

$$D_i = \frac{W^2}{\frac{1}{2}\rho V^2 \pi e b^2}$$

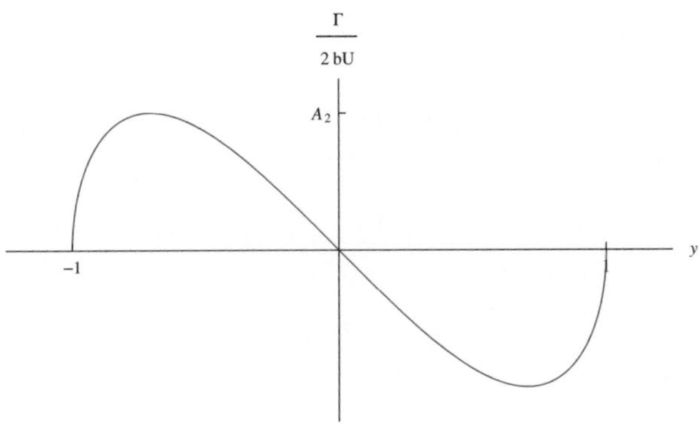

Fig. 15.8 Mode 2 circulation

Application:

$$D_i = 3.6\ N$$

15.3.2.2 Turn

According to the theory, the induced drag coefficient during cruise is given by:

$$C_{Di} = \pi AR\ A_1^2 \left[1 + 3 \left(\frac{1}{9} \right)^2 \right]$$

The sketch of mode 2 is shown in Fig. 15.8.

15.3.2.3 Drag Penalty

The drag during the turn is given by

$$\begin{aligned} C_{Di} &= \pi AR\ A_1^2 \left[1 + 3 \left(\frac{1}{9} \right)^2 + 2 \left(\frac{1}{3} \right)^2 \right] \\ &= \pi AR\ A_1^2 \left[1 + 3 \left(\frac{1}{9} \right)^2 \right] \left[1 + \frac{2(\frac{1}{3})^2}{1 + 3(\frac{1}{9})^2} \right] \end{aligned}$$

The drag penalty is $\frac{\Delta C_{Di}}{C_{Di}} = 0.21$, i.e. a 21 % increase in induced drag.

15.3.3 Glider Equilibrium

Away from high lift, a linear model of the glider in terms of the angle of incidence α (deg) and tail setting angle t_t (deg) is given by:

$$\begin{cases} C_L(\alpha, t_t) = 0.1\alpha + 0.01t_t + 0.7 \\ C_{M,0}(\alpha, t_t) = -0.02\alpha - 0.01t_t - 0.13 \end{cases}$$

15.3.3.1 Definition

The aerodynamic center is the point about which the moment is independent of incidence.

15.3.3.2 Aerodynamic Center

As seen in class, the location of the aerodynamic center satisfies

$$\frac{x_{ac}}{l_{ref}} = -\frac{dC_{M,o}}{d\alpha}/\frac{dC_L}{d\alpha} = 0.02/0.1 = 0.2$$

hence $x_{ac} = 0.3\,\text{m}$.

15.3.3.3 Moment at Aerodynamic Center

Using the transfer of moment formula one finds:

$$C_{M,ac} = C_{M,o} + \frac{x_{ac}}{l_{ref}}C_L = -0.02\alpha - 0.01t_t - 0.13 + \frac{x_{ac}}{l_{ref}}(0.1\alpha + 0.01t_t + 0.7)$$

Using the result for x_{ac} this simplifies to:

$$C_{M,ac} = -0.008t_t + 0.01$$

As expected, it is independent of α.

15.3.3.4 Moment at Center of Gravity

Using the transfer of moment formula one finds:

$$C_{M,cg} = C_{M,ac} - \frac{x_{ac} - x_{cg}}{l_{ref}} C_L = -0.004\alpha - 0.0084 t_t - 0.018$$

The airplane is statically stable because the slope $\frac{dC_{M,cg}}{d\alpha} < 0$.

15.3.3.5 Equilibrium

At equilibrium, $C_{M,cg} = 0$. This yields:

$$\alpha(t_t) = -2.1 t_t - 4.5$$

15.4 Solution to Problem 4

15.4.1 2-D Inviscid, Linearized, Thin Airfoil Theories

15.4.1.1 Incompressible Flow ($M_0 = 0$)

Thickness Effect

In Thin Airfoil theory, the singularity distribution used to represent thickness are sources and sinks.

All the forces and moment coefficients are zero in the symmetric problem associated with thickness, C_l and $C_{m,o}$ are zero by symmetry, and C_d is zero because it is an inviscid flow theory.

The pressure coefficient C_p is affected by thickness, but the lift, drag and moment integrals are not.

Camber Effect

$C_l = 2\pi(\alpha + 2d_m)$ and $C_{m,o} = -\pi(\alpha + 4d_m)/2$.

The angle of zero lift is $(\alpha)_{Cl=0} = -2d_m$.

$$\frac{dC_l}{d\alpha} = 2\pi, \; C_{l0} = 4\pi d_m, \; \frac{dC_{m,o}}{d\alpha} = -\pi/2, \; C_{m,o0} = -2\pi d_m$$

The flow past the mean camber line at $\alpha = 0$ corresponds to the ideal angle of attack or angle of adaptation. The leading edge satisfies a Kutta-Joukowski condition. The flow is symmetric w.r.t $x = c/2$. See Fig. 15.9.

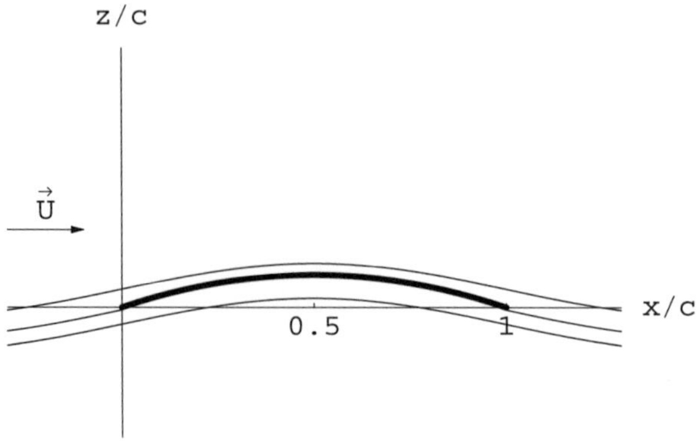

Fig. 15.9 Flow past thin parabolic plate at $\alpha = 0$

Equilibrium About an Axis

At equilibrium the moment about the axis is zero, i.e. $C_{m,o} = 0 \Rightarrow \alpha = -4d_m = -0.344\,\text{rd} = -19.7°$.

The equilibrium is stable because the slope of the moment about the axis, $dC_{m,o}/d\alpha$, is negative.

At equilibrium, the center of pressure is at the nose, since $x_{cp}/c = -C_{m,o}/C_l = 0$. See Fig. 15.10.

15.4.1.2 Supersonic Flow ($M_0 > 1,\ \beta = \sqrt{M_0^2 - 1}$)

Pressure Distributions

$C_p^+ = \frac{2}{\beta}\frac{4d_m}{c}(1 - \frac{2x}{c})$ and $C_p^- = -\frac{2}{\beta}\frac{4d_m}{c}(1 - \frac{2x}{c})$ at $\alpha = 0$. See Fig. 15.11.

The corresponding lift coefficient is $C_l = 0$.

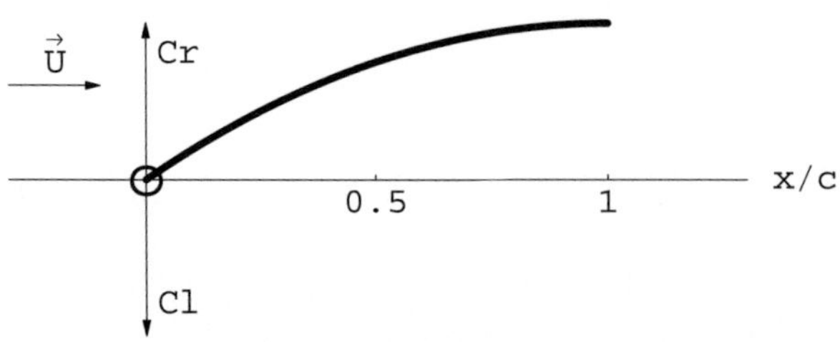

Fig. 15.10 Free body diagram of profile and hinge

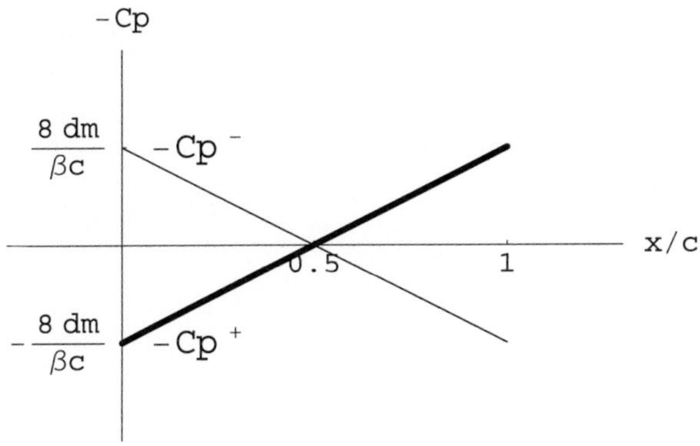

Fig. 15.11 Pressure coefficients

Lift and Moment Coefficients

$$\left(C_{m,o}\right)_{\alpha=0} = \frac{4}{\beta}\int_0^c d'(x)x dx/c^2 = \frac{16}{\beta}d_m \int_0^1 (\xi - 2\xi^2)d\xi = -\frac{8}{3\beta}d_m$$

In the general case, $C_l = \frac{4}{\beta}\alpha$ and $C_{m,o} = -\frac{8}{3\beta}d_m - \frac{2}{\beta}\alpha$ for $\alpha \neq 0$.

Equilibrium About an Axis

The moment about an arbitrary point D along the chord, $C_{m,D} = C_{m,o} + \frac{x_D}{c}C_l = \left(C_{m,o}\right)_{\alpha=0} - \left(1 - \frac{2x_D}{c}\right)\frac{2}{\beta}\alpha$. The condition for the equilibrium to be stable is that $dC_{m,D}/d\alpha < 0$. This will be satisfied provided $x_D/c < 1/2$.

The equilibrium incidence is

$$\alpha_{eq} = \frac{\beta}{2}\left(C_{m,o}\right)_{\alpha=0} / \left(1 - 2x_D/c\right) = -\frac{4}{3}d_m / \left(1 - 2x_D/c\right)$$

α_{eq} is independent of Mach number.

15.4.2 Prandtl Lifting Line Theory

15.4.2.1 Vortex Sheet

The vortex sheet is a stream surface originating at the sharp trailing edge of a finite wing. It is a surface of discontinuity of $\overrightarrow{V}$. It has the following properties:

$< u >= 0$, no jump in u since pressure is continuous at the vortex sheet,

$< w >= 0$, no jump in w because the surface a zero thickness and the fluid is tangent to it (tangency condition)

$< v >\neq 0$, which can be related to the fact that the surface is made of vortex filaments of equation $y = const.,\ z = 0,\ 0 \leq x \leq \infty$.

15.4.2.2 Designing for Tip Vortices

$\Gamma[y(t)] = 2Ub\{A_1 \sin t + A_3 \sin 3t\}$, where $y(t) = -0.5b \cos t,\ 0 \leq t \leq \pi$.

Computing $\Gamma' = (d\Gamma/dt) / (dy/dt)$ shows that as, say, $t \rightarrow 0$ the fraction is of the form $0/0$. However, if $d\Gamma/dt \rightarrow 0$ faster than $\sin t \propto t$, the result will be obtained.

$d\Gamma/dt = 2Ub\{A_1 \cos t + 3A_3 \cos 3t\}$, thus, as $t \rightarrow 0$ the necessary condition is $A_1 + 3A_3 = 0$. With this result, $d\Gamma/dt = 2UbA_1\{\cos t - \cos 3t\} \propto UbA_1t^2$ as $t \rightarrow 0$. The result is confirmed.

One finds $A_1 = (C_L)_{t-o} / (\pi AR)$ and $A_3 = -(C_L)_{t-o} / (3\pi AR)$.

The downwash is given by $w_w[y(t)] = -U\{A_1 + 3A_3 \sin 3t / \sin t\} = -2UA_1(1 - 2\cos^2 t)$. In terms of y this is $w_w(y) = -2UA_1(1 - 2(2y/b)^2)$, a parabolic distribution. See Fig. 15.12.

The induced velocity w_w is negative near the root, but there is upwash near the wing tips.

15.4.2.3 Induced Drag

The induced drag is given by $C_{Di} = \pi AR\{A_1^2 + 3A_3^2\} = (C_{Di})_{elliptic}\{1 + 3\frac{1}{3^2}\} = \frac{4}{3}(C_{Di})_{elliptic}$. There is a 33 % loss compared to the elliptic loading.

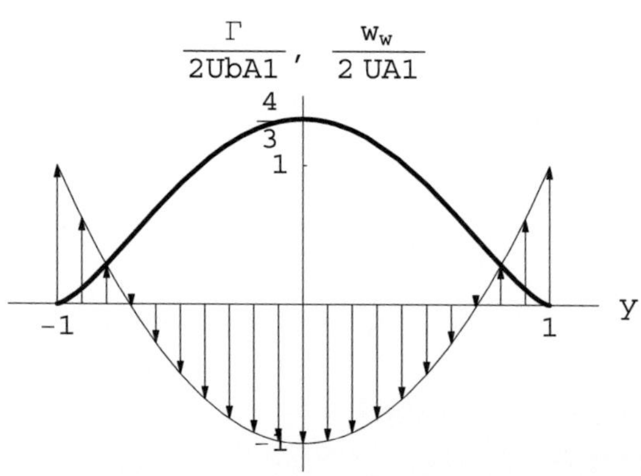

Fig. 15.12 Circulation and downwash for the wing with weak tip vortices

15.4.3 Equilibrium of the Aggie Micro Flyer

The AMF has a span $b_m = 1$ m and a constant chord $c_{xm} = 0.192$ m.

15.4.3.1 Main Wing

The wing aspect ratio is $AR_m = b_m^2/S = b_m/c_{xm} = 5.2$.

The theoretical formula gives $dC_{Lm}/d\alpha = 2\pi/(1 + 2/AR_m) = 4.54 < 4.8701$. This discrepancy is probably due to viscous effects seen in the 2-D viscous polar, in particular a laminar separation bubble near the leading edge on the lower surface that affects the leading edge geometry making the airfoil appear blunter when $\alpha \le 0$. The 2-D lift slope is found to be $(dC_l/d\alpha)_{\alpha=0} = 7.5 > 2\pi$.

15.4.3.2 Lift Curves

The general formula for the lift $C_L(\alpha, t_t)$ of the combination of the wing and tail of areas A_m and A_t, and reference area $A_{ref} = A_m + A_t$ is:

$$C_L = \frac{A_m C_{Lm} + A_t C_{Lt}}{A_m + A_t}$$

Here we get:

$$C_L(\alpha, t_t) = 0.5573 C_{Lm} + 0.4427 C_{Lt} = 3.4374\alpha + 1.5286 t_t + 0.2742$$

15.4.3.3 Moment Curve

The moment coefficient for the complete airplane is given by: $C_{M,o}(\alpha, t_t) = -1.378\alpha - 1.268 t_t - 0.03492$.

The equilibrium equation for the aerodynamic moment is given by:

$$C_{M,cg}(\alpha_{eq}, t_t) = C_{M,o}(\alpha_{eq}, t_t) + \frac{x_{cg}}{l_{ref}} C_L(\alpha_{eq}, t_t) = 0$$

This gives $C_{M,cg}(\alpha_{eq}, t_t) = -0.3248\alpha_{eq} - 0.7996 t_t + 0.0491 = 0$.

Solving for $\alpha_{eq}(t_t) = -2.4618 t_t + 0.1512$.

Application: horizontal flight at $t_t = 1.63° = 0.02845$ rd $\Rightarrow \alpha_{eq} = 0.0812$ rd $= 4.7°$.

15.4.3.4 Take-Off Conditions

At $C_{Lm} = 1.9$, using the wing lift curve one finds $\left(\alpha_{eq}\right)_{t-o} = 0.2464\,\text{rd} = 14.1°$.

The corresponding tail setting angle is given by the equilibrium equation to be $(t_t)_{t-o} = -0.0387\,\text{rd} = -2.216°$. The tail lift curve gives a small positive lift $C_{Lt} = 0.0076$.

15.4.3.5 Extra Credit

The AMF aerodynamic center is located at $x_{ac}/l_{ref} = x_{cg}/l_{ref} + SM = 0.3064 + 0.08 = 0.3864$. The center of gravity is such that $x_{acm}/l_{ref} = 0.27 < x_{cg}/l_{ref} = 0.3064 < x_{ac}/l_{ref} = 0.3864$. At high take-off incidence, the center of pressure of the main wing is close to the aerodynamic center of the main wing, i.e. $x_{cp}/l_{ref} = x_{acm}/l_{ref} + \epsilon \simeq 0.27 < x_{cg}/l_{ref} = 0.3064$. The main wing lift and the weight produce a nose up moment. Hence the tail must have a positive lift to counter that moment. See Fig. 15.13.

15.5 Solution to Problem 5

15.5.1 2-D Inviscid, Linearized, Thin Airfoil Theories

15.5.1.1 Incompressible Flow ($M_0 = 0$)

Definition

In 2-D, the following aerodynamic coefficients C_l, $C_{m,o}$, and C_d are defined as

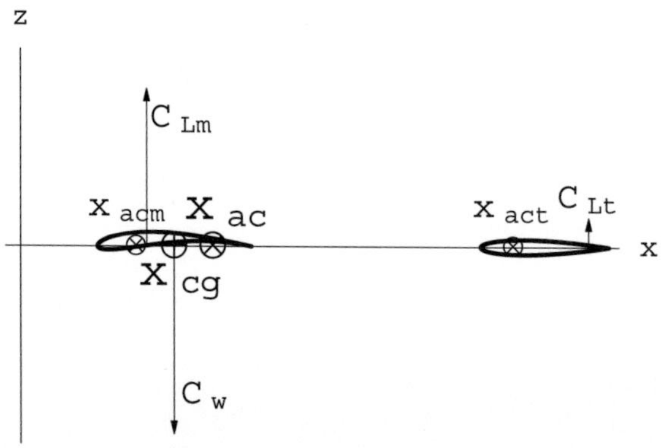

Fig. 15.13 Forces locations at take-off

$$C_l = \frac{L'}{\frac{1}{2}\rho U^2 c}, \quad C_{m,o} = \frac{M'}{\frac{1}{2}\rho U^2 c^2}, \quad C_d = \frac{D'}{\frac{1}{2}\rho U^2 c}$$

The expressions for $C_l(\alpha)$, $C_{m,o}(\alpha)$ and C_d for a symmetric profile are

$$C_l = 2\pi\alpha, \ C_{m,o} = -\frac{\pi}{2}\alpha, \ C_d = 0$$

Suction Force

See the sketch of the flat plate at incidence and of the force due to pressure integration and the suction force on Fig. 15.14.

The suction force coefficient can be obtained since the resulting drag coefficient must be zero. Hence

$$C_s = F'_s / \left(\frac{1}{2}\rho U^2 c\right) = C_l \sin\alpha \cong C_l \alpha = 2\pi\alpha^2$$

The thickness distribution contributes to recuperating the suction force by allowing the boundary layer to remain attached at the rounded leading edge, thanks to the elimination of the singularity at the sharp trailing edge of the plate.

Center of Pressure

Definition: the center of pressure is the point about which the moment of the aerodynamic forces is zero.

As seen in class, the center of pressure for a symmetric profile is at the quarter-chord. This is easily shown by using the transfer of moment formula

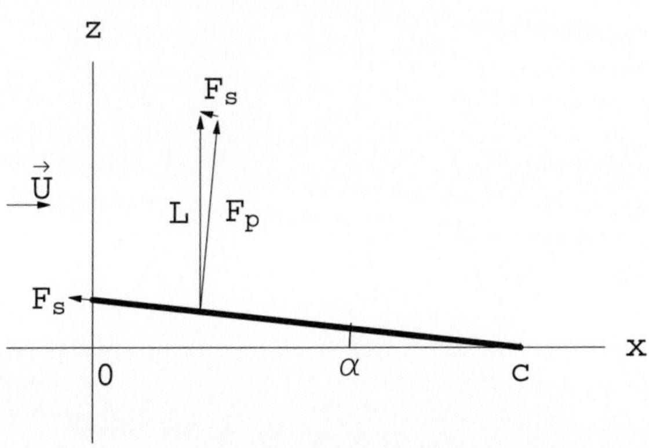

Fig. 15.14 Flat plate at incidence—force due to pressure F_p, suction force F_s, resulting lift force L

$$C_{m,cp} = C_{m,o} + \frac{x_{cp}}{c} C_l = 0$$

Solving for x_{cp} gives

$$\frac{x_{cp}}{c} = -\frac{C_{m,o}}{C_l} = \frac{1}{4}$$

For a symmetric profile, $C_{m,ac} = 0$, since the center of pressure coincides with the aerodynamic center (the formula for the value of $C_{m,ac}$ is the same as that for $C_{m,cp}$.

Computing $C_{m,o}(\alpha)$ from the $C_{m,ac}$, the change of moment formula gives

$$C_{m,o} = C_{m,ac} - \frac{x_{ac}}{c} C_l = -\frac{\pi}{4}\alpha$$

15.5.1.2 Supersonic Flow ($M_0 > 1$, $\beta = \sqrt{M_0^2 - 1}$)

A thin double wedge airfoil equips the fins of a missile cruising at Mach number $M_0 > 1$ in a uniform atmosphere. The chord of the airfoil is c. The profile camber and thickness are:

$$\begin{cases} d(x) = 0 \\ e(x) = \begin{cases} 2\theta x, \ 0 \le x \le c/2 \\ 2\theta(c - x), \ c/2 \le x \le c \end{cases} \end{cases}$$

with $z^{\pm}(x) = d(x) \pm e(x)/2$.

Pressure Distributions

C_p^+ and C_p^- are the same at $\alpha = 0$. One finds

$$\begin{cases} C_p^+ = C_p^- = \frac{2}{\beta}\theta, \ 0 \le x \le c/2 \\ C_p^+ = C_p^- = -\frac{2}{\beta}\theta, \ c/2 \le x \le c \end{cases}$$

See Fig. 15.15.

Lift Coefficient

The lift coefficient $C_l(\alpha)$ for all thin airfoils in supersonic flow is only a function of incidence

$$C_l(\alpha) = \frac{4}{\beta}\alpha$$

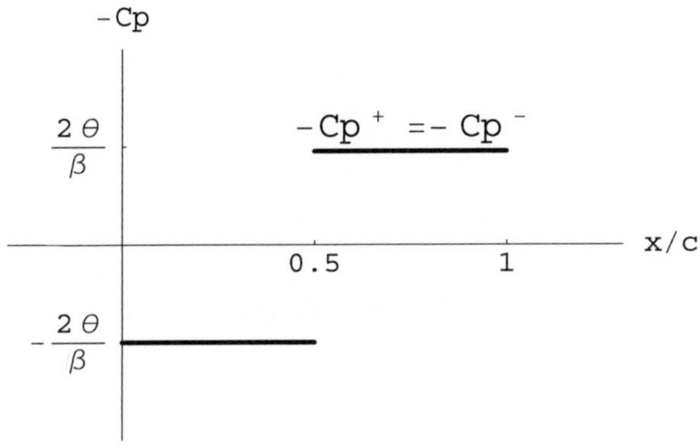

Fig. 15.15 Pressure coefficients at $\alpha = 0$

Drag Coefficient

The drag (wave drag) at zero incidence, $(C_d)_{\alpha=0}$, is

$$(C_d)_{\alpha=0} = \frac{4}{\beta}\int_0^c \left(\frac{e'(x)}{2}\right)^2 \frac{dx}{c} = \frac{4}{\beta}\theta^2$$

C_d for the general case $\alpha \neq 0$ is

$$C_d(\alpha) = \frac{4}{\beta}\theta^2 + \frac{4}{\beta}\alpha^2$$

Moment Coefficient

The zero incidence moment coefficient $\left(C_{m,o}\right)_{\alpha=0} = 0$ since it only depends on camber.

Thus, the values of the coefficient $C_{m,o}$ for the general case $\alpha \neq 0$ is

$$C_{m,o} = -\frac{2}{\beta}\alpha$$

Maximum Finess

The inverse of the finess, $1/f = C_d/C_l$ is

$$1/f = \frac{\theta^2}{\alpha} + \alpha$$

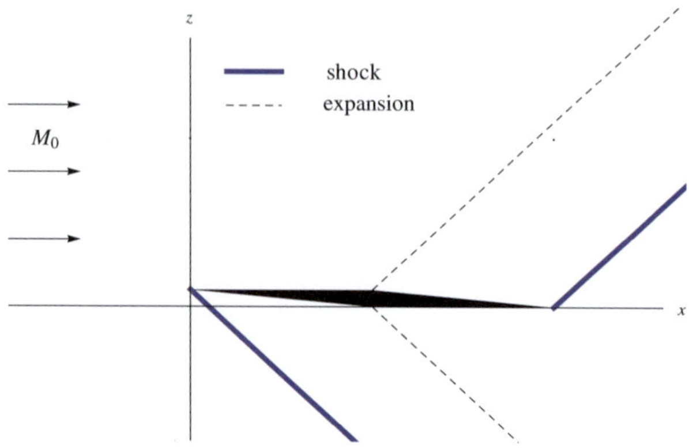

Fig. 15.16 Double wedge at incidence of maximum finess

and the value of α that maximizes f satisfies

$$\frac{d}{d\alpha}\left(\frac{1}{f}\right) = -\frac{\theta^2}{\alpha^2} + 1 = 0$$

The solution is $\alpha = \theta$ ($\alpha = -\theta$ for negative lift). See Fig. 15.16 for remarkable waves (shocks, expansions).

15.5.2 Prandtl Lifting Line Theory

15.5.2.1 Origin of Induced Drag

The remarkable feature, in the flow past a finite wing, is the existence of a vortex sheet, originating at the sharp trailing edge and trailing the wing to downstream infinity (Trefftz plane S_T). In the proximity of the vortex sheet, which is made of vortex filaments parallel to the incoming flow, the flow field is perturbed. Far downstream, in the Trefftz plane, the pressure returns to the undisturbed static pressure, hence $u_T = 0$ (since $C_p = -2u/U$). However, $v_T,\ w_T \neq 0$. An energy balance, applied to a large control volume, Fig. 15.17, shows that the energy in the Trefftz plane is larger that the incoming flow energy. Therefore, some positive work has been done to the fluid. This work is that of a thrust force which balances an equal and opposite drag force (induced drag). It can be shown that

$$A_{ref} C_{Di} = \int_{S_T} \left[\left(\frac{v}{U}\right)^2 + \left(\frac{w}{U}\right)^2\right] dS$$

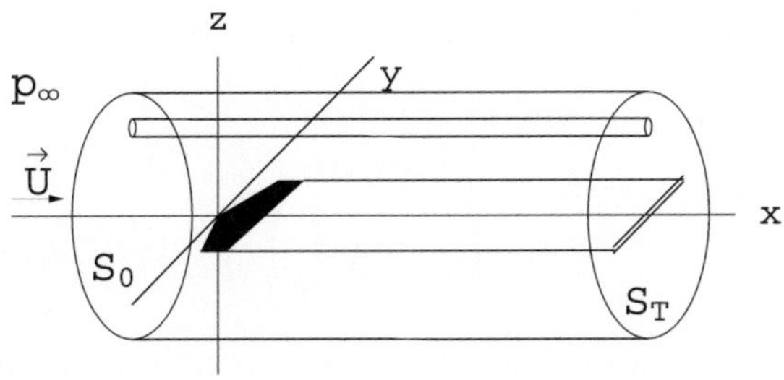

Fig. 15.17 Control volume for the application of the steady energy equation

15.5.2.2 Maximum Lift of a Wing

The first two terms of the series:

$$\Gamma\left[y(t)\right] = 2Ub \sum_{p=0}^{\infty} \frac{2\Gamma_{max}}{\pi U b(2p+1)} \sin(2p+1)t$$

correspond to

$$A_1 = \frac{2\Gamma_{max}}{\pi U b}, \; A_3 = \frac{2\Gamma_{max}}{3\pi U b}$$

See the corresponding distribution of circulation and downwash, Fig. 15.18. There is no upwash, as one finds $w_T(y) = -8UA_1\left(\frac{2y}{b}\right)^2 \le 0$.

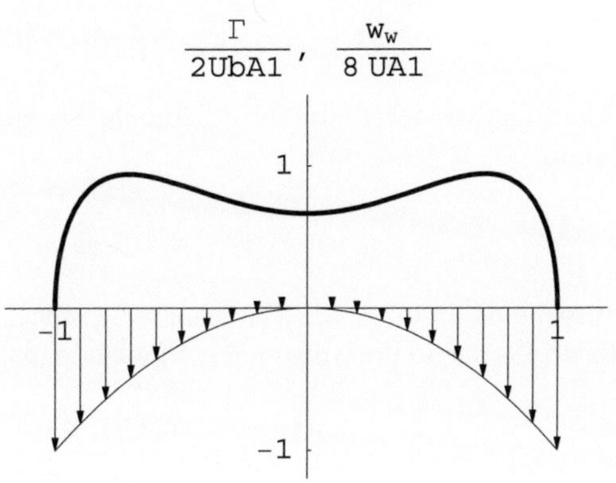

Fig. 15.18 Circulation and downwash

15.5.2.3 Induced Drag

The induced drag can be written

$$C_{Di} = \pi AR\, A_1^2\left(1+3\left(\frac{A_3}{A_1}\right)^2\right) = (C_{Di})_{elliptic}\left(1+\frac{1}{3}\right) = \frac{4}{3}\,(C_{Di})_{elliptic}$$

showing a 33 % loss compared to the elliptic loading.

The induced drag due to two finite strength vortices is obtained when all the terms are added in the infinite series, i.e.

$$(C_{Di})_{finite\ vortices} = (C_{Di})_{elliptic}\left(1+\frac{1}{3}+\frac{1}{5}+\cdots+\frac{1}{2p+1}+\cdots\right) = +\infty$$

This is a diverging series.

15.5.3 Equilibrium of the Aggie Micro Flyer (AMF III)

15.5.3.1 Airplane Lift and Moment Curves

The equilibrium code calculated the lift and moment coefficients for the complete configuration at low incidences to be:

$$\begin{cases} C_L(\alpha, t_t) = 4.47\alpha + 0.37t_t + 1.11 \\ C_{M,o}(\alpha, t_t) = -1.30\alpha - 0.33t_t - 0.28 \end{cases}$$

The effective aspect ratio for the complete configuration *AR* is obtained from the equation

$$\frac{dC_L}{d\alpha} = \frac{2\pi}{1+\frac{2}{AR}} = 4.47$$

corresponding to $AR = 4.9$. The main wing aspect ratio is $AR_m = b_m/c_m = 8$. The difference is due to the wing+tail combination, whereby

$$C_L = \frac{a_m C_{Lm} + a_t C_{Lt}}{a_m + a_t}$$

Therefore, the lift slope and the apparent aspect ratio is a weighted average of the wing and the tail aspect ratios.

As seen in class, the aerodynamic center x_{ac} location is given by

$$\frac{x_{ac}}{l_{ref}} = -\frac{\frac{dC_M}{d\alpha}}{\frac{dC_L}{d\alpha}} = \frac{1.3}{4.47} = 0.29$$

The aerodynamic center is located at $x_{ac} = 0.29 l_{ref} = 0.6\,\text{m}$.

15.5.3.2 Equilibrium Incidence

The center of gravity is located at $x_{cg}/l_{ref} = 0.21$.

The static margin is $SM = (x_{ac} - x_{cg})/l_{ref} = 0.08 = 8\,\%$.

The equilibrium incidence $\alpha(t_t)$ satisfies the equilibrium equation

$$C_M(\alpha_{eq}) + \frac{x_{cg}}{l_{ref}} C_L(\alpha_{eq}) = 0$$

Solving for α_{eq} one obtains

$$\alpha_{eq}(t_t) = -\frac{-(0.33t_t + 0.28) + 0.21(0.37t_t + 1.11)}{-1.3 + 0.21(4.47)} = -0.7t_t - 0.13$$

15.5.3.3 Take-Off Conditions

Substituting the previous result in the lift equation provides the equation for the tail setting at take-off

$$C_{L,to} = 2.0 = 4.47(-0.7t_t - 0.13) + 0.37t_t + 1.11$$

Solving for t_t one finds $t_{t,to} = -0.533\,\text{rd} = -31°$.

The incidence at take-off is therefore $\alpha_{to} = 0.24\,\text{rd} = 14°$.

The tail lift coefficient at take-off is $C_{Lt,to} = 2.49\alpha_{to} + 2.61t_{t,to} - 0.36 = -1.15$.

The lift force (in N) on the tail at take-off is $L_t = \frac{1}{2}\rho U^2 a_t C_{Lt} = -12\ N$.

The force on the tail is down.

15.5.3.4 Extra Credit

Free body diagram at take-off, see Fig. 15.19.

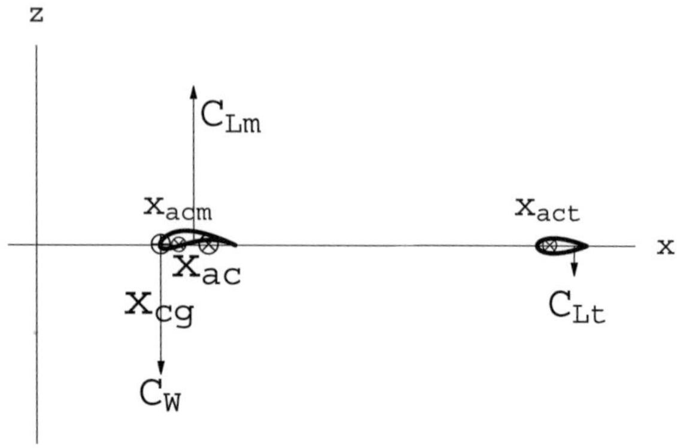

Fig. 15.19 Free body diagram at take-off

15.6 Solution to Problem 6

15.6.1 2-D Inviscid, Linearized, Thin Airfoil Theories

15.6.1.1 Incompressible Flow ($M_0 = 0$)

Airfoil Design

Consider a thin airfoil with *parabolic camberline*. The design requirements are the following:

- the take-off incidence is $(C_l)_{t-o} = 1.8$
- the location of the center of pressure at take-off is $\frac{x_{cp}}{c} = 0.4$

The lift coefficient is $(C_l)_{t-o} = 2\pi(A_0 + \frac{A_1}{2}) = 1.8$

The center of pressure location is $\left(\frac{x_{cp}}{c}\right)_{t-o} = \frac{A_0+A_1-\frac{A_2}{2}}{4(A_0+\frac{A_1}{2})} = 0.4$

A profile with parabolic camberline has only two non-zero Fourier coefficients, A_0 and A_1. The two conditions read

$$\begin{cases} A_0 + \frac{A_1}{2} = \frac{1.8}{2\pi} \\ A_0 + A_1 = \frac{2.88}{2\pi} \end{cases}$$

The solution is $A_0 = 0.114$ and $A_1 = 0.344$.
The take-off incidence is $(\alpha)_{t-o} = A_0 = 0.114\,\text{rd} = 6.5°$.
The profile relative camber is $\frac{d}{c} = \frac{A_1}{4} = 0.086$.
For these conditions, the Selig 1223 can be used.

Lift Curve

The lift slope for all thin airfoils in inviscid, incompressible flow is $\frac{dC_l}{d\alpha} = 2\pi$.

The value of the C_{l0}, corresponding to $\alpha = 0$ is $C_{l0} = \pi A_1 = 1.08$.

The sketch of the lift curve is presented in Fig. 15.20.

Equilibrium at Take-Off with Tail

If the center of gravity of the wing+tail configuration is located at $\frac{x_{cg}}{c} = 0.5$ at take-off, the tail lift coefficient will be positive, $C_l > 0$ since the wing lift and aircraft weight create a nose up moment which must be balanced by the tail.

See Fig. 15.21.

15.6.1.2 Supersonic Flow ($M_0 > 1, \ \beta = \sqrt{M_0^2 - 1}$)

A flat plate equips the fins of a missile cruising at Mach number $M_0 > 1$ in a uniform atmosphere. The chord of the airfoil is c.

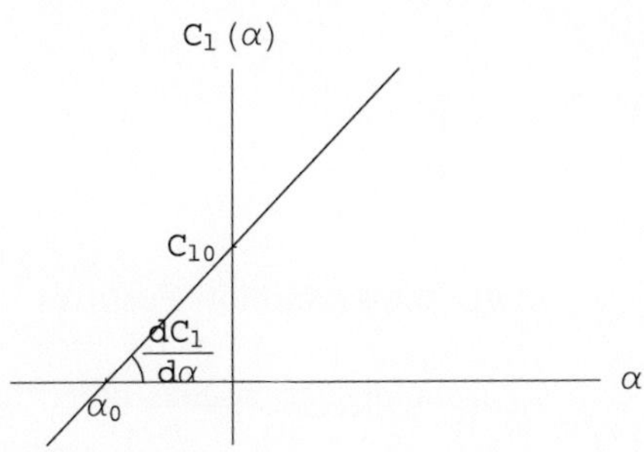

Fig. 15.20 Lift curve $C_l(\alpha)$

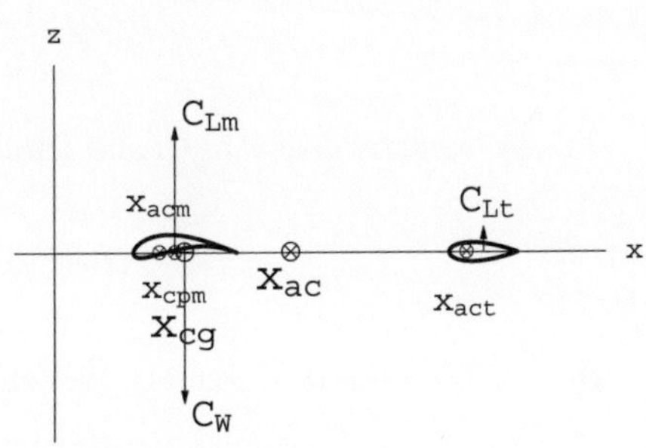

Fig. 15.21 Equilibrium at take-off

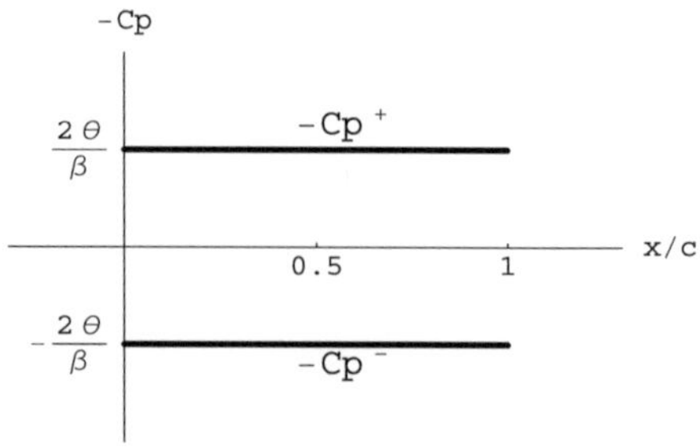

Fig. 15.22 Pressure coefficient distributions

Pressure Distribution and Global Coefficients

The pressure coefficients $-C_p^+$ and $-C_p^-$ versus x for this airfoil at $\alpha > 0$ are simply $\pm\frac{2\alpha}{\beta}$ and are plotted in Fig. 15.22.

From the derivations seen in class

$$C_l(\alpha) = 4\frac{\alpha}{\beta}$$

$$C_{m,o}(\alpha) = -2\frac{\alpha}{\beta}$$

$$C_d(\alpha) = 4\frac{\alpha^2}{\beta}$$

Maximum Finess and Flow Features

With C_d including the viscous drag, the expression of the inverse of the finess, $1/f = C_d/C_l$ is

$$1/f = \frac{C_{d0}}{C_l} + \frac{\beta}{4}C_l$$

Taking the derivative w.r.t C_l and setting the result to zero yield

$$(C_l)^2_{opt} = \frac{4}{\beta}C_{d0}$$

Hence, the optimum C_d and C_l are found to be

$$(C_d)_{opt} = 2C_{d0}, \quad (\alpha)_{opt} = \sqrt{\frac{\beta}{4}C_{d0}}$$

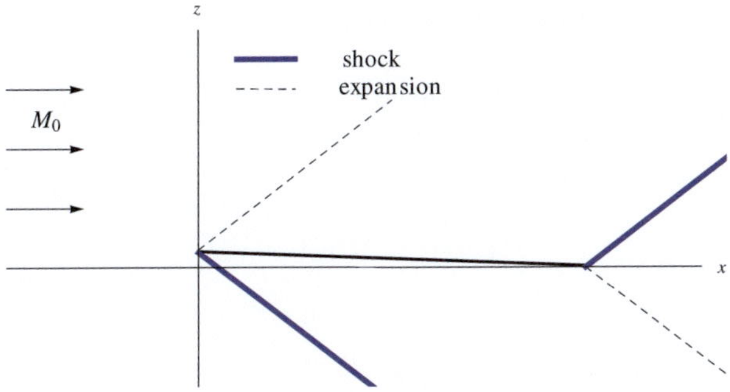

Fig. 15.23 Flow features for supersonic flow past flat plate at incidence

The sketch of the profile at incidence and of the remarkable waves is shown in Fig. 15.23.

15.6.2 Prandtl Lifting Line Theory

15.6.2.1 Downwash Evolution

A semi-infinite vortex sheet trailing a wing and contained in the (x, y) plane induces at a point in that plane a vertical component $w(x, y)$ called downwash. Near the x-axis $(-\frac{b}{2} < y < \frac{b}{2})$, far upstream the effect of the wake decays and the downwash goes to zero. At the lifting line, $w_w(y)$ is the downwash induced by a semi-infinite vortex sheet. Far downstream, in the Trefftz plane, the downwash $w_T(y)$ results from an infinite vortex sheet. Because of the symmetry, the infinite vortex filaments induce velocities that are twice that of semi-infinite ones. As a result, $w_T(y) = 2w_w(y)$.

15.6.2.2 Wing with Ideal Loading

As seen in class, the ideal loading of a wing for minimum induced drag is elliptic, i.e.

$$\Gamma(y) = \Gamma_0\sqrt{1 - \left(\frac{2y}{b}\right)^2}$$

Theory shows that

$$\Gamma_0 = 2UbA_1 = 2Ub\frac{(C_L)_{cruise}}{\pi AR}$$

$$w_w(y) = -UA_1 = -U\frac{(C_L)_{cruise}}{\pi AR} = const.$$

15.6.2.3 Induced Downwash in Manoeuvre

The simplest loading that will contribute to roll is mode 2, which is antisymmetric and does not contribute to lift as any mode $A_n,\ n \geq 2$. The mode 2 circulation and induced downwash are given by

$$\Gamma_2[y(t)] = 2UbA_2 \sin 2t$$

$$w_{w2}[y(t)] = -2UA_2 \frac{\sin 2t}{\sin t} = -4UA_2 \cos t$$

Given that $y(t) = -\frac{b}{2}\cos t,\ 0 \leq t \leq \pi$, the last relation reads

$$w_{w2}(y) = 8UA_2 \frac{y}{b}$$

This is represented in Fig. 15.24.

15.6.2.4 Induced Drag in Manoeuvre

The induced drag is given by

$$C_{Di} = (C_{Di})_{ideal}\left(1 + 2\left(\frac{A_2}{A_1}\right)^2\right) = (C_{Di})_{ideal}\left(1 + \frac{2}{100}\right)$$

The increase in induced drag is 2 %.

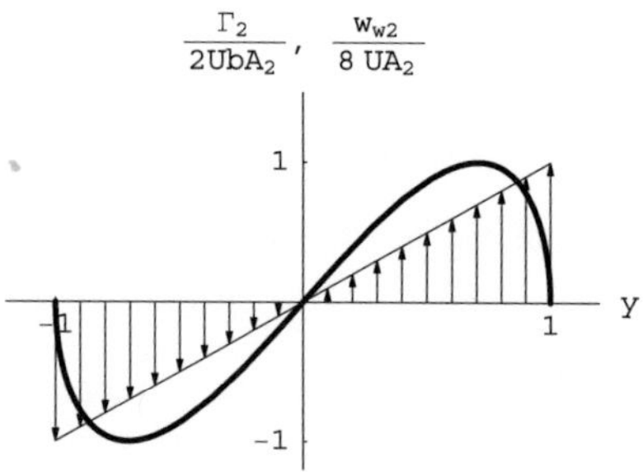

Fig. 15.24 Mode 2 loading for manoeuvre

15.6.3 Equilibrium of the Aggie Micro Flyer

15.6.3.1 Airplane Lift and Moment Curves

The equilibrium code calculates the lift and moment coefficients for the complete configuration at low incidences to be:

$$\begin{cases} C_L(\alpha, t_t) = 4.391\alpha + 0.757t_t + 0.928 \\ C_{M,o}(\alpha, t_t) = -1.733\alpha - 0.686t_t - 0.267 \end{cases}$$

We will use this linear model, even for take-off conditions.

Definition of the *aerodynamic center*:

The aerodynamic center is the point about which the moment is independent of incidence.

As seen in class, the location of the aerodynamic center is given by

$$\frac{x_{ac}}{l_{ref}} = -\frac{\frac{dC_{M,o}}{d\alpha}}{\frac{dC_L}{d\alpha}} = \frac{1.733}{4.391} = 0.395$$

That is $x_{ac} = 0.632\,\text{m}$.

Given that $x_{cg} = 0.503\,\text{m}$, the static margin is

$$SM = \frac{x_{ac} - x_{cg}}{l_{ref}} = 0.08$$

a static margin of 8 %.

15.6.3.2 Take-Off Conditions

The longitudinal equilibrium equation for the moment reads

$$C_{M,cg}(\alpha) = C_{M,o}(\alpha) + \frac{x_{cg}}{l_{ref}} C_L(\alpha) = 0$$

This equation represents the transfer of moment from the nose O of the aircraft to the center of gravity and states that the moment of the aerodynamic forces at the center of gravity must be zero.

The main wing lift curve is given by

$$C_{Lm} = 4.927\alpha + 1.388$$

Since the take-off lift coefficient for the main wing is $(C_{Lm})_{t-o} = 2.7$, the incidence can be found to be $(\alpha)_{t-o} = 0.266\,\text{rd} = 15.3°$.

Find the tail setting angle at take-off is then given by the equilibrium equation as

$$-0.686t_t - 0.728 + 0.314(0.757t_t + 2.096) = 0$$

which gives $t_t = -0.156\,\text{rd} = -8.9°$.

The location of the *center of pressure* is given by

$$\frac{x_{cp}}{l_{ref}} = -\frac{C_{M,o}}{C_L} = \frac{0.621}{1.978} = 0.314$$

Hence, $x_{cp} = 0.502\,\text{m}$.

The global lift coefficient can now be found to be

$$C_L = 4.391\ 0.266 - 0.757\ 0.156 + 0.928 = 1.978$$

By definition we have

$$(a_m + a_t)C_L = a_m C_{Lm} + a_t C_{Lt}$$

Here $a_m = 0.709\,\text{m}^2$ and $a_t = 0.254\,\text{m}^2$. Solving for C_{Lt} gives

$$C_{Lt} = -0.04$$

The tail lift coefficient at take-off is close to zero (within our linear model accuracy).

15.7 Solution to Problem 7

15.7.1 2-D Inviscid, Linearized, Thin Airfoil Theories

15.7.1.1 Incompressible Flow ($M_0 = 0$)

Cambered Plate Geometry

Given the thin cambered plate equation

$$d(x) = Ac\frac{x}{c}\left(1 - 2\frac{x}{c}\right)\left(1 - \frac{x}{c}\right)$$

one calculates $d'(x)$ to be

$$d'(x) = A\left(1 - 6\frac{x}{c} + 6\frac{x^2}{c^2}\right) = A(1 - 6\xi + 6\xi^2)$$

where $\xi = \frac{x}{c}$. $d'(x) = 0$ for $\xi = \frac{x}{c} = \frac{3\pm\sqrt{3}}{6}$. The plate is depicted in Fig. 15.25.

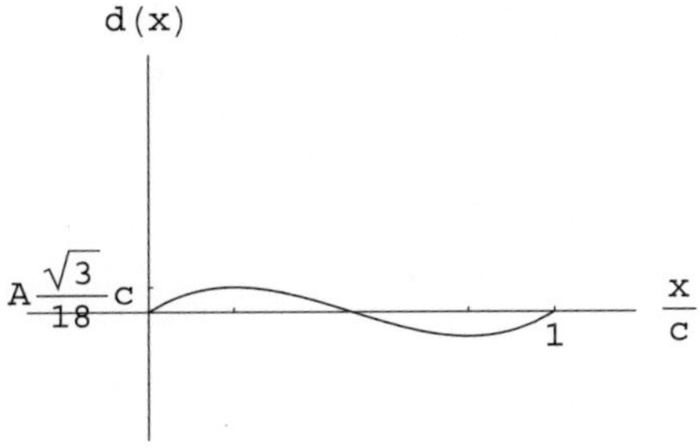

Fig. 15.25 Thin cambered plate

Fourier Coefficients

The Fourier coefficients $A_0, A_1, A_2, \ldots, A_n$ for this thin cambered plate are obtained from the equation

$$d'[x(t)] - \alpha = -A_0 + \sum_{n=1}^{\infty} A_n \cos nt$$

Changing variable from x to t in $d'(x)$ yields

$$d'[x(t)] = A\left(1 - 3(1 - \cos t) + \frac{3}{2}(1 - \cos t)^2\right)$$
$$= A\left(-\frac{1}{2} + \frac{3}{2}\cos^2 t\right) = \frac{A}{4}(1 + 3\cos 2t)$$

where we have made use of the identity $\cos^2 t = (1 + \cos 2t)/2$. Substitution in the above equation provides by simple identification all the Fourier coefficients

$$\frac{A}{4}(1 + 3\cos 2t) - \alpha = -A_0 + \sum_{n=1}^{\infty} A_n \cos nt$$

$$A_0 = \alpha - \frac{A}{4}, \quad A_1 = 0, \quad A_2 = \frac{3}{4}A, \quad A_3 = \cdots = A_n = 0, \quad n \geq 3$$

From the result for A_0 one finds $\alpha_{adapt} = \frac{A}{4}$.
Sketch of the flow at the incidence of adaptation is in Fig. 15.26.

Definition of Aerodynamic Center

The aerodynamic center is the point about which the moment of the aerodynamic forces is independent of incidence.

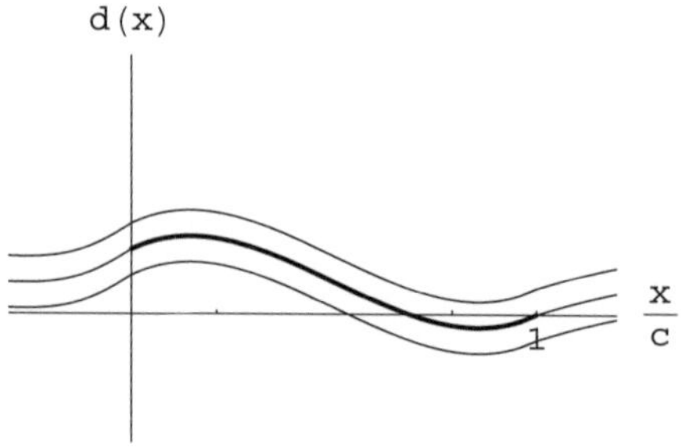

Fig. 15.26 Thin cambered plate at incidence of adaptation and some streamlines

Aerodynamic Coefficients

The aerodynamic coefficients $C_l(\alpha)$, $C_{m,o}(\alpha)$ and $C_{m,a.c.}$ are all expressed in terms of the Fourier coefficients, i.e.

$$C_l(\alpha) = 2\pi\left(A_0 + \frac{A_1}{2}\right) = 2\pi\left(\alpha - \frac{A}{4}\right)$$

$$C_{m,o}(\alpha) = -\frac{\pi}{2}\left(A_0 + A_1 - \frac{A_2}{2}\right) = -\frac{\pi}{2}\left(\alpha - \frac{A}{4} - \frac{3A}{8}\right) = -\frac{\pi}{2}\left(\alpha - \frac{5A}{8}\right)$$

$$C_{m,a.c.} = -\frac{\pi}{4}(A_1 - A_2) = \frac{3\pi}{16}A$$

Static Equilibrium About an Axis

The change of moment formula gives the aerodynamic moment coefficient at the mid-chord, $\frac{x}{c} = \frac{1}{2}$

$$C_{m,\frac{1}{2}}(\alpha) = C_{m,o}(\alpha) + \frac{1}{2}C_l(\alpha) = -\frac{\pi}{2}\left(\alpha - \frac{5A}{8}\right) + \frac{1}{2}2\pi\left(\alpha - \frac{A}{4}\right) = \frac{\pi}{2}\left(\alpha + \frac{A}{8}\right)$$

The equilibrium angle α_{eq} is such that $C_{m,\frac{1}{2}}(\alpha_{eq}) = 0$. Hence

$$\alpha_{eq} = -\frac{A}{8}$$

The equilibrium is unstable because $\frac{dC_{m,\frac{1}{2}}}{d\alpha} > 0$.

15.7.1.2 Supersonic Flow ($M_0 > 1,\ \beta = \sqrt{M_0^2 - 1}$)

Pressure Distribution and Flow Features

The pressure coefficients are given by the Ackeret formulae

$$C_p^+ = \frac{2}{\beta}\left(A\left(1 - 6\frac{x}{c} + 6\frac{x^2}{c^2}\right) - \alpha\right)$$

$$C_p^- = -\frac{2}{\beta}\left(A\left(1 - 6\frac{x}{c} + 6\frac{x^2}{c^2}\right) - \alpha\right)$$

The plot of $-C_p^+$ and $-C_p^-$ versus x for this airfoil at $\alpha = 0$ is seen in Fig. 15.27.
See sketch of the flow at $\alpha = 0$ (shocks, characteristic lines, expansion shocks), Fig. 15.28.

Static Equilibrium About an Axis

$C_{m,\frac{1}{2}}$ is given by

$$C_{m,\frac{1}{2}}(\alpha) = \left(C_{m,o}\right)_{\alpha=0} - \frac{2\alpha}{\beta} + \frac{1}{2}C_l(\alpha)$$

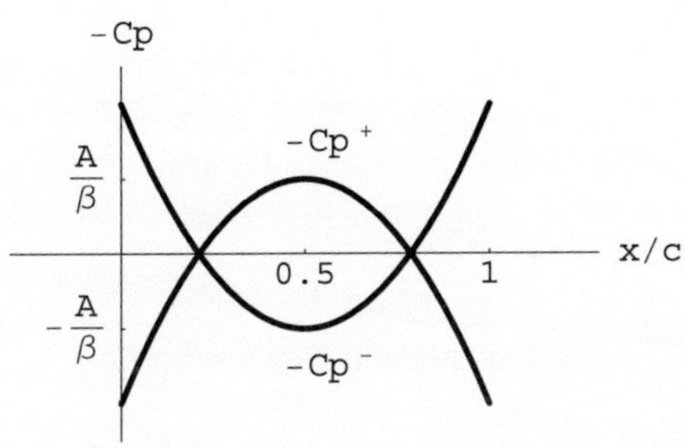

Fig. 15.27 Pressure coefficients on the thin cambered plate at zero incidence

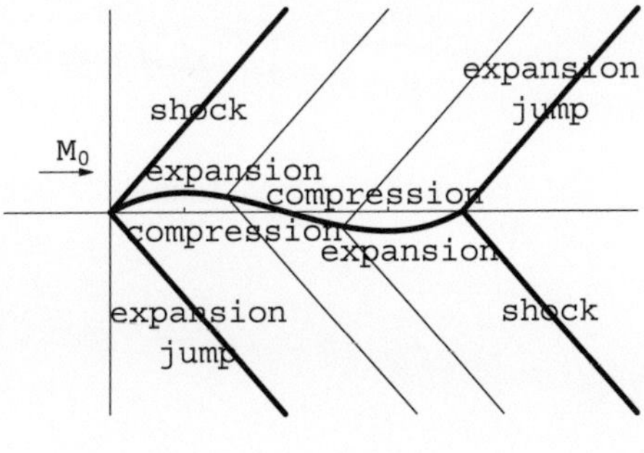

Fig. 15.28 Flow field past the thin cambered plate at zero incidence

The coefficient $\left(C_{m,o}\right)_{\alpha=0}$ needs to be calculated as

$$\left(C_{m,o}\right)_{\alpha=0} = \frac{4}{\beta}\int_0^c d'(x)\frac{x}{c}\frac{dx}{c} = \frac{4A}{\beta}\int_0^1 (1-6\xi+6\xi^2)\xi d\xi = 0$$

Hence the moment about the mid-chord reduces to

$$C_{m,\frac{1}{2}}(\alpha) = -\frac{2\alpha}{\beta} + \frac{1}{2}\frac{4\alpha}{\beta} = 0$$

Any value of α satisfies the equilibrium requirement. The equilibrium is neutral.

15.7.2 Prandtl Lifting Line Theory

15.7.2.1 Vortex Sheet Characteristics

The conditions $< u >=< w >= 0$ on the vortex sheet behind a finite wing result from

- the continuity of pressure across the vortex sheet (a fluid surface),
- the tangency condition along the surface, respectively.

The existence of induced drag for a wing in incompressible, inviscid flow results from the non-zero components of v and w near the vortex sheet in the Trefftz plane, far downstream, where $u = 0$ (the pressure is back to the undisturbed value). The kinetic energy associated with v and w is not recoverable. There is more kinetic energy out of the control volume than into the control volume as shown by the application of the Steady Energy Equation. This excess kinetic energy comes from the work of the thrust that balances the drag.

15.7.2.2 Circulation Representation

The circulation is represented by a Fourier series

$$\begin{cases} \Gamma[y(t)] = 2Ub\sum_{n=1}^{\infty} A_n \sin nt \\ y(t) = -\frac{b}{2}\cos t, \ 0 \le t \le \pi \end{cases}$$

The first three modes with unit coefficients $A_1 = A_2 = A_3 = 1$ are shown in Fig. 15.29.

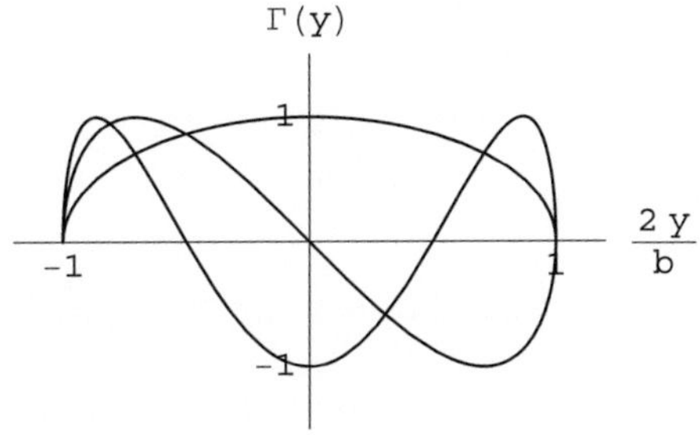

Fig. 15.29 First three modes of circulation representation

15.7.2.3 Efficiency Factor

As seen in class, the definition of the Efficiency Factor e in terms of the Fourier coefficients is

$$\frac{1}{e} = 1 + 2\left(\frac{A_2}{A_1}\right)^2 + \cdots + n\left(\frac{A_n}{A_1}\right)^2 + \cdots$$

Given that the wing loading is symmetrical, only the odd modes are present. Since the "simplest" wing is assumed for the AMAT09 wing, only modes 1 and 3 are non-zero

$$\frac{1}{e} = 1 + 3\left(\frac{A_3}{A_1}\right)^2 = \frac{1}{0.95} = 1.053, \Rightarrow \frac{A_3}{A_1} = 0.132$$

Furthermore, the mode 1 amplitude is related to the lift coefficient by

$$C_L = \pi AR\, A_1, \Rightarrow A_1 = \frac{C_L}{\pi AR}$$

The results for the three phases of flight are

- take-off: $\alpha = 18°$, $C_L = 2.5, \Rightarrow A_1 = 0.141$, $\Rightarrow A_3 = 0.019$
- top speed: $\alpha = -3°$, $C_L = 1.0$, $\Rightarrow A_1 = 0.056$, $\Rightarrow A_3 = 0.007$
- power-off descent: $\alpha = 4°$, $C_L = 1.6$, $\Rightarrow A_1 = 0.09$, $\Rightarrow A_3 = 0.012$

15.7.3 Equilibrium of the AMAT09

15.7.3.1 Airplane Aerodynamic Center and Static Margin

The *aerodynamic center* is given by

$$\frac{x_{a.c.}}{l_{ref}} = -\frac{\frac{\partial C_{M,o}}{\partial \alpha}}{\frac{\partial C_L}{\partial \alpha}} = \frac{1.124}{3.912} = 0.287 = 28.7/100$$

The *static margin SM* is

$$SM = \frac{x_{a.c.}}{l_{ref}} - \frac{x_{c.g.}}{l_{ref}} = 0.287 - 0.227 = 0.06 = 6/100$$

15.7.3.2 Equilibrium Condition and Static Stability

The moment coefficient at the center of gravity, $C_{M,c.g.}(\alpha, t_t)$ is given by

$$C_{M,c.g.}(\alpha, t_t) = C_{M,o}(\alpha, t_t) + \frac{x_{c.g.}}{l_{ref}} C_L(\alpha, t_t)$$

$$C_{M,c.g.}(\alpha, t_t) = -1.124\alpha - 0.625t_t - 0.138 + 0.227(3.912\alpha + 0.626t_t + 0.844)$$

One finds

$$C_{M,c.g.}(\alpha, t_t) = -0.236\alpha - 0.483t_t + 0.054$$

The slope of the $C_{M,c.g.}$ is negative. The equilibrium is stable.
At equilibrium $C_{M,c.g.} = 0$, hence

$$\alpha_{eq}(t_t) = -2.064t_t + 0.229$$

15.7.3.3 Top Speed

The top speed is obtained for $t_t = 7.6° = 0.1326\,\text{rd}$. The corresponding value of α_{eq} is

$$\alpha_{eq} = -2.06\,0.1326 + 0.229 = -0.0448\,\text{rd} = -2.6°$$

The top speed, given that $A_{ref} = 2.225\,\text{m}^2$, $\rho = 1.2\,\text{kg/m}^3$ is such that the lift balances the weight

$$\frac{1}{2}\rho U_{eq}^2 A_{ref} C_{L,eq} = Mg$$

But $C_{L,eq} = 3.912(-0.045) + 0.626(0.1326) + 0.844 = 0.751$
One finds

$$U_{eq}^2 = \frac{30\,9.81}{0.5\,1.2\,2.225\,0.751} = 293.5, \Rightarrow U_{eq} = 17.1\,\text{m/s}$$

15.8 Solution to Problem 8

15.8.1 2-D Inviscid, Linearized, Thin Airfoil Theories

15.8.1.1 Incompressible Flow ($M_0 = 0$)

Profile Geometry

Consider a half-double wedge profile of chord c of equation

$$\begin{cases} f^+(x) = d(x) + \frac{1}{2}e(x) = \Theta x, \ 0 \le x \le \frac{c}{2} \\ f^+(x) = d(x) + \frac{1}{2}e(x) = \Theta(c-x), \ \frac{c}{2} \le x \le c \end{cases}$$

$$f^-(x) = d(x) - \frac{1}{2}e(x) = 0, \ 0 \le x \le c$$

See Fig. 15.30. The camber $d(x)$ and thickness $e(x)$ distributions are given by

$$\begin{cases} d(x) = \frac{1}{2}\left(f^+(x) + f^-(x)\right) = \frac{1}{2}\Theta x, \ 0 \le x \le \frac{c}{2} \\ d(x) = \frac{1}{2}\left(f^+(x) + f^-(x)\right) = \frac{1}{2}\Theta(c-x), \ \frac{c}{2} \le x \le c \end{cases}$$

$$\begin{cases} e(x) = f^+(x) - f^-(x) = \Theta x, \ 0 \le x \le \frac{c}{2} \\ e(x) = f^+(x) - f^-(x) = \Theta(c-x), \ \frac{c}{2} \le x \le c \end{cases}$$

Fourier Coefficients

The expressions of the Fourier coefficients A_0 and A_n in the expansion of the vorticity for an arbitrary profile are

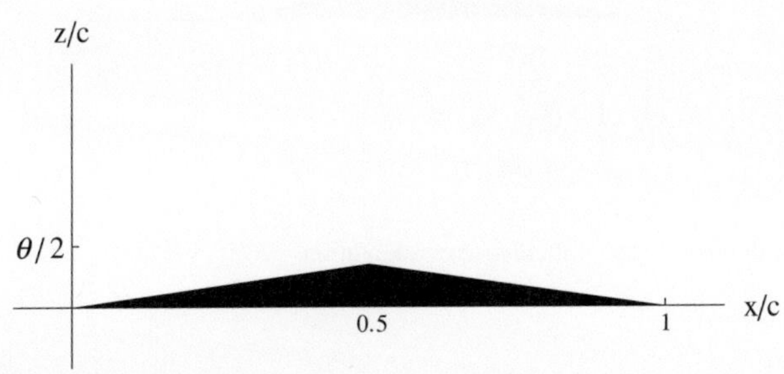

Fig. 15.30 Half double-wedge geometry

$$A_0 = \alpha - \frac{1}{\pi}\int_0^{\pi} d'[x(t)]dt, \quad A_n = \frac{2}{\pi}\int_0^{\pi} d'[x(t)]cosntdt$$

A_0 and A_2 is given by

$$A_0 = \alpha - \frac{1}{\pi}\left\{\int_0^{\frac{\pi}{2}} \frac{\Theta}{2}dt + \int_{\frac{\pi}{2}}^{\pi}\left(-\frac{\Theta}{2}\right)dt\right\} = \alpha$$

A_1 is given by

$$A_1 = \frac{2}{\pi}\left\{\int_0^{\frac{\pi}{2}} \frac{\Theta}{2}costdt + \int_{\frac{\pi}{2}}^{\pi}\left(-\frac{\Theta}{2}\right)costdt\right\} = \frac{2}{\pi}\frac{\Theta}{2}\{1-(-1)\} = \frac{2}{\pi}\Theta$$

A_2 is given by

$$A_2 = \frac{2}{\pi}\left\{\int_0^{\frac{\pi}{2}} \frac{\Theta}{2}cos2tdt + \int_{\frac{\pi}{2}}^{\pi}\left(-\frac{\Theta}{2}\right)cos2tdt\right\} = \frac{2}{\pi}\frac{\Theta}{2}\{0-(-0)\} = 0$$

The incidence of adaptation α_{adapt} is such that $A_0 = 0$. Here we have $\alpha_{adapt} = 0$. See Fig. 15.31.

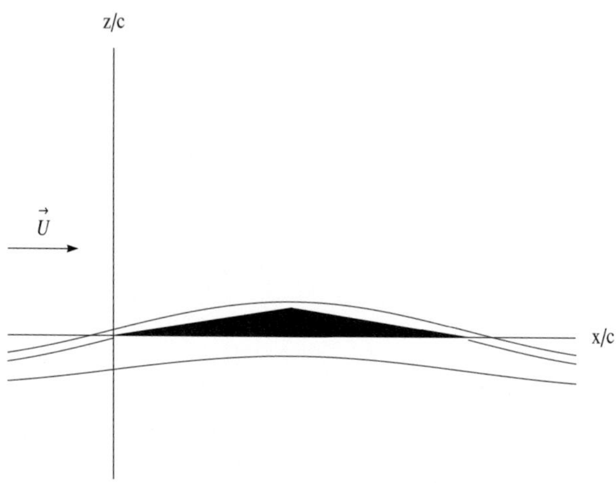

Fig. 15.31 Half double wedge at incidence of adaptation

Definition of Aerodynamic Center

The aerodynamic center is the point about which the moment of the aerodynamic forces is independent of α.

Aerodynamic Coefficients

The aerodynamic coefficients $C_l(\alpha)$ and $C_{m,o}(\alpha)$ are given in terms of the Fourier coefficients

$$C_l(\alpha) = 2\pi\left(A_0 + \frac{A_1}{2}\right) = 2\pi\left(\alpha + \frac{\Theta}{\pi}\right)$$

$$C_{m,o}(\alpha) = -\frac{\pi}{2}\left(A_0 + A_1 - \frac{A_2}{2}\right) = -\frac{\pi}{2}\left(\alpha + 2\frac{\Theta}{\pi}\right)$$

15.8.1.2 Supersonic Flow ($M_0 > 1$, $\beta = \sqrt{M_0^2 - 1}$)

The same profile equips the wing of an airplane cruising at Mach number $M_0 > 1$ in a uniform atmosphere.

Pressure Distribution and Flow Features

Using Ackeret formula one finds $C_p^+ = \pm 2\Theta/\beta$ and $C_p^- = 0$. This is shown in Fig. 15.32 at $\alpha = 0$. The flow features are displayed in Fig. 15.33 (shocks, characteristic lines, expansion shocks).

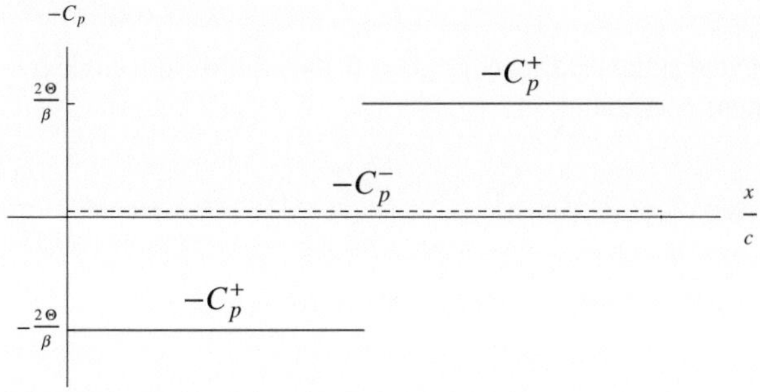

Fig. 15.32 C_p distribution on the half double wedge at $\alpha = 0$

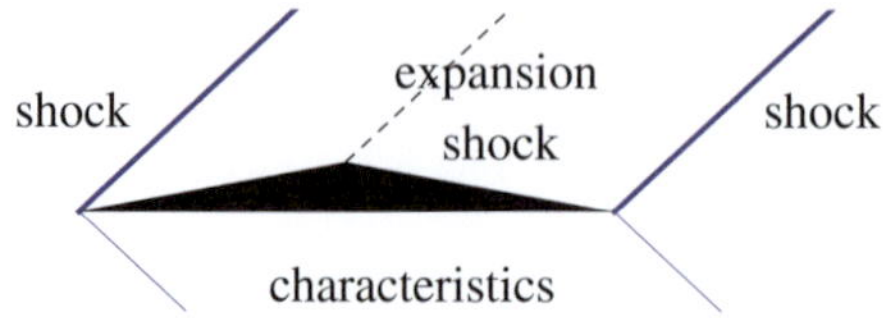

Fig. 15.33 Flow features in supersonic flow

Aerodynamic Coefficients

At $\alpha = 0$, the drag coefficient $(C_d)_{\alpha=0}$ is given by

$$(C_d)_{\alpha=0} = \frac{4}{\beta}\int_0^c \left\{(d'(x))^2 + \left(\frac{e'(x)}{2}\right)^2\right\}\frac{dx}{c}$$

$$= \frac{4}{\beta}\left[\int_0^{\frac{c}{2}} \left\{\left(\frac{\Theta}{2}\right)^2 + \left(\frac{\Theta}{2}\right)^2\right\}\frac{dx}{c} + \int_{\frac{c}{2}}^c \left\{\left(-\frac{\Theta}{2}\right)^2 + \left(-\frac{\Theta}{2}\right)^2\right\}\frac{dx}{c}\right] = 2\frac{\Theta^2}{\beta}$$

and moment coefficient $\left(C_{m,o}\right)_{\alpha=0}$ by

$$\left(C_{m,o}\right)_{\alpha=0} = \frac{4}{\beta}\int_0^c d'(x)\frac{x}{c}\frac{dx}{c} = \frac{4}{\beta}\left\{\int_0^{\frac{1}{2}} \frac{\Theta}{2}\xi d\xi + \int_{\frac{1}{2}}^1 \left(-\frac{\Theta}{2}\right)\xi d\xi\right\}$$

$$= \frac{4}{\beta}\frac{\Theta}{4}\left\{\left(\frac{1}{2}\right)^2 - \left(1 - \left(\frac{1}{2}\right)^2\right)\right\} = -\frac{\Theta}{2\beta}$$

Static Equilibrium About an Axis

If an axis is located at the leading edge, $x = 0$, the equilibrium angle α_{eq} corresponds to a zero moment coefficient, i.e.

$$C_{m,o}(\alpha_{eq}) = \left(C_{m,o}\right)_{\alpha=0} - 2\frac{\alpha_{eq}}{\beta} = 0, \quad \Rightarrow \quad \alpha_{eq} = -\frac{\Theta}{4}$$

The equilibrium is stable because $dCm, o/d\alpha < 0$.

15.8.2 Prandtl Lifting Line Theory

15.8.2.1 Circulation Representation

A paper airplane wing is made of manila folder paper by cutting an ellipse of large aspect ratio (b, c_0). Then the wing is given some small camber such that the relative camber is constant $(d/c = const)$ along the span, and there is no twist, $t(y) = 0$. The circulation for an arbitrary wing is represented by a Fourier series

$$\begin{cases} \Gamma[y(\theta)] = 2Ub\sum_{n=1}^{\infty} A_n \sin n\theta \\ y(\theta) = -\frac{b}{2}\cos\theta,\ 0 \le \theta \le \pi \end{cases}$$

The chord distribution of the wing can be expressed as $c[y(\theta)] = c_x sin\theta$ with the above change of variable.

Prandtl Integro-differential equation reads

$$\Gamma(y) = \pi U c(y)\left\{\alpha + \alpha_i(y) + 2\frac{d(y)}{c(y)} + t(y)\right\}$$

where $\alpha_i(y) = w_w(y)/U$ is the induced angle of attack. Taking into account the wing characteristics, i.e. the constant relative camber, the absence of twist and the elliptic chord distribution, as seen in class, the above equation is satisfied identically with an elliptic loading since the corresponding downwash or induced incidence is constant as well. The relationship between the root circulation Γ_0, the incidence α, the constant induced incidence α_i and the relative camber d/c reduces to

$$\Gamma_0 sin\theta = \pi U c_0 sin\theta\left\{\alpha + \alpha_i + 2\frac{d}{c}\right\}, \quad \Rightarrow \quad \Gamma_0 = \pi U c_0\left\{\alpha + \alpha_i + 2\frac{d}{c}\right\}$$

15.8.2.2 Ideal Angle of Attack

From the previous result, one can calculate the lift coefficient, starting with

$$\Gamma_0 = 2\pi U c_0\frac{d}{c}, \quad \Rightarrow C_L = \frac{2\Gamma_0}{Uc_0} = 4\pi\frac{d}{c}$$

The geometric incidence, α_{ideal} that will make the effective incidence zero, i.e. $\alpha_{eff} = \alpha_{ideal} + \alpha_i = 0$ is such that

$$C_L = \frac{2\pi}{1+\frac{2}{AR}}\left(\alpha_{ideal} + 2\frac{d}{c}\right) = 4\pi\frac{d}{c}, \quad \Rightarrow \alpha_{ideal} = \frac{4}{AR}\frac{d}{c}$$

See Fig. 15.34.

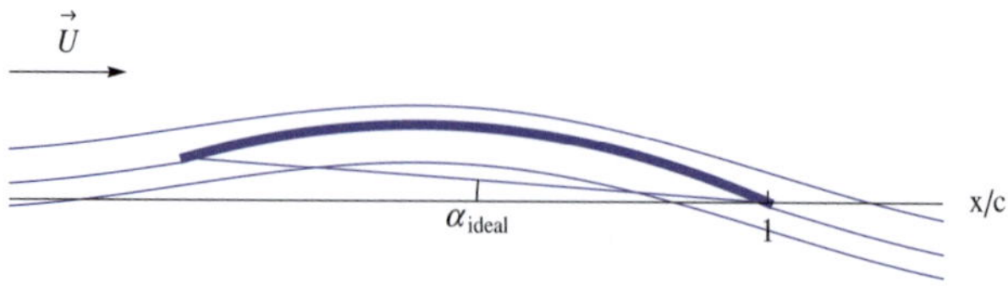

Fig. 15.34 Cross section of elliptic wing at ideal incidence

15.8.2.3 Including Twist

The paper wing is actually slightly warped with a linear twist $t(y) = -2t_x y/b$, where t_x is a small positive number that represents the tip twist.

Revisiting Prandtl Integro-differential equation and substituting in the left-hand-side the two Fourier modes gives

$$2Ub\left\{A_1 sin\theta + A_2 sin(2\theta)\right\} = \pi U c_0 sin\theta \left\{\alpha + \alpha_i + 2\frac{d}{c} + t_x cos\theta\right\}$$

where we have made use of $y(\theta) = -\frac{b}{2}\cos\theta$. From the previous result, the first mode balances the first three of the terms in the right-hand-side and there remains only

$$2UbA_2 sin(2\theta) = 4UbA_2 sin\theta cos\theta = \pi U c_0 sin\theta\, t_x cos\theta = \pi U t_x c_0 sin\theta cos\theta$$

resulting in

$$A_2 = \frac{\pi}{4}\frac{c_0}{b}t_x$$

15.8.2.4 Efficiency Factor

The warped wing loading is represented by the first two modes in the Fourier series. Given that $A_2 = A_1/10$, one can calculate the induced drag of the warped wing as

$$C_{Di} = \pi AR\sum_{n=1}^{\infty} nA_n^2 = \pi AR\left(A_1^2 + 2A_2^2\right) = \pi AR\, A_1^2\left(1 + 2\left(\frac{A_2}{A_1}\right)^2\right)$$

$$= (C_{Di})_{elliptic}\,(1+\delta)$$

The increase in induced drag is $\delta = 0.02 = 2\,\%$.

The efficiency factor is $e = 1/(1+\delta) = 0.98$.

15.8.3 Equilibrium of the AMAT2010

The AMAT10 has a rectangular main wing with span $b_m = 3.6$ m and constant chord $c_{xm} = 0.35$ m. The tail is also defined with a 33 % moving flap. The equilibrium code provides the aircraft characteristics and a maximum take-off mass $M = 24$ kg.

15.8.3.1 Airplane Aerodynamic Center and Static Margin

The equilibrium code calculates the lift and moment coefficients for the complete configuration at low incidences to be:

$$\begin{cases} C_L(\alpha, t_f) = 4.479\alpha + 0.808t_f + 0.9314 \\ C_{M,o}(\alpha, t_f) = -1.469\alpha - 0.7479t_f - 0.1565 \end{cases}$$

where α is the geometric incidence (in radians, measured from the fuselage axis) and t_f is the tail flap setting angle (in radians). $C_{M,o}$ is the aerodynamic moment about the origin of the coordinate system (located at the nose O). We will use this linear model for simplicity.

The center of gravity is located at $x_{cg}/l_{ref} = 0.268$.

The *aerodynamic center* has been defined above. Using the change of moment formula, it has been shown that

$$\frac{x_{a.c.}}{l_{ref}} = -\frac{dC_{M,o}}{d\alpha}/\frac{dC_L}{d\alpha} = \frac{1.469}{4.479} = 0.328$$

The *static margin SM* in % of l_{ref} is

$$\frac{x_{a.c.}}{l_{ref}} - \frac{x_{c.g.}}{l_{ref}} = 0.328 - 0.268 = 0.06$$

That represents a 6 % static margin.

15.8.3.2 Equilibrium Condition and Static Stability

The moment coefficient at the center of gravity, $C_{M,c.g.}(\alpha, t_f)$ is given by the equation

$$C_{M,c.g.}(\alpha) = C_{M,o}(\alpha) + \frac{x_{c.g.}}{l_{ref}} C_L(\alpha)$$

Substitution of the linear model yields

$$C_{M,c.g.}(\alpha) = -1.469\alpha - 0.7479t_f - 0.1565 + 0.268\left(4.479\alpha + 0.808t_f + 0.9314\right)$$

$$C_{M,c.g.}(\alpha) = -0.2686\alpha - 0.5314t_f + 0.0931$$

The equilibrium is stable since $dC_{M,c.g.}/d\alpha < 0$.

The condition for equilibrium is $C_{M,c.g.}(\alpha_{eq}) = 0$. Solving for α_{eq} gives

$$\alpha_{eq}(t_f) = -1.9784t_f + 0.3466$$

15.8.3.3 Take-Off Conditions

The take-off speed of $U = 11.49\,\text{m/s}$ is obtained for $t_f = 7.13° = 0.1244\,\text{rd}$. Hence, the incidence at take-off is $\alpha_{eq} = 0.1\,\text{rd} = 5.75°$.

The tail aerodynamic lift curve is given by

$$C_{Lt} = 3.032\alpha + 2.886t_f - 0.3259$$

The lift coefficient of the tail at take-off is $C_{Lt} = 0.336$. The force on the tail, given that the tail reference area is $A_t = 0.49\,\text{m}^2$, and $\rho = 1.2\,\text{kg/m}^3$ will be

$$L_t = \frac{1}{2}\rho U^2 A_t C_{Lt} = 13.0\,N$$

The force is up. The tail is lifting.

15.9 Solution to Problem 9

15.9.1 2-D Inviscid, Linearized, Thin Airfoil Theories

15.9.1.1 Incompressible Flow ($M_0 = 0$)

Aerodynamic Center

The aerodynamic center is the point about which the moment of the aerodynamic forces is independent of the incidence α.

The aerodynamic center is located at the quarter-chord for all thin airfoils at low speeds.

Second Mode only Airfoil

Consider a thin cambered plate such that the vorticity distribution is given by the second mode with $A_2 > 0$ as

$$\begin{cases} \Gamma'[x(t)] = 2U\left\{A_0\frac{1+\cos t}{\sin t} + A_2\cos 2t\right\} \\ x(t) = \frac{c}{2}(1-\cos t),\ 0 \le t \le \pi \end{cases}$$

As seen in class

$$d'[x(t)] = \alpha - A_0 + \sum_{n=1}^{\infty} A_n \cos nt = \alpha - A_0 + A_2 \cos 2t$$

Integrating in t yields (using $d(0) = d(c) = 0$)

$$\begin{aligned} d(c) &= \int_0^{\pi} (\alpha - A_0 + A_2\cos 2t)\,\frac{c}{2}\sin t\,dt \\ &= \frac{c}{2}\int_0^{\pi}\left(\alpha - A_0 + A_2(2\cos^2 t - 1)\right)\sin t\,dt \\ &= (\alpha - A_0 - A_2)\frac{c}{2}[-\cos t]_0^{\pi} + A_2 c\left[-\frac{\cos^3 t}{3}\right]_0^{\pi} \\ &= (\alpha - A_0 - A_2)c + 2\frac{A_2}{3}c = 0 \end{aligned}$$

This determines $A_0 = \alpha - A_2/3$. The incidence of adaptation is

$$\alpha_{adapt} = A_2/3$$

If one eliminates A_0 in the equation of the profile slope, one gets

$$d'[x(t)] = \frac{A_2}{3}(1+3\cos 2t) = \frac{A_2}{3}(6\cos^2 t - 2) = 4\frac{A_2}{3}\left(1 - 6\frac{x}{c} + 6\frac{x^2}{c^2}\right) = d'(x)$$

and upon integration the profile shape as

$$d(x) = 4\frac{A_2}{3}c\left(\frac{x}{c} - 3\frac{x^2}{c^2} + 2\frac{x^3}{c^3}\right) = 4\frac{A_2}{3}c\frac{x}{c}\left(1 - 2\frac{x}{c}\right)\left(1 - \frac{x}{c}\right)$$

which is the equation of a cubic. See Fig. 15.35.

Aerodynamic Coefficients

The lift coefficient is

$$C_l(\alpha) = 2\pi\left(A_0 + \frac{A_1}{2}\right) = 2\pi\left(\alpha - \frac{A_2}{3}\right)$$

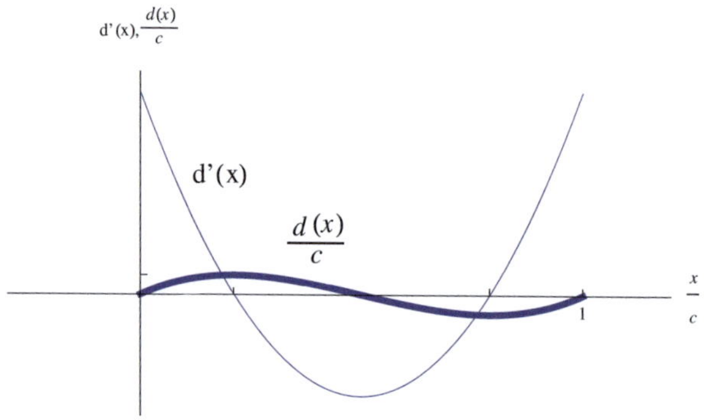

Fig. 15.35 $d'(x)$ and $d(x)/c$ for the cambered plate

The moment coefficient

$$C_{m,o}(\alpha) = -\frac{\pi}{2}\left(A_0 + A_1 - \frac{A_2}{2}\right) = -\frac{\pi}{2}\left(\alpha - 5\frac{A_2}{6}\right)$$

The moment coefficient, at the aerodynamic center, is

$$C_{m,a.c.} = -\frac{\pi}{4}(A_1 - A_2) = \frac{\pi}{4}A_2$$

The drag coefficient is $C_d = 0$.

Moment About an Axis

The aerodynamic moment about the mid-chord is calculated, using the change of moment formula

$$C_{m,c/2} = C_{m,o}(\alpha) + \frac{1}{2}C_l(\alpha) = \frac{\pi}{2}\left(\alpha + \frac{A_2}{6}\right)$$

If the profile is allowed to rotate without friction about an axis located at mid-chord, the equilibrium incidence, α_{eq}, corresponds to $C_{m,c/2} = 0$, i.e.

$$\alpha_{eq} = -\frac{A_2}{6}$$

No, the equilibrium is unstable because $dC_{m,c/2}/d\alpha = \pi/2 > 0$.

15.9.1.2 Supersonic Flow ($M_0 > 1,\ \beta = \sqrt{M_0^2 - 1}$)

Consider the cubic plate of equation

$$d(x) = \frac{4}{3}Ac\frac{x}{c}\left(1 - 2\frac{x}{c}\right)\left(1 - \frac{x}{c}\right),\ A > 0$$

The slope is given by

$$d'(x) = \frac{4}{3}A\left(1 - 6\frac{x}{c} + 6\frac{x^2}{c^2}\right)$$

This plate equips the fins of a supersonic rocket.

Pressure Distribution and Flow Features

At $\alpha = 0$ the $-C_p^+$ and $-C_p^-$ are

$$-C_p^+ = -\frac{2}{\beta}d'(x) = -\frac{8}{3\beta}A\left(1 - 6\frac{x}{c} + 6\frac{x^2}{c^2}\right)$$

$$-C_p^- = \frac{2}{\beta}d'(x) = \frac{8}{3\beta}A\left(1 - 6\frac{x}{c} + 6\frac{x^2}{c^2}\right)$$

See Fig. 15.36.

The flow features (shocks, characteristic lines, expansion shocks are shown in Fig. 15.37).

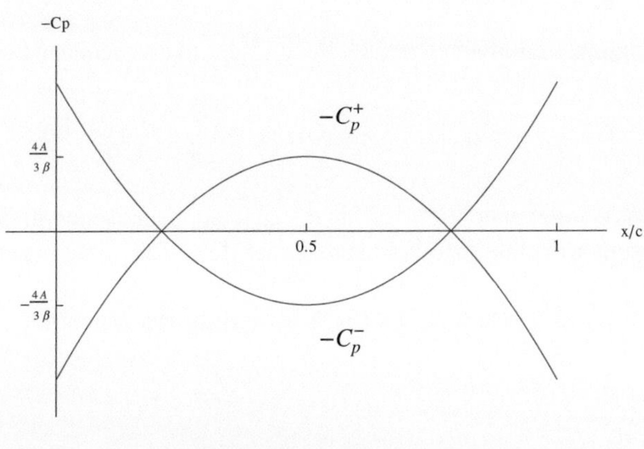

Fig. 15.36 Cubic plate pressure coefficients at $\alpha = 0$

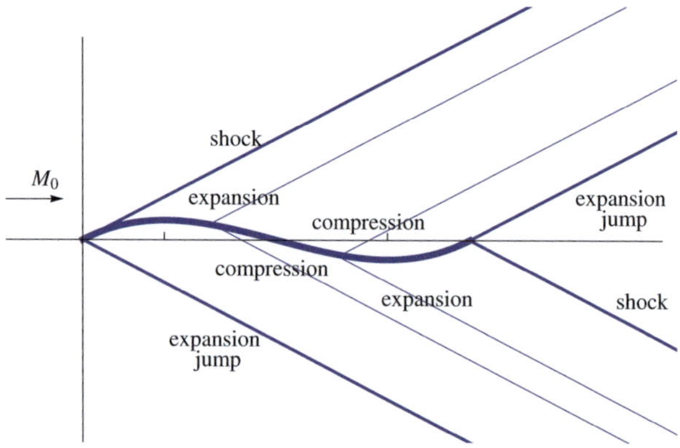

Fig. 15.37 Flow feature for the cubic plate in supersonic flow at $\alpha = 0$

Moment Coefficient

The moment coefficient $\left(C_{m,o}\right)_{\alpha=0}$

$$\left(C_{m,o}\right)_{\alpha=0} = \frac{4}{\beta}\int_0^c d'(x)\frac{x}{c}\frac{dx}{c} = \frac{16A}{3\beta}\int_0^1 \left(1 - 6\xi + 6\xi^2\right)\xi\, d\xi$$

$$= \frac{16A}{3\beta}\left[\frac{\xi^2}{2} - 2\xi^3 + 3\frac{\xi^4}{2}\right]_0^1 = 0$$

For arbitrary α the moment coefficient reads

$$C_{m,o}(\alpha) = \left(C_{m,o}\right)_{\alpha=0} - 2\frac{\alpha}{\beta} = -2\frac{\alpha}{\beta}$$

Using the change of moment formula, the mid-chord moment can be evaluated as

$$C_{m,c/2} = C_{m,o}(\alpha) + \frac{1}{2}C_l(\alpha) = -2\frac{\alpha}{\beta} + \frac{1}{2}4\frac{\alpha}{\beta} = 0$$

The moment is independent of α. This is expected, because the mid-chord is the aerodynamic center of thin airfoils in supersonic flow. The value is zero because it is the same as $\left(C_{m,o}\right)_{\alpha=0}$.

There is equilibrium at any value of α. The equilibrium is neutral.

15.9.2 Prandtl Lifting Line Theory

The WWII Spitfire was designed with an elliptic wing of span $b = 11.2$ m and wing area $S = 22.5\,\text{m}^2$. It is equipped with a NACA2209.4 profile of 2 % relative camber ($d/c = const = 0.02$). The top speed in cruise is $V = 170\,\text{m/s}$ (378 mph) with a take-off mass of $M = 3{,}000$ kg.

15.9.2.1 Vortex Sheet

The vortex sheet is a surface of discontinuity for the velocity, that is generated at the sharp trailing edge of a wing, where a Kutta-Joukowski condition holds. In small disturbance theory and for a large aspect ratio wing, one assumes that the vortex sheet is in the base surface generated by straight lines touching the lifting line and parallel to the incoming velocity vector. The rolling-up of the sheet edges is neglected. The u-component of velocity is continuous ($< u >= 0$) across the vortex sheet, since the pressure is continuous and $C_p = -2u/U$. The w-component is also continuous ($< w >= 0$) since the vortex sheet is a stream surface of zero thickness and the fluid is tangent to it (tangency condition imposes w). Only the v-component has a jump ($< v > \neq 0$) at the vortex sheet. At a point, this jump is interpreted as the jump due to a vortex filament passing through that point. The vortex sheet is made of vortex filaments that induce a down-wash (or normal wash), responsible for the induced drag. The fundamental effect of the vortex sheet is to produce an inviscid drag, balanced by a propulsive force, whose work adds irreversible kinetic energy to the flow field.

15.9.2.2 Lift Coefficient

By definition, the lift coefficient C_L is given by

$$C_L = \frac{W}{\frac{1}{2}\rho U^2 S} = \frac{3,000\,9.81}{0.5\,1.2\,170^2 22.5} = 0.075$$

15.9.2.3 Elliptic Loading

The aspect ratio of the wing is $AR = b^2/S = 5.58$. The induced drag C_{Di} is

$$C_{Di} = \frac{C_L^2}{\pi AR} = 0.00032$$

The first mode amplitude A_1 in the Fourier Series expansion is

$$A_1 = \frac{C_L}{\pi AR} = 0.00428$$

The equation for the lift coefficient C_L in terms of aspect ratio *AR*, geometric incidence α and relative camber d/c reads

$$C_L(\alpha) = \frac{2\pi}{1 + \frac{2}{AR}} \left(\alpha + 2\frac{d}{c}\right)$$

In cruise, the incidence α will satisfy the above equation: $\alpha = -0.0238$ rd $= -1.36°$

15.9.2.4 Added Twist

Adding the third mode only with amplitude $A_3 = -0.002$ increases the induced drag according to

$$C_{Di} = (C_{Di})_{elliptic} \left\{1 + \sum_{n=1}^{\infty} n \left(\frac{A_n}{A_1}\right)^2\right\} = (C_{Di})_{elliptic} \left\{1 + 3\left(\frac{A_3}{A_1}\right)^2\right\}$$

$$C_{Di} == (C_{Di})_{elliptic} \{1 + 0.655\} = 0.00053$$

The percent change in induced drag is 65.5 %.
The corresponding Efficiency factor $e = 1/(1 + 0.655) = 0.604$

15.9.3 Equilibrium of the AMAT11

The AMAT11 has a rectangular main wing with span $b_m = 2.1$ m and constant chord $c_{xm} = 0.3$ m. The tail is also rectangular with span $b_t = 1.0$ m and chord $c_{xt} = 0.3$ m. The equilibrium code provides the aircraft characteristics and a maximum take-off mass $M = 19$ kg. The reference area is $A_{ref} = A_m + A_t$.

15.9.3.1 Aspect Ratios of Lifting Elements—Global Lift Slope

The aspect ratio of the wing, AR_m and of the tail, AR_t are computed as

$$AR_m = \frac{b_m}{c_{xm}} = 7.0, \quad AR_t = \frac{b_t}{c_{xt}} = 3.33$$

Given that

$$\frac{dC_{Lm}}{d\alpha} = 4.675, \quad \frac{dC_{Lt}}{d\alpha} = 2.696$$

the global lift slope $dC_L/d\alpha$ for the wing+tail configuration will be

$$\frac{dC_L}{d\alpha} = \left(A_m \frac{dC_{Lm}}{d\alpha} + A_t \frac{dC_{Lt}}{d\alpha} \right) / (A_m + A_t) = 4.037$$

15.9.3.2 Airplane Center of Gravity

The aerodynamic center is located at $x_{a.c.}/l_{ref} = 0.3$.

The center of gravity $x_{c.g.}/l_{ref}$, given a 6% *static margin* (*SM*) will be such that

$$\frac{x_{c.g.}}{l_{ref}} = \frac{x_{a.c.}}{l_{ref}} - 0.06 = 0.24$$

15.9.3.3 Equilibrium Condition and Static Stability

The equilibrium code calculates the linear model for lift and moment coefficients for the complete configuration, at low incidences, to be:

$$\begin{cases} C_L(\alpha, t_f) = 3.955\alpha + 0.984 t_f + 0.712 \\ C_{M,o}(\alpha, t_f) = -1.188\alpha - 0.907 t_f - 0.008 \end{cases}$$

The moment coefficient at the center of gravity, $C_{M,c.g.}(\alpha, t_f)$ is

$$C_{M,c.g.}(\alpha, t_f) = C_{M,o}(\alpha, t_f) + \frac{x_{c.g.}}{l_{ref}} C_L(\alpha, t_f) = -0.239\alpha - 0.671 t_f + 0.163$$

The condition for equilibrium is that the moment at the center of gravity vanishes

$$C_{M,c.g.}(\alpha_{eq}, t_f) = 0$$

The equilibrium is stable because $dC_{M,c.g.}/d\alpha < 0$.

Solving for the equilibrium incidence gives

$$\alpha_{eq}(t_f) = -2.808 t_f + 0.682$$

15.9.3.4 Take-Off Conditions

The take-off speed of $U = 13.11$ m/s is obtained for $t_f = 9.2°$.

α_{eq} at take-off is $\alpha_{eq} = 0.231$ rd $= 13.2°$.

The lift coefficient of the tail at take-off , given that

$$C_{Lt} = 2.696\alpha + 2.705t_f - 0.347$$

is $C_{Lt}(\alpha_{eq}) = 0.711$.

The force on the tail in (N) is $L_t = 0.5\,\rho U^2 A_t C_{Lt} = 22\,N$.

The force is up.

15.10 Solution to Problem 10

15.10.1 2-D Inviscid, Linearized, Thin Airfoil Theories

15.10.1.1 Incompressible Flow ($M_0 = 0$)

Profile Camber Estimation

Thin parabolic plate has only two non-zero Fourier coefficients:

$$A_0 = \alpha, \quad A_1 = 4\frac{d_m}{c}$$

In thin airfoil theory, the lift coefficient is given by

$$C_l = 2\pi\left(A_0 + \frac{A_1}{2}\right)$$

Solving for A_1, given C_l and α reads

$$A_1 = 2\left(\frac{C_l}{2\pi} - \alpha\right) = 2\left(\frac{2.1878}{6.2832} - 0.06981\right) = 0.5568$$

The Fourier coefficients are

$$A_0 = \alpha, \; A_1 = 0.5568, \quad A_2 = A_3 = \cdots = A_n = 0, \; n \geq 2$$

The relative camber d_m/c of the thin parabolic plate is $A_1/4$

$$\frac{d_m}{c} = 0.14$$

a 14 % relative camber!

Take-Off Incidence

For a lift coefficient $C_l = 2.5$, the incidence needed is

$$\alpha = \frac{C_l}{2\pi} - 2\frac{d_m}{c} = 0.1195\,\text{rd} = 6.85^\circ$$

Nose Pitching Moment Coefficient

In thin airfoil theory, the pitching moment coefficient $C_{m,o}(\alpha)$ is given by

$$C_{m,o}(\alpha) = -\frac{\pi}{2}\left(A_0 + A_1 - \frac{A_2}{2}\right) = -\frac{\pi}{2}(\alpha + A_1)$$

The nose pitching moment coefficient of the wing profile $\left(C_{m,0}\right)_{profile}$ at $\alpha = 4^\circ$ is predicted to be

$$\left(C_{m,0}\right)_{profile} = -\frac{3.1416}{2}(0.06981 + 0.5568) = -0.984$$

not a bad approximation to the calculated result of -0.935.

Aerodynamic Center Pitching Moment Coefficient

Definition: the aerodynamic center is the point about which the moment of the aerodynamic forces is independent of incidence.

As seen in class, the pitching moment coefficient about the aerodynamic center $C_{m,a.c.}$ is

$$C_{m,a.c.} = -\frac{\pi}{4}(A_1 - A_2) = -\frac{\pi}{4}A_1 = -0.4373$$

15.10.1.2 Supersonic Flow ($M_0 > 1,\ \beta = \sqrt{M_0^2 - 1}$)

Consider the parabolic plate of equation

$$d(x) = 4d_m\frac{x}{c}\left(1 - \frac{x}{c}\right)$$

The slope is given by

$$d'(x) = 4\frac{d_m}{c}\left(1 - 2\frac{x}{c}\right)$$

where the relative camber is given to be $d_m/c = 0.14$.

Lift Coefficient

The lift coefficient $C_l(\alpha)$ for all thin airfoils is

$$C_l(\alpha) = 4\frac{\alpha}{\beta}$$

Drag Coefficient

The drag coefficient $(C_d)_{\alpha=0}$

$$(C_d)_{\alpha=0} = \frac{4}{\beta}\int_0^c \left(d'(x)\right)^2 \frac{dx}{c} = \frac{4}{\beta}\left(4\frac{d_m}{c}\right)^2 \int_0^1 (1-2\xi)^2\, d\xi$$

$$= \frac{64}{\beta}\left(\frac{d_m}{c}\right)^2 \left[\xi - 2\xi^2 + \frac{4}{3}\xi^3\right]_0^1 = \frac{64}{3\beta}\left(\frac{d_m}{c}\right)^2$$

The expression of $C_d(\alpha)$ for this airfoil is

$$C_d(\alpha) = \frac{64}{3\beta}\left(\frac{d_m}{c}\right)^2 + 4\frac{\alpha^2}{\beta}$$

Pitching Moment Coefficient

The moment coefficient $\left(C_{m,o}\right)_{\alpha=0}$

$$\left(C_{m,o}\right)_{\alpha=0} = \frac{4}{\beta}\int_0^c d'(x)\frac{x}{c}\frac{dx}{c} = \frac{4}{\beta}4\frac{d_m}{c}\int_0^{c1} (1-2\xi)\,\xi d\xi$$

$$= \frac{16}{\beta}\frac{d_m}{c}\left[\frac{\xi^2}{2} - \frac{2}{3}\xi^3\right]_0^1 = -\frac{8}{3\beta}\frac{d_m}{c}$$

The expression of $C_{m,o}(\alpha)$ for this airfoil is

$$C_{m,o}(\alpha) = -\frac{8}{3\beta}\frac{d_m}{c} - 2\frac{\alpha}{\beta}$$

Equilibrium About an Axis

The moment at the quarter-chord is

$$C_{m,\frac{c}{4}}(\alpha) = C_{m,o}(\alpha) + \frac{1}{4}C_l(\alpha) = -\frac{8}{3\beta}\frac{d_m}{c} - 2\frac{\alpha}{\beta} + \frac{\alpha}{\beta} = -\frac{8}{3\beta}\frac{d_m}{c} - \frac{\alpha}{\beta}$$

The equilibrium incidence corresponds to $C_{m,\frac{c}{4}}(\alpha) = 0$, which gives $\alpha_{eq} = -0.3733\,\text{rd} = -21.4°$.

The equilibrium is stable, since the moment slope is negative.

15.10.2 Prandtl Lifting Line Theory

15.10.2.1 Lifting Line Theory

The downwash results from the presence of a vortex sheet, downstream of the sharp trailing edge of a finite wing. The sheet is made of line vortices which induce a negative vertical velocity component when the lift is positive, as required from a balance of momentum.

Downwash is associated with induced drag, whereas upwash will produce thrust (negative drag).

There is no net drag benefit to design a wing with upwash. The wing with minimum induced drag for a given lift has elliptic circulation and constant downwash. Any other induced velocity distribution, including with upwash, will produce more induced drag.

15.10.2.2 Design of a Wing

The circulation distribution is represented by the Fourier series

$$\begin{cases} \Gamma[y(t)] = 2Ub\sum_{n=1}^{\infty} A_n \sin nt \\ y(t) = -\frac{b}{2}\cos t,\ 0 \leq t \leq \pi \end{cases}$$

Combining modes 1 and 3 and taking the derivative w.r.t. t yields

$$\frac{d\Gamma[y(t)]}{dt} = 2Ub\,(A_1 \cos t + 3A_3 \cos 3t) = 0, \quad at\ t = 0\ and\ \pi$$

requiring

$$A_1 + 3A_3 = 0 \quad \Rightarrow \quad A_3 = -\frac{A_1}{3}$$

Using the identity: $\cos 3t = \cos t\left(4\cos^2 t - 3\right)$ one finds

$$\frac{d\Gamma[y(t)]}{dt} = 8UbA_1 \cos t\left(1 - \cos^2 t\right) = 8UbA_1 \sin^2 t \cos t$$

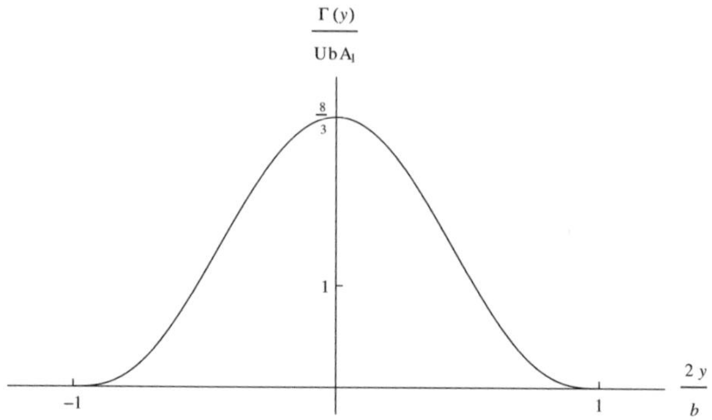

Fig. 15.38 Circulation distribution for zero tip vortex strength

and the derivative w.r.t. y reads

$$\frac{d\Gamma(y)}{dy} = \frac{d\Gamma[y(t)]}{dt} / \frac{dy(t)}{dt} = 16UA_1 \sin t \cos t = 8UA_1 \sin 2t = 0, \quad at\ t = 0\ and\ \pi$$

The requirement is satisfied for zero strength tip vortices.
With this choice, the circulation distribution becomes

$$\Gamma\{y(t)\} = 2UbA_1 \left(\sin t - \frac{1}{3} \sin 3t \right) = \frac{8}{3} UbA_1 \sin^3 t = \frac{8}{3} UbA_1 \left[1 - \left(\frac{2y}{b} \right)^2 \right]^{\frac{3}{2}}$$

where we used the identity: $\sin 3t = \sin t (4 \cos^2 t - 1)$. The circulation distribution $\Gamma[y(t)]$ is shown in Fig. 15.38.

15.10.2.3 Induced Velocity

From the result derived in class

$$w_w[y(t)] = -U \sum_{n=1}^{\infty} n A_n \frac{\sin nt}{\sin t}$$

the distribution of induced velocity reads

$$w_w[y(t)] = -UA_1 \left(1 - \frac{\sin 3t}{\sin t} \right) = -2UA_1 \left(1 - 2\cos^2 t \right)$$

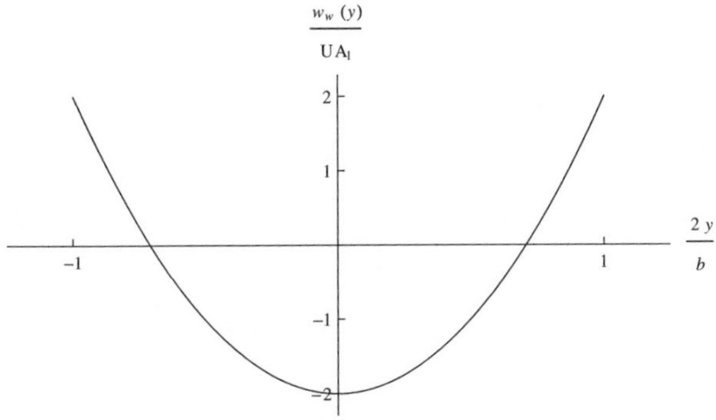

Fig. 15.39 Induced velocity distribution for zero tip vortex strength

$$= -2UA_1\left(1 - 2\left(\frac{2y}{b}\right)^2\right)$$

The parabolic distribution of $w_w[y(t)]$ is shown in Fig. 15.39.
The wing experiences upwash for $|2y/b| \geq 1/\sqrt{2}$.

15.10.2.4 Induced Drag

The induced drag of the wing without tip vortices is given by

$$C_{Di} = (C_{Di})_{elliptic}\left(1 + \sum_{n=2}^{\infty} n\left(\frac{A_n}{A_1}\right)^2\right) = (C_{Di})_{elliptic}\left(1 + 3\left(\frac{1}{3}\right)^2\right)$$

$$= (C_{Di})_{elliptic}\left(1 + \frac{1}{3}\right)$$

The percent change in induced drag is 33 % increase compared to the elliptic loading. The corresponding efficiency factor is $e = 0.75$.

15.10.3 Equilibrium of the Glider

The AMAT12 has a rectangular main wing with span $b_m = 2.1$ m and constant chord $c_{xm} = 0.3$ m. The tail is also rectangular with span $b_t = 1.0$ m and chord $c_{xt} = 0.338$ m. The equilibrium code provides the aircraft aerodynamic characteristics and a maximum take-off mass $M = 18$ kg. The reference area is $A_{ref} = A_m + A_t$. The reference length is $l_{ref} = 2.12$ m. The engine is turned off. The global aerodynamic lift and nose pitching moment coefficients are found to be

$$\begin{cases} C_L(\alpha, t_t) = 3.853\alpha + 0.902t_t + 0.716 \\ C_{M,o}(\alpha, t_t) = -1.336\alpha - 0.894t_t - 0.06 \end{cases}$$

where α represents the geometric incidence and t_t the tail setting angle in rd.

15.10.3.1 Equilibrium About the Center of Gravity

The pitching moment about the center of gravity $C_{M,c.g.}(\alpha, t_t)$ is given by

$$C_{M,c.g.}(\alpha, t_t) = C_{M,o}(\alpha, t_t) + \frac{x_{c.g.}}{l_{ref}} C_L(\alpha, t_t)$$

$$C_{M,c.g.}(\alpha, t_t) = -1.336\alpha - 0.894t_t - 0.06 + 0.2868\,(3.853\alpha + 0.902t_t + 0.716)$$

$$C_{M,c.g.}(\alpha, t_t) = -0.2367\alpha - 0.6367t_t + 0.1442$$

The airplane is statically stable since the slope $\partial C_{m,c.g.}/\partial\alpha$ of the moment curve $C_{M,c.g.}(\alpha, t_t)$ is negative.

The aerodynamic center of the glider is at $x_{a.c.} = 0.735$ m. The static margin SM in %

$$SM = 100\left(\frac{x_{a.c.}}{l_{ref}} - \frac{x_{c.g.}}{l_{ref}}\right) = 100\,(0.3467 - 0.2868) = 6\,\%$$

15.10.3.2 Equilibrium Incidence

The equilibrium condition is given by $C_{M,c.g.}(\alpha_{eq}, t_t) = 0$ and can be solved for $\alpha_{eq}(t_t)$

$$\alpha_{eq}(t_t) = -2.69t_t + 0.609$$

15.10.3.3 Trimming for Maximum Distance

The equilibrium code calculates the tail setting angle for the maximum $C_L/C_D = 9.3$ to be $t_t = 12° = 0.2094$ rd.

Substitution of the value of t_t in the equilibrium equation gives $\alpha_{eq} = -2.69$ $0.2094 + 0.609 = 0.0457$ rd $= 2.6°$.

The corresponding values of C_L and C_D, as well as the descent angle β are: $C_L = 1.08$, $C_D = 0.116$ and $\beta = -C_D/C_L = -0.107$ rd $= -6.15°$.

The lift forces (in N) on the main wing and the tail, given that the velocity is $U = 17.3$ m/s and the air density $\rho = 1.1214$ kg/m^3 are given by

$$L_m = \frac{1}{2}\rho U^2 A_m C_{Lm}, \quad L_t = \frac{1}{2}\rho U^2 A_t C_{Lt}$$

The wing and tail lift curves are given respectively by

$$\begin{cases} C_{Lm}(\alpha) = 4.664\alpha + 1.312 = 1.525 \\ C_{Lt}(\alpha, t_t) = 2.342\alpha + 2.583t_t - 0.396 = 0.25 \end{cases}$$

The results are: $L_m = 161.2\,N$, $L_t = 14.18\,N$.

Appendix A
Special Techniques

A.1 Methods of Complex Variables

Solutions of two-dimensional, incompressible, potential flow problems can be obtained using functions of complex variables. In the following we follow Moran [1] and others [2].

A complex number z consists of two real numbers x and y, where

$$z = x + iy, \; \bar{z} = x - iy, \quad i = \sqrt{-1} \tag{A.1}$$

or

$$z = re^{i\theta}, \; \bar{z} = re^{-i\theta}, \quad r = \sqrt{x^2 + y^2}, \; e^{i\theta} = \cos\theta + i\sin\theta \tag{A.2}$$

Notice, the inverse of a complex number is a complex number since

$$\frac{1}{z} = \frac{1}{x + iy}\frac{x - iy}{x - iy} = \frac{x}{x^2 + y^2} - i\frac{y}{x^2 + y^2} \tag{A.3}$$

A function w of complex variables

$$w(z) = w(x + iy) = \phi(x, y) + i\psi(x, y) \tag{A.4}$$

is called the complex potential.

The derivative of a function of a complex variable is defined as

$$\frac{dw}{dz} = \lim_{\Delta z \to 0} \frac{w(z + \Delta z) - w(z)}{\Delta z} \tag{A.5}$$

For the derivative of w to exist, it is necessary that the limit of the difference quotient in z be the same no matter how Δz approaches zero.

If $\Delta z = \Delta x$, then

$$\frac{dw}{dz} = \frac{\partial\phi}{\partial x} + i\frac{\partial\psi}{\partial x} \tag{A.6}$$

J.J. Chattot and M.M. Hafez, *Theoretical and Applied Aerodynamics*,
DOI 10.1007/978-94-017-9825-9

On the other hand, if $\Delta z = i\Delta y$, then

$$\frac{dw}{dz} = -i\frac{\partial \phi}{\partial y} + \frac{\partial \psi}{\partial y} \tag{A.7}$$

Hence,

$$\frac{\partial \phi}{\partial x} + i\frac{\partial \psi}{\partial x} = -i\frac{\partial \phi}{\partial y} + \frac{\partial \psi}{\partial y} \tag{A.8}$$

or

$$\frac{\partial \phi}{\partial x} = \frac{\partial \psi}{\partial y}, \quad \frac{\partial \phi}{\partial y} = -\frac{\partial \psi}{\partial x} \tag{A.9}$$

The two conditions of Eq. (A.9) are known as the Cauchy-Riemann equations. In general, since $x = (z + \bar{z})/2$ and $y = (z - \bar{z})/2i$, $w(x, y) = w(z, \bar{z})$. In fact

$$\frac{\partial w}{\partial \bar{z}} = \frac{1}{2}\left(\frac{\partial \phi}{\partial x} - \frac{\partial \psi}{\partial y}\right) + \frac{i}{2}\left(\frac{\partial \phi}{\partial y} + \frac{\partial \psi}{\partial x}\right) \tag{A.10}$$

Therefore, for analytic functions, $\partial w/\partial \bar{z} = 0$, i.e. w is a function of z only.

If $w(z)$ possesses a derivative at $z = z_0$ and at every point in some neighborhood of z_0, then $w(z)$ is said to be analytic.

Let

$$u = \frac{\partial \phi}{\partial x} = \frac{\partial \psi}{\partial y}, \quad v = \frac{\partial \phi}{\partial y} = -\frac{\partial \psi}{\partial x} \tag{A.11}$$

then

$$\frac{dw}{dz} = u - iv \tag{A.12}$$

is the complex velocity.

It will be shown later that the derivative of an analytic function is also analytic function and hence an analytic function possesses derivatives of all orders and can be represented by convergent Taylor series at any point in the region where the function is analytic (the radius of convergence is determined by the closest point where the function fails to be analytic, i.e. the closest singular point). It follows then for any analytic function w

$$\frac{\partial^2 w}{\partial \bar{z} \partial z} = 0 \tag{A.13}$$

However,

$$\frac{\partial^2 w}{\partial z \partial \bar{z}} = \frac{1}{4}\left(\frac{\partial^2 \phi}{\partial x^2} + \frac{\partial^2 \phi}{\partial y^2}\right) + \frac{i}{4}\left(\frac{\partial^2 \psi}{\partial x^2} + \frac{\partial^2 \psi}{\partial y^2}\right) = 0 \tag{A.14}$$

hence, ϕ and ψ satisfy Laplace equation.

Since, dw/dz is also analytic function, then Eq. (A.12) implies

$$\frac{\partial u}{\partial x} = -\frac{\partial v}{\partial y}, \quad \frac{\partial u}{\partial y} = \frac{\partial v}{\partial x} \tag{A.15}$$

The first of Eq. (A.15) is the continuity equation and the second is the irrotationality condition. Hence ϕ and ψ are the potential and stream functions of two-dimensional incompressible, irrotational flows and $w(z) = \phi(x, y) + i\psi(x, y)$ is called complex potential. Both the real and the imaginary parts of the above analytic function have continuous second partial derivatives satisfying Laplace's equation

$$\frac{\partial^2 \phi}{\partial x^2} + \frac{\partial^2 \phi}{\partial y^2} = 0, \quad \frac{\partial^2 \psi}{\partial x^2} + \frac{\partial^2 \psi}{\partial y^2} = 0 \tag{A.16}$$

Moreover, the curves of the family $\phi(x, y) = constant$ are the orthogonal trajectories of the curves of the family $\psi(x, y) = constant$, (and vice versa if ϕ and ψ satisfy Laplace equation, then $w(z)$ is an analytic function).

Examples of complex potential are

(i)

$$w = Ue^{-i\alpha}z = U(x \cos \alpha + y \sin \alpha) + iU(y \cos \alpha - x \sin \alpha) \tag{A.17}$$

The above expression represents the potential and stream functions of a uniform flow making an angle α with the x-axis.

(ii)

$$w = \frac{1}{2\pi}(Q + i\Gamma) \ln z \tag{A.18}$$

This w represents the complex potential of a source of strength Q and a potential vortex of strength Γ at the origin.

(iii)

$$w = z + \frac{a^2}{z} = x\left(1 + \frac{a^2}{r^2}\right) + iy\left(1 - \frac{a^2}{r^2}\right) \tag{A.19}$$

This w represents potential flow and stream functions of flow at unit speed around a cylinder of radius a.

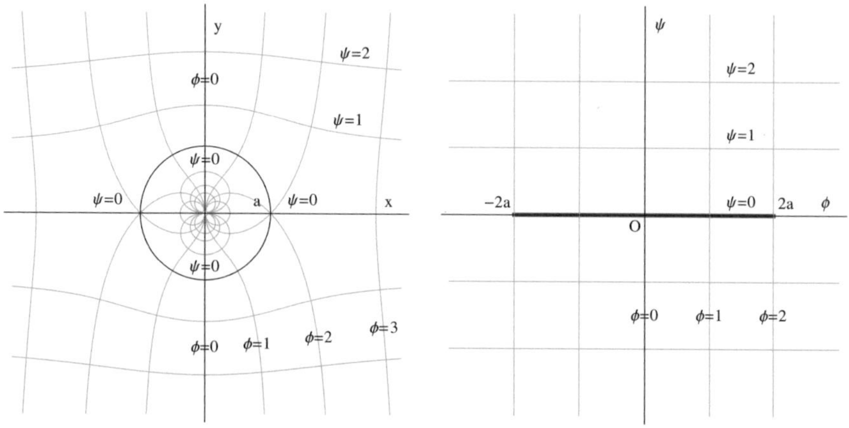

Fig. A.1 Z-plane (*left*); w-plane (*right*)

In the following, the applications of functions of complex variables in aerodynamics will be discussed, in particular, conformal mapping and Blasius theorems.

Conformal mapping

The function $W(Z) = \Phi(X, Y) + i\Psi(X, Y)$ defines a mapping or a transformation of the Z-plane onto the w-plane or z-plane with $x = \Phi$ and $y = \Psi$. Figure A.1.

To study when the transformation has a single-valued inverse, let

$$dx = \frac{\partial \Phi}{\partial X}dX + \frac{\partial \Phi}{\partial Y}dY, \quad dy = \frac{\partial \Psi}{\partial X}dX + \frac{\partial \Psi}{\partial Y}dY \tag{A.20}$$

The Jacobian of the transformation is

$$J = \begin{vmatrix} \frac{\partial \Phi}{\partial X} & \frac{\partial \Phi}{\partial Y} \\ \frac{\partial \Psi}{\partial X} & \frac{\partial \Psi}{\partial Y} \end{vmatrix} = \frac{\partial(x, y)}{\partial(X, Y)} \tag{A.21}$$

Using the Cauchy-Riemann relations, J becomes

$$J = \begin{vmatrix} \frac{\partial \Phi}{\partial X} & -\frac{\partial \Psi}{\partial X} \\ \frac{\partial \Psi}{\partial X} & \frac{\partial \Phi}{\partial X} \end{vmatrix} = \begin{vmatrix} \frac{\partial \Psi}{\partial Y} & \frac{\partial \Phi}{\partial Y} \\ -\frac{\partial \Phi}{\partial Y} & \frac{\partial \Psi}{\partial Y} \end{vmatrix}$$

$$= \left(\frac{\partial \Phi}{\partial X}\right)^2 + \left(\frac{\partial \Psi}{\partial X}\right)^2 = \left(\frac{\partial \Phi}{\partial Y}\right)^2 + \left(\frac{\partial \Psi}{\partial Y}\right)^2 = \left|\frac{dW}{dZ}\right|^2 = \left|\frac{dz}{dZ}\right|^2 \tag{A.22}$$

Hence, if $W(Z)$ is analytic, $W = W(Z)$ will have a single-valued inverse in the neighborhood of any point where the derivative of the mapping function is different from zero.

Now let $\Delta Z = |\Delta Z|\, e^{i\theta}$ and $\Delta W = |\Delta W|\, e^{i\omega}$, hence

$$\frac{dW}{dZ} = \lim_{\Delta Z \to 0} \frac{\Delta W}{\Delta Z} = \lim_{\Delta Z \to 0} \left(\left| \frac{\Delta W}{\Delta Z} \right| e^{i(\theta - \omega)} \right) \tag{A.23}$$

Therefore

$$|\Delta W| = \left| \frac{dW}{dZ} \right| |\Delta Z|, \quad \arg \Delta W = \arg \left(\frac{dW}{dZ} \right) + \arg \Delta Z \tag{A.24}$$

In analytic mapping, the length of infinitesimal segments are altered by a factor $\left|\frac{dW}{dZ}\right|$ which depends only on the point from which the segments are drawn (regardless of the direction). The infinitesimal areas are then magnified by the factor $\left|\frac{dW}{dZ}\right|^2$, i.e. by J.

Also, two infinitesimal segments forming an angle will both be rotated in the same direction by the same amount. A transformation which preserves the sense and the magnitudes of angles is said to be conformal. Hence, any analytic function $W = W(Z)$ defines conformal transformation (as long as $\frac{dW}{dZ} \neq 0$).

An important theorem of conformal mappings is the following: Laplace equation is transformed to a Laplace equation, for example, consider the one-to-one transformation $z = z(Z)$, its inverse $Z = Z(z)$ and the complex potential $W(Z) = \Phi(X, Y) + i\Psi(X, Y) = w(z) = \phi(x, y) + i\psi(x, y)$
hence,

$$\frac{\partial^2 \Phi}{\partial X^2} + \frac{\partial^2 \Phi}{\partial Y^2} = 0, \quad \frac{\partial^2 \Psi}{\partial X^2} + \frac{\partial^2 \Psi}{\partial Y^2} = 0 \tag{A.25}$$

in the Z-plane, are transformed to

$$\frac{\partial^2 \phi}{\partial x^2} + \frac{\partial^2 \phi}{\partial y^2} = 0, \quad \frac{\partial^2 \psi}{\partial x^2} + \frac{\partial^2 \psi}{\partial y^2} = 0 \tag{A.26}$$

in the z-plane (where $\frac{dz}{dZ} \neq 0$). Moreover,

$$\phi(x, y) = \Phi(X, Y), \quad \psi(x, y) = \Psi(X, Y), \quad w(z) = W(Z) \tag{A.27}$$

The proof is based on the fact that an analytic function of an analytic function is analytic, see [3].

It can be also shown that the kinetic energy of the motion remains constant under conformal transformation, see [2].

Let the derivatives of the complex potentials with respect to z and Z be dw/dz and dW/dZ.

Since $W = w$,

$$\frac{dW}{dZ} = \frac{dw}{dz}\frac{dz}{dZ}, \quad \frac{dw}{dz} = \frac{dW}{dZ}\frac{dZ}{dz} \tag{A.28}$$

The complex velocity in the Z-plane is defined as

$$\frac{dW}{dZ} = U - iV \tag{A.29}$$

Hence, using the chain rule

$$\frac{dW}{dZ} = u - iv \tag{A.30}$$

The equality of the moduli yields

$$\left|\frac{dw}{dz}\right| = \left|\frac{dW}{dZ}\right|\left|\frac{dZ}{dz}\right|, \quad \Rightarrow \quad |u - iv| = |U - iV|\left|\frac{dZ}{dz}\right| \tag{A.31}$$

hence the velocity magnitude can be found in the physical plane. Since the ratio of two incremental areas in Z- and z-planes is equal to the Jacobian, hence the integral over corresponding domains of the kinetic energy for constant density flows is preserved. Using Bernoulli's law, this relation can be used to find the pressure distribution anywhere in the physical plane.

The simplest example of conformal transformation is mapping the circle to a slit, given by

$$z = Z + \frac{a^2}{Z}, \quad x = X\left(1 + \frac{a^2}{X^2 + Y^2}\right), \quad y = Y\left(1 - \frac{a^2}{X^2 + Y^2}\right) \tag{A.32}$$

The general Joukowski transformation is given by

$$z = Z + \frac{a^2}{Z}, \quad r_0^2 = (a + \epsilon)^2 + \delta^2, \quad \epsilon \geq 0 \tag{A.33}$$

The cylinder of radius r_0, centered at $(-\epsilon, \delta)$ in the Z-plane, passing by the point B $(a, 0)$, maps onto a Joukowski profile, with cusped trailing edge at point b, Fig. A.2. Special cases correspond to $\epsilon = 0$ and/or $\delta = 0$. In the Z-plane, the flow at angle α with the x-axis, about the circular cylinder, is given by the complex potential

$$W = U\left((Z + \epsilon - i\delta)e^{-i\alpha} + \frac{r_0^2}{(Z + \epsilon - i\delta)e^{-i\alpha}}\right) + i\frac{\Gamma}{2\pi}\ln\left(\frac{(Z + \epsilon - i\delta)e^{-i\alpha}}{a}\right) \tag{A.34}$$

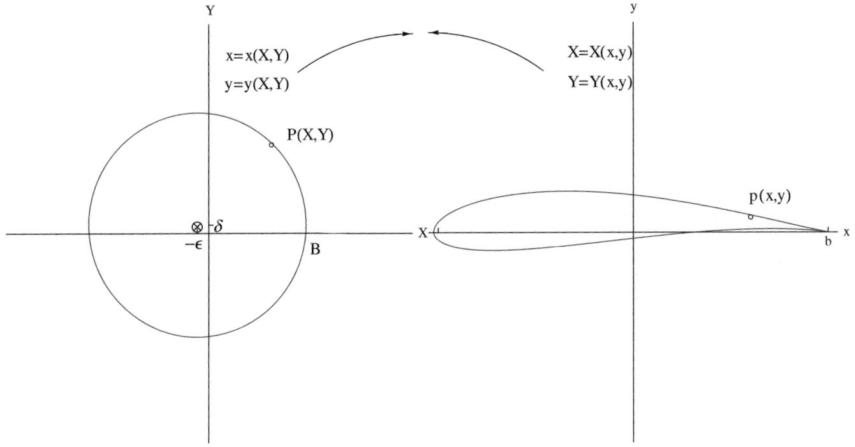

Fig. A.2 Z-plane (*left*); z-plane (*right*)

The complex velocity in the physical plane is given by

$$u - iv = \frac{dW}{dZ}\frac{dZ}{dz} = \left[U\left(e^{-i\alpha} - \frac{r_0^2 e^{i\alpha}}{(Z+\epsilon-i\delta)^2}\right) + i\frac{\Gamma}{2\pi}\frac{1}{Z+\epsilon-i\delta}\right] / \left[1 - \frac{a^2}{Z^2}\right] \tag{A.35}$$

There is a singularity at $Z = \pm a$. The singularity at $Z = a$ on the cylinder, is needed to have a cusped airfoil, while the singularity at $Z = -a$ inside the cylinder, is excluded from the domain of interest.

The Kutta-Joukowski condition requires that velocity to be finite at the trailing edge, hence the numerator should vanish at $Z = a$, meaning that the point is a stagnation point for the flow past the circular cylinder. The condition reads

$$\Gamma = 4\pi U\left((a+\epsilon)\sin\alpha + \delta\cos\alpha\right) \tag{A.36}$$

Note that the condition obtained is the same for the real and imaginary parts of the complex equation $dW(a)/dZ = 0$ because the flow is constrained to be tangent to the cylinder. After some algebra one finds

$$u - iv = U\frac{(Z-a)e^{-i\alpha} + 2\left\{(a+\epsilon)\cos\alpha - \delta\sin\alpha\right\}}{(Z+a)}\frac{Z^2}{(Z+\epsilon-i\delta)^2} \tag{A.37}$$

There are two important special cases:
(i) $\epsilon = \delta = 0$, for a flat plate at angle of attack. In this case, $r_0 = a$, $z = ae^{i\theta}$ and

$$\Gamma = 4\pi U a\sin\alpha \tag{A.38}$$

(ii) $\epsilon = 0$, $\delta \neq 0$ (say $\delta > 0$), for a circular arc at angle of attack. In this case, $r_0 = \sqrt{a^2 + \delta^2}$, and

$$\Gamma = 4\pi U \left(a \sin \alpha + \delta \cos \alpha\right) \tag{A.39}$$

Cauchy Integral Theorem

The Cauchy integral theorem states that the line integral of analytic function in a simply connected region S around a closed curve C in this region, is zero, i.e.

$$\oint_C w(z)dz = 0 \tag{A.40}$$

To prove the theorem, let $w(z) = \phi + i\psi$ and $z = x + iy$, hence

$$w(z)dz = (\phi + i\psi)(dx + idy) \tag{A.41}$$

The integral can be expanded as

$$\oint_C w(z)dz = \oint_C (\phi dx - \psi dy) + i \oint_C (\psi dx + \phi dy) \tag{A.42}$$

Using Green's theorem with domain D inside curve C

$$\oint_C (\phi dx - \psi dy) = -\int\int_D \left(\frac{\partial \phi}{\partial y} + \frac{\partial \psi}{\partial x}\right) dxdy = 0 \tag{A.43}$$

$$\oint_C (\psi dx + \phi dy) = -\int\int_D \left(\frac{\partial \psi}{\partial y} - \frac{\partial \phi}{\partial x}\right) dxdy = 0 \tag{A.44}$$

An extension of Cauchy theorem for multiply connected domain where C_1 and C_2 are closed curves in the region that can be continuously deformed into each other without leaving the region, is given by

$$\oint_{C1} w(z)dz - \oint_{C2} w(z)dz = 0 \tag{A.45}$$

See Fig. A.3.

Based on Cauchy theorems, one can show that

$$w(z_0) = \frac{1}{2\pi i} \oint \frac{w(z)}{z - z_0} dz \tag{A.46}$$

where z_0 is any point in the interior of the simply connected region R whose boundary is sectionally smooth. Moreover, the derivative of $w(z)$ at point z_0 is given by

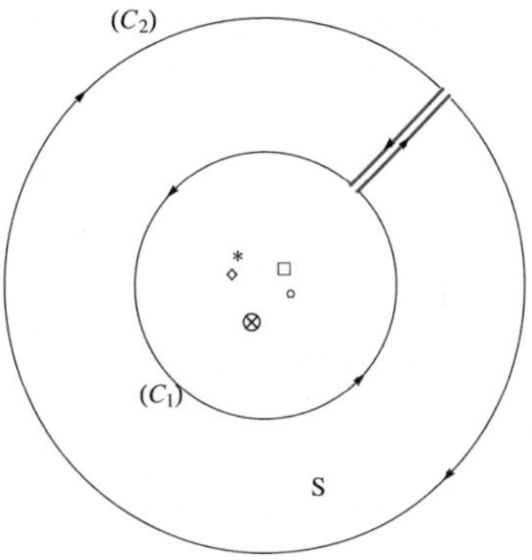

Fig. A.3 Contour for multiply connected region

$$w'(z_0) = \frac{1}{2\pi i} \oint \frac{w(z)}{(z - z_0)^2} dz \tag{A.47}$$

Higher derivatives can be obtained formally by repeated differentiation with respect to z_0. Finally, the converse of Cauchy theorem is true, namely if $w(z)$ is a continuous function in a region R and if $\oint w(z)dz = 0$ for every simple closed curve C which can be drawn in R, then $w(z)$ is analytic in R, see Ref. [4].

Blasius Theorems

The pressure force acting on an element of surface is pds. Assuming unit width, the components in x and y directions are

$$dF_x = -pdy, \quad dF_y = pdx \tag{A.48}$$

Hence, if one defines $dF = dF_x - idF_y$, then

$$dF = -ip(dx - idy) \tag{A.49}$$

According to Bernoulli formula, $p = p_\infty + \rho U^2/2 - \rho V^2/2$ and $V^2 = (u - iv)(u + iv)$. On a closed contour, the constant pressure term $p_\infty + \rho U^2/2$ does not contribute to the force. Therefore, the elementary force due to pressure reduces to

$$dF = i\frac{\rho}{2}(u - iv)(u + iv)(dx - idy) \tag{A.50}$$

But, the complex velocity is related to the complex potential $w(z)$ by

$$\frac{dw}{dz} = u - iv \tag{A.51}$$

Hence

$$dw = d\phi + id\psi = (u - iv)(dx + idy) \tag{A.52}$$

The body contour is a streamline along which dw is real since $d\psi = 0$. Consequently, dw is equal to its complex conjugate and can be written as

$$dw = (u + iv)(dx - idy) \tag{A.53}$$

Hence

$$dF = dF_x - idF_y = i\frac{\rho}{2}\frac{dw}{dz}dw = i\frac{\rho}{2}\left(\frac{dw}{dz}\right)^2 dz \tag{A.54}$$

The elementary moment about the origin can also be found from

$$dM, o = -\Re\{izdF\} \tag{A.55}$$

with the moment counted positive clockwise.

Upon integration around the obstacle one obtains

$$F = i\frac{\rho}{2}\oint\left(\frac{dw}{dz}\right)^2 dz \tag{A.56}$$

$$M_{,o} = \frac{\rho}{2}\Re\left\{\oint z\left(\frac{dw}{dz}\right)^2 dz\right\} \tag{A.57}$$

These integrals of analytic functions along closed contour can be evaluated with the theorem of residues.

Far from the obstacle, the velocity perturbation dies out as the flow returns to uniform flow condition. The complex potential can be expanded in a series of inverse powers of z, as

$$\left(\frac{dw}{dz}\right)^2 = A_0 + \frac{A_1}{z} + \frac{A_2}{z^2} + ... + \frac{A_n}{z^n} + ... \tag{A.58}$$

Using as integration contour a large circle of radius R, centered at the origin, one obtains

$$F = i\frac{\rho}{2}\oint\left(A_0 + \frac{A_1}{z} + \frac{A_2}{z^2} + ... + \frac{A_n}{z^n} + ...\right)dz \tag{A.59}$$

Since $z = Re^{i\theta}$, then $dz = iRe^{i\theta}d\theta$, this reads

$$F = i\frac{\rho}{2}\int_0^{2\pi}\left(iA_0Re^{i\theta} + iA_1 + i\frac{A_2}{R}e^{-i\theta} + ... + i\frac{A_n}{R^{n-1}}e^{-i(n-1)\theta} + ...\right)d\theta \tag{A.60}$$

Integrating term by term one obtains

$$F = i\frac{\rho}{2}\left[A_0 Re^{i\theta} + iA_1\theta - \frac{A_2}{R}e^{-i\theta} + \ldots - \frac{A_n}{(n-1)R^{n-1}}e^{-i(n-1)\theta} + \ldots\right]_0^{2\pi} = -\pi\rho A_1 \tag{A.61}$$

The integral of moment expands as

$$M_{,o} = \frac{\rho}{2}\Re\left\{\oint\left(A_0 z + A_1 + \frac{A_2}{z} + \ldots + \frac{A_n}{z^{n-1}} + \ldots\right)dz\right\}$$

$$= \frac{\rho}{2}\Re\left\{\int_0^{2\pi}\left(iA_0R^2e^{i2\theta} + iA_1Re^{i\theta} + iA_2 + \ldots + i\frac{A_n}{R^{n-2}}e^{i(n-2)\theta} + \ldots\right)d\theta\right\} \tag{A.62}$$

resulting in

$$M_{,o} = \frac{\rho}{2}\Re\left\{\left[\frac{A_0}{2}R^2e^{i2\theta} + A_1Re^{i\theta} + iA_2\theta + \ldots + \frac{A_n}{(n-2)R^{n-2}}e^{i(n-2)\theta} + \ldots\right]_0^{2\pi}\right\}$$

$$= \rho\Re\left\{i\pi A_2\right\} \tag{A.63}$$

where A_1 and A_2 are, in general complex.

Application of the above formulas, for example, to steady flow around a circular cylinder with circulation (positive clockwise)

$$w(z) = U\left(z + \frac{a^2}{z}\right) + i\frac{\Gamma}{2\pi}\ln z, \quad \frac{dw}{dz} = U\left(1 - \frac{a^2}{z^2}\right) + i\frac{\Gamma}{2\pi z} \tag{A.64}$$

yields

$$A_1 = i\frac{U\Gamma}{\pi}, \quad A_2 = -\left(2a^2U^2 + \frac{\Gamma^2}{4\pi^2}\right) \tag{A.65}$$

Hence, $F_x = 0$, $F_y = \rho U\Gamma$, and $M_{,o} = 0$. See Ref. [5] for more details.

Problem

The objective is to find the potential and stream functions for flow around a corner.

Consider the transformation

$$w(z) = Az^n \tag{A.66}$$

where A is real and $n = \pi/\alpha,\ 0 < \alpha < 2\pi$.

Using polar coordinate

$$w(re^{i\theta}) = Ar^{\pi/\alpha}\left(\cos\frac{\pi}{\alpha}\theta + i\sin\frac{\pi}{\alpha}\theta\right) \tag{A.67}$$

Therefore

$$\phi = Ar^{\pi/\alpha} \cos \frac{\pi}{\alpha}\theta, \quad \psi = Ar^{\pi/\alpha} \sin \frac{\pi}{\alpha}\theta \tag{A.68}$$

The complex velocity is given by

$$\frac{dw}{dz} = u - iv = \frac{\pi}{\alpha} Az^{(\pi/\alpha - 1)} \tag{A.69}$$

The streamline $\psi = 0$ is given by $\theta = 0$ and $\theta = \alpha$. The flow between two planes at angle α are shown in Fig. A.4.

Another example is the flow near the sharp leading edge of a flat plate. In this case the complex potential is $w(z) = A\sqrt{z}$, see Fig. A.5.

Notice that at the origin dw/dz is zero for $\alpha < \pi$ and dw/dz is infinite for $\alpha > \pi$. The origin is a singular point in both cases, see Ref. [5].

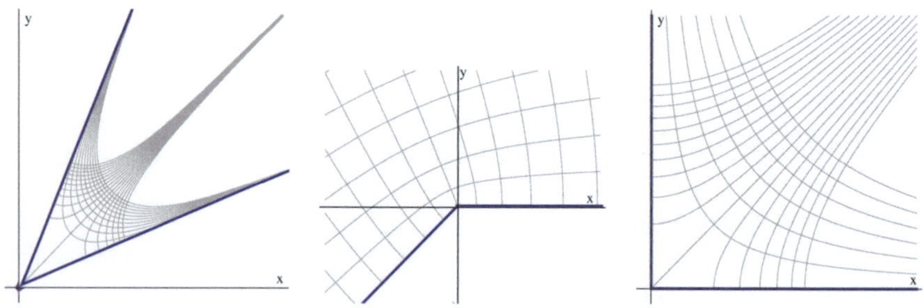

Fig. A.4 Flow between two planes at an angle: $\alpha = 45°$ (*left*); $\alpha = 135°$ (*center*); $\alpha = 90°$ (*right*)

Fig. A.5 Flow around the leading edge of a plate

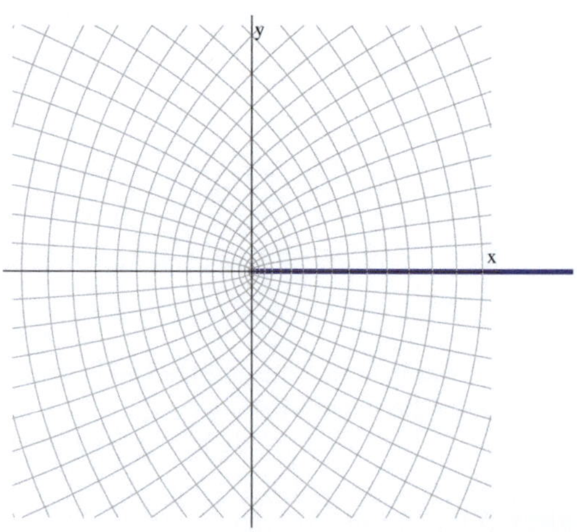

A.2 Method of Characteristics—Prandtl-Meyer Expansion

Introduction

The governing equations for steady, two-dimensional, inviscid, adiabatic, smooth (without shock), perfect gas flows, in Cartesian coordinates, are given by

$$\frac{\partial \rho u}{\partial x} + \frac{\partial \rho v}{\partial y} = 0 \tag{A.70}$$

$$-\frac{\partial u}{\partial y} + \frac{\partial v}{\partial x} = \omega = 0 \tag{A.71}$$

$$\frac{p}{p_0} = \left(\frac{\rho}{\rho_0}\right)^{\gamma} \tag{A.72}$$

$$e + \frac{p}{\rho} + \frac{1}{2}\left(u^2 + v^2\right) = H \tag{A.73}$$

and $p = \rho RT$, $e = c_v T$, $R = c_p - c_v$ and $\gamma = c_p / c_v$.

u, v are the velocity components, p, ρ, T are pressure, density and temperature, e is the specific internal energy, H is the total enthalpy, c_p, c_v are the specific heats under constant pressure and volume, respectively, γ is the ratio of specific heats and R is the gas constant. Notice, ω is the vorticity (twice the angular velocity of an element around its center) and it vanishes identically under the above assumptions. Equation (A.73) is the compressible Bernoulli's equation (including the contribution of internal energy). Combining equations (A.72) and (A.73), one obtains the density relation in terms of the magnitude of the velocity

$$\frac{\rho}{\rho_0} = \left\{1 + \frac{\gamma - 1}{2} M_0^2 \left(\frac{u^2 + v^2}{U^2} - 1\right)\right\}^{1/(\gamma-1)} \tag{A.74}$$

where $M_0 = U/a_0$, the free stream Mach number, and a_0 is the speed of sound at the free stream conditions.

Equations (A.70) and (A.71) can be rewritten in terms of potential or stream function, where

$$u = \frac{\partial \phi}{\partial x} = \frac{1}{\rho}\frac{\partial \psi}{\partial y}, \quad v = \frac{\partial \phi}{\partial y} = -\frac{1}{\rho}\frac{\partial \psi}{\partial x} \tag{A.75}$$

The full potential and stream function equations read

$$\left(1 - \frac{u^2}{a^2}\right)\frac{\partial^2 \phi}{\partial x^2} - 2\frac{uv}{a^2}\frac{\partial^2 \phi}{\partial x \partial y} + \left(1 - \frac{v^2}{a^2}\right)\frac{\partial^2 \phi}{\partial y^2} = 0 \tag{A.76}$$

and

$$\left(1-\frac{u^2}{a^2}\right)\frac{\partial^2\psi}{\partial x^2}-2\frac{uv}{a^2}\frac{\partial^2\psi}{\partial x\partial y}+\left(1-\frac{v^2}{a^2}\right)\frac{\partial^2\psi}{\partial y^2}=0 \tag{A.77}$$

where $a^2=\gamma p/\rho=\gamma RT$, therefore

$$\frac{a^2}{\gamma-1}+\frac{u^2+v^2}{2}=H=\frac{a_0^2}{\gamma-1}+\frac{U^2}{2} \tag{A.78}$$

Equation (A.77) can be obtained directly from the quasi linear form of equations (A.70) and (A.71), namely

$$\frac{\partial \mathbf{f}}{\partial \mathbf{w}}\frac{\partial \mathbf{w}}{\partial x}+\frac{\partial \mathbf{g}}{\partial \mathbf{w}}\frac{\partial \mathbf{w}}{\partial y}=\mathbf{0} \tag{A.79}$$

where

$$\mathbf{w}=\begin{pmatrix}u\\ v\end{pmatrix},\quad \mathbf{f}=\begin{pmatrix}\rho u\\ v\end{pmatrix},\quad \mathbf{g}=\begin{pmatrix}\rho v\\ -u\end{pmatrix} \tag{A.80}$$

and the Jacobian matrices are

$$\frac{\partial \mathbf{f}}{\partial \mathbf{w}}=\begin{pmatrix}\rho\left(1-\frac{u^2}{a^2}\right) & -\rho\frac{uv}{a^2}\\ 0 & 1\end{pmatrix},\quad \frac{\partial \mathbf{g}}{\partial \mathbf{w}}=\begin{pmatrix}-\rho\frac{uv}{a^2} & \rho\left(1-\frac{v^2}{a^2}\right)\\ -1 & 0\end{pmatrix} \tag{A.81}$$

The nonlinear transonic small disturbance equation in terms of the perturbation potential $\varphi(x, y)$ is given by

$$\left(1-M^2\right)\frac{\partial^2\varphi}{\partial x^2}+\frac{\partial^2\varphi}{\partial y^2}=0 \tag{A.82}$$

where $1-M^2\simeq(1-M_0^2)-(\gamma+1)M_0^2\frac{\partial\varphi}{\partial x}$.

The linearized equation becomes

$$\left(1-M_0^2\right)\frac{\partial^2\varphi}{\partial x^2}+\frac{\partial^2\varphi}{\partial y^2}=0 \tag{A.83}$$

For incompressible flows, $M_0=0$ and $\rho=\rho_0$, the governing equations reduce to the Cauchy-Riemann equations for u and v and Laplace equation for ϕ and ψ.

Also, for linearized subsonic flows, the governing equations are similar to the incompressible flow equations (via Prandtl-Glauert transformation). On the other hand, for supersonic flows, the linearized equations can be written in the form

$$\begin{pmatrix} \frac{\partial v}{\partial y} \\ \frac{\partial u}{\partial y} \end{pmatrix} = \begin{pmatrix} 0 & M_0^2 - 1 \\ 1 & 0 \end{pmatrix} \begin{pmatrix} \frac{\partial v}{\partial x} \\ \frac{\partial u}{\partial x} \end{pmatrix} \tag{A.84}$$

Calculation of the Characteristics

The eigenvalues of the matrix in Eq. (A.84) are $\lambda = \pm\sqrt{M_0^2 - 1}$. They are real for $M_0 > 1$ and imaginary for $M_0 < 1$.

Let $\beta = \sqrt{M_0^2 - 1}$, equation (A.84) can be rewritten as

$$\begin{cases} \frac{\partial}{\partial x}(\beta u) - \frac{1}{\beta}\frac{\partial}{\partial y}(v) = 0 \\ \frac{\partial}{\partial x}(v) - \frac{1}{\beta}\frac{\partial}{\partial y}(\beta u) = 0 \end{cases} \tag{A.85}$$

Adding and subtracting the above two equations yields

$$\begin{cases} \frac{\partial}{\partial x}(\beta u + v) - \frac{1}{\beta}\frac{\partial}{\partial y}(\beta u + v) = 0 \\ \frac{\partial}{\partial x}(\beta u - v) + \frac{1}{\beta}\frac{\partial}{\partial y}(\beta u - v) = 0 \end{cases} \tag{A.86}$$

The two equations in (A.86) are uncoupled.

Let $Q = \beta u + v$ and $R = \beta u - v$, hence

$$\begin{cases} \frac{\partial Q}{\partial x} - \frac{1}{\beta}\frac{\partial Q}{\partial y} = 0 \\ \frac{\partial R}{\partial x} + \frac{1}{\beta}\frac{\partial R}{\partial y} = 0 \end{cases} \tag{A.87}$$

Since

$$\frac{dS}{dx} = \frac{\partial S}{\partial x} + \frac{dy}{dx}\frac{\partial S}{\partial y} \tag{A.88}$$

hence, $dQ = 0$, i.e. $Q = constant$ along the line of slope $dy/dx = -1/\beta$. Similarly, $R = constant$ along the line of slope $dy/dx = 1/\beta$. Q and R are called the Riemann invariants and $dy/dx = \pm 1/\beta$ are called the characteristic directions. In this case, the characteristic directions are the Mach lines.

Let $\xi = x + \beta y$ and $\eta = x - \beta y$, one can show that, with this change of variables, the governing equation (A.83) is transformed into

$$\beta^2 \frac{\partial^2 \varphi}{\partial x^2} - \frac{\partial^2 \varphi}{\partial y^2} = 4\frac{\partial^2 \varphi}{\partial \xi \partial \eta} = 0 \tag{A.89}$$

Therefore, the general solution is of the form

$$\varphi(x, y) = F(\xi) + G(\eta) = F(x + \beta y) + G(x - \beta y) \tag{A.90}$$

Equation (A.90) is the d'Alembert solution of the wave equation.

Notice, the wave equation (A.83), can be factored with real coefficients as

$$\left(\beta\frac{\partial}{\partial x}-\frac{\partial}{\partial y}\right)\left(\beta\frac{\partial}{\partial x}+\frac{\partial}{\partial y}\right)\varphi=0 \tag{A.91}$$

while the Laplace equation can be factored with complex coefficients

$$\left(\beta\frac{\partial}{\partial x}-i\frac{\partial}{\partial y}\right)\left(\beta\frac{\partial}{\partial x}+i\frac{\partial}{\partial y}\right)\varphi=0 \tag{A.92}$$

As shown in the complex variables methods

$$\frac{\partial^2\varphi}{\partial x^2}+\frac{\partial^2\varphi}{\partial y^2}=4\frac{\partial^2\varphi}{\partial z\partial\bar{z}}=0 \tag{A.93}$$

where $z=x+iy$ and $\bar{z}=x-iy$ and $\varphi(x,y)=F(z)+G(\bar{z})$. An analytic solution is a function of z only.

For systematic treatment of the characteristics, consider the system

$$\begin{cases}\frac{\partial\bar{u}}{\partial x}dx+\frac{\partial\bar{u}}{\partial\bar{y}}d\bar{y}=d\bar{u}\\ \frac{\partial v}{\partial x}dx+\frac{\partial v}{\partial\bar{y}}d\bar{y}=dv\\ -\frac{\partial\bar{u}}{\partial x}+\frac{\partial v}{\partial\bar{y}}=0\\ \frac{\partial v}{\partial x}-\frac{\partial\bar{u}}{\partial\bar{y}}=0\end{cases} \tag{A.94}$$

where $\bar{u}=\beta u$ and $\bar{y}=\beta y$, see Ref. [6].

In a matrix form, one obtains

$$\begin{pmatrix} dx & d\bar{y} & 0 & 0\\ 0 & 0 & dx & d\bar{y}\\ -1 & 0 & 0 & 1\\ 0 & -1 & 1 & 0\end{pmatrix}\begin{pmatrix}\frac{\partial\bar{u}}{\partial x}\\ \frac{\partial\bar{u}}{\partial\bar{y}}\\ \frac{\partial v}{\partial x}\\ \frac{\partial v}{\partial\bar{y}}\end{pmatrix}=\begin{pmatrix}d\bar{u}\\ dv\\ 0\\ 0\end{pmatrix} \tag{A.95}$$

Using Gaussian elimination or Cramer's rule, one obtains

$$\frac{\partial\bar{u}}{\partial x}=-\frac{d\bar{u}dx+dvd\bar{y}}{dxdx-d\bar{y}d\bar{y}} \tag{A.96}$$

Along the characteristic directions, $d\bar{y}/dx=\pm 1$, $\frac{\partial\bar{u}}{\partial x}$ is not defined (i.e. indeterminate) if $d\bar{u}/dv=-d\bar{y}/dx$. The latter condition is called compatibility relation.

Another way to obtain the last relation is to diagonalize the matrix A of the governing equations

$$\begin{pmatrix} \frac{\partial \bar{u}}{\partial x} \\ \frac{\partial v}{\partial x} \end{pmatrix} = A \begin{pmatrix} \frac{\partial \bar{u}}{\partial \bar{y}} \\ \frac{\partial v}{\partial \bar{y}} \end{pmatrix} = \begin{pmatrix} 0 & 1 \\ 1 & 0 \end{pmatrix} \begin{pmatrix} \frac{\partial \bar{u}}{\partial \bar{y}} \\ \frac{\partial v}{\partial \bar{y}} \end{pmatrix} \tag{A.97}$$

The eigenvalues are $\lambda = \pm 1$. The two eigenvectors are the columns of the matrix

$$T = \begin{pmatrix} 1 & 1 \\ -1 & 1 \end{pmatrix} \tag{A.98}$$

The inverse of T is

$$T^{-1} = \frac{1}{2} \begin{pmatrix} 1 & -1 \\ 1 & 1 \end{pmatrix} \tag{A.99}$$

Hence, $T^{-1}AT$ is

$$T^{-1}AT = \begin{pmatrix} -1 & 0 \\ 0 & 1 \end{pmatrix} = \Lambda \tag{A.100}$$

and

$$T^{-1} \begin{pmatrix} \frac{\partial \bar{u}}{\partial x} \\ \frac{\partial v}{\partial x} \end{pmatrix} = \Lambda T^{-1} \begin{pmatrix} \frac{\partial \bar{u}}{\partial \bar{y}} \\ \frac{\partial v}{\partial \bar{y}} \end{pmatrix} \tag{A.101}$$

or

$$\begin{cases} \frac{\partial \bar{u}}{\partial x} - \frac{\partial v}{\partial x} = -\left(\frac{\partial \bar{u}}{\partial \bar{y}} - \frac{\partial v}{\partial \bar{y}} \right) \\ \frac{\partial \bar{u}}{\partial x} + \frac{\partial v}{\partial x} = \frac{\partial \bar{u}}{\partial \bar{y}} + \frac{\partial v}{\partial \bar{y}} \end{cases} \tag{A.102}$$

Equation (A.102) are the same as Eq. (A.86). Hence, the compatibility relations are obtained by multiplying the governing equations by the matrix T^{-1}. See Ref. [7].

In the following, the characteristics of the nonlinear small disturbance equation and the full potential equation are derived.

In the first case

$$\frac{dy}{dx} = \pm \frac{1}{\sqrt{M^2 - 1}} \tag{A.103}$$

where $M = u/a$ is the local Mach number.

The small disturbance equation is a special case of the full potential equation with $v \ll u$.

The full potential (or stream function) equation can be rewritten as a system of two equations in u and v

$$\left(1-\frac{u^2}{a^2}\right)\frac{\partial u}{\partial x}-\frac{uv}{a^2}\left(\frac{\partial v}{\partial x}+\frac{\partial u}{\partial y}\right)+\left(1-\frac{v^2}{a^2}\right)\frac{\partial v}{\partial y}=0 \tag{A.104}$$

$$\frac{\partial v}{\partial x}-\frac{\partial u}{\partial y}=0 \tag{A.105}$$

Equations (A.104) and (A.105) are cast in the form

$$\begin{pmatrix} a & b \\ 0 & 1 \end{pmatrix}\begin{pmatrix} \frac{\partial u}{\partial x} \\ \frac{\partial v}{\partial x} \end{pmatrix}=-\begin{pmatrix} b & c \\ -1 & 0 \end{pmatrix}\begin{pmatrix} \frac{\partial u}{\partial y} \\ \frac{\partial v}{\partial y} \end{pmatrix} \tag{A.106}$$

where

$$a=1-\frac{u^2}{a^2},\quad b=\frac{uv}{a^2},\quad c=1-\frac{v^2}{a^2} \tag{A.107}$$

$$\begin{pmatrix} \frac{\partial u}{\partial x} \\ \frac{\partial v}{\partial x} \end{pmatrix}=-\begin{pmatrix} a & b \\ 0 & 1 \end{pmatrix}^{-1}\begin{pmatrix} b & c \\ -1 & 0 \end{pmatrix}\begin{pmatrix} \frac{\partial u}{\partial y} \\ \frac{\partial v}{\partial y} \end{pmatrix}$$

$$=-\frac{1}{a}\begin{pmatrix} 1 & -b \\ 0 & a \end{pmatrix}\begin{pmatrix} b & c \\ -1 & 0 \end{pmatrix}\begin{pmatrix} \frac{\partial u}{\partial y} \\ \frac{\partial v}{\partial y} \end{pmatrix}=-\frac{1}{a}\begin{pmatrix} 2b & c \\ -a & 0 \end{pmatrix}\begin{pmatrix} \frac{\partial u}{\partial y} \\ \frac{\partial v}{\partial y} \end{pmatrix} \tag{A.108}$$

The characteristic equation is

$$a\lambda^2-2b\lambda+c=0 \tag{A.109}$$

The eigenvalues are

$$\lambda_{1,2}=\frac{b\pm\sqrt{b^2-ac}}{a} \tag{A.110}$$

or

$$\lambda_{1,2}=\frac{uv\pm a^2\sqrt{M^2-1}}{u^2-a^2} \tag{A.111}$$

Let, $u=U\cos\theta$ and $v=U\sin\theta$, $\tan\theta=v/u$ and $\tan\mu=1/M$, where μ is the Mach angle. Then it can be shown that

$$\left(\frac{dy}{dx}\right)_{1,2}=\lambda_{1,2}=\tan\left(\theta\pm\mu\right) \tag{A.112}$$

To find the compatibility relations along the characteristics, let

$$\begin{pmatrix} dx & dy & 0 & 0 \\ 0 & 0 & dx & dy \\ a & b & b & c \\ 0 & -1 & 1 & 0 \end{pmatrix} \begin{pmatrix} \frac{\partial u}{\partial x} \\ \frac{\partial u}{\partial y} \\ \frac{\partial v}{\partial x} \\ \frac{\partial v}{\partial y} \end{pmatrix} = \begin{pmatrix} du \\ dv \\ 0 \\ 0 \end{pmatrix} \tag{A.113}$$

Again, using Cramer's rule, $\frac{\partial u}{\partial x}$ is not defined along the characteristics if

$$\left(\frac{du}{dv}\right)_{1,2} = -\left(\frac{dy}{dx}\right)_{1,2} \tag{A.114}$$

Equation (A.114) indicates that

$$\left(\frac{dy}{dx}\right)_1 \left(\frac{dv}{du}\right)_1 = -1, \quad \left(\frac{dy}{dx}\right)_2 \left(\frac{dv}{du}\right)_2 = -1 \tag{A.115}$$

These relations have practical implications for the graphical construction of the characteristic nets as will be discussed later.

Alternatively, one can obtain the same compatibility relations by multiplying equation (A.106) by T^{-1}, where T consists of the eigenvectors corresponding to the eigenvalues given by equation (A.111).

One can show that (see Ref. [8])

$$T^{-1} = \begin{pmatrix} 1 & 1 \\ c\ \cot(\theta + \mu) & c\ \cot(\theta - \mu) \end{pmatrix} \tag{A.116}$$

and the compatibility relation is given by

$$du_{\pm} + \lambda_{\pm} \mp dv_{\pm} = 0 \tag{A.117}$$

Geometrical Interpretation of Characteristic Directions and Surfaces

If a simple wave solution of the form

$$\mathbf{w} = \mathbf{w}_0 e^{(ipx+iqy)}, \quad i = \sqrt{-1} \tag{A.118}$$

is sought for the quasi linear equation

$$J_1 \frac{\partial \mathbf{w}}{\partial x} + J_2 \frac{\partial \mathbf{w}}{\partial y} = \mathbf{0} \tag{A.119}$$

The determinant of the combined matrix must vanish, i.e.

$$|J_1 p + J_2 q| = 0 \tag{A.120}$$

Equation (A.120) is the characteristic equation. The direction normal to the characteristic surface $S(x, y) = constant$, $\mathbf{n}$, is given by

$$\frac{dy}{dx} = -\frac{p}{q} = -\frac{n_x}{n_y} \tag{A.121}$$

The compatibility relations contain only derivatives along the surface S. For hyperbolic equations all the eigenvectors are linearly independent. For more details, see [7].

Notice, supersonic flows are inherently non analytic in character and characteristic curves are loci of possible discontinuities in derivatives of fluid properties, see [6].

Application of Method of Characteristics to Steady, Two-dimensional, Isentropic, Irrotational Supersonic Flows

In this case, Riemann invariants exist and it is recommended to use the natural coordinate system. Let (ω, θ) be the velocity magnitude and direction, while (s, n) are the streamline direction and normal to it.

The governing equations of continuity and irrotationality are (see Ref. [9])

$$\frac{\cot^2\mu}{\omega}\frac{\partial\omega}{\partial s} - \frac{\partial\theta}{\partial n} = 0 \tag{A.122}$$

$$\frac{1}{\omega}\frac{\partial\omega}{\partial n} - \frac{\partial\theta}{\partial s} = 0 \tag{A.123}$$

where $\cot^2\mu = M^2 - 1$.

Introducing the function ν, defined by

$$\nu = \int \frac{\cot^2\mu}{\omega} d\omega \tag{A.124}$$

The governing equations can be rewritten as

$$\frac{\partial}{\partial s}(\nu - \theta) + \tan\mu \frac{\partial}{\partial n}(\nu - \theta) = 0 \tag{A.125}$$

$$\frac{\partial}{\partial s}(\nu + \theta) - \tan\mu \frac{\partial}{\partial n}(\nu + \theta) = 0 \tag{A.126}$$

Hence, there are two functions $Q = \nu + \theta$ and $R = \nu - \theta$ which are invariant along the characteristic directions. In the following Fig. A.6, the method of calculation is demonstrated.

Fig. A.6 Riemann invariants

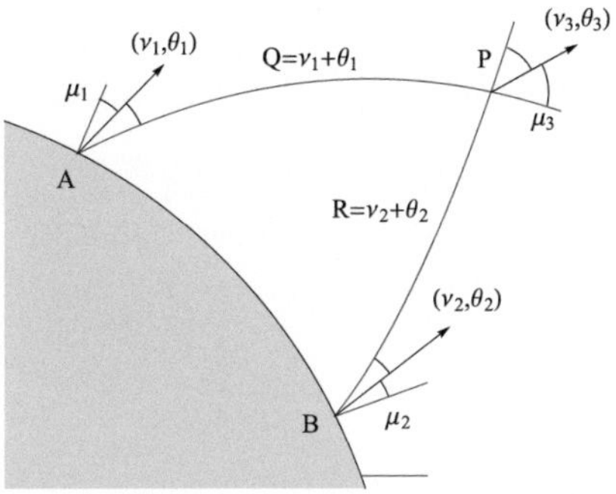

Fig. A.7 Prandtl-Meyer expansion around a corner

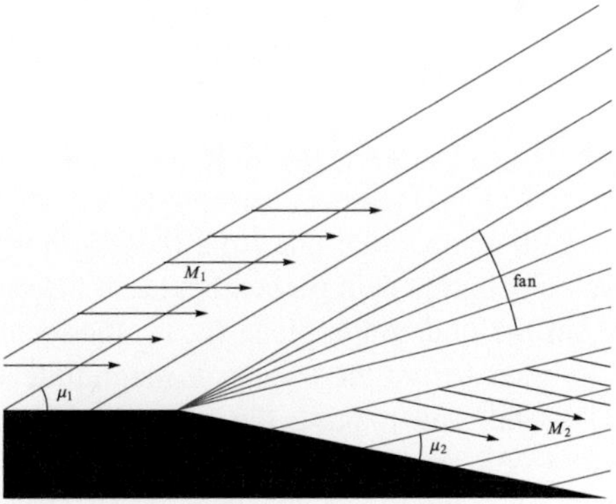

Given the initial data on a non-characteristic curve AB (i.e. a curve not tangent to any of the characteristic lines), the conditions at point P, which is the intersection of the two characteristics, can be found. Notice $Q_3 = Q_1$ and $R_3 = R_2$. In terms of the invariants, $\nu_3 = (Q_3 + R_3)/2$ and $\theta_3 = (Q_3 - R_3)/2$.

If A is close to B, the mesh sides may be approximated by straight line segments. The accuracy of the calculations depend on the mesh size. Iterative methods can be used to improve the procedure. Special treatments are needed for points at solid body or free surface.

Notice for the case of axially symmetric, rotational or non isentropic flows, no Riemann invariants exist and the compatibility relations along the characteristics must be solved numerically (for example using finite differences). See Ref. [9].

Supersonic Flow Around a Corner—Prandtl-Meyer Expansion Fan

The change of speed due to a deflection (Fig. A.7) is given by

$$\frac{\Delta\omega}{\omega} = -\frac{\Delta\theta}{\sqrt{M_1^2 - 1}} \tag{A.127}$$

Taking the limit

$$\frac{d\omega}{\omega} = -\frac{d\theta}{\sqrt{M^2-1}}, \quad \Rightarrow \quad -d\theta = \sqrt{M^2-1}\frac{d\omega}{\omega} = d\nu \tag{A.128}$$

Using the definition of Mach number and Bernoulli's law, then

$$\frac{d\omega}{\omega} = \frac{1}{1+\frac{\gamma-1}{2}M^2}\frac{dM}{M} \tag{A.129}$$

hence

$$\nu(M) = \int_1^M \frac{\sqrt{M^2-1}}{1+\frac{\gamma-1}{2}M^2}\frac{dM}{M}$$

$$= -\sqrt{\frac{\gamma+1}{\gamma-1}}\arctan\sqrt{\frac{\gamma-1}{\gamma+1}}\left(M^2-1\right) + \arctan\sqrt{M^2-1} \tag{A.130}$$

where $\nu(M=1)=0$. As $M \to \infty$, $\nu \to -130.45°$. At this angle, the velocity is maximum and temperature and density vanish (vacuum).

Notice the radial line from the origin is a Mach wave (or characteristic) and the velocity component perpendicular to the radial line is sonic. See Fig. A.7.

In the hodograph plane (u, v), the locus of the extremities of the velocity vectors issued from the origin, of magnitude V/a^* normalized with the critical speed of sound, and making the angle ν with the u-axis, is an epicycloid. An epicycloid is a curve generated by a point on the circumference of a circle, rotating without slipping, on an other circle. Here, the point is on the circle of diameter $2R = \left(\sqrt{(\gamma+1)/(\gamma-1)} - 1\right)$ rotating on the sonic circle (unit circle) and initially located at $(1, 0)$. In Fig. A.8, $\phi = R\psi$ is the contact angle of the two circles, M describes the epicycloid. In complex form, the equation for point M reads

$$u + iv = (1+R)e^{iR\psi} - Re^{i(1+R)\psi} \tag{A.131}$$

The mirror image epicycloid corresponds to a counter clockwise Prandtl-Meyer expansion.

Prandtl-Meyer expansion fan can be derived also as a self similar solution, where all the dependent variables are function of the polar angle and not the radial distance from the corner.

Subsonic Flow Around Sharp Corner

Starting with subsonic flow, a Prandtl-Meyer expansion around a corner begins at $M = 1$. Figure A.9.

The problem is complicated due to the formation of a shock emanating from the corner, see Ref. [10].

Notice, for viscous flow, a vortex exists followed by a separated region, Ref. [10].

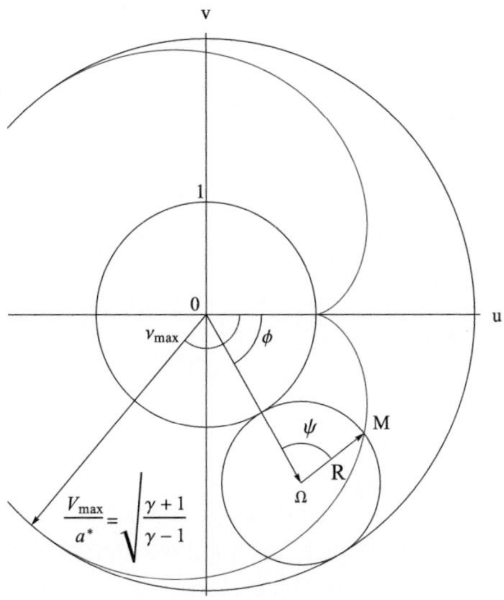

Fig. A.8 Epicycloid and its mirror image

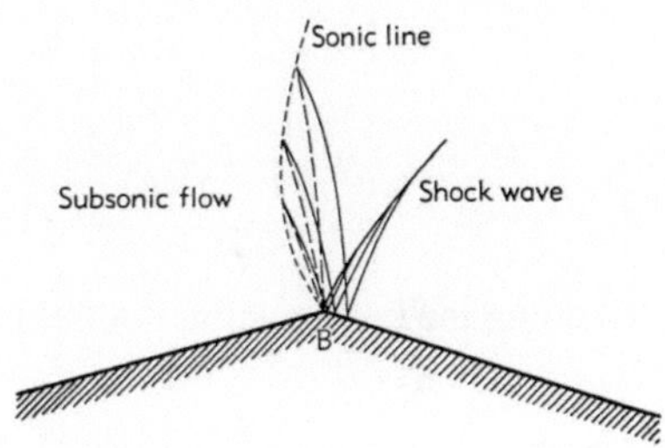

Fig. A.9 Expansion around a corner of a subsonic flow (from General Theory of High Speed Aerodynamics, with kind permission of Princeton University Press)

A.3 Conservation Laws, Shock Waves and Contact Discontinuities

Introduction

The classical theory for a scalar and a system of conservation laws of unsteady one dimensional flows is discussed first, followed by the application to steady two-dimensional supersonic flows.

Following Lax [11] and also Chattot [12], consider the partial differential equation

$$\frac{\partial u}{\partial t} + \frac{\partial f(u)}{\partial x} = 0, \quad \frac{df(u)}{du} = a(u), \quad \Rightarrow \quad \frac{\partial u}{\partial t} + a(u)\frac{\partial u}{\partial x} = 0 \qquad \text{(A.132)}$$

where $u(x, 0) = u_0(0)$ is given as initial condition.

Comparing the above equation with the expression of a total derivative and using chain rule, one obtains

$$\Delta u = \frac{\partial u}{\partial t}\Delta t + \frac{\partial u}{\partial x}\Delta x, \quad \Rightarrow \quad \frac{\Delta u}{\Delta t} = \frac{\partial u}{\partial t} + \frac{\partial u}{\partial x}\frac{\Delta x}{\Delta t} \tag{A.133}$$

Taking the limit, Eq. (A.133) implies that u is constant along the curve of slope (characteristic)

$$\frac{dx}{dt} = a(u) \tag{A.134}$$

Notice that since u is constant along the characteristics, the characteristics have constant slope, hence they are straight lines.

The general solution of Eq. (A.132) is

$$u(x, t) = F\left(x - a(u)t\right) \tag{A.135}$$

where F is an arbitrary function of a single variable $\xi = x - a(u)t$. It is easily verified that this is indeed the general solution, as

$$\frac{\partial u}{\partial t} = F'\left\{-a(u) - a'(u)\frac{\partial u}{\partial t}t\right\}, \quad \Rightarrow \quad \frac{\partial u}{\partial t} = -\frac{a(u)F'}{1 + a'(u)tF'} \tag{A.136}$$

$$\frac{\partial u}{\partial x} = F'\left\{1 - a'(u)\frac{\partial u}{\partial x}t\right\}, \quad \Rightarrow \quad \frac{\partial u}{\partial x} = \frac{F'}{1 + a'(u)tF'} \tag{A.137}$$

Adding the last equation multiplied by $a(u)$ to the previous one yields

$$\frac{\partial u}{\partial t} + a(u)\frac{\partial u}{\partial x} = \frac{a(u)F' - a(u)F'}{1 + a'(u)tF'} = 0 \tag{A.138}$$

F is determined by the initial condition, here $F \equiv u_0$

$$u(x, t) = u_0\left(x - a(u)t\right) \tag{A.139}$$

Assume $a'(u) > 0$, then if F' (or u_0') < 0 at some point, both partial derivatives will tend to infinity as $1 + a'(u)tF' \to 0$.

Consider two points x_1 and x_2 such that $x_1 < x_2$ and $u_1 = u_0(x_1) > u_0(x_2) = u_2$ (since $u_0' < 0$) and $a_1 = a(u_1) > a(u_2) = a_2$ (since $a'(u) > 0$), so the characteristics originating from x_1 and x_2 intersect at time $t = (x_2 - x_1)/(a_1 - a_2)$. At the point of intersection, u is multi-valued and the need for a shock wave arises.

As an example, let $a(u) = u$, the Burgers-Hopf equation reads

$$\frac{\partial u}{\partial t} + \frac{\partial}{\partial x}\left(\frac{u^2}{2}\right) = \frac{\partial u}{\partial t} + u\frac{\partial u}{\partial x} = 0 \tag{A.140}$$

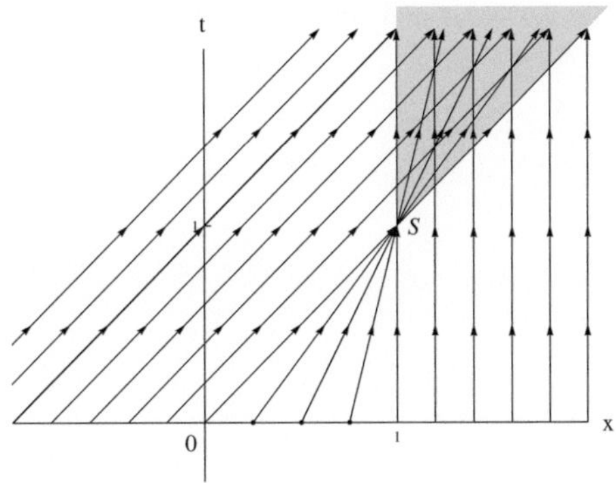

Fig. A.10 Multi-valued solution to initial value problem

and, initial condition

$$u_0(x) = \begin{cases} 1, & -\infty < x \le 0 \\ 1 - x, & 0 \le x \le 1 \\ 0, & 1 \le x < +\infty \end{cases} \tag{A.141}$$

The solution is shown in Fig. A.10.

The solution is single-valued for $t \le 1$ and triple-valued for $t > 1$ (in the gray wedge subdomain).

On the other hand, let

$$u_0(x) = \begin{cases} 0, & -\infty < x < 0 \\ 1, & 0 < x < +\infty \end{cases} \tag{A.142}$$

The solution is given in Fig. A.11.

In the wedge-like subdomain, $0 \le x \le t$, a self-similar solution exists, as

$$u(x, t) = \frac{x}{t}, \quad 0 \le \frac{x}{t} \le 1 \tag{A.143}$$

that produces a continuous transition between two states, an expansion fan.

To study the discontinuous solutions of the conservation law given by Eq. (A.132), let's integrate the two terms from x_1 to x_2 at time t, where the shock $x = x_s(t)$ intersects the interval $x_1 \le x_s \le x_2$. The integral form of the equation reads

$$\frac{d}{dt}\int_{x1}^{x2} u\,dx = -\int_{x1}^{x2} \frac{\partial f(u)}{\partial x} dx = f_1 - f_2 \tag{A.144}$$

Notice

$$\frac{d}{dt}\int_{x1}^{x2} u\,dx = \frac{d}{dt}\left(\int_{x1}^{xs} u\,dx + \int_{xs}^{x2} u\,dx\right) \tag{A.145}$$

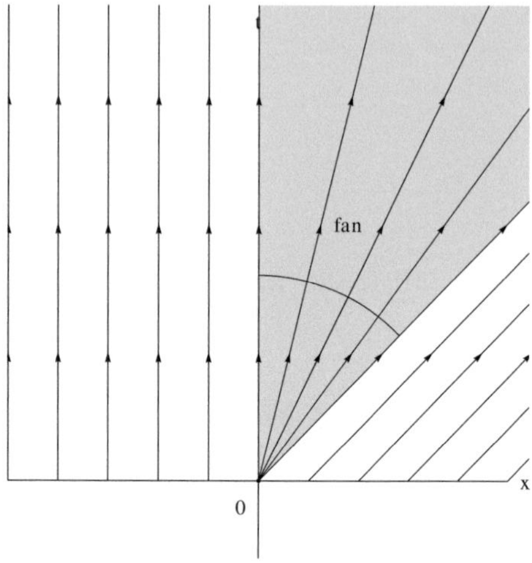

Fig. A.11 Solution with expansion fan

Therefore

$$f_1 - f_2 = \int_{x1}^{xs} \frac{\partial u}{\partial t} dx + u_l \frac{dx_s}{dt} + \int_{xs}^{x2} \frac{\partial u}{\partial t} dx - u_r \frac{dx_s}{dt} = f_1 - f_l + u_l s - f_2 + f_r - u_r s \tag{A.146}$$

Hence

$$f_r - f_l = (u_r - u_l)s \tag{A.147}$$

where $s = dx_s/dt$.

In the above example, where $f = u^2/2$, one finds $s = (u_l + u_r)/2 = \bar{u}$ and $x_s(t) = (u_l + u_r)t/2 = \bar{u}t$. The solution jumps from u_l to u_r along the line of compression shock ($u_l > u_r$).

For the case where $u_l < u_r$, an expansion fan given by $u = x/t$ for $u_l t \leq x \leq u_r t$ provides a continuous solution. Expansion shock solution can be eliminated via an "entropy" condition

$$a(u_l) = \left(\frac{df}{du}\right)_l > s > a(u_r) = \left(\frac{df}{du}\right)_r \tag{A.148}$$

or, the characteristic lines must go into the shock curve as t increases. Such a condition renders the generalized solution of Eq. (A.132) unique.

Notice, the conservation form itself is not unique. Following Strang [3], multiply Eq. (A.140) by u, hence

$$u\frac{\partial u}{\partial t} + u^2 \frac{\partial u}{\partial x} = 0 \tag{A.149}$$

The above equation can be rewritten in conservation form as

$$\frac{\partial}{\partial t}\left(\frac{u^2}{2}\right)+\frac{\partial}{\partial x}\left(\frac{u^3}{3}\right)=0 \tag{A.150}$$

Let

$$E=\int_{x1}^{x2}\frac{u^2}{2}dx \tag{A.151}$$

then, upon integration in x of Eq. (A.150), accounting for a discontinuity at x_s, $x_1 \le x_s \le x_2$

$$\frac{dE}{dt}+\frac{u_2^3}{3}-\frac{u_1^3}{3}=-\frac{dx_s}{dt}\left(\frac{u_r^2}{2}-\frac{u_l^2}{2}\right)+\frac{u_r^3}{3}-\frac{u_l^3}{3} \tag{A.152}$$

The speed of discontinuity associated with the conservation law Eq. (A.150) is given by setting the right-hand-side to zero

$$\frac{u_r^3}{3}-\frac{u_l^3}{3}=\left(\frac{u_r^2}{2}-\frac{u_l^2}{2}\right)s \tag{A.153}$$

and can be obtained after some algebra to be different from the original speed $\bar{u}$

$$s=\bar{u}+\frac{1}{12}\frac{<u>^2}{\bar{u}}\neq\bar{u} \tag{A.154}$$

However, if the shock speed is required to correspond to the original conservation law, i.e. $s=\bar{u}$, then the right-hand-side of Eq. (A.152) is no longer zero, but can be evaluated after some algebra to be

$$\frac{dE}{dt}+\frac{u_2^3}{3}-\frac{u_1^3}{3}=\frac{<u>^3}{12}\le 0 \tag{A.155}$$

This provides an entropy inequality, since the velocity must decrease across a shock, hence a negative result for $<u>^3$ or $<u>$.

The choice of the conservation law should be based on the physics of the problem and not the differential equation, as will be discussed in the next section together with the entropy function and the inequality it satisfies.

The solution of Eq. (A.132) with shock satisfying the entropy condition is called a weak solution.

Multiplying the equation by any smooth test function $v(x,t)$ that vanishes at infinity and via integration by parts, one obtains

$$\int_0^\infty \int_{-\infty}^\infty \left(u\frac{\partial v}{\partial t} + f(u)\frac{\partial v}{\partial x} \right) dxdt = -\int_{-\infty}^\infty u_0 v(x,0)dx \tag{A.156}$$

In the above equation, u is called a distribution solution. Such solution can be constructed as the limit of u_ϵ of the parabolic equation as $\epsilon \to 0$

$$\frac{\partial u_\epsilon}{\partial t} + \frac{\partial f(u_\epsilon)}{\partial x} = \epsilon \frac{\partial^2 u_\epsilon}{\partial x^2}, \quad \epsilon > 0 \tag{A.157}$$

Based on maximum principle, the above equation has at most one solution. The proof that u_ϵ converges to a limit u is omitted.

Systems of Conservation Laws

Following Lax [11], consider a first order system of conservation laws in time and single space dimension

$$\frac{\partial u_i}{\partial t} + \frac{\partial f_i(u)}{\partial x} = 0, \quad i = 1, ..., n \tag{A.158}$$

where $f = \{f_i\}$ is a vector function of $u = \{u_i\},\ i = 1, .., n$.

The above equations are related to the quasi-linear system

$$\frac{\partial u}{\partial t} + A\frac{\partial u}{\partial x} = 0 \tag{A.159}$$

where $A = \partial f/\partial u$ is the Jacobian matrix. System Eq. (A.159) is assumed to be totally hyperbolic, i.e. A has n real eigenvalues λ_k with the corresponding distinct right and left eigenvectors, r_k, l_k. It is also assumed that (normalization)

$$r_k.\nabla\lambda_k = 1 \tag{A.160}$$

Each of the n conservation laws must satisfy the condition

$$< f_k > = < u_k > s, \quad k = 1, ..., n \tag{A.161}$$

where s is the speed of propagation of the discontinuity.

The entropy condition requires that, for a single conservation equation, the characteristics on either side of a discontinuity run into the line of the discontinuity, therefore

$$\lambda(u_l) = a(u_l) > s > \lambda(u_r) = a(u_r) \tag{A.162}$$

For systems, it is required that for some index $k,\ 1 \le k \le n$

$$\lambda_k(u_l) > s > \lambda_k(u_r), \quad \lambda_{k-1}(u_l) < s < \lambda_{k+1}(u_r) \tag{A.163}$$

Therefore, $n-k+1$ characteristics impinge on the line of discontinuity from the left and k from the right, a total of $n+1$.

The n relations, Eq. (A.161) can be written

$$< f >= A(u_l, u_r) < u >= s < u > \tag{A.164}$$

showing that s is an eigenvalue of the matrix $A(u_l, u_r)$ and $< u >$ is a right eigenvector. Given the left state u_l, it is possible to find the right state u_r for the shock speed s, see Roe [13].

For continuous solutions of centered rarefaction waves at the origin, let $u(x,t) = w(x/t) = w(\xi)$ and substitute in the quasi-linear form to obtain

$$-\frac{x}{t^2}w' + \frac{1}{t}Aw' = 0, \quad \Rightarrow \quad (A - I\xi)\, w' = 0 \tag{A.165}$$

hence, ξ is an eigenvalue of A and $w' \propto r$ is proportional to the right eigenvector. In such a case, only one family of waves corresponding to index k carries information, while the $n-1$ other characteristics are not transporting any information, hence there must be constant quantities associated with this simple wave flow. One can show that there are $n-1$ functions $\Psi(w)$ that remain constant along those characteristics, called Riemann invariants. It can be shown that the unique solution is $w(\xi) = u_l + \int_{\xi l} K(\xi) r(\xi) d\xi,\ \lambda_k(u_l) \leq \xi \leq \lambda_k(u_r)$, where K is a scalar function of ξ.

As in a single conservation law, the limit solution of equations augmented with artificial viscosity, satisfies the entropy condition, namely u is the limit of u_ϵ as $\epsilon \to 0$, where

$$\frac{\partial u_\epsilon}{\partial t} + A\frac{\partial u_\epsilon}{\partial x} = \epsilon\frac{\partial^2 u_\epsilon}{\partial x^2}, \quad \epsilon > 0 \tag{A.166}$$

Applications to Compressible Flows

Following Chattot [12], the unsteady one dimensional flow equations are given by conservation of mass, momentum and energy

$$\frac{\partial \rho}{\partial t} + \frac{\partial \rho u}{\partial x} = 0 \tag{A.167}$$

$$\frac{\partial \rho u}{\partial t} + \frac{\partial (\rho u^2 + p)}{\partial x} = 0 \tag{A.168}$$

$$\frac{\partial \rho E}{\partial t} + \frac{\partial \rho u H}{\partial x} = 0 \tag{A.169}$$

where

$$E = \frac{p}{(\gamma - 1)\rho} + \frac{u^2}{2}, \quad H = \frac{\gamma p}{(\gamma - 1)\rho} + \frac{u^2}{2} \tag{A.170}$$

This is a hyperbolic system of equations. The eigenvalues are $\lambda_1 = u - a$, $\lambda_2 = u$ and $\lambda_3 = u + a$. There are three Riemann invariants. Entropy remains constant along $\lambda_2 = u$ characteristic. The compatibility relations read

$$\frac{\partial}{\partial t}\left(u - 2\frac{a}{\gamma - 1}\right) + (u - a)\frac{\partial}{\partial x}\left(u - 2\frac{a}{\gamma - 1}\right) = 0 \tag{A.171}$$

$$\frac{\partial S}{\partial t} + u\frac{\partial S}{\partial x} = 0, \quad S = \frac{p}{\rho^\gamma} \tag{A.172}$$

$$\frac{\partial}{\partial t}\left(u + 2\frac{a}{\gamma - 1}\right) + (u + a)\frac{\partial}{\partial x}\left(u + 2\frac{a}{\gamma - 1}\right) = 0 \tag{A.173}$$

The above relations are used as non-reflecting boundary conditions (see Chap. 5).

For shocks, by convention, the states 1 and 2 correspond to the orientation of the normal to the discontinuity, i.e. $\vec{n}$ pointing from 1 to 2.

The jump conditions are

$$< \rho > n_t + < \rho u > n_x = 0 \tag{A.174}$$

$$< \rho u > n_t + < \rho u^2 + p > n_x = 0 \tag{A.175}$$

$$< \rho E > n_t + < \rho u H > n_x = 0 \tag{A.176}$$

The speed of the discontinuity is $dx/dt = -n_t/n_x$.

The entropy inequality is given by

$$\frac{\partial \rho S}{\partial t} + \frac{\partial \rho u S}{\partial x} \geq 0, \quad at\ the\ jump \quad < \rho S > n_t + < \rho u S > n_x \geq 0$$

where $S = p/\rho^\gamma$ is the entropy function.

There are however two types of jump lines, those crossed by the fluid and those along a particle path.

If the fluid crosses a jump line, i.e. $u_1, u_2 > -n_t/n_x$ or $u_1, u_2 < -n_t/n_x$, there is a discontinuity in the velocity, density and pressure: the discontinuity is a shock wave.

If the fluid does not cross the discontinuity, i.e. $u_1 = u_2 = -n_t/n_x$, the pressure is continuous across the jump line and the jump in density cannot be determined from the jump conditions: the discontinuity is a contact discontinuity.

Steady Two Dimensional Supersonic Flows

In Cartesian coordinates, the governing equations are

$$\frac{\partial \rho u}{\partial x} + \frac{\partial \rho v}{\partial y} = 0 \tag{A.177}$$

$$\frac{\partial \rho u^2 + p}{\partial x} + \frac{\partial \rho u v}{\partial y} = 0 \tag{A.178}$$

$$\frac{\partial \rho u v}{\partial x} + \frac{\partial \rho v^2 + p}{\partial y} = 0 \tag{A.179}$$

$$\frac{\partial \rho u H}{\partial x} + \frac{\partial \rho v H}{\partial y} = 0 \tag{A.180}$$

where

$$H = \frac{\gamma p}{(\gamma - 1)\rho} + \frac{u^2 + v^2}{2} \tag{A.181}$$

The jump conditions are

$$< \rho u > n_x + < \rho v > n_y = 0 \tag{A.182}$$

$$< \rho u^2 + p > n_x + < \rho u v > n_y = 0 \tag{A.183}$$

$$< \rho u v > n_x + < \rho v^2 + p > n_y = 0 \tag{A.184}$$

$$< \rho u H > n_x + < \rho v H > n_y = 0 \tag{A.185}$$

Introduce the mass flux across the discontinuity as

$$\rho V_n = \rho u n_x + \rho v n_y \tag{A.186}$$

Then Eq. (A.182) becomes simply

$$(\rho V_n)_2 = (\rho V_n)_1 = \rho V_n \tag{A.187}$$

The x- and y-momentum can be rewritten as

$$\rho V_n < u > + < p > n_x = 0 \tag{A.188}$$

$$\rho V_n < v > + < p > n_y = 0 \tag{A.189}$$

and the energy jump condition as

$$\rho V_n < H >= 0 \tag{A.190}$$

The component of velocity tangential to the discontinuity can be defined as

$$V_t = u n_y - v n_x \tag{A.191}$$

The two momentum equations can be combined to give

$$\rho V_n < V_t >= 0, \quad and \quad \rho V_n < V_n > + < p >= 0 \tag{A.192}$$

If $\rho V_n \neq 0$ then $< V_t >= 0$, the tangential component of velocity is continuous. But $< V_n >\neq 0$, else there would be no discontinuity in velocity nor in density and the pressure would also be continuous (trivial situation). Hence, in this case, the pressure is discontinuous ($< p >\neq 0$) as well as density ($< \rho >\neq 0$) and normal velocity ($< V_n >\neq 0$), but the total enthalpy is continuous ($< H >= 0$). If the flow is isoenergetic, then $H = constant$ everywhere. The discontinuity is a shock wave.

If $\rho V_n = 0$, then $< V_t >, < \rho >, < H >$ are undetermined. The pressure jump is zero ($< p >= 0$). Such a discontinuity is a contact discontinuity.

The entropy inequality is given by

$$\frac{\partial \rho u S}{\partial x} + \frac{\partial \rho v S}{\partial y} \geq 0 \tag{A.193}$$

and its jump condition counterpart

$$< \rho u S > n_x + < \rho v S > n_y = \rho V_n < S > \geq 0 \tag{A.194}$$

For details of shock relations, see Liepmann and Roshko [9].

Attached shocks around wedges can be easily calculated, see Fig. A.12.

The solution is not unique. Weak and strong shocks satisfy all required conditions. For $\theta > \theta_{critical}$, no attached shock solution exists and the shock is detached, see Ref. [9].

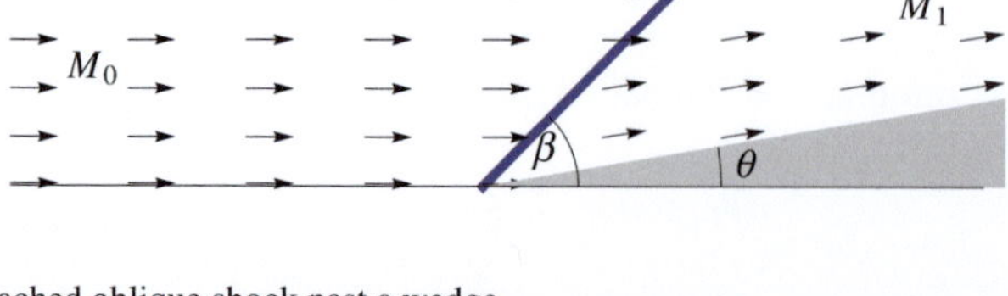

Fig. A.12 Attached oblique shock past a wedge

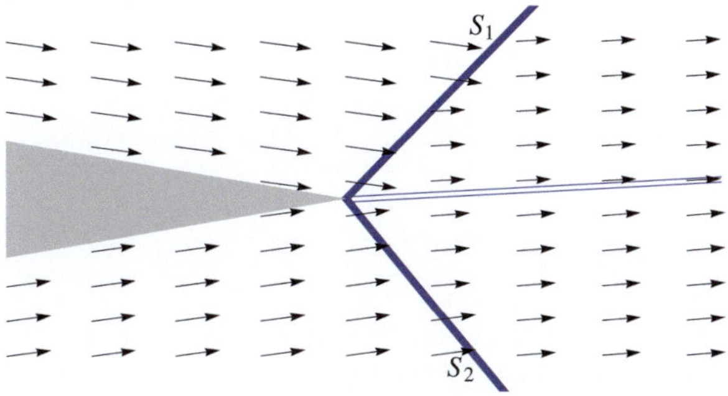

Fig. A.13 Contact discontinuity from intersection of two shock waves

As an example of contact discontinuity, consider the intersection of two oblique shocks of different strengths, say at the trailing edge. A vortex sheet or slip surface will develop across which the velocity jumps, while the pressure is continuous. To calculate the pressure and the flow direction, the shock jump relations must be used and again the solution, in general, is not unique.

Notice, the temperature and density on either side of the contact surface are different due to the fact that the entropy on either side is different, Fig. A.13.

Only compression shocks are allowed according to the entropy inequality. If θ in Fig. A.12 is negative, Prandtl-Meyer expansion is used instead.

For potential, isentropic flows, the conservation laws are

$$\frac{\partial \rho u}{\partial x} + \frac{\partial \rho v}{\partial y} = 0 \tag{A.195}$$

$$\frac{\partial v}{\partial x} - \frac{\partial u}{\partial y} = 0 \tag{A.196}$$

and $\rho = \rho(u^2 + v^2)$.

There is a mathematical entropy condition, allowing only compression shocks. For details, see Chattot [14].

Notice that momentum is not conserved across the shock wave in potential flow calculations, and the drag is related to the difference in the momentum across the shock. The mathematical entropy inequality is implying that the wave drag is positive (negative drag is thrust).

Cole and Cook [15] constructed a divergence form for transonic small disturbance equation by multiplying the continuity equation by u and the irrotationality condition by v

$$(K - (\gamma + 1)u)\, u\frac{\partial u}{\partial x} + u\frac{\partial v}{\partial y} = 0 \tag{A.197}$$

$$v\frac{\partial v}{\partial x} - v\frac{\partial u}{\partial y} = 0 \tag{A.198}$$

By combining the above equations one can obtain a conservation form

$$\frac{\partial}{\partial x}\left(K\frac{u^2}{2} - (\gamma-1)\frac{u^3}{3} - \frac{v^2}{2}\right) + \frac{\partial uv}{\partial y} = 0 \tag{A.199}$$

The above form is not however conserved across the shock. It can be shown that the wave drag across the shock is related to

$$\begin{aligned} C_{dw} &= -\int_{shocks}\left\{< uv > dx - < K\frac{u^2}{2} - (\gamma-1)\frac{u^3}{3} - \frac{v^2}{2} > dy\right\} \\ &= -\frac{\gamma+1}{12}\int_{shocks} < u >^3 dy \end{aligned} \tag{A.200}$$

The last expression is consistent with Oswatitsch formula for the wave drag

$$D_w = \rho_0 T_0 \int_{shocks} < S > dy \tag{A.201}$$

where $< S >$ is the entropy jump for small disturbance Euler equations (see Ref. [15]).

Shock Viscous Layer

For inviscid, adiabatic flow, the shock is considered a discontinuity. In reality, it is a thin layer of finite thickness, of order of a few mean free paths. Although the validity of the continuum model is questionable, the results of the Navier-Stokes equations are in good agreement with experiments, particularly for moderate Mach numbers.

The governing equations for a normal shock in steady one dimensional flows are (see [16], [17])

$$\frac{d\rho u}{dx} = 0 \tag{A.202}$$

$$\frac{d(\rho u^2 + p)}{dx} = \frac{d\bar{\tau}}{dx} \tag{A.203}$$

$$\frac{d\rho u H}{dx} = \frac{d(\bar{\tau}u - \bar{q})}{dx} \tag{A.204}$$

where $\bar{\tau} = \bar{\mu}du/dx, \bar{q} = -\bar{k}dT/dx$.

Notice, $\bar{\mu} \neq \mu$, but $\bar{k} = k$. Also $\bar{\tau}$, the compressive stress is different from the shear stress τ.

Upon integration once, one obtains

$$\rho u = \rho_1 u_1 \tag{A.205}$$

$$\rho u^2 + p = \rho_1 u_1^2 + p_1 + \bar{\tau} \tag{A.206}$$

$$\rho u H = \rho_1 u_1 H_1 + \bar{\tau} u - \bar{q} \tag{A.207}$$

where subscript 1 denotes the conditions upstream, where both the shear and heat transfer vanish.

Notice, across the shock region, the jump conditions are satisfied, namely

$$\rho_2 u_2 = \rho_1 u_1 \tag{A.208}$$

$$\rho_2 u_2^2 + p_2 = \rho_1 u_1^2 + p_1 \tag{A.209}$$

$$H_2 = H_1 \tag{A.210}$$

if $\bar{P}_r = \bar{\mu} c_p / k = 1$, an energy integral exists and $H = constant$ throughout the shock layer. In this case, the momentum equation can be rewritten as

$$\frac{\bar{\mu}}{\rho_1 u_1} \frac{d\theta}{dx} = \frac{\gamma + 1}{2\gamma} \frac{(\theta - \theta_1)(\theta - \theta_2)}{\theta} \tag{A.211}$$

where

$$\theta = \frac{\rho_1 u_1^2}{\rho_1 u_1^2 + p_1} \frac{1}{\rho} \tag{A.212}$$

The solution of the above ODE is (see [16, 17])

$$\frac{\rho_1 u_1}{\bar{\mu}} x = \frac{2\gamma}{\gamma + 1} \frac{\theta_1 \ln(\theta_1 - \theta) - \theta_2 \ln(\theta - \theta_2)}{\theta_1 - \theta_2} The \tag{A.213}$$

entropy difference across the shock can be obtained using the second law of thermodynamics, where

$$\frac{ds}{dx} = \frac{1}{T} \frac{dh}{dx} - \frac{1}{\rho} \frac{dp}{dx}, \quad or \quad m \frac{ds}{dx} = \frac{\bar{\tau}}{T} \frac{du}{dx} - \frac{1}{T} \frac{d\bar{q}}{dx} \tag{A.214}$$

Via integration, one obtains

$$m(s_2 - s_1) = \int_1^2 \frac{\bar{\tau}}{T}\frac{du}{dx}dx - \int_1^2 \frac{1}{T}\frac{d\bar{q}}{dx}dx \tag{A.215}$$

or

$$m(s_2 - s_1) = \int_1^2 \frac{\bar{\mu}}{T}\left(\frac{du}{dx}\right)^2 dx + \int_1^2 \frac{1}{T}\frac{d}{dx}\left(k\frac{dT}{dx}\right)dx \tag{A.216}$$

The last integral can be integrated by parts to yield

$$m(s_2 - s_1) = \int_1^2 \frac{\bar{\mu}}{T}\left(\frac{du}{dx}\right)^2 dx + \int_1^2 \frac{k}{T^2}\left(\frac{dT}{dx}\right)^2 dx \tag{A.217}$$

Thus, the entropy increases due to the viscous dissipation and heat transfer.

In general, it can be shown that the entropy profile is monotonic, but it has a maximum at sonic condition.

The particles flowing through the shock pass from one state of thermodynamic equilibrium to another. Since entropy is a state variable, the difference of entropy across the shock is independent of viscosity and heat transfer and the wave drag can be estimated from solution of the Euler equations in terms of entropy jump (Oswatitsch formula [16]).

Viscous layer of Parallel Streams

Unlike shock waves, in contact discontinuities the pressure is continuous with a jump in velocity (even for incompressible flows).

To examine the viscous effects, consider the layer between two parallel, laminar streams which move at different speeds, U_1 and U_2.

In the mixing zone, the transverse velocity component is everywhere small, compared to the longitudinal velocity u. The boundary layer equations can be used, with uniform pressure, to analyze this problem (see [17]).

A self similar solution can be obtained in terms of a nondimensional variable $\eta = y\sqrt{U_1/(\nu x)}$and stream function $\psi = \sqrt{U_1 \nu x}\, f(\eta)$. Assuming $u/U_1 = f'$, the governing equation is

$$ff'' + 2f''' = 0 \tag{A.218}$$

with the boundary conditions

$$\begin{cases} f' = 1 \quad as \quad \eta \to +\infty \\ f' = U_2/U_1 = \lambda \quad as \quad \eta \to -\infty \end{cases} \tag{A.219}$$

At the plane of separation, $f = 0$ at $\eta = 0$.

Fig. A.14 Viscous layer of parallel streams

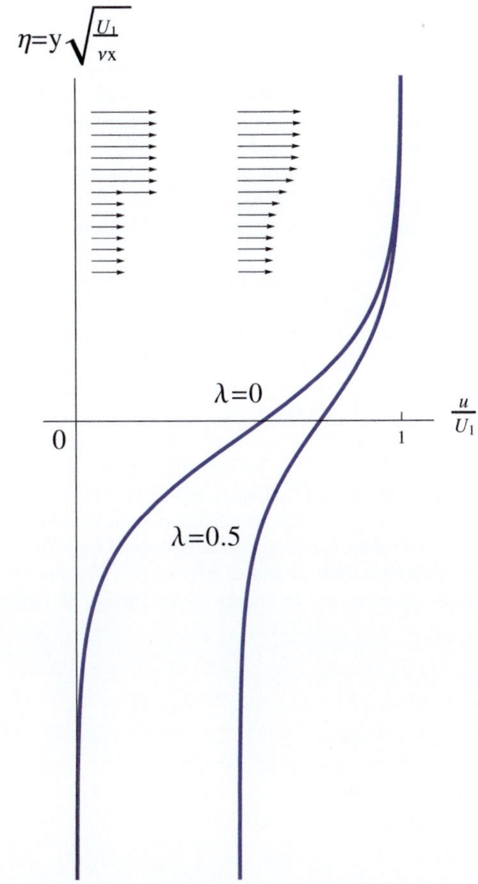

Lock [18] obtained solutions for several ratios λ as shown in Fig. A.14.

Improved numerical solution was reported by Christian [19]. The case of compressible flow is solved in Ref. [20]. For more details, see Ref. [21].

References

1. Moran, J.: Theoretical and Computational Aerodynamics. Wiley, New York (1984)
2. Robertson, J.: Hydrodynamics in Theory and Application. Prentice-Hall, Englewood Cliffs (1965)
3. Strang, G.: Introduction to Applied Mathematics. Wellesley-Cambridge Press, MA (1986)
4. Wyllie, C.R.: Advanced Engineering Mathematics. McGraw Hill, New York (1966)
5. Streeter, V.: Fluid Dynamics. McGraw Hill, New York (1948)
6. Shapiro, A.: The Dynamics and Thermodynamics of Compressible Fluid Flow, vol. I and II. Ronal Press, New York (1953)

7. Hirsch, C.: Numerical Computation of Internal and External Flows, vol. I and II. Wiley, New York (1990)
8. Abbott, M.: An Introduction to the Method of Characteristics. Thames and Hudson, UK (1965)
9. Liepmann, H., Roshko, A.: Elements of Gas Dynamics. Wiley, New York (1957)
10. Ferri, A.: Supersonic flows with shock waves. In: General Theory of High Speed Aerodynamics. Princeton University Press, New Jersey (1955)
11. Lax, P.D.: Hyperbolic Systems of Conservation Laws and the Mathematical Theory of Shock Waves. SIAM, Philadelphia (1973)
12. Chattot, J.-J.: Computational Aerodynamics and Fluid Dynamics—An Introduction. Springer, New York (2002)
13. Roe, P.: Approximate Riemann Solvers, parameter vectors and difference schemes. J. Comput. Phys. **43**(2), 357–372 (1981)
14. Chattot, J.-J.: Uniqueness Condition for Irrotational Compressible Flows of Perfect Fluid (Comptes Rendus de l'Academie des Sciences, t. 286-A), pp. 111–113 (1978)
15. Cole, J.D., Cook, P.: Transonic Aerodynamics. North Holland, Amsterdam (1986)
16. Oswatitsch, K.: Gas Dynamics. Academic Press, New York (1956)
17. Thompson, P.: Compressible Fluid Dynamics. McGraw Hill, New York (1972)
18. Lock, R.C.: The velocity distribution in the laminar boundary layer between parallel streams. Q. J. Mech. Appl. Math. **4**, 42–63 (1951)
19. Christian, W.: Improved Numerical Solution of Blasius Problem with Three Point Boundary Conditions. J. Aerosp. Sci. **28**, 911–912 (1961)
20. Chapman, D.R.: Laminar Mixing of a Compressible Fluid. NACA TN 1800 (1949)
21. Schlichting, H.: Boundary Layer Theory. McGraw Hill, New York (1979)

Index

J.J. Chattot and M.M. Hafez, *Theoretical and Applied Aerodynamics*,
DOI 10.1007/978-94-017-9825-9

MIX
Papier aus verantwortungsvollen Quellen
Paper from responsible sources
FSC® C105338

If you have any concerns about our products,
you can contact us on
ProductSafety@springernature.com

In case Publisher is established outside the EU,
the EU authorized representative is:
Springer Nature Customer Service Center GmbH
Europaplatz 3, 69115 Heidelberg, Germany

Printed by Libri Plureos GmbH
in Hamburg, Germany